IDENTIFICATION OF MATERIALS
VIA PHYSICAL PROPERTIES
CHEMICAL TESTS AND MICROSCOPY

IDENTIFICATION OF MATERIALS

VIA PHYSICAL PROPERTIES CHEMICAL TESTS AND MICROSCOPY

BY

The Late A. A. BENEDETTI-PICHLER
QUEENS COLLEGE OF THE
UNIVERSITY OF THE CITY OF NEW YORK

WITH 2 PLATES AND 75 FIGURES

SPRINGER-VERLAG NEW YORK INC.

SECOND PRINTING, DECEMBER, 1965

ALLE RECHTE, INSBESONDERE DAS DER ÜBERSETZUNG
IN FREMDE SPRACHEN, VORBEHALTEN
OHNE SCHRIFTLICHE GENEHMIGUNG DES VERLAGES
IST ES AUCH NICHT GESTATTET, DIESES BUCH ODER TEILE DARAUS
AUF PHOTOMECHANISCHEM WEGE (PHOTOKOPIE, MIKROKOPIE)
ODER SONSTWIE ZU VERVIELFÄLTIGEN

ALL RIGHTS INCLUDING TRANSLATION INTO OTHER LANGUAGES RESERVED
NO PART OF THIS BOOK MAY BE REPRODUCED IN ANY FORM,
BY PHOTOSTAT, MICROFILM, OR ANY OTHER MEANS, WITHOUT WRITTEN
PERMISSION FROM THE PUBLISHERS

© 1964 BY SPRINGER-VERLAG/WIEN

LIBRARY OF CONGRESS CATALOG CARD NUMBER 64-16107

Printed in the United States of America

Preface

This book has been written for the practicing chemist whose occasional task may be qualitative analysis. It deals with the investigation of things as they are without any limitations to the scope. It emphasizes the identification of materials – inorganic, organic, organized (biological), common, rare, described or not described in the accessible literature – as they actually occur in nature and industry, or are met in the investigation of mishaps and crime.

The description of techniques – macro to submicro – and the practice exercises have been included since the teaching of these arts is rarely a part of academic curricula and it happens with increasing frequency that chemists have to acquire them "on the job".

In the systematic procedure given, emphasis is placed upon the investigation of minute specimens and upon acute reasoning that continuously weighs all accumulating evidence. The work begins with the consideration of the history of the material under investigation. Especially when specks of an organic substance shall be identified, it should be realized that the discovery of the source – and consequently of the possibilities involved – may be the most valuable clue to an efficient solution of the problem.

Because of the increasing practical importance of the identification of small specimens, the systematic scheme starts with the application of non-destructive tests. Inspection of solids under the microscope reveals material of biological origin, and references to the literature are given for a continuation of the search. The appearance under the microscope may also permit to recognize man-made materials (artifacts), and the microscope may be used as an aid in studying color, fluorescence, hardness, refraction, and crystallographic data. Tests for radio activity and ferromagnetism are applied, and the density may be determined or estimated.

If the definitely non-destructive tests do not permit recognition, the solid is first heated to 300° C and then to higher temperatures. Its behavior and the testing of products of decomposition will usually reveal whether the material is inorganic or organic. The observed transition temperatures of organic substances may permit prompt identification by the performance of "eutectic" and "mixed" melting points requiring very little material. Simple inorganic substances may be recognized sufficiently by the accumulating evidence to permit arriving at final decisions by decisive confirmatory tests.

At the same time, enough information has been obtained to decide whether instrumental or chemical methods are best suited to continue the investigation. Advice is supplied to aid in arriving at an intelligent decision, and directions are given for the chemical analysis of inorganic materials. Instructions for the separation of analytical groups are outlined. The separation of the tantalum group of NOYES and BRAY is included, but otherwise it is assumed that the isolation of a group will suffice to identify the members – rare as well as common – by specific tests.

It is the opinion of the author that the book may also serve as a basis for an interesting laboratory course. The traditional course in qualitative analysis, which usually gravitates toward mechanical, lengthy, and tedious separations, is getting discarded to the detriment of the training in inorganic chemistry. The new approach can bring new life into the laboratory practice of qualitative analysis since it is based upon nicety of observation and penetrating reasoning which often lead to surprisingly quick identification. The recommended frequent use of the microscope – and a very inexpensive one will do – will contribute to the enjoyment of the work and familiarize the student of chemistry with a very valuable tool.

The book is based upon a critical evaluation of the experience gained in 40 years of teaching and occasional practice of qualitative analysis, micro analysis, and chemical microscopy. It makes use of the contributions of about three generations of analytical chemists, and the more glaring appropriations are duly acknowledged by references to the literature. The basic idea for the systematic scheme of identification stems from a recollection of an ingenious and simple course in qualitative analysis given to us by Professor KARL WITMANN, around 1910 at the *1. Staatsrealschule* in Graz, Austria. For inspiration, I am primarily in debt to him and to my teacher FRIEDRICH EMICH.

Furthermore, I am indebted to my late friend NICHOLAS D. CHERONIS for encouragement; to the analytical staffs of the Technical Service Laboratory of the Socony Mobil Oil Company and the Technical Center of General Motors for being kept aware of practical needs and procedures; and certainly also to my associates, coworkers, and students for criticism, suggestions, help, and the testing of procedures.

Special gratitude goes to the publisher and to all those who had a part in the production of the book for the understanding shown and the special care taken.

New York, December 1963

A. A. Benedetti-Pichler

Contents

	Page
Introduction	1
The Task of Qualitative Analysis	1
Principles and Definitions	2
Specificity and Sensitivity	4
The Method of Qualitative Analysis	6
The Minimum Size of the Sample for Chemical Analysis	7
Properties Affected by the Size of the System	9
Apparatus, Technique, and Scale of Work	11
Observation of Properties	16
Selection of Procedure	18
Part I: Technique of Observation and Manipulation	20
Use of Optical Aids	20
The Microscope	20
Experiment 1. Inspecting and Cleaning the Microscope	28
Illumination of the Specimen	34
Experiment 2. Illumination of Microscopical Specimens	35
Experiment 3. Calibration of Eyepiece Micrometer; Working Distances and Fields of Vision	37
The Immersion Method for the Determination of Refraction	39
Experiment 4. Phenomena Caused by Differences in Refraction	41
Observation of Schlieren	44
Experiment 5. Visual Observation of Schlieren	45
Use of Polarized Light	47
Experiment 6. Testing and Adjusting a Polarizing Microscope	50
Experiment 7. Isotropic and Anisotropic Substances in Polarized Light	52
Experiment 8. Determination of Vibration Directions in Relation to Profile	55
Experiment 9. Use of Compensators	57
Experiment 10. On the Determination of the Refractive Indices of Anisotropic Materials	60
Experiment 11. Observation of Pleochroism	61
Experiment 12. Observation of Axial Figures	61
Experiment 13. Transition Phenomena in Polarized Light	63
Experiment 14. Preserving Microscopical Preparations	65
Technique of Experimentation and Observation	67
Basic Rules	67
Work on the Gram Scale	68
Experiment 15. Phenomena Observed Upon Heating in an Inert Atmosphere	75

Contents

Experiment 16. Phenomena Observed Upon Heating in a Stream of Air 77
Experiment 17. Heating Upon the Charcoal Block 78
Experiment 18. Performance of Flame Tests 79

Work on the Centigram Scale 81
Experiment 19. Preparation of Capillaries, Capillary Pipets, and Microburners 99
Experiment 20. Emich's Method of Fractional Distillation .. 101
Experiment 21. Emich's Method for the Determination of Boiling Points 102

Work on the Milligram Scale 104
Experiment 22. Calibration of Capillary Pipets and Centrifugal Pipets................................... 125
Experiment 23. Preparation and Calibration of Platinum Loops and Hooks 127

Techniques of the Submilligram Scale 128
  Spot Tests.................................... 128
  Experiment 24. Test for Mercuric Mercury and Lead 131
  Experiment 25. Chromate Test for Silver 132
  Experiment 26. Tests for Bismuth and Antimony.......... 132
  Experiment 27. Test for Copper, Nickel, and Cobalt 134
  Experiment 28. Test for Cadmium 134
  Experiment 29. Precipitation of Silver Arsenate 135
  Experiment 30. Molybdenum Blue Test for Tin 135
  Slide Tests..................................... 136
  Experiment 31. Silver Dichromate 139
  Experiment 32. Recrystallization of Silver Chloride 139
  Experiment 33. Lead Iodide 140
  Experiment 34. Potassium-Lead-Copper Nitrite 141
  Experiment 35. Cesium Iodobismuthite and Cesium Iodoantimonite 142
  Experiment 36. Bismuth Cobalticyanide Pentahydrate...... 143
  Experiment 37. The Mercurithiocyanates of Copper, Zinc, Cadmium, and Cobalt 144
  Experiment 38. Test for Cadmium with Brucine and Bromide 146
  Experiment 39. Test for Mercury with Zinc, Copper, and Thiocyanate 146
  Experiment 40. Magnesium-Ammonium Arsenate and Silver Arsenate 147
  Experiment 41. Rubidium Chlorostannate 148
  Experiment 42. Estimation of the Relative Quantities of the Metals in a Slurry Containing Arsenic, Antimony, Copper, and Silver 149
  Working Upon the Surface of a Slide..................... 150
  Experiment 43. Conversion of Silver Chloride to Silver Dichromate. Handling Precipitates and Solutions, Fusion, and Electrolytic Reduction .. 151
  Experiment 44. Separation of Bismuth and Lead. Evaporation, Extraction of "Invisible" Residues ... 153

Experiment 45. Separation of Silver, Lead, and Mercurous Mercury, Sublimation, Extraction with Boiling Solutions 156
Experiment 46. Test for Ammonium Ion, Use of the Gas Reaction Cell 158

Working in Capillaries 159
    Experiment 47. Lead Sulfate, Triple Nitrite, Lead Chromate. The Capillary as Adjunct to Working Upon the Microscope Slide 160
    Experiment 48. Recrystallization of Lead Iodide 161
    Experiment 49. Isolation of Metallic Mercury, Conversion to Iodide 162
    Experiment 50. Bettendorff's Test for Arsenic............... 164
    Experiment 51. Oxidation of Arsenic to Arsenic Acid. Carius' Treatment in Capillaries 165
    Experiment 52. Conversion of Aniline to Acetanilide 167
    Experiment 53. Conversion of Urea to Symmetrical Diphenyl Urea...................................... 170
    Experiment 54. Purification of Benzene; Separation by Partial Melting in the Capillary 171
    Experiment 55. Cupric Ammonia Complex, Observation of Color in the Capillary 175

Working in Filter Paper 177
    Experiment 56. Ring-Oven Technique for Extraction and Evaporation in Paper 180

Particles of Ion Exchange Resins as Reaction Media 182

Working on Textile Fibers and Wires 183
    Experiment 57. Turmeric Test for Boric Acid 188
    Experiment 58. Test for Bismuth, Precipitation of the Sulfide Upon the Fiber and Conversion to Sulfate, Chromate, and Elemental Bismuth 190
    Experiment 59. Bead Test for Cobalt 191
    Experiment 60. Luminescence Test for Bismuth, Antimony, and Manganese 193

Work on the Microgram Scale................................ 193
  Apparatus 194
  Technique 202
    Experiment 61. Mechanical Separation of the Components of a Powder 219
    Experiment 62. Precipitation of Silver Dichromate Upon the Platform of the Condenser Rod 221
    Experiment 63. Estimation of the Quantities of Arsenic and Antimony in a Solution of Unknown Concentration 222
  Additional Practice Experiments for the Chosen Scale of Work 224
    Experiment 64. Study of Chemical Behavior 224
    Experiment 65. Analysis of Two Unknown Solutions 225
    Experiment 66. Identification of Simple Compounds of the Common Metals of the Hydrogen Sulfide Group 225

VIII   Contents

    Experiment 67. Identification of Simple Inorganic Compounds 226
    Experiment 68. Identification of Simple Compounds 227
    Experiment 69. Identification of Simple Compounds 227
    Experiment 70. Identification of Materials as They Occur in Nature, Industry, and Research 227

Part II: Systematic Analysis 228
    Choice of Materials and Cleaning 228
Sampling for Analysis 230
Systematic Procedure of Analysis 234
    The History of the Sample................................. 234
    Description of Sample and Record of Investigation............ 235
Preliminary Inspection 235
    The Sample is a Liquid.................................... 236
    Identification of Organized Matter 240
    Identification of Artifacts 242
    Well-Developed Crystals 243
    Solids of Random Shape and Structure 244
Non-Destructive Testing... 244
    Action Upon Light, Color 244
    Investigation of Crystals and Crystal Fragments............... 247
    Testing for Radioactive Decay 253
    Testing for Ferromagnetism 254
    Odor 256
    Hardness 258
    Refractive Index 260
    Density 262
Classification Tests 268
    Observation of Transition Points Below $350°$ C 269
    Ignition Above $300°$ C 274
    Solubility 285
        Performance of Solubility Tests 287
    Review: Inorganic Substances 291
              Organic Substances................................. 294
    Elemental Analysis of Organic Substances 299
    Testing with Dilute Sulfuric Acid........................... 301
    Test with Concentrated Sulfuric Acid 309
    Flame Tests 313
    Bead Tests 315
    Review of Findings.. 317
    Dissolution of the Sample.................................. 320
    Treatment of Substances Insoluble in Acids 322
Confirmatory Tests 324
    Group I A: Alkali Metals.................................. 325
        No. 3: Lithium, 6.939................................. 325
        No. 11: Sodium, 22.9898 326
        No. 19: Potassium, 39.102 326
        No. 37: Rubidium, 85.47 327
        No. 55: Cesium, 132.905 328

Contents

Group II A: Alkaline Earths 329
   No. 4: Beryllium, 9.0122 329
   No. 12: Magnesium, 24.312 329
   No. 20: Calcium, 40.08 330
   No. 38: Strontium, 87.62 330
   No. 56: Barium, 137.34 331
Group III B: Scandium Group 331
   No. 21: Scandium, 44.956 331
   No. 39: Yttrium, and the Lanthanides, Nos. 57 to 71 332
   No. 58: Cerium, 140.12 332
   No. 63: Europium, 151.96 333
   No. 70: Ytterbium, 173.04 333
Group IV B: Titanium Group............................. 333
   No. 22: Titanium, 47.90............................... 333
   No. 40: Zirconium, 91.22 334
   No. 90: Thorium, 232.038 335
Group V B: Vanadium Group............................. 335
   No. 23: Vanadium, 50.942 335
   No. 41: Niobium, 92.906 335
   No. 73: Tantalum, 180.948 336
Group VI B: Chromium Group........................... 336
   No. 24: Chromium, 51.996............................ 336
   No. 42: Molybdenum, 95.94.......................... 337
   No. 74: Wolfram (Tungsten), 183.85................... 337
   No. 92: Uranium, 238.03 338
Group VII B: Manganese Group.......................... 338
   No. 25: Manganese, 54.9380 338
   No. 75: Rhenium, 186.2 340
Group VIII: Fe-Ni Triad 340
   No. 26: Iron, 55.847 340
   No. 27: Cobalt, 58.9332 341
   No. 28: Nickel, 58.71 342
Group VIII: Ru-Pd Triad 343
   No. 44: Ruthenium, 101.07 343
   No. 45: Rhodium, 102.905............................ 344
   No. 46: Palladium, 105.4 344
Group VIII: Os-Pt Triad 345
   No. 76: Osmium, 190.2 345
   No. 77: Iridium, 192.2 346
   No. 78: Platinum, 195.09.............................. 346
Group I B: Copper Group 347
   No. 29: Copper, 63.54................................. 347
   No. 47: Silver, 107.870 347
   No. 79: Gold, 196.967 348
Group II B: Zinc Group 349
   No. 30: Zinc, 65.37 349
   No. 48: Cadmium, 112.40............................. 351
   No. 80: Mercury, 200.59 351
Group III A: Boron-Thallium Group 351
   No. 5: Boron, 10.811.................................. 351

Contents

No. 13: Aluminium, 26.9815 352
No. 31: Gallium, 69.72 353
No. 49: Indium, 114.82 354
No. 81: Thallium, 204.37 355
Group IV A: Carbon-Lead Group 355
No. 6: Carbon, 12.01115 355
No. 14: Silicon, 28.086 362
No. 32: Germanium, 72.59 363
No. 50: Tin, 118.69 364
No. 82: Lead, 207.19 365
Group V A: Nitrogen-Bismuth Group 365
No. 7: Nitrogen, 14.0067 365
No. 15: Phosphorus, 30.9738 369
No. 33: Arsenic, 74.9216 373
No. 51: Antimony, 121.75 374
No. 83: Bismuth, 208.980 375
Group VI A: Oxygen-Polonium Group 375
No. 8: Oxygen, 15.9994 375
No. 16: Sulfur, 32.064 377
No. 34: Selenium, 78.96 382
No. 52: Tellurium, 127.60 383
Group VII A: Halogen Group 385
No. 9: Fluorine, 18.9984 385
No. 17: Chlorine, 35.453 387
No. 35: Bromine, 79.909 391
No. 53: Iodine, 126.9044 393

Separations 395
Systematic Schemes for the Detection of Cations............. 396
The Classical Scheme 396
Outline for the Separation of the Analytical Groups........ 397
Observations and Notes 400
Analysis of Metals and Alloys Attacked by Nitric Acid...... 403
Separation of the Analytical Groups of NOYES and BRAY... 404
Systematic Search for Anions 422
I. Only Alkali Metals or (and) Ammonium Are Present 424
II. Nonmetallic Materials Readily Dissolved or Decomposed
by Water or Acids 427
III. Nonmetallic Refractory Materials 427
Final Review of Observations and Report 428

Appendix 430

Test Solutions... 430
Preparation of Unknowns 431
Reagents 432
Table 1. Color of Some Inorganic Substances 432
Table 2. Substances Crystallizing in the Cubic System 433
Elements 433
Inorganic Compounds 433
Organic Compounds 435

Table 3. Substances Crystallizing in the Hexagonal System 435
            Elements 435
            Inorganic Compounds 436
            Organic Compounds................................. 437
Table 4. List of Common Inorganic Compounds in the Order of Their Melting Points 438
Table 5. Inorganic Substances that Sublime, Arranged According to Color 457
Table 6. Inorganic Solids which Burst into Flame when Heated in Air, Ignition Temperatures in Centigrades 458
Table 7. List of Solids which Explode on Heating 459
            Inorganic Compounds 459
            Organic Compounds................................. 460
Table 8. Inorganic Solids Moderately Soluble in Water at Room Temperature 461

Literature 463
    General Reference Books 463
    Theory of Chemical Analysis 463
    Reagents 464
    Standard Tests and Procedures of Qualitative Analysis 464
    Chromatography and Ion Exchange 465
    Instrumental Methods 466
    Chemical Microscopy 466
    Slide Tests and Spot Tests 468
    Micro Analysis and Microtechnique 469
    Miscellaneous Applications of Microtechnique 470
    Mineralogy.. 471
    Journals 472
    Reports... 483
    Theses 483
    Unpublished Experiments 483
    Private Communications...................................... 483
    Demonstrations of Microgram Technique by Dr. M. CEFOLA 484
    Meetings.. 484
    Addresses 484

Subject Index... 485

Introduction

The Task of Qualitative Analysis

The purpose of qualitative analysis is the discovery of the essential nature of the object under investigation. A procedure of qualitative analysis, which shall be useful in actual analytical practice, must be generally applicable. It must give information without regard whether the object is a mixture or a pure substance; rare or common; simple or complex; organized, organic or inorganic; and whether it is known and described or as yet "unknown" and not yet mentioned in the literature. In addition to this, the procedure must be applicable to very small amounts of sample since qualitative analysis is frequently needed to determine the nature of minute objects.

It is impossible to ever present a complete practical scheme of qualitative analysis between the covers of a book since the practicing analyst may have to draw on any and every part of our knowledge of material things. A sensible approach, however, will give some useful information under all conditions and at the same time provide a high probability that simple problems will be solved satisfactorily and efficiently. The amount of information that can be gained with a suitable scheme will depend on the nature of the object. Obviously, qualitative identification must fail whenever distinguishing features are missing, and it may be impossible to indicate more than the general class of matter (lanthanides, mixture of fatty acids, proteid substance).

It is also plain that lack of material may seriously limit the scope of investigation and the amount of information that can be gained. Assuming, however, the existence of suitable apparatus and technique (145), one may say that the amount of material available for testing is only a minor factor in estimating the difficulty of the task. The time and labor needed depend in the first place upon the amount of information available concerning the sample and upon the amount of knowledge to be gained by the analysis. Under otherwise equal conditions, it will depend mainly upon the complexity of the object of investigation and upon the number and kind of separations required, whereby the emphasis should be on the latter since the number of constituents in the sample is by far not as important as their nature and their mass ratios.

If the object is a substance which is not yet described in the literature (146), qualitative analysis may or may not reveal this fact. It may furnish descriptive material, possibly some data concerning physical and chemical properties, the elemental composition as to kind, and perhaps evidence concerning functional groups and the general class of substances to which it belongs. As a rule, however, precise determination of percentage composition, number of functional groups, molecular weight, etc., are required to give a new substance its proper name, i. e., to assign it the correct niche in our classification of matter. If some unique behavior, physical or chemical, is discovered in the course of the investigation, this will facilitate the recognition of the substance when it occurs again.

This book outlines a plan of investigation for the use of the practicing chemist. Emphasis is placed upon a proper start which assures accumulation of useful information with small expenditure of sample material and the quick identification of common simple materials. References to the literature serve to widen the field of investigation in all directions toward the limits of our knowledge. Descriptions of technique are included to permit the investigation not only of small samples but of truly minute objects.

Principles and Definitions

In the strict sense, analysis ($\dot{\alpha}\nu\dot{\alpha}\lambda\nu\sigma\iota\varrho$ = resolution) should apply only to the method of taking apart or breaking down things for the purpose of investigation. Destruction of the sample, however, is not necessarily a part of qualitative analysis.

The modifier *qualitative*, too, should be accepted with a rather loose meaning. It indicates that no effort will be made toward a precise determination of the mass ratios if it should turn out that the sample is a mixture of several chemical specimens. On the one hand, the purely "qualitative" finding that the sample consists of sodium chloride implies the quantitative facts that it is composed of 60.67 per cent chlorine and 39.33 per cent sodium and represents practically 100 per cent of the combination of the two. On the other hand, a statement that the sample consists of sodium, calcium, sulfate, chloride, and water would not permit identifying the object of analysis. It could be gypsum contaminated by some rock salt, or table salt contaminated with gypsum, or some hydrate of calcium chloride mixed with some sodium sulfate, or a concentrated or dilute solution of some combination of the four possible salts, etc. Obviously, an approximate statement of the ratio of the ingredients is required before the object of investigation can be given a name; a merely qualitative statement does not suffice in such instances where several chemical specimens are found in the sample.

The primary task of qualitative analysis, the disclosure of the general nature of the sample, is usually accomplished by a rather crude classification according to mass ratio. Customarily one distinguishes *majors* (more than

5 per cent of the whole), *minors* (between 0.1 to 5 per cent), and *traces* (less than 0.1 per cent of the whole).

Whereas observations of mass ratios are frequently required, the performance of laborious quantitative determinations is avoided since the *general nature* of the object, as a rule, may be recognized without precise knowledge of ratios. In most instances it will suffice to know the major and minor constituents. A more exacting qualitative task, the *identification* of a special steel, of a special batch of copper, or of a reagent taken from a certain bottle, etc., may be solved by a study of the trace constituents or may require precise quantitative determinations.

There is no sharp dividing line between analysis and micro analysis. One speaks of **micromethods** whenever the experimentation proceeds with significantly less material than is customarily used for the purpose. If "significantly" is translated to mean ten times, there still remains the interpretation of "customarily", which will change with time and circumstance. In addition, there is the fact of the continuously varying amount of matter (sample) taken for testing. Schemes of separation for inorganic qualitative analysis usually start with one decigram to one gram of sample, but only milligrams and fractions of milligrams are used for the final confirmatory tests; the quantities decrease even to micrograms when flame tests, spot tests, fluorescence phenomena, etc., are included among the final tests.

By **sample** is always meant that part of the object of investigation, which is being treated or observed at the particular stage of experimentation. It is always assumed that it is truly representative of the whole object.

Because of the impossibility of giving a precise meaning to the term micro and derived terms, EMICH suggested to indicate the **scale of work** by stating the approximate mass of the sample used at the time. The magnitude of mass is indicated with the use of the prefixes of the metric system. Accordingly one may state that schemes of separation in inorganic analysis are customarily started on a decigram or gram scale, whereas preliminary tests and confirmatory tests are carried out on either the centigram, milligram, or submilligram scales. If the total amount of sample required or the amount of sample required for the separation of cations (these two quantities are in the approximate ratio of two to one) is taken for criterion, the terms semimicro, micro, and ultramicro may be interpreted as refering to inorganic qualitative analysis with decigram, milligram, and microgram samples, respectively.

The subdivision into inorganic and organic analysis complies with the fact that the limitations of human capability call for specialization. Since the objects of analysis are not necessarily labelled organic or inorganic, a procedure must be followed that leads to the recognition of organic substances in time so that the task may be turned over to a specialist if the already established facts and some additional simple tests, which are recommended, do not suffice for a solution of the problem. The practical analyst cannot adopt the rules of game of the typical school texts on qualitative analysis, which restrict the search to a small selection of inorganic cations and anions and may even require that the sample for analysis is an aqueous solution of a certain concentration range.

Specificity and Sensitivity

So far, chemical reactions cannot be observed as such. The objects of observation are the physical properties of the reacting system before, during, and after the reaction. The reactions used for recognizing substances produce either one or both of the two following effects:

a) Changes affecting the transmission of light by the appearance or disappearance of phases (boundary lines): changes from homogeneity to heterogeneity and vice versa. Into this category belong precipitation, evolution of gas, dissolution, condensation, evaporation, transition to a different state of aggregation, etc.

b) Distinct changes affecting the absorption or emission of light, i. e., changes of color, fluorescence, phosphorescence.

The practical value of an analytical test or procedure depends mainly upon the nicety of discrimination, its selectivity, and upon the limiting proportions. Tests (procedures, reagents) to which only one substance responds are called **specific**. Tests which are given only by a small number of substances are classified as **selective** (1241).

Sensitivity of a test (procedure, reagent) is to be described by listing the limiting concentration, the limit of identification, and the limiting proportions (850). The first two usually apply only in the absence of interfering substances and become less favorable in the presence of such.

The **limiting concentration** (L. C.) is defined as the lowest concentration of substance which always produces a positive test. Its dimension is mass per volume (856, 859).

The **limit of identification** (L. I.) is the smallest absolute quantity (mass or volume) which always gives a positive test (850).

The **limiting proportions** (L. P.) are the smallest mass ratios (substance sought/substance interfering) which still always permit to get a positive test (1137).

If perceptibility, the limitation imposed by our senses, is not considered, one may assume that limiting concentrations and limiting proportions are determined by essentially chemical phenomena: rate of reaction, shift of equilibrium, side reactions, establishment of metastable equilibria, complex formation, catalysis, solubility, coprecipitation, etc. Thus it is reasonable to expect that limiting concentrations and the important limiting proportions will not depend upon the size of the reacting system.

This assumption is supported by the available evidence, and it may be expected that the limiting proportions (and limiting concentrations) determined for gram samples will also hold on the microgram or nanogram scales and *vice versa*. The limits of identification, of course, depend upon the absolute size of the sample or the volume of solution taken for the test and may be improved by reducing the scale of work. The relation L. I. = = L. C. × volume should hold within a considerable range.

As a rule, the limiting proportions are less favorable than the limits of identification or limiting concentrations which apply only after the substance has been isolated in a more or less pure state. It is important to realize that negative confirmatory tests must be interpreted with reference to the sensitivity criteria. A negative test does not prove more than that the amount of substance, if it should be present, is less than indicated by the limit of identification or limiting proportions. If separations have preceded the test, the criterion is provided by the operation with the least favorable limiting proportion. Absence in an absolute sense cannot be proven by an analytical test; at times, it may be concluded from the history of the object under investigation.

Limiting concentrations, limits of identification, and limiting proportions cannot be determined with high accuracy since the conditions which always give a positive test and those which always lead to a negative result have no sharp borderline. They are separated by a range of conditions, which produces uncertainty as to the outcome: some trials give positive and some negative results (22, 150, 1580). This implies that the limiting conditions "always" leading to a positive result must be established by a sufficient number of experiments.

Suggestions for the brief notation of sensitivity criteria originated with GILLIS (899), FLASCHKA (341), and MALISSA (895, 898). For this book, the simple notation according to HAHN (962) has been adopted:

a) If a solid sample is being tested, the exponent (negative logarithm) of the limit of identification given in gram, pL. I., is listed in parenthesis: (6.0).

b) If a solution is being tested, the parenthetical notation indicates the subtraction pL. C. − pL. I.: (4.0 − 7.0), whereby the L. C. is given in gram per milliliter and the L. I. in gram. Performance of the subtraction leads to the logarithm of the volume of test solution given in milliliter: 0.001 ml = $1\,\mu$l. The notation consequently reveals also the scale of performance.

c) To indicate the limiting proportions, the interfering substances are listed with the logarithm of their limiting mass per unit mass of substance tested for: (6.0; Fe, Co, Ni: 2; U: 1) means that $1\,\mu$g of X may still be detected in presence of $100\,\mu$g of Fe, Co, or Ni and in presence of $10\,\mu$g of U.

(4.2 — 7.5; Ca: 2.5) indicates that $0.03 \,\mu g$ X, L. C. $= 60 \,\mu g$ X/ml, may be detected in a drop of $0.5 \,\mu l$ in presence of $9 \,\mu g$ Ca.

HAHN suggests to give the figures with one decimal.

The Method of Qualitative Analysis

Two *principal steps* may be recognized when considering the performance of a qualitative analysis (463): the *preliminary treatment* of the sample material and the *performance of the tests* that identify the substances which are present. This classification applies whether all tests are performed on one portion of the sample, or various tests require separate portions of the sample; in either instance, the material taken for analysis usually must undergo some treatment before the decisive tests may be applied.

The final tests, which are known as *confirmatory tests*, obviously must be based upon the observation of some phenomenon that, under the prevailing conditions, is produced only by the substance under consideration. The observed phenomenon may originate with a chemical, physical, or biological property that is characteristic of the substance (145); the intensity of the phenomenon is usually directly related to the mass of substance so that a crude estimation of quantity becomes possible. As a rule, confirmatory tests as well as estimations of quantity may be performed quite rapidly and without difficulty, once substances have been obtained in a reasonably pure state.

The *treatment* of the sample material *preliminary to* the *confirmatory tests* may involve transfers, disintegration, mixing, aliquot partitions, preparation of solution(s), and involved procedures for the separation of constituents. It may require nothing beyond transfer of the sample or part of it to some apparatus for treatment or observation if either the sample has a simple composition, or very specific and sensitive tests are available for obtaining the desired knowledge. It should be considered in this connection that the task may require merely a decision on absence or presence of a certain substance.

Whenever little or nothing is known concerning the nature of the sample, its composition is more or less complex, and sensitive as well as specific tests are missing, then it becomes necessary that isolation precede the identification of the various constituents. Isolation implies separation and transfer in space, i. e., mechanical operations which require that the substances to be separated have been collected in different phases.

Separable phases may exist to begin with. As a rule, however, they must be established for the purpose of separation. Only three ways are available for the creation of a phase: (1) the phase may be added mechanically; (2) it may be caused to appear as a consequence of a change in temperature or (and) pressure; and (3) it may be made to separate as the product of a chemical reaction proceeding in the system. Obviously, one will try to obtain a new phase of such a kind that some of the substances to be separated collect in it in their entirety, whereas the others remain altogether outside of it. Noticeable deviations from this ideal are quite common as a consequence of finite distribution ratios as well as adsorption phenomena and are responsible for the difficulty of getting perfect separations.

Apparatus and manipulative technique for the purely mechanical part of the separations must depend upon the nature of the phases to be separated. The three states of matter give five combinations which represent the natural divisions for a discussion of mechanical separations: solid-solid, solid-liquid, solid-gas, liquid-liquid, and liquid-gas. When suitable apparatus and proper techniques are applied, the purely mechanical part of the separations should not introduce errors or cause losses of a magnitude that would render them significant in qualitative analysis.

The Minimum Size of the Sample for Chemical Analysis

The various species of matter are identified by their properties. Obviously, the microchemist must be concerned about the effects of diminishing size upon the properties of matter and the means for their observation.

It follows from the atomistic concept of matter that its properties should remain unchanged down to the size of the fundamental units. These already contain the correct kind, ratio, and arrangement in space of the building stones, that determine the particular behavior. Therefore, the fundamental unit should represent the size of the smallest sample suitable for the study of physical properties since the particular material and the properties belonging to it cannot exist in a still smaller quantity of matter. Masses larger than the fundamental unit are simply accumulations of the latter and should not present any novel features excepting incidental phenomena such as size and shape, that can be assumed by any material.

The size of the fundamental unit that still permits qualitative identification on the basis of the typical properties depends greatly upon the type of material. In the instance of a coarse conglomerate, it may take a piece of several kilograms to reveal the essential characteristics; with crystalline rocks, a fraction of a gram may be required. The fundamental unit of mixtures must represent a reasonably close approximation of the composition of the whole (13, 14). Of suspensions, emulsions, colloidal sols, and molecular dispersions, a volume will be required which includes a fairly characteristic percentage of the dispersed matter. In the instance of chemical specimens (substances) the fundamental unit may be identified with the molecule or the unit cube of a crystal lattice.

When dealing with chemical specimens (solids and liquids), the fundamental unit representing the critical limit to identity and identification can hardly be imagined to exceed 1000 atoms corresponding to a mass of $1000 \times 1.7 \times 10^{-24} W$ gram or 3.4×10^{-19} gram if an average molecular weight is assumed such as $W = 200$.

After remembering that the energy content of one molecule or one fundamental unit need not represent the temperature of the bulk and that the collision of light with just one fundamental unit may not lead to the commonly accepted value for the refraction, one may be inclined to take into account the essentially statistical nature of the phenomena exhibited by matter. Accordingly one may adopt a safety factor of 1000 and expect that this number of fundamental units would give a satisfactorily close approach to the average effects observed with large quantities of matter and thus permit identification by observation of the characteristic behavior with about 10^{-15} to 10^{-16} g (0.003 pg) of substance (145).

There is no doubt that such and smaller quantities will suffice for identification if the phenomena are striking and originate in the atom or in the molecule. Thus it has been assumed long ago (1149) that quantitative determinations via the radiation count are possible with 10^{-15} g of highly radioactive elements, which implies that one hundredth of this amount or less (10^{-17} to 10^{-18} g) should suffice for qualitative identification. EMICH (150) demonstrated that 10^{-15} g hydrogen are sufficient to produce the line H_α, and PANETH and PETERS (652, 1201) devised gasvolumetric apparatus that permitted to identify 2×10^{-14} g helium by means of the glow discharge. With the use of the microscope, fluorescence may be studied on tiny objects; thus taking hemispherical drops of 1-μm (μ) diameter, one should be able to recognize 2.5×10^{-16} g fluorescein in 0.1 per cent solution.

It appears safe to assume that general theoretical reasoning permits reducing the scale of chemical experimentation to 0.001 pg = 10^{-15} g of mass and possibly less. One may expect with quantities down to this limit that matter still has the customarily accepted values for density and specific volume, refraction, optical rotation, absorption of radiation, magnetic susceptibility, conductivity, electrode potential, etc. One may also expect that the customarily known equilibria will establish themselves in reaction mixtures down to this size. This has important practical consequences.

In most microchemical work, one is very far from approaching the *critical size limit*. Therefore one may feel certain that there will be no fundamental change in physical properties and in chemical equilibria when the available quantity shrinks to milligrams, micrograms, nanograms, or picograms. Since equilibria are determined, aside from temperature and pressure, solely by the concentration of the reactants, working *directions* may be simply *transposed* from anyone scale to any other by multiplying all weights and volumes with the same factor and retaining concentrations, temperatures, and pressures without change. Any test or scheme of separation may be carried out with micrograms exactly as described for the gram scale by just reducing all weights and volumes to one millionth, by taking micrograms instead of grams, nanograms instead of milligrams, and nanoliters in place of milliliters. Experiences gained on any scale hold for every other scale of work; the utilization depends merely upon application of proper techniques in observation and experimentation. The "chemistry" and the theory of qualitative analysis may be applied as found in the general literature (13—82).

It has been demonstrated by RACHELE (424) that the critical size limit may be closely approached by the suitable performance of common test tube tests. Droplets of test solution and reagent solution were deposited by means of a micromanipulator in a film of paraffin oil that had been deposited upon the underside of the top plate (cover slip) of a moist chamber. The operation was performed in the field of a high-power microscope so that it was possible to measure the diameter of the spherical drop of test solution and to ascertain that both drops were clear. Drops of different size merged spontaneously if deposited close to each other, whereupon the separation of a precipitate or the development of a coloration was observed. Thus the limit of identification was reached with 10^{-14} g of barium ion supplied in a drop of 1 pl (12 μm diameter) of 0.001 per cent solution. With magnifications of about 400 diameters, a fine crystalline precipitate could be observed with transmitted light as well as with darkfield illumination. The Prussian blue test was reliably obtained with 0.4 pg of ferric ion supplied by drops of 4-μm diameter (0.04 pl volume) of 1 per cent solution. In a like manner, ANNA G. LOSCALZO (1261) could identify 0.02 pg silver as silver chloride, 0.2 pg cobalt as cobalt-mercuric thiocyanate, and 0.5 pg iron as ferric thiocyanate.

In all instances it was felt that the phenomena were probably obtained with still smaller quantities, but that it became impossible to observe them with the means employed.

Properties Affected by the Size of the System

It is to be expected that all properties that do not reside in the fundamental unit may vary with the amount of matter. Thus all statistically established equilibria must be affected even though our insufficiently precise methods of observation fail to reveal the differences.

Most of all one may expect to find deviations from the customary experience with properties or equilibria acting on or across phase boundaries. This is to be expected because the relative amount of surface must increase when mass and volume diminish.

It may be readily demonstrated that the **specific surface area,** defined as surface area per unit volume, A/v, must be inversely proportional to the cube root of the volume (or mass) provided that the shape and its proportions are retained. In the instances of the tetrahedron, the cube, and the sphere, the equations are $A/v = 7.2/v^{1/3}$ (tetrahedron), $A/v = 6/v^{1/3}$ (cube), and $A/v = 4.8/v^{1/3}$ (sphere).

The decrease of mass and volume and the corresponding increase of specific surface area affect the surface tension and as a consequence the vapor tension, solubility, melting point, and boiling point. The changes are insignificant until the particle diameters drop below $1\,\mu$m, i. e., the volume below 10^{-12} ml and the mass below 10^{-11} to 10^{-12} g. Noticeable increases of vapor tension and solubility are expected with decidedly smaller amounts of matter.

In the instance of barium sulfate, DUNDON (22) estimates a nearly hundredfold increase of the solubility when the diameter of the particles dropped to $0.4\,\mu$m corresponding to a mass of a cube-shaped particle of 3×10^{-13} g. The solubility should have been still higher in the experiments of RACHELE (424) where only about 2×10^{-14} g of barium sulfate were available at the limit of identification. The published photomicrograph indicates at least four large particles, and if five of them were present, their diameter could not have exceeded $0.1\,\mu$m. RACHELE, however, assumes that these particles were aggregates and that a large number of still finer particles collected at the interface with the paraffin oil. The findings are not necessarily in contradiction. Aside from the fact that there is no definite assurance concerning the absence of finer particles in the experiments of DUNDON, excess of sulfate ion in RACHELE's tests may have more than compensated for increased solubility. Finally, somewhere near the critical size limit, there must be some reversal in the solubility–particle diameter relation to permit the formation and growth of nuclei.

In addition to these effects, a large specific surface area may have a noticeable *influence* upon *the composition* of matter. In the instance of a palpable collection of matter (crystalline or liquid), the mass of the substances adsorbed on the surface may be alltogether negligible when compared with the bulk. The ratio of adsorbed foreign matter to the mass of the bulk must increase, however, when the latter diminishes. If the mass decreases to include just a small number of unit cubes or molecules, one may very well imagine that the amount of adsorbed matter could assume such a proportion as to change the identity of the accumulation. If no precautions are taken, adsorption might materially affect the performance of chemical separations on a very small scale. Fortunately, however, adsorbed matter is, as a rule, readily removed by washing or displacement.

So far, microchemical experimentation has produced no evidence of a change of chemical or physical behavior as compared with that customarily observed. This is probably due to the fact that most microchemical work has been performed with quantities far above the critical size limit. It means that our chemical knowledge applies without change down to at least the picogram scale and that any deviations observed are probably due to faulty technique. On the other hand, it will be wise to be on the lookout for changes in the behavior of matter when entering the critical size range of less than 10^{-14} g $= 0.01$ pg.

Apparatus, Technique, and Scale of Work

Whereas it may be assumed that the properties of matter and equilibria remain essentially unchanged throughout the whole size range of applied micro analysis, it is quite obvious that the technique of experimentation must be varied depending upon the amount of matter to be dealt with. Apparatus and procedures suitable for work with milligrams of material may become quite impractical when adopted for use on either the gram or the microgram scale. When the inescapable necessity of adjusting apparatus and technique to the amount of matter is compared with the universal validity of our chemical knowledge over the whole field of applied micro experimentation, it appears only proper to substitute terms like **microtechnique** of chemical experimentation and microtechnique of chemical analysis for the established but misleading composites microchemistry and micro analysis. The expression **microchemistry** might be more properly reserved for the behavior of matter below the critical size range, where the customary statistical equilibria are no longer attained.

Whatever the scale of work may be, proper apparatus and technique will give efficiency. In other words, they will assure attainment of the purpose with a minimum expenditure of skill, effort, and time. In addition to these requirements, the micro analyst must be concerned with the economical use of the sample material. In this respect it may be desirable to have the possibility of recovering the material under investigation in some form, but even this is not enough. It is important to keep this material always **confined** in a known small space so that its manifestations give an intensity to the phenomena assuring their reliable recognition by the striking contrast with the behavior of the surroundings. Accordingly apparatus and technique must be designed to retain the material under investigation in the smallest practicable and, if possible, approximately equidimensional space. The success of this reasoning is demonstrated by the various techniques for the performance of slide tests in very small areas: on the plateau of 0.2-mm diameter and 0.03-mm² area with a L. I. of 10 ng (p. 217); on the Brenneis electrode of 0.025-mm diameter and 0.0005-mm² area with a L. I. of about 1 ng; and in the confined field of the electron microscope of 0.0001-mm² area with a L. I. of 50 pg (934, 964).

It is in keeping with this same principle that centrifugal force is widely used for collecting matter under investigation into a small space. Filtration is usually replaced by decantation; this avoids exposure of liquids to the large surface of filter mats and does away with the often quite difficult task of separating the collected solid from the filter material. Consequently, centrifuge tubes of 3-ml to 1-μl capacity are widely used and are standard equipment for work on the centigram, milligram, submilligram, and microgram scales.

Of particular interest are capacity and shape of that part of the apparatus, in which the matter under investigation is collected. The small volume assures high concentration and corresponding intensity of action; approach of the occupied space to the shape of a sphere reduces the area of contact with surrounding matter and the extent of undesirable consequences such as adhesion, exchange of matter, and adsorption. Exposure to air would invite vaporization of volatile matter, absorption of water vapor and laboratory fumes, and contamination by dust. Mechanical collection would become difficult and inefficient if the area of contact with the apparatus were large. Extensive contact with the wall of the apparatus would also increase the amount of contamination by solvent action and magnify losses due to adsorption and ion exchange.

A suitable approach to the spherical shape is obtained by filling a circular cylinder to a height equal to its diameter, D. The sphere and such a cylinder give the same specific surface area, $6/D$, which may be readily tested with the use of the equations for surface area and volume of the sphere, πD^2 and $\pi D^3/6$, and of the equidimensional cylinder, $1.5 \pi D^2$ and $\pi D^3/4$. For obvious reasons, cylindrical apparatus will be given a height exceeding its diameter, but the latter should be such that the contents will approach the shape of the equidimensional cylinder. To obtain this condition, it is necessary to select the bore of the cylindrical apparatus according to the scale of work. To this end, Table I lists the diameters of the equidimensional cylinders as a function of their volumes. The working scales are named after the approximate mass of solid sample available. The corresponding volumes in the second column are the volumes of solution occurring in the course of a qualitative analysis. If one gram of sample is available, the largest volume of up to 300 ml may occur during the precipitation with hydrogen sulfide, and confirmatory tests may be performed with 1-ml aliquots. The corresponding cylindrical apparatus would be beakers of 7.5-cm diameter and test tubes of 11 mm bore. The table shows that D becomes inconveniently small for volumes of 1 nl = $= 10^{-6}$ ml corresponding to test-tube tests on the microgram scale.

Manipulation in the apparatus is materially facilitated by giving it the shape of a cone which may be considered to represent an infinite number of circular cylinders of zero height and continuously decreasing diameter.

Table I. *Dimensions of Apparatus as a Function of Its Capacity*

Scale of Work	Volume v	Diameter (and Height) of Equidimensional Circular Cylinder $1.08\,v^{1/3}$	Height of Circular Cone of 20 Degrees $3.13\,v^{1/3}$
Gram Scale	300 ml	7.4 cm	21 cm
	100 ml	5.1 cm	14.5 cm
	10 ml	2.4 cm	6.7 cm
	3 ml	1.6 cm	4.5 cm
Centigram Scale	1 ml	11 mm	31 mm
	0.3 ml	7.4 mm	21 mm
	0.1 ml	5 mm	15 mm
Milligram Scale	0.03 ml	3.4 mm	10 mm
	10 µl	2.4 mm	6.8 mm
	3 µl	1.6 mm	4.5 mm
	1 µl	1.1 mm	3.1 mm
Microgram Scale	0.3 µl	0.74 mm	2.1 mm
	100 nl	0.5 mm	1.5 mm
	1 nl	0.11 mm	0.31 mm
Nanogram Scale	0.3 nl	0.074 mm	0.21 mm
	1 pl	0.011 mm	0.03 mm

The cone has also the same advantage as a staircase of cylinders of diminishing diameter: that a separating phase of small volume may occupy an approximately equidimensional space, while this condition also holds for the whole bulk of the system. This simultaneity cannot be obtained with cylindrical apparatus truncated at a right angle; but it could be rather closely approached with a cone having an angle of 60 degrees.

The taper of centrifuge tubes usually has an angle of less than 60 degrees, which has the advantage of decreasing the specific surface area,

$$3(1 + \sin \alpha/2)/r \cos \alpha/2 = 3.05(1 + \sin \alpha/2)/v^{1/3}(\operatorname{tg} \alpha/2)^{1/3} \cos \alpha/2$$

which for a given volume v reaches infinity for $\alpha = 0°$ and $\alpha = 180°$ and becomes a minimum of $6.11\,v^{-1/3}$ for $\alpha = 35°$. The relative surface varies from $6.48\,v^{-1/3}$ for $20°$ to $6.35\,v^{-1/3}$ for $60°$, and the angle of the cone may be suitably chosen between these limits. Of course, the specific surface area will increase directly with $v^{-1/3}$ when the volume of the system becomes smaller.

The right-hand column of Table I lists the heights to which a circular cone having an angle of 20 degrees must be filled to represent the listed volumes v. The computed dimensions of the table show that, in general, the shapes of centrifuge tubes have been selected with good sense. A taper of 20- to 25-mm length fitted to a cylindrical tube of 4- to 6-mm bore is appropriate for the needs of work on the milligram scale. For ease of manipulation, especially cleaning, the tip of centrifuge cones is usually rounded off as indicated in Fig. 22. This is no longer feasible for cones of the microgram or nanogram scales, and consequently one may prefer to give their tapers somewhat wider angles.

Very obvious general observations concerning type of apparatus may be made with regard to the handling of substance of high **vapor tension**. When dealing with small amounts of volatile matter, loss will result in a short time if evaporation is not prevented; volatile solvents may vaporize so fast from small drops of solutions that working becomes impossible. Naturally, the vapor pressure that may be tolerated depends upon the scale of work. In the instance of aqueous solutions, the volatility of the solvent gives little concern on the milligram scale; only when slide tests are performed in a dry atmosphere, the rapid drying up of test drops may cause some annoyance. On the microgram scale, where the volumes range from 1 to 300 nl, experimentation with aqueous solutions becomes impossible if evaporation is not prevented.

If the material under investigation is volatile, work on a small scale requires that the volatile material is enclosed in a restricted space. The size of this space may be readily estimated by means of the gas law. One unit of volume of a liquid or solid substance of molecular weight M, density d, and vapor tension p mm will occupy $62\,000\, d\, T/p\, M$ units of volume at T Kelvin or $18 \cdot 10^6\, d/p_{20}\, M$ units of volume at 20° C. If not more than one per cent shall be lost by vaporization at 20° C, 1 μl (1 cubic millimeter) of the liquid or solid substance must be confined into a space of not more than $180\,000\, d/p_{20}\, M\, \mu$l. The volumes are for

carbon disulfide, chloroform	10 μl
ethyl alcohol	70 μl
acetic acid	0.25 ml
water	0.57 ml
iodine, camphor, valeric acid	3 ml
naphthalene	30 ml

The corresponding volume for water at boiling temperature is 16 μl, and it follows that water and dilute aqueous solutions will not show a decrease of volume when heated to 100° C if they are sealed in a container and the gas space is not more than 16 times of the volume of the liquid (uniform capillary of 17-mm length and 1-mm height of liquid column).

The evaporation of solutions may be prevented also by maintaining an atmosphere that is saturated with the vapor of the solvent. For work under the microscope, this is simply accomplished by using a moist chamber, Fig. 60, and entirely satisfactory on the microgram scale, i. e., solution volumes down to $1 \text{ nl} = 0.001\,\mu\text{l}$. RACHELE (424) found, however, that drops of a few micrometers in diameter (0.01 to 0.1 pl or 10^{-7} to $10^{-8}\,\mu\text{l}$) evaporated within a few seconds even when the moist chamber was completely sealed and contained air so saturated with water vapor that the inside surfaces of the chamber were covered with condensate. Obviously the equilibrium with the large drops of the condensate was still altogether insufficient to prevent the rapid evaporation of the far smaller drops of dilute salt solutions. The phenomenon was suppressed by increasing the electrolyte concentration of the small drops to 1 per cent or adding 0.6 to 2.5 per cent of glycerol, which gave a sufficient reduction of the vapor pressure of the small drops to establish equilibrium with the atmosphere of the cell. Interesting is the fact that this precaution had to be retained even when the small drops were suspended inside a film of paraffin oil mounted in the moist chamber.

The feature of micromanipulation that is most apt to attract the attention of the casual observer is connected with the increase of **surface forces** on the small scale. Pipets and siphons operate without use of suction; containers may be inverted or even dropped without spilling of contents; etc. In general, the capillary attraction frequently facilitates manipulation, and the tendency toward the spherical shape dictated by surface tension is an aid in keeping matter confined into a small space. The relatively large fraction of liquid left behind when emptying glass capillary tubing is mostly a consequence of the large specific surface area ($= 4/D$) of the cylinder surface, which increases proportionally with the shrinkage of the diameter D. The amount of residual liquid approaches a minimum when the outflow is slow, a fact that is well established for the customary pipets and burets of the analytical laboratory. Even with very slow delivery, the hold-up in glass capillary tubing may become very large and cause losses that could become inconvenient even for qualitative work. Experiments of LOSCALZO (437) indicate that the fraction of residual liquid over total liquid reaches a maximum of about 0.25 when the inner diameter of the tube is between 0.2 and 0.3 mm. When the meniscus travelled between 0.005 to 0.08 mm per second, the following percentages of residual liquid were determined in experiments with water: diameter of bore, $D = 0.26$ mm, 24% residue; $D = 0.22$ mm, 24%; $D = 0.16$ mm, 20%; $D = 0.14$ mm, 13%; $D = 0.12$ mm, 12%; and $D = 0.11$ mm, 10%. The maximum for a bore of about 0.25 mm may be explained by a decrease of the thickness of the residual film in narrower capillaries. It should be obvious that, on emptying, the amount of residual liquid will be of similar magnitude with small apparatus of any shape; one might expect it to become most unfavorable when collecting liquid that has spread on a plane surface.

The above statements apply to clean glass surfaces. The amount of residual liquid as well as that of solid particles suspended in it will greatly depend upon the nature of the surface of the container and the presence or absence of surface active substances in the liquid. The technique of using apparatus must be adjusted depending upon the behavior of liquids upon its surface. Vice versa, the surface of apparatus may have to be given a special treatment to assure the success of the technique to be applied. So far, relatively little use has been made of the advantages offered by surfaces that repel liquids, but their importance will probably increase with the refinement of technique. The spreading of drops on the surface of clean glass slides may be merely annoying when working on the milligram scale; it must be prevented by the use of a repelling barrier when working with micrograms, p. 201; and it would lead to complete loss of material on a still smaller scale of work. Since very small apparatus (cylindrical, conical, etc.) would not admit liquid that is repelled by its surface, one will have to work with more or less spherical drops that either adhere to a repelling surface or float in an inert medium, the techniques tried by RACHELE (424) on the subpicogram scale.

Observation of Properties

Common sense indicates that it will be the easier to observe a property, the larger the object; it follows that all available material should be collected into one body to give as much bulk as can be obtained for the observation. All phases of microtechnique are designed to serve this end. Spot tests and fiber tests are performed so that the colored matter is concentrated into a compact body. Precipitations on the slide are made to take place under conditions favoring the separation of few large crystals, which facilitates the observation of their shape as well as optical and other properties. The principle may be quite rigorously stated for the observation of the absorption of radiation and color.

The law of BOUGUER-LAMBERT-BEER gives the absorbance A as a function of absorptivity a, breadth of layer b, and concentration of the absorbing matter c.

$$A = a\,b\,c.$$

Since concentration is mass m over volume v and the volume may be expressed as the product of breadth times cross-section area α, it follows that

$$A = m\,a/\alpha.$$

For a given test and method of observation, there will be a smallest absorbance effect A_0 that still may be recognized with certainty. The amount of matter, m_0, corresponding to that intensity of the phenomenon is given by

$$m_0 = \alpha\, A_0/a$$

and represents the limit of identification, L. I. Obviously the test may be refined, i. e., the L. I. reduced in direct proportion to the reduction of the cross-section area occupied by the absorbing matter. Use of this relation is made wherever absorption of radiation serves the identification: spot tests, bead tests, coloriscopic capillary, microspectrophotometry, micropolarimetry, and fiber tests.

The equation may also be used for an estimation of the sensitivity of tests based upon the perception of color or characteristic absorption. If two color tests use the same conditions of observation (same cross-section area), one may expect that the limits of identification will be in a ratio inverse to that of the respective absorptivities

$$(m_0)_x : (m_0)_y = a_y : a_x$$

since the limiting absorbances A_0 will be similar. Obviously, this will hold for all types of radiation that obey BEER's law, and it will be the more nearly correct, the smaller the frequency range of the radiation.

Of special importance for the microscopic observation of color is the fact that absorbance does not vary with the intensity of the light. To obtain a bright image of an absorbing object, it is necessary to increase the intensity of the illumination in proportion to the square of the lineal magnification, but this does not affect the contrast of the color of object and surroundings. This makes it possible to collect just enough coloring matter into an area of microscopic dimensions to obtain a perceptible color effect and then to use the microscope or microprojection to get a greatly enlarged image of that area without diminishing the color contrast. It is assumed that the intensity of the illumination can be increased without changing the composition of the radiation.

In reducing the scale of working, apparently unsurmountable difficulties have been met in the observation of phenomena. The equation above indicates that a certain minimum mass of colored substance has to act on the beam of light, i. e., is required per unit of cross-section area if the color shall become visible. As a consequence it is found that faint colorations which are readily visible in thick layers of matter cannot be seen in thin sheets or small droplets of the same. If dye indicators shall be used, they must be added in much higher concentrations than customarily tolerated (437). Furthermore, when working with droplets of 7- to 14-μm diameter floating in a film of paraffin oil, RACHELE (424) found that the observation of heterogeneity as well as that of color was hampered by refractive phenomena. With transmitted light, the drops were transparent but surrounded by a heavy dark rim which closed up with very small drops so that these appeared as black disks. When changing to darkfield illumination, drops smaller than 7 μm in diameter became bright disks and the recognition of particles of a precipitate in the interior became uncertain. Removing this difficulty would require the discovery of an otherwise suitable liquid with a refractive index close to that of water.

Selection of Procedure

An ideal collection of microtechniques would offer apparatus and manipulative procedures for the efficient performance of any desired operation and on any desired scale. At this time, only the framework of the structure is available, and the experimenter must generally be on the lookout for possible improvements of the described technique. Deviations from tested techniques should not be made, however, without careful consideration of all possible consequences.

Usually there are good reasons for every detail of the directions. Thus, whereas the order in which reagents are taken into a capillary may be immaterial from the point of view of the chemical equilibrium, it may have been chosen to prevent contamination of a particular reagent, to avoid side reactions, to facilitate or to prevent mixing, or merely to assure proper rinsing of the tip before sealing it.

Concerning the choice of method, there are usually various chemical approaches for the solution of an analytical task, and they may require different operations. Depending upon the scale of work, certain operations may be awkward and inefficient, and it will be wise to avoid chemical procedures that require such. In general, it will be best to keep away also from schemes which need a large amount of manipulation that has to be done by a skilled or experienced operator and to favor procedures using operations that proceed spontaneously without much manipulation and supervision.

Evaporation, distillation, sublimation, and diffusion consume much time on a large scale and are therefore avoided in qualitative analysis. When working on a small scale, they require little time and become eminently practical, for they proceed spontaneously with little supervision. On the centigram scale, it may still be practical to neutralize a solution by adding small portions of acid and base, but the "dropwise" addition of reagent becomes rather inconvenient already on the milligram scale that would require "drops" of 0.05 μl volume if a macroprocedure is to be faithfully copied. On the small scale, it is far more efficient to adjust pH by exposing the solution to an atmosphere containing a known concentration of hydrogen chloride or ammonia gas, obtained by equilibrating air with a large volume hydrochloric acid or ammonia solution of known concentration (immersion of a test drop into the air space of the reagent bottle will often suffice). If this is not practical, it is usually possible to evaporate the solution to dryness after making certain that it contains an excess of volatile acid or base and to dissolve the residue, possibly after ignition, in a medium of the desired composition and pH. This latter procedure has the additional advantage of preventing the accumulation of salts in the solution.

Careful adjustment of equilibria, i. e., control of concentrations and measuring of reagents, is essential regardless of the scale of work. Special attention is required when working on an unfamiliar scale since the intuitive choice of correct proportions becomes unreliable under the conditions. The measuring of test substance, solutions, and reagents is not only necessary for the proper adjustment of the conditions, it is also required for the estimation of the quantities of substances found.

The obvious choice is the operationally simplest procedure which has the required sensitivity. From a purely technical view point, the simplest procedure may be the one with the smallest numbers of transfers. Especially the transfer of solids is fraught with the danger of loss. As far as the sensitivity is concerned, it is necessary to recall that it may be determined by some step in the separation preceding the final test. Of course, only an approximate estimate may be possible at this time since the limiting proportions depend upon the composition, essential facts of which may still be unknown.

Part I

Technique of Observation and Manipulation

Use of Optical Aids

The Microscope

The microscope is what its name implies, an instrument for viewing small things. It usually has the following parts:

I. the **Stand** consisting of,

Base, usually horseshoe-shaped, which supports the

Pillar or **Column**, often joined by a

Tilting Mechanism to the

Arm which carries the substage, the stage, and the body tube so that the distance between the two may by changed by rack-and-pinion (coarse adjustment) and possibly also micrometer motion (fine adjustment);

II. the **Body tube** containing the optical system that produces the enlarged image of the object and consequently represents the microscope proper consisting of:

Objective giving an enlarged image of the object in the plane of the field diaphragm of the eyepiece,

Eyepiece which further enlarges the image given by the objective, and

Drawtube permitting variation of the distance between eyepiece and objective;

III. the **Stage** supporting the object under investigation and needed auxiliary devices; and

IV. the **Substage** which usually may be moved mechanically and carries the apparatus for illumination with transmitted light, which may consist of:

Mirror with one plane and one concave side,

Condenser, a single lens or a lens combination, and

Diaphragm.

The formation of the image is purely schematically illustrated since presentation with correct proportions leads to fine detail which is difficult to discern.

In the left upper corner of Fig. 1 is indicated that the light emerging from the eyepiece of any optical apparatus forms a double cone, of which the cross section of narrowest diameter is known as RAMSDEN disk. Every observer automatically places his eye so that the iris of the eye is around the RAMSDEN disk, because failure to do this has the unpleasant effect that part of the light rays are cut off by the iris and consequently only

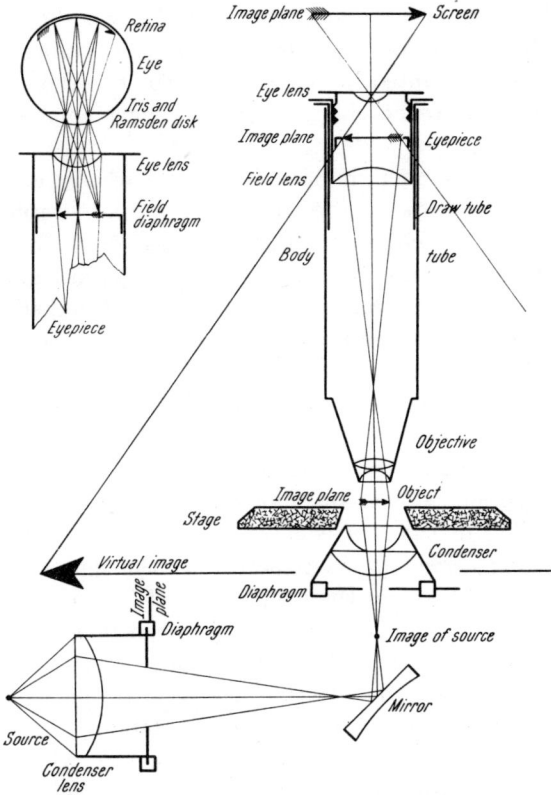

Fig. 1. Action of the Compound Microscope

part of the image is received by the retina and visible. This annoying phenomenon may be experienced with high-power eyepieces having the RAMSDEN disk close to the eye lens so that the eye touches the latter before the iris arrives in the desired position; it also may happen with auxiliary devices (color filter or cap analyzer) on top of the eyepiece, which prevent the eye from attaining the proper position. With properly designed eyepieces, the eye is held at a definite distance from the eye lens, which not only fixes the focal length of the rigid combination of eye lens and eye, but also determines the distance of the receiving screen (retina) from the eye lens. Since focal length and distance of the receiving screen are fixed, also the location of the object is determined: it is the conjugated image plane in the eyepiece half way between eye lens and field lens and indicated in the upper right corner of Fig. 1.

The location of this image plane will vary somewhat depending upon the characteristics of the eye. This does not impair the action of a field diaphragm in the eyepiece, set into an image plane corresponding to a "normal" eye. On the other hand, whenever the image plane is accurately predetermined by placing a scale, cross hairs, or other rulings in the eyepiece, then the eye lens is mounted in the eyepiece so that the distance between eye lens and rulings can be changed until a sharp image of the latter is received upon the retina.

Obviously, the reasoning changes very little when the image given by the eye lens is received upon a screen or upon a photographic plate instead of upon the retina. In either instance, the action of the eye is omitted and the magnification may be greatly varied by changing the distance between the eye lens and the receiving surface. Furthermore, in microprojection and photomicrography the final image is received on a plane surface, whereas the retina is curved. Especially in photomicrography it may become desirable to use specially corrected objectives and eyepieces for the flattening of the final image: Homals of ZEISS (88).

When rulings in the image plane of the eyepiece are once focused upon the retina, projection screen, or photographic plate, it is obvious that anything appearing in the image plane of the eyepiece will be portrayed upon the receiving screen (retina, plate). If the lenses of the objective together with the field lens of the eyepiece produce an image of the object in the image plane of the eyepiece, a conjugated image of the object will also appear upon the screen. This will happen during focusing when a distance between object and objective has been reached which places the inverted image of the object into the image plane of the eyepiece.

To prevent undesirable contact of the objective with the object, the latter is first closely approached with the objective while watching from the side. Then the eye is brought to the eyepiece or the screen is observed while the distance between object and objective is increased. An infinitely large image is formed at infinity when the object arrives at the principal focus of the objective combination. After this, the image is rapidly pulled in from infinity into the body tube while the distance from object to objective is increased to twice the focal length. Especially with high-power objectives of short focal length, the image must move large distances with very high speed while the object—objective distance grows by fractions of a millimeter, and the image may appear in the conjugated image plane of the eyepiece and on the retina or screen for such a short time that it cannot be perceived.

When the object is in focus, the plane of the object, the plane of the rulings in the eyepiece, and the plane in which the final image appears (retina, screen, plate) become conjugated image planes, i. e., any object in one of these planes has an image in all others, provided that light proceeds in both directions. If the object is located outside the principal focus of the condenser, there must be another conjugated image plane on the other side of the condenser lenses (Fig. 1 assumes that it is the plane of the iris diaphragm of the illuminating lamp), and any objects or structures in this plane will give images in the object plane and consequently in the image plane of the eyepiece and on the retina, screen, or photographic plate. Provided that the lenses of the microscope condenser are highly corrected, this makes it possible to include into the final image of the small object also reduced likenesses of relatively large objects (pointers, scales, rulings) placed into the conjugated image plane on the far side of the microscope condenser (88).

Fig. 1 shows that, because of double inversion, the image received on retina, screen, or photographic plate is right side up. For this reason, the image received upon the retina is interpreted corresponding to an inverted virtual image.

As an aid to better understanding, one may copy Fig. 1 on a large sheet of paper by actual construction of the images using the rules of elementary optics and neglecting the effect of the field lens of the eyepiece.

Magnification. The **total magnification** of a microscope is the *product of the individual magnifications of the objective and the eyepiece* being used; it is the ratio of the diameter of the virtual image (Fig. 1), imagined 250 mm in front of the eye of the observer, divided by the diameter of the object.

The magnification of objectives and eyepieces is usually indicated on the mounting. If the focal length is given, the magnification may be estimated as follows,

$$\text{magnification of objective} = \frac{\text{tube length}}{\text{focal length of objective}},$$

$$\text{magnification of eyepiece} = \frac{250 \text{ mm}}{\text{focal length of eyepiece}}.$$

Most manufacturers use a standard tube length of 160 mm. If the microscope is provided with a draw tube, usually permitting a variation of the tube length from 140 mm to 180 mm, the tube length may be read off a scale engraved upon the outside of the draw tube.

Total magnifications of less than 100 are considered "low", and of more than 500, "high"; magnifications of 100 to 500 are called "medium". Micro analysis very rarely requires magnifications above 250, and, consequently, the following suggestions for the use and care of the microscope consider mostly low-power microscopy.

Eyepiece. The eyepieces shown in Fig. 1 are of the Huygenian type suited for use with common achromatic objectives. Micrometer eyepieces with adjustable eye lens are preferable for most chemical work.

Eyepieces are readily exchanged by slipping them in and out of the draw tube or the body tube of the microscope. The eyepiece remains safely in the microscope as long as the instrument is *not inverted*. Eyepieces which are not in use should be kept in the rack provided in every microscope case. One eyepiece must always remain in the tube of the microscope to keep the dust out.

Objective. The quality of the final image depends largely upon the perfection of the objective, and consequently it is imperative to handle objectives with care and to keep them clean at all times. When not attached to the microscope, they must be kept in the containers provided by the manufacturer.

Dry objectives with magnifications from 2 to 20 will suffice for most chemical work. They usually have several lenses mounted in a metal tube with a screw thread for attaching to the body tube or objective changing devices.

Objective Changers. The component parts of objective changers, such as the clamp changing device (clutch) or the more precisely working sliding objective changer, may be attached to objectives and body tubes of standard make. Use of a revolving nose piece has the disadvantage that all attached objectives are continuously exposed to the objectionable agents frequently present in chemical work (fumes, heat).

Aperture. The angle (hatched in Fig. 2b) between the most divergent rays used for image formation is called the **angular aperture**, A. A., of the objective. The **numerical aperture**,

$$N. A. = n \cdot \sin (A. A./2)$$

is computed from the angular aperture and the refractive index n of the medium occupying the space between the preparation and the front lens of the objective.

Aperture determines the amount of light which an objective may grab and consequently the brightness of the final image. It also determines the **power of resolution** of the objective. Obviously, a small fraction of the light originating from a gross feature of the object will suffice to give an image of it. It is necessary, however, to employ the widest possible cone of rays emanating from a minor heterogeneity in the object in the hope of revealing its existence.

Front Lenses are those lenses of objective and condenser which are closest to the object on the stage. Their exposed surfaces, which are always plane, must be carefully guarded since they are most likely to become soiled, corroded, or mechanically damaged by contact with or proximity of the object.

Inspection of Fig. 2 shows that front lenses of objectives having low magnification must have relatively large diameters to get a reasonable aperture. This provides a simple means for recognizing low-power objectives.

Working Distance of an objective is the distance in millimeters between the front lens and the object when the latter is in focus. The working distance of a low-power objective is somewhat less than its focal length; that of a high-power objective, only a fraction of the focal length. Obviously, long working distances for the investigation of the interior of thick objects or the convenient operation of tools between object and front lens can be obtained only with low-power objectives.

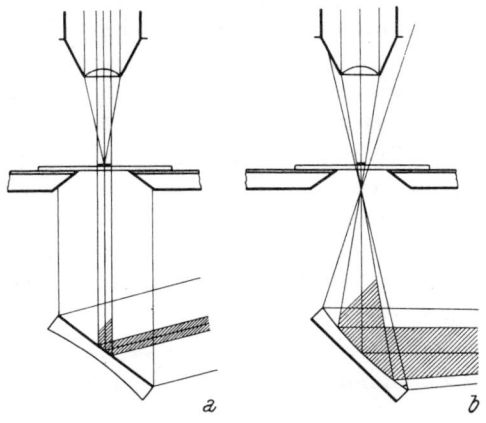

Fig. 2. Use of Plane and Concave Mirrors. The diameters of the preparations and the angular apertures of the objectives are indicated. The shaded areas show the bundles of light that are used for the formation of the image

Field Diaphragms are used to limit the visible area without affecting its brightness. They are placed in the conjugated image planes to obtain sharp images of their edges giving a sharp circular boundary to the microscopic field.

Aperture Diaphragms are placed into a focal plane of their respective lens system so that they uniformly restrict the amount of light used in forming an image without affecting the size of the field.

The aperture diaphragm in the lower focal plane of the condenser (Fig. 1) serves to regulate the brightness of the image. A conjugated image of this diaphragm forms in the upper focal plane of the objective, where another aperture diaphragm may be used for improving the quality of the image formed by the objective. Images of both aperture diaphragms are finally formed in the RAMSDEN disk and may be observed there with the use of a low-power magnifying lens.

Bastard Diaphragms and Baffles. The former are located in positions intermediate to focal planes and image planes. They affect both field and aperture, and they should be avoided since they do not permit independent regulation of these two factors. Baffles may be located wherever they serve best in the elimination of stray light, but their openings should be so large that they permit passage of all image-forming rays and cannot act as diaphragms.

Selection of Microscopes. Whereas the simplest toy microscope will suffice for many chemical tasks, efficient performance of work suggests the use of several instruments selected according to the special needs of the laboratory.

A chemical microscope as specified by CHAMOT and MASON (88) will serve for most work of a routine nature. A very desirable companion is a binocular microscope of the Greenough type with objective magnifications from about one to eight, which may be used on a microscope stand or held by some type of universal stand to serve as binocular magnifier.

Very satisfactory is a large stand with wide body tube and calibrated rotating stage, as used in biological research. If properly chosen, it will permit photography and projection as well as the use of polarizing equipment and a great variety of illuminating devices. It should be possible to exchange the stage for a rotating stage with built-in mechanical movements vertical to the axis of the microscope, to use a spectroscopic eyepiece, or to change to binocular viewing.

A simple microscope may be set aside for use with a heating stage and provided with an illuminating device that meets the requirements of the heating stage.

For convenience, microscope tables may be designed which hold the instrument so that the stage appears at the level of the table top and which are equipped with push buttons and voltage regulators or rheostats to provide the various types of illumination for direct observation, photography, and projection.

Microprojection and Photomicrography. Specially corrected lenses which flatten the image are desirable in photomicrography; the use of reasonably monochromatic (green) light may do away with the need for elaborate correction for chromatic aberration. The common problem of projection and micrography is the need for strong illumination and simultaneous exclusion of stray light. It is obvious that the latter will befog photographic emulsions. Only the experiment seems to be able to bring the realization that projection (especially with polarized light) requires a completely dark room if high magnifications shall be used.

The source should be small to permit efficient collection of the emitted light with condenser lens (and concave mirror). A tungsten arc or an incandescent lamp with concentrated filament furnish steady illumination for photomicrography. They are also satisfactory for microprojection if an image of only 10- to 25-cm diameter is needed. The crater of the positive electrode of a d. c. carbon arc is most suitable for the projection of large images. A close approach to a point source of high brilliancy are the Concentrated-(zirconium)-Arc lamps which are offered with 0.075- to 2.5-mm diameter of the arc and 2 to 300 Watt power.

Stray light is effectively reduced by using **critical illumination** as schematically indicated in Fig. 1. The light of the source is gathered by a condenser with iris diaphragm. The substage condenser is focused to obtain an image of the opening of the diaphragm of the auxiliary condenser in the plane of the object, whereupon the iris diaphragm is closed until only the field of vision receives light. This reduces the width of the illuminating beam of light and, consequently, the stray light to the unavoidable minimum. In practice proceed as follows.

Place upon the stage a preparation of good transparency (thin layer of thymol crystals from Expt. 13). Adjust the source, condenser, and mirror with all diaphragms wide open until the light beam passes the preparation and enters the objective so that a circular image appears upon the screen (ground glass plate). By the standard procedure (p. 30), focus the preparation to obtain a sharp image upon the screen and, if necessary, readjust mirror, source, condenser, and substage condenser until the field is evenly illuminated. Close the diaphragm of the auxiliary condenser to leave an opening of about 1-cm diameter and focus with the substage condenser until a sharp image of the opening of this diaphragm is obtained upon the screen; be certain not to change the focus of the microscope. The diaphragm of the auxiliary condenser is now in a conjugated image plane, and its sharp image appears in the object plane and the image plane in the eyepiece as well as upon the screen. Open and close the diaphragm and see its image changing upon the screen. Move the mirror until the image of the opening of the diaphragm is concentric with the circular boundary of the screen image. Finally open the diaphragm of the auxiliary condenser wide enough so that the image of its opening is just outside the field. If rulings contained in the eyepiece shall be used, focus with the eye lens until they appear on the screen. Then again focus the preparation and readjust the position of the substage condenser if this should be necessary to restore critical illumination.

Since the size of the field of vision depends upon the objective magnification, it is obvious that critical illumination must be restored after each change of objective. To maintain an image of a given size, the amount of light needed for its satisfactory illumination must be concentrated into an area (field of vision) which becomes the smaller, the higher the objective magnification. Very soon, the intensity of the unavoidable infrared radiation becomes so high that the object of investigation is endangered; solids may explode, melt, sublime, char, or burn; liquid preparations start to boil. To avoid this, cells filled with 5% solution of copper sulfate pentahydrate or an acidified saturated solution of ferrous sulfate are inserted between source and mirror of the microscope; plates of special infrared absorbing glass may be used in addition.

Consult the special literature for details (88, 106, 1240); ROYER (708, 1240) writes on the determination of exposure for color photography; COURTNEY-PRATT and HUGGINS (1075) on 100000 frames per second cinematography.

Microscope Lamps. A microscope lamp for chemical use should permit quick adjustment in height so that one may readily change from work by transmitted to work by reflected light. It should permit focusing a bright small spot of light upon the object and have a slot for the insertion of ground-glass plate and color (daylight) filters, singly and in pairs. It should be light in weight, small in size, and it should not heat up so that it becomes difficult to handle. A telescoping tube containing a low-wattage bulb and a condenser lens could be mounted in a well-balanced manner on a ball-and-socket universal joint which may be moved along a vertical rod.

As a substitute, desk lamps with gooseneck arm are quite satisfactory. The use of spot lamps for automobiles and motorcycles as well as of electric bicycle headlights may be considered. A rather efficient and handy microscope lamp may be obtained with a lens bulb (Mazda 222, 0.55 watt) for pen lights. The glass at the top of the tiny bulb is fused into a bead which acts as a powerful condenser lens. Potential of 2.2 volts may be supplied to several of such bulbs via an ordinary 2.5-volt filament transformer and a 10-ohm variable rheostat in series with the bulbs.

Experiment 1

Inspecting and Cleaning the Microscope

Lens paper; camel's-hair brush, round, bathe in 1 ml of acetone and allow to dry; cotton and tooth picks; clean rag.

Remove the microscope from its case and place it on a clean table top. When **carrying a microscope,** grasp its arm with the right hand and *support the base* of the stand with the palm of the left hand. This prevents accidents such as smashing the lower part of the instrument into the edge of a table top.

Inspect inside and outside of the case. Check the inventory for objectives, containers, eyepieces, racks, and other accessories, and then proceed to the testing of the microscope.

1. *Adjustment of Eyepiece, Method of Viewing.* Micrometer and crosshair eyepieces should have an adjustable eye lens. Remove the eyepiece from the microscope, hold it against a suitable source of light (try a window, a white wall, lamps) and move the eye lens until a sharp image of the rulings is obtained; use spectacles if they correct for astigmatism. Keep

the eye lens in this position and return the eyepiece into the tube of the microscope.

In the instance of **binocular** microscopes and binocular eyepiece attachments, at least one of the paired eyepieces should have an adjustable eye lens. In addition, there must be some means for adjusting the distance between the eyes (interpupillary distance). Use spectacles to correct for astigmatism. Cover the eyepiece having the adjustable eye lens and focus as directed under (5) and (6). When sharp focus has been obtained, transfer the cover to the other eyepiece and, without changing the position of the microscope tube, focus with the eye lens of the eyepiece until again a sharp image is obtained. It is understood that the left eye always is before the left eyepiece and the right, before the right. Finally remove the cover and view through both eyepieces. In general, two circular images will be seen, that overlap for the most part. Operate the adjustment of the interpupillary distance until the two images fuse into one.

Binocular vision facilitates the recognition of three-dimensional structures and prevents abuse of the eyes by permitting them to function as intended by nature. Since both eyes must be kept centered with respect to two eyepoints, head and neck must be held very rigid, which may cause muscular fatigue (88).

2. For *monocular observation* make it a principle to frequently alternate the use of left and right eye and to relax the eyes at frequent intervals by looking at distant objects. Both eyes should be kept open while looking into the microscope; the unoccupied eye might be protected by a black shade, which may prevent development of the unfortunate habit that one eye goes automatically blind when viewing through an instrument and binocular vision with instruments becomes impossible.

When viewing through the eyepiece of any optical instrument, relax the eye so that it remains focused at infinity and make no effort to "see". Let the instrument (eye lens) do the focusing. If very much time will be spent with looking into an instrument, obtain a binocular eyepiece attachment or use projection of the image upon a screen, which last solution is most convenient in every respect.

3. Clean both sides of the mirror with lens paper.

4. *Preliminary Adjustment of Illumination.* If the substage has a diaphragm, open it wide. Place a lamp in front of the microscope and adjust the concave mirror until the light is sent through the hole in the stage to the front lens of the objective. If a microscope lamp is not available, use sunlight reflected from white clouds or bright walls, or the light from ceiling lamps. In such instances, try the plain and the concave mirror, and use the one which gives the better illumination.

5. Place a *clean* slide on the stage so that its edge runs across the center of the hole in the stage, and then *focus with the standard procedure*:

a) Estimate the working distance of the attached objective from the data available. Information on focal lengths and magnifications is often posted inside the carrying case.

Observe the objective from the **side** with the eyes held at the level of the microscope stage. Lower the tube with the coarse adjustment until the front lens of the objective has approached the preparation (top surface of the slide) to approximately half of the estimated working distance. With a high-power objective, go as near as possible to the preparation **without** touching it.

b) Look into the eyepiece, and slowly **raise** the tube by means of the coarse adjustment until a sharp image of the upper edge of the slide is obtained. If no image appears, one of three things may have happened: the tube has been raised too fast, and the image has been missed; the front lens of the objective never was below the working distance; or the slide is not in the proper position on the stage. Make the necessary adjustments, and focus again as directed under (a) and (b).

It may happen to a beginner that he raises the tube so high that rack and pinion disengage (properly designed microscopes have a stop to prevent this). The meshing of the gears must be done very carefully to avoid stripping.

The fine adjustment should not be used for focusing with low-power objectives, but reserved for small final adjustments in the study of detail and for the focusing with high-powers. It should be kept engaged near the center of the micrometer motion to provide a maximum of adjustment in both directions. Lost motion and lag indicate wear or poor workmanship. The fine adjustment should be graduated for the measurement of vertical displacement.

6. *Final Adjustment of Illumination.* Adjust the mirror and try the plane and the concave surface until the most effective and even illumination of the whole field is obtained. If the illumination is too strong, place a ground-glass plate or a sheet of paper in front of the mirror or move a lamp farther away from the microscope.

If the substage carries a condenser and diaphragm, move the condenser up and down to find the position which gives the best even illumination of the whole field. Finally adjust the brightness of the field by opening or closing the substage diaphragm.

7. *Examination of the Image.* The edge of the slide should appear sharply defined; otherwise the field should be empty. *Films of dirt* on the lenses of the microscope proper will impair the sharpness of the image; the edge of the slide will appear blurred or clouded by a haze, and refocusing will give no improvement. *Particles* adhering to lens surfaces are indicated by the presence of more or less sharply defined spots in the field of vision. The location of the imperfections may be found as follows.

a) Move the **slide**. Images of dust particles and fingerprints adhering to the slide move simultaneously. Clean the slide until these imperfections disappear.

b) Rotate the **eyepiece**; a cross-hair eyepiece must be slightly lifted to make this possible. This imparts rotary motion to images produced by dust or dirt on the glass surfaces of the eyepiece.

c) Move the mirror slightly. If a condenser is used, move it up and down. This will reveal images of irregularities in the **light source** (the condenser may be focused upon the frosted glass or the filament of the bulb, etc.) or images of foreign matter adhering to the **condenser** lenses. Eliminate the former by changing the distance of the light source from the microscope; the latter, by cleaning the lenses of the condenser and the aperture of the substage diaphragm.

d) Imperfections not located in either of the aforementioned parts of the optical system must originate in the objective. Corrosion, dirt, and dust on the lenses of the objective destroy the sharpness and clarity of the image of the object, but they are only vaguely or not perceived when looking into the eyepiece of the microscope.

e) Imperfections which are in sharp focus simultaneously with the object (edge of the slide) must be located in one of the **conjugated image planes**: plane of the field diaphragm of the eyepiece (micrometer scale or cross hairs), plane of the preparation viewed, source of illumination or some plane between microscope condenser and lamp.

Repeat the tests with the other objectives and eyepieces belonging to the microscope. Finally inspect and, if necessary, clean the parts as directed in the following. Start by carefully washing and drying the hands.

8. *Eyepiece.* Remove the eyepiece from the drawtube of the microscope, unscrew the mountings of eye lens and field lens, and place the parts upon a sheet of clean paper. Inspect all lens surfaces for corrosion, scratches, and dirt by holding them so that light is reflected from them to the eye. A perfect surface looks like a clean, faultless mirror. Rotate each lens around its optic axis during inspection. Imperfections will rotate with the lens, whereas images and reflections will remain more or less stationary.

Clean lens surfaces by first brushing or blowing off dust to make certain that abrasive particles cannot make scratches when wiping. Then breathe upon them and wipe off the condensed moisture with doubled layers of lens paper. Do not apply much pressure, or oil from the skin will go through the paper and collect on the lens surfaces. Inspect the lenses again and, if they are clean, assemble the eyepiece without delay.

Micrometer eyepieces. Unscrew also in the middle to get access to the micrometer plate. When unscrewing, hold the eyepiece right side up so that the micrometer plate will not be dropped. Often it will suffice to clean the upper surface of the plate with a camel's-hair brush without removing it from the mounting. It may be taken out, however, and inspected and cleaned as described for lenses.

Cross hairs are either real hairs mounted on a ring or engravings on a glass plate. In the former instance, cleaning is rarely necessary, but may be done by gently blowing against the hairs which must not be touched, for they are easily broken. In the latter instance, proceed as with eyepiece micrometer scales.

9. *Objective*. Inspection and cleaning of lenses has been outlined under the caption "Eyepiece". **Standard objectives** must **not be taken apart,** and cleaning that requires the separation of lenses should be done by the manufacturer. Thus restrict cleaning to the lower surface of the front lens and the upper surface of the top lens. For cleaning the latter, use a swab of cotton on a toothpick and finish with a strong blast of clean air.

The exposed surface of the front lens needs special and continuous attention. Clean it immediately if it should have had contact with a preparation or been exposed to a preparation giving off vapors of any kind. As the first step in cleaning, the surface of the lens and the surrounding part of the mounting may be carefully rinsed with tap water and wiped dry with lens paper. Alcohol and other organic solvents are best avoided since they may act upon a cement used by the manufacturer.

An inexpensive microscope may have an objective of the separable type, which provides two or three magnifications. In such instances, inspect and clean each section. When using it, keep in mind that such an objective has as many front lenses as magnifications and that the perfection of each of these front lenses must be continuously guarded.

10. *Coarse Adjustment*. Test for excess play and lost motion. The body tube carrying the full equipment should remain in every position into which it is brought. Check the directions of the manufacturer concerning tightening and loosening the rack-and-pinion motion.

Microscopes may still be found, that lack a stop preventing the pinion from overriding the rack, and consequently their coarse adjustment may become damaged by inexperience, haste, or carelessness in meshing the gears. For inspection, raise the body tube to its highest position with the pinion head. Note that the gears are disengaged, and then grasp the body tube and pull it up to remove it from the stand for inspection of the rack. Rotate the pinion head while inspecting the teeth of the pinion. Any imperfections of the teeth of either part of the mechanism must be corrected immediately to prevent its complete destruction.

For the **meshing of rack and pinion,** insert the rack of the body tube into the pinion slot and push the tube down until the rack just touches the pinion. Then impart short turns to the pinion head and a very gentle downward pressure to the tube until proper meshing occurs so that the gear works smoothly. Avoid haste and the use of force. Follow the procedure recommended for meshing whenever it happens that the pinion overrides the rack.

11. *Condenser.* With some microscopes, the substage is moved by a rack-and-pinion arrangement and may be removed from the stand like the body tube. Others have a quick-acting screw which permits swinging the condenser out of the optic axis. The condenser proper is usually held in a ring clamp which may be provided with a set screw; usually, the uppermost lens of the condenser may be removed by unscrewing the mounting. In the instance of petrographic stands, provision may be made allowing the uppermost lens of the condenser to be swung aside by lever action. An iris diaphragm and polarizing equipment may be attached below the condenser lenses. The mirror is attached to either substage or stand and may be removed by pulling the carrying rod out of its socket. Usually the parts of the substage are more conveniently accessible after tilting the microscope into a horizontal position.

Swing out the condenser or remove it completely from the microscope. Dust and dirt collect mostly on the top surfaces of the front lenses. Inspect and clean the accessible lens surfaces as directed above. Dust will also collect on the top surface of a diaphragm; remove it with a brush (working on the inverted apparatus) or a blast of clean air. Close and open an iris diaphragm, and inspect the blades for absence of corrosion, kinks, and bends.

If the microscope is equipped with objectives of magnification 20 and higher, the centration of the corresponding condensers should be checked. Do this, after focusing objective and condenser upon a preparation, by removing the eyepiece and looking down the tube at the back aperture of the objective. When the substage diaphragm is nearly closed, its small opening should be seen at the *center* of the back aperture. A thoroughly satisfactory substage will permit adjustment. If none is provided, however, the centering may be done at the objective or the objective changing device; this assumes that a rotating stage is also equipped with centering screws.

12. *Stage.* Inspect for corrosion and damage by heat. Sockets for attaching clips or other apparatus should not be clogged. A central opening of 25 mm or more in diameter will permit the use of all kinds of condensers.

A rotating stage should be provided with graduations around the circumference and a vernier; a locking device is desirable. The stage should rotate smoothly without any indication of lateral motion. Inspection should show whether or not it is readily removable for lubrication and adjustment. Center the stage (Expt. 6) and make certain that all necessary accessories are supplied.

13. *Final Inspection.* Assemble the microscope after cleaning and repeat the tests given under (7). If necessary, repeat the cleaning and adjustment. Permanent imperfections of lenses can be removed by the manufacturer.

Illumination of the Specimen

Whether or not the eye is aided by a magnifying lens or a microscope, the basic requirement for the observation of the structure or color of a specimen is that the latter is able to properly act upon light of the right kind. Light travelling through the specimen will be affected differently from light reflected from the surface of the specimen. Light transmitted through the specimen will show the color resulting from selective absorption (body color) and will reveal the shape by means of refraction rather than reflection. Light reflected from the surface of the specimen will show the color resulting from selective reflection (surface color) and reveal the structure by reflection rather than refraction. Light passing in such a direction through the specimen that it cannot reach the eye but for gross deflection may reveal heterogeneity and fluorescence. In addition and quite generally, no matter what the effect of the specimen, it will be the more readily observed the more it differs from the surroundings.

It follows for all visual observation, with and without use of magnifying devices, that **illumination** (kind of light and direction of incidence upon the specimen in relation to the location of the eye) **and background** will decide what may be seen. Magnifying devices only aid in resolving the detailed information already contained in the light coming from the specimen.

As a rule, white light similar in composition to sunlight is preferred for illumination so that colors appear as in the natural light of the day. There is never any harm in trying ultraviolet which may reveal tell-tale fluorescence. Monochromatic light is useful in photomicrography and in the determination of optical constants. Polychromatic illumination gives special darkfield effects.

The angle of incidence may be varied through 180 degrees in an attempt of finding all available information. By all means, one should try transmitted light as well as reflected light with backgrounds of various colors. The detection of heterogeneity (turbidity, opalescence) and fluorescence calls for Tyndall illumination (incidence vertical to the line of vision and black background) or darkfield illumination. One might keep in mind that appearances revealed by reflected light are more readily interpreted since one is accustomed to see by reflected light. The wide use of transmitted light in microscopy is made possible by the fact that many small objects are transparent.

The methods of illumination used in microscopy are briefly outlined in Table II. The variety is explained by the desire of seeing small detail. Very slight gradations of optical density must be brought out by the use of special diaphragms and other devices in the image-forming system and in the illuminating apparatus: *Schlieren* microscope and phase contrast microscopy (88).

Table II. *Types of Illumination Used in Visual Microscopy*

Transmitted Light	Reflected Light
Axial: a) *Parallel:* beam of light parallel to the optic axis of the microscope; *source of light at a great distance and plane mirror in the optic axis,* Fig. 2a. b) *Convergent:* pencil of rays along the optic axis; *concave mirror, Fig. 2b, or condenser,* Fig. 3. **Oblique:** a) *Parallel:* beam of light inclined to the optic axis; *plane mirror swung to one side and no condenser.* b) *Convergent:* pencil of rays inclined to the optic axis; *concave mirror swung to one side and no condenser; condenser used and cardboard inserted to cover all but one side; condenser with small opening of its diaphragm off center.* c) *Annular:* dark-field condenser; two-color illumination of Wright and Rheinberg used in Zeiss' Mikropolychromar (88).	**Axial:** beam or pencil of light parallel to the optic axis of the microscope; *transparent reflector (cover slip inclined at 45 degrees) between objective and specimen so that a horizontal beam of light is reflected along the optic axis upon the surface under examination; vertical illuminator for medium- and high-power objectives* (metallography). **Inclined:** beam or pencil of rays coming from above the stage and inclined to the optic axis; *direct illumination from light source located above the stage.* **Annular Oblique:** *Silverman illuminator for low magnifications; Epi-Condenser W* (Zeiss) or *Ultropak* (Leitz) *for medium and high magnifications.*

Experiment 2

Illumination of Microscopical Specimens

Slide with ground-glass surface; squares of white, black, and colored paper, about 3 cm × 3 cm; cedarwood oil.

To obtain a suitable specimen, draw a line with a silver coin through the center of the rough surface of the ground-glass slide. Parallel to this line and about 1 mm away from it, make another streak with a penny. With a soft black pencil draw a line vertically across both streaks.

Place the slide on the stage of the microscope so that the intersections of the lines appear at the center of the opening. Use the weakest objective and remove the condenser. Adjust the illumination and focus upon the rough surface of the slide as directed in Expt. 1 (5). Try sunlight or the light from a distant lamp with the plane mirror, and also try the microscope lamp with plane and with concave mirror to obtain even illumination of satisfactory brightness over the whole field of vision. Note that the choice between plane and concave mirror is determined by the size of the field of vision, the intensity and distance of the source of light, and the needed brightness of the image, Fig. 2. Observe that the rough surface of the slide interferes greatly with the recognition of the particles left behind by coins and pencil.

Cut off the light coming from the mirror, slip a square of white paper underneath the slide, and direct the illuminating beam of light down upon the preparation at an angle of about 45 degrees. To this end, either raise the microscope lamp, or use a special arc lamp or sunlight which may be guided with the use of a mirror (plane or concave). Note the appearance of the rough glass surface and the visibility of graphite and the metallic particles.

Place a fragment of a cover slip upon the part of the slide where the streaks are located, and (with capillary or glass rod) add to the edge of the cover slip a droplet of cedarwood oil so that the latter is drawn by capillarity into the space between slide and cover. Since the cedarwood oil has nearly the same refractive index as the glass, the particles of the streak now appear imbedded in a sheet of clear glass. Note the visibility and color of the particles by reflected light when using white, black, and colored paper as background. Finally turn back to transmitted light and again note the color of the particles. When doing this, be certain to exclude reflected light by cupping the hands around preparation and objective or enclosing the stage by a black curtain. Try the plane mirror as well as the convex one. Note that transmitted light cannot show anything beyond the silhouette of an opaque object.

Change to an objective of magnification 20 or 10, focus with the standard procedure, Expt. 1 (5), and try to obtain satisfactory illumination by transmitted light with the use of plane and concave mirror. Finally insert the condenser below the stage and try for even illumination of the whole field and satisfactory brightness by using the condenser once with the top lens in place and once with the top lens removed. In each instance, have the substage diaphragm wide open and move the condenser up and down until the whole field is brightly illuminated; then reduce the intensity by closing the substage diaphragm until the image can be viewed with comfort.

Since the opening of the field diaphragm of the eyepiece is practically constant, the field of vision (the **actual** area of the object which is reproduced inside this diaphragm and consequently in the final image) must decrease at the rate at which the square of the magnification of the objective grows. Since the final image on the retina, screen, or photographic plate retains its size and always needs the same amount of light, it becomes necessary to concentrate the needed light into a rapidly contracting area of the object when proceeding to higher objective magnifications. This is done by collecting the light by means of the substage condenser into the field of vision, and the smaller the latter becomes, the more powerful must be the condenser. The focal length of the condenser should be properly related to that of the objective, and for the most efficient use of the available light, both should have the same aperture, Fig. 3a. Removal of the uppermost lens reduces the focal length and the aperture of the condenser so that it becomes better suited for use with objectives of moderate magnifying power. In spite of this, because of the difference of aperture, only a small portion (shaded in Fig. 3b) of the light may be used for image formation. Under these conditions, unconventional use of the concave side of the mirror and empirical readjustment of the position of the condenser may improve the illumination.

Again cut off the light coming from the mirror and try observation by reflected light with backgrounds of different color. Note the effectiveness

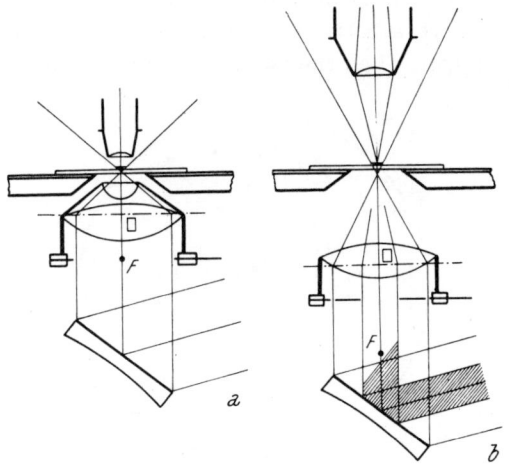

Fig. 3. Use of Condensers. The shaded area indicates the pencil of light actually used for image formation

of the various kinds of illumination in the differentiation of the three types of particles and in the observation of the structure of the rough glass surface. If available, try also other types of illumination.

Experiment 3

Calibration of Eyepiece Micrometer; Working Distances and Fields of Vision

Stage micrometer; square of millimeter graph paper.

The center of the stage micrometer carries an accurately divided scale produced by a photographic process or by engraving. The value of the divisions is indicated by the manufacturer. If necessary, the stage micrometer may be cleaned with lens paper, but care should be taken to avoid damage to the cover slip protecting the micrometer scale.

First decide upon the tube length to be used and adjust it accordingly. Then carefully focus the eyepiece micrometer scale with the eye lens.

Place the square of graph paper upon the stage, attach the objective of lowest magnification, and by either transmitted or reflected light focus upon the ruling. Obtain the diameter of the field of vision in millimeters by counting the number of millimeter spaces included. Get the working distance of the objective (i. e., the distance from the preparation to the front lens of the objective focused upon the preparation) with a millimeter rule held parallel to the optic axis of the microscope.

Comparison of Scales. Test for absence of parallax by slightly moving the eye in front of the eyepiece. If the rulings seem to change their relative positions when the eye is moved, repeat the focusing of the eyepiece micrometer scale and of the ruling on the graph paper until a movement of the eye will no longer give a relative displacement of the images of the two scales. This will happen when the two scales are truly located in conjugated

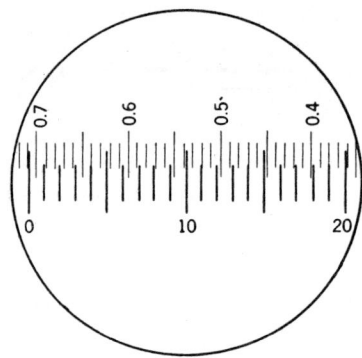

Fig. 4. Calibration of Eyepiece Micrometer. Using the coincidences at 4 and 14 of the eyepiece scale and the right edges of these lines, one finds 10 div. = 0.17 mm and 1 div. = 17 μm

image planes. Obviously, a micrometer eyepiece in conjunction with an image forming system (objective of microscope or telescope, lens of refractometer) may be generally used for the exclusion of parallactic errors by the projection of the image of the object into the plane of the rulings.

Rotate the eyepiece or the stage so that the lines of both scales become parallel. Then select two points, not too far from the center of the field and approximately one-third of the diameter of the field apart, where lines of the eyepiece scale coincide with lines of the graph paper. It may be necessary to move the graph paper to obtain coincidence. The lines of the graph paper may appear so wide that it becomes necessary to decide whether the center or an edge of the lines shall be used.

Determine and record the distance between the two points of coincidence in units of both scales. Compute and record the value of a division of the eyepiece micrometer scale for the particular objective and tube length. If a stage micrometer scale is not available, repeat with all other objectives and collect all data in a table.

If a stage micrometer scale is available, use it for repeating the comparison of scales. The direct focusing of any small object such as a stage micrometer scale is rather trying, especially when a strong objective is being used. It is difficult to place the slide so that the small object gets within the minute field of vision, the location of which may only be guessed. In such instances, focus first upon some coarse structure located in approximately the same level as the object: the edge of a cover slip or label; or the upper edge of the slide, etc. Focus by the standard procedure, Expt. 1 (5), and then move the slide until the object (micrometer scale) appears in the field of vision.

Starting with the test for absence of parallax, repeat the comparison of scales as directed above. From the known value of the division of the stage micrometer scale compute the value of the division of the eyepiece micrometer scale for the particular objective and tube length in absolute measure (mm or μm). If the tenths of the divisions have been estimated, three to four figures may be significant. Since the eyepiece scale is now accurately calibrated, compute the distance between the lines of the graph paper in millimeters.

Repeat the whole experiment with all objectives supplied and collect all data (eyepiece and tube length; objective: focal length, working distance, magnification, diameter of field of vision, value of division of micrometer scale, and possibly distance determined for one interval of the graph paper) in form of a table which may be posted inside the microscope case.

The Immersion Method for the Determination of Refraction

The outlines of an object become visible if the object and its surroundings behave differently to light. They may differ in absorption or rate of transmission (refractive index) or both. Selective absorption in the visible range imparts color; complete absorption, opacity of a homogeneous object. Differences in the rate of transmission lead to refraction, reflection, and diffraction phenomena at the boundaries, which render them visible.

When the outlines of a colorless transparent solid are invisible after immersion into a colorless clear liquid, one may assume that their refractive indices are identical. If the refraction of either one of them may be found, that of the other is already determined. Series of liquids as well as solids have been collected to serve as standards of known refractive index and may be obtained from laboratory supply houses. A series of solid standards, supplied in powder form, has been developed by KOFLER (98, 163); it starts with cryolite ($n = 1.3400$), fluorite (1.4339), and vitreous silica (1.4584) and continues with the use of twenty different glasses up to $n = 1.6718$. R. FISCHER adds sodium fluoride (1.3255) and lithium fluoride (1.3918). A list of liquid standards is assembled in Table III. Since even the most stable substances may slowly undergo changes, the refractive index of liquid standards should be occasionally checked with the refractometer. Gaps in refraction may be bridged by mixing adjacent liquids until the outlines of the solid disappear, whereupon the index of the mixture is determined with the ABBE refractometer; to this end, it is necessary that the liquids are miscible, and it is desirable that they have about the same vapor tension (boiling point) to prevent fractionation during use. Since the solid must be insoluble in the liquid medium, it is necessary to have alternatives, and aqueous solutions of NaCl, KI, and K_2HgI_4 will provide standards for the range from 1.34 (water) to 1.72.

Table III. *Liquids for Use as Standards in Refractive Index Determination*

	n	b. pt. °C		n	b. pt. °C
A			**B**		
Methanol	1.329	65	*Mixtures of* SPANGEN-		
Water	1.333	100	BERG (1225)		
Acetone	1.359	56	1.47 to 1.62		
Ethanol	1.37	78	Glycerol and	1.47	290
n-Hexane	1.37	69	Quinoline	1.624	237
Ethyl acetate	1.373	77	1.54 to 1.62		
n-Heptane	1.388	98	Diethylaniline and	1.542	218
n-Butanol	1.399	118	Quinoline	1.624	237
n-Pentanol	1.40	138	1.59 to 1.74		
n-Butyl chloride	1.402	79	Bromoform and	1.596	150
p-Dioxane	1.422	101	Methylene iodide	1.740	180
Ethylene glycol	1.431	197	1.62 to 1.66		
Ethyl citrate	1.442		Quinoline and	1.624	237
Chloroform	1.45	61	α-Bromonaphthalene	1.658	281
Cyclohexanone	1.450	156	**C**		
Cyclohexanol	1.466	161	*Aqueous Solutions*		
p-Cymene	1.490	177	Sodium chloride	1.33 to 1.37	
Benzene	1.501	80	Potassium iodide	1.33 to 1.50	
Ethyl iodide	1.513	72	Potassium tetra-		
Anisole	1.517	154	iodomercurate	1.50 to 1.72	
Chlorobenzene	1.525	132			

The dependence of the refractive index upon the frequency of the light and the temperature should be considered. Both factors affect the refractive index of liquids much more than that of solids. In the instance of liquids, the refractive index decreases by about 0.0015 for an increase of 10 nm (100 Å) in the wave length and by about 0.0004 per degree Celsius (centigrade) rise of temperature. For solids, the **dispersion** due to difference in wave length is about half that given for liquids, and the temperature effect is about one-tenth of that for liquids.

It follows that the boundaries cannot be made to vanish when white light is being used since the refractive indices of solid and liquid may be matched for only one color of light at a time; when this happens, the boundary still remains outlined in the colors of those parts of the spectrum for which balance has not been achieved. Use of monochromatic light overcomes this difficulty and permits estimation of the index to one unit in the third decimal place (for a given wave length and temperature); if the monochromatic light is chosen so that it is not noticeably absorbed by the two media, the immersion method may be extended to objects with body color. In addition, small differences in refraction may be bridged by varying either the wave length by means of an adjustable monochromator or the temperature with a heating stage (98, 163); both are adjustable with EMMONS' double variation apparatus consisting of arc lamp, monochromator, polarizing microscope equipped with special universal stage, ABBE refractometer, and system for circulating water of controlled temperature (113).

The effect of differences in refractive index is already known from the experiment (2) with the ground-glass surface. It may be imagined that the difference of refractive index may be estimated to the nearest tenth of the unit by mere comparison with a scale of specimens exhibiting differences of 0.0, 0.1, 0.2, and 0.3. A systematic procedure will start by comparing with a standard of $n = 1.5$ and use a test that indicates whether the unknown has the higher or lower refractive index. The next comparison will accordingly be made with a standard of about 1.4 or 1.6. Two more trials should reduce the gap to about 0.02, whereafter it should be possible to decide upon the procedure for the final matching. Tests for determining the direction of the refraction difference are described in Expt. 4.

For lack of suitable immersion liquids, the immersion method becomes impractical for solids with refractive indices above 1.75.

Experiment 4

Phenomena Caused by Differences in Refraction

Ethanol, ethylene bromide, quinoline, and cedarwood oil.

The following experiments suggest also types of illumination for getting a good image of objects immersed in a medium of similar refractive index.

A. Clean three slides, and place upon the center of each a drop of about $0.5\,\mu l$ of 1% NaCl solution. Put the slides aside until the drops have evaporated completely, and use the time for starting part B of the experiment *(see below)*.

When evaporation is complete, drop on each residue a clean cover slip. Transfer one of the preparations to the stage of the microscope and focus with an objective of magnification 10 or somewhat stronger. If the residue is small or difficult to see, start by focusing the edge of the cover glass.

Most of the residue may form irregular clusters of crystals, but usually it will be possible to find some well developed single crystals with square or rectangular outline. Use such crystals for the following observations.

With a stirring rod, transfer a drop of ethanol to the edge of the cover slip so that the liquid is drawn into the space between slide and cover glass. Focus upon a simple crystal and note that the outlines are now less pronounced than before adding the ethanol. Vary the intensity of the illumination by opening and closing the substage diaphragm, raising and lowering the condenser, or changing the distance between mirror and light source. Note that the *outlines become weaker when the intensity of the illumination is increased*. The difference of refractive indices is now $1.54 - 1.36 = 0.18$. The direction of the difference is found by the following two tests of about equal sensitivity, ± 0.001 with monochromatic and ± 0.005 with white light. Frequently, it will be wise to try both tests to find which gives the more conclusive evidence.

The *Becke Test* is most suitable for thin objects bounded by surfaces about parallel to the optical axis of the microscope. It may not be reliable with rounded grains.

Move the slide to bring the test crystal into the center of the field. Use weak axial illumination, either just the plane mirror with the source at a distance or the condenser with the diaphragm only slightly open.

Focus the crystal to get it bounded by a single dark line. Continuously looking at the crystal, *slowly* raise the body tube a *short* distance. Note that a fine bright line appears parallel to the dark outline. This "halo", Fig. 5, is the Becke line. *Slowly* lower the body tube and note that the Becke line moves toward the black outline and disappears when the crystal is sharply in focus again. Lower the body tube some more and observe the appearance of the Becke line on the *other* side of the black outline. If the phenomenon is not distinctly visible, repeat the experiment with a thinner (smaller) crystal.

For interpreting the Becke test, remember the rule: on **lowering** the body tube, the Becke line moves into the medium of **lower** refractive index; on *raising* the tube, the Becke line moves in the direction in which the refraction *rises*.

Oblique Illumination. Retain in the center of the field the crystal which gave a good Becke test. If a substage condenser is being used, bring it into the highest position that gives a good illumination; this is necessary to avoid an otherwise possible reversing of the shadow effect. Open the substage diaphragm wide and obtain oblique illumination by

Expt. 4 The Immersion Method for the Determination of Refraction

covering first all but the extreme left side of the condenser opening with a strip of black cardboard, about 4-cm wide, and then moving it to the left until only the extreme right side of the condenser opening admits light. Move the strip back-and-forth while watching the crystals through the microscope for changes in the distribution of shadow and light.

If the substage condenser is mounted to permit lateral movements, repeat the experiment by using a narrow opening of the substage diaphragm and moving the condenser-diaphragm combination back-and-forth between the extreme positions on the left and on the right.

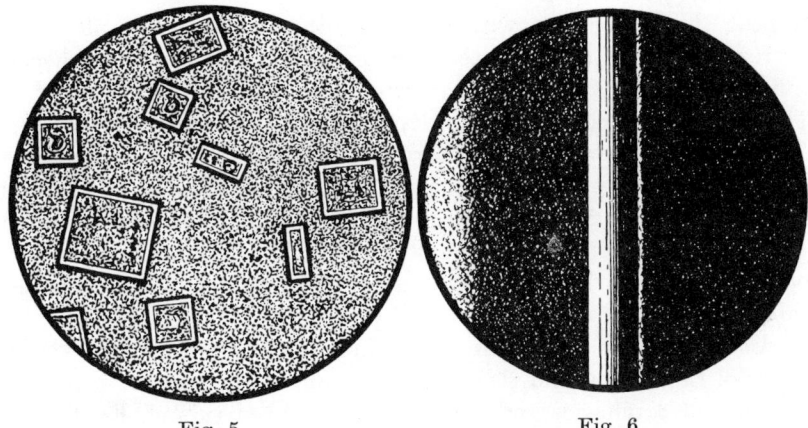

Fig. 5.   Fig. 6

Fig. 5. Sodium Chloride Crystals Immersed in Alcohol. Microscope is raised above the position of sharp focus

Fig. 6. Glass Thread Immersed in Water. Mirror of the microscope is swung to the left-hand side

Finally remove the condenser, open wide a remaining diaphragm, and use the plane side of the mirror. Start with the mirror in the optic axis of the microscope, and adjust it so that the best illumination is obtained. Then, while observing the crystals through the microscope, swing the mirror to the left side until a shadow enters the field of vision and its edge almost arrives at the opposite boundary of the field, Fig. 6. While watching for changes in the distribution of shadow and light in the images of the crystals, swing the mirror through the center to the right side—and then back again, etc.

This test is best suited for rounded objects and may be disappointing with the sodium chloride crystals. The phenomenon may be improved by using an objective of lower aperture, and a special objective equipped with iris diaphragm is recommended for the investigation of very small particles (88). The interpretation of the observed phenomenon must consider the method of illumination in use.

With the condenser in high position so that it focuses above the preparation, the heavy *shadow* in the image of the object will move with the source of the light (opening in the diaphragm) from the left to the right and will face the shaded side of the field if the *object has the higher* refractive index than the surrounding medium. (If the condenser is lowered to focus below the preparation, the phenomenon is reversed and also less distinct.)

If the test is made with the mirror and without use of a condenser, the *bright* spot or line in the image of the object will follow the motions of the mirror from left to right if the *object has the higher refractive* index than the surrounding medium.

Repeat the Becke test with the two other sodium chloride residues after treating one with ethylene bromide ($n = 1.54$) and the other with quinoline ($n = 1.6$) and applying cover slips.

The effect of ethylene bromide depends upon the purity of the reagent and upon the temperature. As a rule, the outlines of the crystals will still be visible when the intensity of illumination is reduced to a practical minimum. If color fringes appear, identity of refraction may be assumed for that part of the spectrum that does not show up in the fringes. If the proper monochromatic light were used, the outlines would completely disappear.

B. Use glass wool or obtain glass thread by drawing out a rod to a diameter of first 1 mm and then 0.1 to 0.2 mm in the manner used for the preparation of capillaries and capillary pipets from tubing, Expt. 19.

Mount a 1-cm piece of fine glass thread in water between slide and cover slip. Place the preparation on the stage so that the image of the thread crosses the center of the field going from 6 to 12 o'clock. (Treating the field of vision like the face of a clock, one finds it easy to specify locations such as "two-thirds of the way from the center to eight thirty".)

Apply the Becke test and oblique illumination, and repeat this with cedarwood oil ($n = 1.515$) for immersion liquid. Note that both tests are satisfactory with this latter medium.

C. Aerate saliva in the mouth; transfer a droplet of the foaming liquid to a slide, and cover it with a glass slip at once. The bubbles of air ($n = 1.00$) retain their spherical shape long enough to permit use of the Becke test and of oblique illumination.

Observation of Schlieren

TOEPLER defined *schlieren* (pr. shleeren) as regions of changing refraction and devised a highly sensitive method for their detection by an arrangement of diaphragms which eliminate the participation in the image formation of those rays which pass through the carefully corrected central portions of the lenses (170, 1014). Rather simple combinations of lenses and diaphragms (606, 760, 1014, 1220, 1221) suffice to demonstrate the hot air rising from a human hand, to detect inhomogeneity in a transparent plastic sheet (1055), or to photograph the compression waves surrounding a projectile flying through air. EMICH (1015, 1017, 1018, 1157) adopted the technique for the study of chemical problems and demonstrated that it offers a very sensitive means for detecting small differences in refraction, density, or both and consequently for recognizing hidden circumstances that bring such differences about: different composition, different concentration, temperature differences, diffusion or (and) chemical action affecting concentration or temperature, etc.

Testing identity and purity are foremost among the analytical applications. To these ends, the least sensitive methods of *schlieren* observation will usually suffice. Because of the effects of the particular impurities, they will often allow to identify a certain batch of some "pure" solvent.

For such comparison, one permits a slow, fine stream of the unknown liquid to enter a small sample of the standard which must have the same temperature. If a suitable method of observation is used and the two liquids differ sufficiently in refraction or dispersion, the stream of the liquid becomes visible and the distribution of light and shade indicates the medium of higher refractive index. At the same time, the direction of flow shows which medium has the higher density. Miscible liquids of the same optical density may give *schlieren* nevertheless as a consequence of phenomena occurring in the zone of contact (chemical reaction or diffusion leading to the establishment of concentration gradients), but this cannot happen with substances which are closely similar such as isomers.

The visual method for the observation of *schlieren* (1155) used in the following experiment permits detecting differences of 0.0001 to 0.0002 in the refractive indices and in the specific gravities of two liquids. The magnitude of the difference in refraction may be crudely estimated by comparison with *schlieren* given by known refraction differences and identical optical conditions (position of *schlieren*, eye, etc.). Observed should be the width and the intensity of the shadow portion of the *schlieren*. The differences may be expressed in empirical numbers by using some simple device permitting measurement of that displacement of some part of the optical system, which represents a sufficient deviation from the most sensitive adjustment to render the *schlieren* invisible.

Experiment 5

Visual Observation of Schlieren

Cells. The cell, Fig. 7, formed at one end of a standard microscope slide, may be less than 1 mm thick and may be given the outline of a V or of an amphora to reduce the capacity to 0.1 ml and less. The glass parts may be formed by cutting or grinding microscope slides. For use with aqueous solutions, they may be cemented together with Canada balsam or any suitable plastic cement. For use with organic liquids, the parts may be polished so that they form a tight cell when pressed together in a suitable metallic frame. It is also possible to buy cells made entirely of glass.

Capillary Pipets. For the preparation see Expts. 19 and 22. Determine the diameter of the opening, 0.1 to 0.2 mm, as directed in Expt. 22. The wide part should be 1 mm or more in diameter.

Background. A suitable background is obtained by mounting a frosted 60-watt bulb behind a ground-glass plate, approximately 25 cm by 25 cm. The left half of the glass plate is made opaque by covering with a metal sheet, a black cardboard, or painting with black lacquer. It is important that the dark part of the field is set off by a sharp perpendicular boundary against the bright half.

Procedure. Fast emergence from a wide tip does not only waste sample, it also gives "false" *schlieren* because of the friction which raises the temperature of the flowing liquid. Thus, select a pipet with a bore of the tip not exceeding 0.2 mm. Insert the tip into 1% NH_4Cl solution and apply gentle suction with the mouth until about 4 cm of the length of the wide capillary are filled.

Stretch a rubber band around the upper part of a clean glass cell as shown in Fig. 7, and with a pipet add water to fill three fourth of the cell. Mount the capillary pipet on the slide of the cell as shown by the illustration, but do not insert the tip of the pipet into the cell liquid. Place the cell assembly for a few minutes upon a metal object (stand of a microscope) to make certain that the liquids acquire the same temperature.

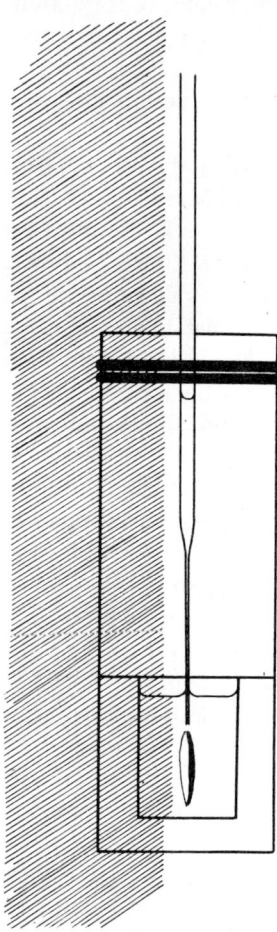

Fig. 7. Observation of *Schlieren*; nat. size

Grasp the cell at the upper left-hand corner of the slide and push the capillary pipet down so that its tip enters the liquid in the cell and the outflow starts. Because of the narrow bore of the tip of the pipet, the flow of the ammonium chloride solution continues for several minutes giving ample time for the observation of the phenomenon.

Hold the cell about 120 cm in front of the ground-glass plate so that the perpendicular boundary of the dark half of the field is seen near to the opening of the pipet, Fig. 7. Note that the NH_4Cl solution flows to the bottom of the cell and that the **shaded half** of the *schlieren* **faces away from the dark half of the background.** The phenomenon corresponds to differences of 0.001 in refraction and 0.003 in density, both being higher for the solution flowing from the pipet.

Clean and dry the cell as directed for centrifuge cones (p. 124) by washing it with distilled water and then drawing air through it. Then repeat the *schlieren* observation by using 1% NH_4Cl in the cell and water in the capillary pipet. This time insert the opening of the pipet deeper into the cell liquid than before. Note that the *schlieren* ascends with the shaded part of it facing the dark half of the background.

Again clean the cell and repeat the experiment. This time use water in both, the cell and the pipet; no *schlieren* should be observed.

Empirical data on *schlieren* strength may be obtained by mounting the apparatus on a board of about 1.5-m length so that the relative positions of lamp, background, cell, and eye are fixed (1155). Mount the ground-glass plate in a vertical position so that it may be moved sideways and its position may be read off a scale. Locate the cell 120 cm in front of the background, and mount 25 cm in front of the cell an eyepiece consisting of a blackened tube with a vertical slit (1 to 2 mm wide) at the far end. First adjust the set-up to obtain maximum sensitivity for the detection of refraction differences. To obtain estimates of *schlieren* strength, move the ground-glass plate to the left until the *schlieren* disappears (dark shadow close to the opening of the pipet vanishes) and then without delay to the right until the *schlieren* again becomes invisible and record the distance between the two points. Absolute data may be obtained by preparing a calibration curve with liquids of known refractive indices.

Use of Polarized Light

All use of polarized light is based upon the fact that properties of anisotropic matter (all crystallized matter except cubic crystals; cubic crystals and glasses under strain; and liquids having the molecules more or less uniformly oriented) are a function of the direction of action with regard to the internal structure. When the interdependence of kind of substance, internal structure, and variability of properties is once known, a study of the latter may be used to get information on the internal structure and on the chemical nature.

The action of anisotropic matter upon light may be briefly summarized as follows. In most directions, only plane polarized light is transmitted, and only two certain planes of vibration which always are vertical to one another are available. Both planes of vibration may be used simultaneously, and light entering the anisotropic material is resolved into the two components using the two available planes of vibration in that ratio which makes the most efficient use of the nature of the incoming light; ordinary light and plane polarized light vibrating at an angle of 45 degrees to both offered planes of vibration are distributed with equal intensity; plane polarized light which has one of the possible planes of vibration continues through the anisotropic material with that plane of vibration, and the other component will not form; plane polarized light with the plane of vibration inclined at an angle different from zero and 45 to the offered planes of vibration is very unequally distributed so as to favor that plane which is closer to the plane of vibration of the incoming light. As a rule, the anisotropic material acts more or less differently upon the two components of light, vibrating in planes perpendicular to one another, even when they travel along the same path. As a rule, the two components travel with different speed, which implies different refractive indices, different wave lengths, and acquisition of a phase difference on the path through the anisotropic matter; the two components may also experience different selective absorption and acquire different "color".

Any anisotropic material produces a mixture of equal parts of two components of plane polarized light when illuminated with ordinary light. Polarizing equipment, regardless whether it is used for the investigation of large objects or whether it is combined with a microscope for the study of small objects, always has the purpose of getting plane polarized light of *just one* plane of vibration, and this may be utilized in two ways. Obviously it may serve for isolating either one component of plane polarized light emerging from the object so that the properties of the object (refraction, absorption) may be studied in the light of one component at a time (determination of refractive index, observation of pleochroism). Most frequently it is used for combining the two components of plane polarized light emerging from the object so that they assume identical planes of vibration and consequently interfere. The phenomena resulting from interference permit conclusions concerning type and orientation of the structure of the object.

In principle, the polarizing apparatus proper is extremely simple. Its two parts, the polarizer and the analyzer, are identical and interchangeable in essence. Both are devices for obtaining plane polarized light of only one plane of vibration and usually consist of some anisotropic medium (crystal or aggregate of crystals having identical orientation) that unavoidably converts ordinary light into two components of plane polarized light, but has the exceptional property of removing one of the components by either turning it away on a different path (Nicol prism or nicol for short) or by absorbing it (various types of Polaroid). One of these devices (polars) is called polarizer and is mounted below the stage to supply to the preparation plane polarized light with one definite plane of vibration. The other is called analyzer and is mounted above the preparation to get all light emerging from the preparation to vibrate in one prescribed plane.

Any conventional microscope may be quite readily converted to a polarizing microscope with the use of accessories having Polaroid films and retardation plates (mica or selenite) in circular mountings between glass plates. The polarizer may be inserted into the stage aperture or into a slot above or beneath the substage diaphragm. The (cap) analyzer is usually placed upon the top of the eyepiece. Retardation plates may be obtained in mountings that fit on top of the polarizer mount. A substitute for a rotating stage may be obtained by fitting a circular (graduated) disk around a short piece of metal tubing that slips into the opening of the stationary microscope stage.

A polarizing microscope (petrographic microscope) may use Nicol prisms for polarizer and analyzer. The polarizer is usually mounted below the diaphragm of the substage so that it may be rotated around the optical axis of the microscope. The direction of the plane of vibration (or that perpendicular to it) is usually indicated upon the mounting, and the 6 to 12 o'clock position of that direction may be indicated by a stud falling into a notch of the mounting. The analyzer is frequently mounted above the objective so that it may be quickly slid in and out of the tube of the microscope. Below it, a slot is provided in the tube for the insertion of retardation plates and quartz wedge. The rotating stage is graduated and provision is made for aligning axis of rotation of the stage with the optic axis determined by the objective. The centering may be done with the stage, with the objective changing device, or with individual objectives, and an objective changing device is used, which assures that objectives are always returned into the same position.

Essential for a polarizing microscope is furthermore that the cross-hair eyepieces carry a stud to engage in a notch of the draw-tube so that the hairs indicate the directions 6 to 12 and 3 to 9 o'clock, which are also the directions of vibration in polarizer and analyzer in the positions parallel and crossed. Obviously, the draw tube must be keyed to the body tube so that it cannot be rotated in the latter. Customarily, a divisible Abbe condenser with iris diaphragm is provided so that it may be focused mechanically or completely swung aside together with the polarizer. A Bertrand lens, an auxiliary lens for the observation of axial figures, is frequently found mounted in a metal frame so that it may be slid into the body tube above the analyzer.

The interpretation of the phenomena observed with a polarizing microscope easily leads to wrong conclusions and will give a meager yield if not based upon a rather thorough study of crystal optics (89, 93, 96, 98, 107, 110).

Experiment 6
Testing and Adjusting a Polarizing Microscope

The inspection of the stand and the microscope proper has been treated in Expt. 1. The following additional tests apply to the polarizing equipment and accessories.

1. *Approximate Vibration Directions of Polarizer and Analyzer.* The following procedure is frequently needed just for checking the meaning of marks on the mountings.

Take either the analyzer or the polarizer out of the microscope, hold it close to the eye, and observe through it the light reflected from some plane horizontal surface of dark color such as a black glass plate, table top, or some surface coated with black enamel or lacquer. Most of the reflected light is plane polarized and vibrates horizontal, i. e., in a direction parallel to the reflecting surface. Rotate the polarizing device to find the position in which it transmits least of the reflected light. In this position, the diameter produced by a perpendicular cut through the mounting represents the vibration direction of the transmitted polarized light. (The horizontal diameter represents the vibration direction of the component swallowed by the device.)

2. *Action of Polarizer and Analyzer.* Place upon the stage of the microscope a slide with a cover slip on it. Focus upon the edge of the cover slip and adjust the illumination to get bright illumination of the whole field when both, analyzer and polarizer, are removed. Insert the polarizer below the stage so that its vibration direction goes from 6 to 12 o'clock (in microprojection, it may be preferable to rotate the polarizer into the position in which it will transmit any plane polarized light already obtained by reflection on the surface of a mirror or prism). As a rule, it is impossible to perceive that the intensity of the illumination is cut into half by the insertion of the polarizer.

Insert the analyzer and rotate it (or if this cannot be done, rotate the polarizer) until no more light is transmitted and the field appears dark when looking into the eyepiece. This happens when the light transmitted by one polarizing device has that plane of vibration which makes it the component that is swallowed up by the other. The planes of vibrations of the components that are transmitted by the devices are **crossed** at right angles.

The vibration direction of one polar is already known; that of the other must now be perpendicular to it. If the adjustments are properly carried out, the vibration directions should be parallel to the cross hairs in the eyepiece (6 to 12 and 3 to 9 o'clock). A more precise test follows under (6).

The position *parallel* is obtained by rotating the analyzer or, if this cannot be done, the polarizer through 90 degrees so that the vibration directions of their transmitted components arrive in the same plane.

Rotate one of the polars a few times through 360 degrees and observe that the positions of minimum and maximum brightness corresponding to **crossed** and **parallel** occur twice during a full turn and that the blackening out occurs quite suddenly and at a precisely adjustable position, whereas the position of maximum brightness is only vaguely defined. If the rotating device has a graduated scale, record the average position of minimum brightness and its standard error computed from several settings. If the analyzer is in fixed position and the polarizer is being rotated, minimum brightness should occur when the stud clicks into the notch (readings zero and 180). If the polarizer is set to zero and the analyzer is rotated, darkness should occur at 90 and 270 if polarizer and analyzer circles have the zero at 12 o'clock.

Test all objectives by focusing on the edge of the cover slip in a bright field and then crossing the polars to the position of minimum brightness. Even with very strong illumination, the field should become reasonably and evenly dark. If fringes or spots of color or brightness are noticed, move the preparation, rotate the objective (and move the lenses of the condenser if this is possible) in an attempt to find the source of the imperfection. Slides and lenses may become anisotropic due to mechanical stress or heating and rapid cooling. A small amount of light is always derived from the partial depolarization on refraction and is most noticeable with high-power objectives having several strongly curved lenses (88).

3. *Centering.* Add some glycerol to the cover slip to hold it in place. Remove the analyzer. With a cross-hair eyepiece in place, focus a corner of the cover slip with the objective having the lowest magnification. Move the slide to get the point of the corner to coincide with the intersection of the cross hairs and, if necessary, use a stage clip to hold the slide in place. While looking into the microscope, rotate the stage. To this end, cup the hands around the stage so that they make contact with its outer edge but do not touch the top surface of the stage. Estimate the center of the circle which is described by the image of the point of the cover slip and mark it by moving the slide until the point of the slip is located there. Using only the centering screws, bring the point to the intersection of the cross hairs. Again test by rotating the stage and repeat the process of adjusting and testing until the mark represented by the corner of the cover glass remains at the intersection of the cross hairs when the stage is rotated.

If the centering is done at the objective, repeat it for each objective belonging to the microscope. If the centering is done at the stage or at the tube of the microscope, repeat it with the strongest objective and convince yourself that the setting is satisfactory for the objectives of lower power also.

Finally check the centration of the condenser with the top lens in place and the strongest objective as directed in Expt. 1.

4. *Checking the Graduation of the Stage.* Align one edge of the cover slip with one of the cross hairs. Use a stage clip if necessary. Read the position of the stage, and then rotate the stage to turn the image upside down so that the edge is again lined up with the cross hair. Again read the position of the stage, which should have changed by 180 degrees. If it appears advisable, repeat the test by using another part of the scale (turn the slide).

5. *Checking the Angle of the Cross Hairs.* Align the edge of the cover glass first with one cross hair, and read the position of the stage. Rotate the stage until the edge is lined up with the other cross hair, and again read the position. The readings should differ by 90 degrees.

6. *Checking the Position of the Cross Hairs.* The cross hairs are supposed to indicate the vibration directions in the polarizing devices in the positions crossed and parallel.

Place a large drop of nitrobenzene upon the center of a microscope slide. Add a small amount of anthraquinone which must be crystallized as fine orthorhombic needles. Stir and place a cover slip upon the mixture. Use an objective of magnification 10 and focus first upon the edge of the cover slip; then select a perfect, long needle, and move the slide until one of its long edges goes through the intersection of the cross hairs.

Secure strong, even illumination of the field. Insert the analyzer, and carefully cross analyzer and polarizer so that their vibration directions presumably take the marked positions (click) 6 to 12 and 3 to 9 o'clock, while the *field acquires minimum brightness*. Rotate the stage while cupping the hands around it until the needle is in the position of minimum brightness (appears as dark as the field). Then turn or remove the analyzer and determine the angle formed by edge of the needle and the cross hair. Return the analyzer into the crossed position, turn the stage through 90 degrees, and again bring the needle in the position of minimum brightness with the hands cupped around the stage. Turn or remove the analyzer and observe the inclination of the edge of the needle to the second cross hair.

Both times, the cross hairs should be closely parallel to the edge of the crystal. Otherwise they do not correctly indicate the vibration directions of the polarizing devices and must be aligned by either rotating the cross hairs or both nicols.

Reserve the preparation for Expt. 10.

Experiment 7

Isotropic and Anisotropic Substances in Polarized Light (96, 110)

Red first order selenite (gypsum) plate, quartz wedge as supplied with polarizing microscopes. Substitutes may be made of thin cellophane sheet: cut a ribbon, 5 to 10 mm wide, and cut this into squares; mount between slides and cover slips (a) a sandwich of three or four squares in the *same* orientation and slightly displaced so that one side forms a stair with steps of less than 1-mm width, (b) two squares, one rotated against the other through 90 degrees, side by side, but slightly overlapping to give a double layer of less than 1-mm width. — Crystals of $Na_2S_2O_3 \cdot 5\ H_2O$.

Using the objective of lowest power and a cross-hair eyepiece, remove the analyzer and focus upon an edge of the gypsum (monoclinic, for D line: $\alpha = 1.5205$, $\beta = 1.5226$, $\gamma = 1.5296$) plate or the cellophane square. Note the relation of this edge to the vibration direction of the slower component of light usually indicated upon the mounting of the selenite plate (if the mounting does not permit focusing upon an edge of the selenite plate, use India ink or some marking pen to draw a parallel to the direction indicated upon the glass cover cemented over the selenite plate and focus upon this line).

Insert the analyzer and cross the polars. Rotate the stage and confirm that the anisotropic gypsum plate is always the same hue of red excepting when in the four positions of extinction ocurring during a full turn of 360 degrees. Note that the parts of the field that are not taken up by the gypsum plate and contain only isotropic material (glass, immersion liquid, clear cement, air) always remain dark. Confirm that, in the positions of complete extinction, the cross hairs indicating the vibration directions in polarizer and analyzer also **indicate the vibration directions in the specimen,** the gypsum plate (the one marked upon the mounting and the other vertical to it). Try to explain why darkness must prevail with crossed polarizers when the specimen is isotropic or when an anisotropic specimen has the vibration directions coinciding with those of the polars; why is there brightness and sometimes color when an anisotropic specimen is not in the extinction position? Why is the hue of the color always the same when the stage is rotated?

Make the polars parallel and again rotate the stage. Observe that the phenomena are complementary to those observed with crossed polars: the isotropic media remain always bright and appear in the same color as in ordinary light; the gypsum plate is colorless bright in the position of extinction and green in the 45 degree position (its vibration directions at 45 degrees to those of the polars). Note that the green color fades rapidly as soon as the specimen is rotated out of the 45 degree position and that the plate is white most of the time when the stage is rotated. Try to explain this as well as the green color in the 45-degree position.

Substitute the quartz wedge (stair) for the gypsum plate and, using crossed polars, select a portion that shows bands of different color. Repeat the procedure followed with the selenite plate as directed above. Realize that the differences in color must be due to differences in thickness since all other conditions are equal. Note that all colors retain their hue during the rotation of the stage and that all of them change to complementary colors when proceeding from crossed to parallel polars. Also observe that the colors are in general more brilliant with crossed than with parallel polars.

Cross the polars and bring the quartz wedge into a 45 degree position. Keeping it in this position, push it through the field of vision while observing through the microscope. Pay attention to the succession of colors. Note that (a) the thin edge appears black; (b) the appearance changes to gray, lavender, yellow, orange, red, violet, blue, green, yellow, orange, red, etc. as the thickness increases; (c) that the series of colors is repeated, but that the reds, blues, etc. of the successive series (orders) differ in hue so that a certain color and hue corresponds to a definite thickness; (d) that the colors lose purity and brilliance and assume pastel hues as order and thickness increase; (e) that it is possible to distinguish and recognize the colors of the various orders either by memory or by comparison with the color series of a wedge.

Make the polars parallel and repeat the observation of the succession of colors (it may be easier to keep the wedge in the 45 degree position by using the slot in the body tube). Note that the series starts at the thin edge with white, yellow, orange, and red of first order and then continues as with crossed polars.

Bring the band of the green of second order, i. e., the first green from the thin end, into the center of the field and then change to crossed polars. Confirm that the red obtained as complementary color is the red of first order by slowly moving the thin edge of the wedge into the field and observing the succession of colors. Note that crossed polars require a thicker specimen (longer path through the specimen) than parallel polarizers to produce the same interference color, and that the sequence of color bands moves closer to the thin edge when changing from crossed to parallel.

Repeat all the experiments with the quartz wedge by using monochromatic light (sodium lamp or color filter) for illumination, and try to explain the phenomena. Observe that the dark bands move toward the thin edge when changing from crossed to parallel. Change again to white light for the last part of the experiment, which follows.

Place upon the center of a slide a small clear crystal of $Na_2S_2O_3 \cdot 5\ H_2O$ and fuse it by placing the slide upon a steam bath. Without delay, place a cover slip on the drop of melt. The melt should fill the entire space between the glass surfaces. If too much thiosulfate was used, remove the excess by touching the edge of filter paper to the edge of the cover slip until only a very thin film of liquid is left. Start crystallization by touching the melt at the edge of the cover glass with a needle that has been in contact with a crystal of the pentahydrate.

Remove the analyzer and use an objective of power 10 for focusing the preparation. Note that the outlines of the thin monoclinic plates are not very distinct and are much better seen with polarized light. Use crossed polars and rotate the stage. Note that: (a) air bubbles enclosed by the crystals remain always black; (b) each plate retains its color during the revolution of the stage; (c) not all plates have the same color in spite of the fact that they consist of the same materials and must have the same thickness; explanation? (d) that extinction does not occur simultaneously for all crystals in the field since they are differently oriented.

Observe the inclination of the long edges of the plates to the cross hairs when they are in the position of extinction, and repeat the observation with different sets of plates. Search for a center (point of seeding?) from which crystals radiate in all directions, and bring it into the center of the field. Rotate the stage. Observe and explain the more or less stationary black cross. Prepare a crude pencil sketch of some part of the preparation, and indicate by arrows the vibration directions in the various crystals.

Select a very colorful portion of the preparation, and then rotate the analyzer and observe the change to complementary colors when going from crossed to parallel and back. Finally place the analyzer into position parallel, rotate the stage, and observe that the color display is far less dazzling than with crossed polarizers.

Save the preparation for Expts. 8 and 13.

If a spectroscopic eyepiece is available, use it to analyze the compositions of colors of increasing order as exhibited by the quartz wedge and also of crystals showing the white of higher order. For an alternative, the image of the crystals may be projected onto the slit of a standard spectroscope (1283).

Experiment 8

Determination of Vibration Directions in Relation to Profile (96, 110)

Three 1-inch watch glasses, filter paper; saturated solutions of $Na_2S_2O_3 \cdot 5\ H_2O$, $BaCl_2 \cdot 2\ H_2O$, and $CuSO_4 \cdot 5\ H_2O$.

Of each of the three reagent solutions take one drop of $10\ \mu l = 0.01$ ml, transfer it to the center of a microscope slide, and spread it to a circle of not more than 1-cm diameter. Put the slides aside, but inspect them from time to time. When crystals begin to separate, start observing them with the lowest magnification available. When well developed crystals of 0.1- to 1-mm length or diameter have been obtained, stop the evaporation by inverting small watch glasses over the drops or by drawing off the mother liquor by touching filter paper to it. The latter alternative has the disadvantage that annoying crusts of tiny crystals may form on evaporation of residual liquid. These may be dissolved by breathing upon the slide and quickly covering the residue with a small watch glass.

The outcome of crystallization experiments is quite unpredictable, but one may hope that some suitably developed crystals will be obtained.

If crystallization is too copious or too rapid, dilute some of each solution with an equal volume of water, and repeat the crystallization with drops of the diluted solutions.

Perform the following procedure with all preparations which appear promising and try it also with the thiosulfate crystals of Expt. 7.

Remove the analyzer and use the lowest available magnification to start the search for suitably developed crystals. Large thin plates with clearly defined edges are most desirable.

When a crystal has been selected, move the slide to bring it into the center of the field. If necessary, use a stronger objective so that the outstanding dimension of the crystal becomes more than one-fourth of the diameter of the field. Then use the polarizing equipment in position crossed, and rotate the stage to bring the crystal in the position of complete extinction. Remove the analyzer and observe the position of the outstanding dimension of the crystal (long edge, outstanding diagonal) in relation to the cross hairs which give the vibration directions for the given view of the crystal.

The extinction is called **parallel** *(symmetric)*, if the outstanding dimension (edge or diagonal) is parallel and vertical to the cross hairs. It is **oblique,** if the outstanding dimension is inclined to the cross hairs. The **extinction angle** is the angle enclosed by the outstanding direction and the nearest cross hair.

Try to record all pertinent facts without worrying about the correct crystallographic interpretation. Prepare a sketch of the crystal and indicate the vibration directions. Record all angle measurements in the sketch.

To measure a *silhouette or profile angle*, move the slide to bring the point of the angle to the intersection of the cross hairs. Rotate the stage to align one side of the angle with one of the cross hairs and read the position of the stage with the use of the vernier; it is preferable to have the point of the angle slightly off the center of revolution so that the lines become parallel when they are very close to one another without actually coinciding. Record the position and then rotate the stage so that the aligned cross hair sweeps through the angle. Align it with the other side of the angle in like manner as before, and again record the position of the stage. Repeat the process and compute averages for both readings. Their difference represents the angle.

Determine the *angle of extinction* as follows. Move the slide so that the outstanding direction of the crystal passes through the intersection of the cross hairs and is parallel and very close to one of the cross hairs. Read and record the position of the stage. Cross polarizer and analyzer and rotate the stage with the hands cupped around it until the crystal is in the position of extinction that is closest to the starting position. Again read the position of the stage. Remove the analyzer, return to the starting position, and repeat the determination. The difference of the averages of the readings represents the extinction angle which is recorded in the sketch.

If possible, repeat the angle measurements with several crystals of each substance and record the results with the use of drawings. Occasional large discrepancies should not surprise. Profile angles depend upon the orientation of the crystal, and the same holds for extinction angles. The true outstanding directions may be obscured by irregular development of crystals.

If a rotating stage is not available, angle measurements may be performed with a goniometer eyepiece, or the image may be projected or photographed and a protractor used on the screen image, tracings, or a (possibly enlarged) photomicrograph.

Save all preparations containing well-shaped large crystals. $Na_2S_2O_3 \cdot 5 H_2O$ and $BaCl_2 \cdot 2 H_2O$ are monoclinic; $CuSO_4 \cdot 5 H_2O$ is triclinic.

Experiment 9

Use of Compensators (96, 110)

First–order red gypsum plate; quarter wave length mica plate; quartz wedge. — Saturated solutions of $NH_4H_2PO_4$ and NH_4ClO_4; the latter may be prepared by bringing together solutions of equivalent amounts of $Ba(ClO_4)_2$ and $(NH_4)_2SO_4$.

Proceed as in Experiment 8 to get well-developed crystals of $NH_4H_2PO_4$ and NH_4ClO_4.

The first part of the experiment requires superimposing the quartz wedge upon a retardation (gypsum or mica) plate. If the microscope tube has a slot, use it for the quartz wedge and place the retardation plate upon the stage; if not, try to insert the retardation plate into the substage and use the wedge upon the stage; if this too cannot be done, clamp the retardation plate to the stage and use the wedge on top of it.

Cross the polars and focus on the top surface of the device upon the stage. Take the quartz wedge out of the path of light and rotate the gypsum plate into that 45-degree position of maximum brightness for which the slower component (arrow) vibrates in the direction 7:30 to 1:30 o'clock. Superimpose the quartz wedge with the same vibration direction of the slower component; with the thin edge leading, advance it slowly into the field and observe the shift of the color bands. If it is possible, rearrange the position of the gypsum plate so that it occupies only the upper half of the field and the other half of the field shows the color bands as they occur in the wedge without the addition of the gypsum plate. Note that the adding of the gypsum plate, which by itself produces the red of first order, advances the colors of the wedge by one order. The gypsum plate **adds** to the thickness of the wedge so that its very edge shows already the red of first order.

Rotate the gypsum plate through 90 degrees so that the vibration direction of its slower component is parallel to the vibration direction of the faster component in the wedge. If possible, again move the gypsum plate so that it appears only in the upper half of the field. Note that the action of the gypsum plate **subtracts** from the thickness of the wedge. Where the red of first order is located in the wedge, the combination gives a black band indicating zero thickness or complete **compensation**. The optical thickness increases on both sides of the black band. Going toward the thick end of the wedge, the colors of first, second, third order appear displaced by the space of one order; toward the thin edge, the optical effect of the gypsum plate increases from the gray to the red of first order, which latter appears at the edge where the effect of the quartz has become zero.

Obviously, the quartz wedge may be used to determine the order of an interference color by superimposing it upon the specimen so that it compensates and then advancing it until the compensation becomes complete and the specimen blacks out. This will happen at that point where the optical effect (thickness) of the wedge is equal to that of the specimen and, consequently, where the wedge exhibits the same color as the specimen. It then remains to identify the order of the color by counting the number of orders (red bands) between the color giving complete compensation and the thin edge of the compensator. The experiment also shows that the gypsum plate exhibits the red of first order and may serve as a corresponding standard of retardation or optical thickness.

The quartz wedge as well as any other anisotropic plate (gypsum, mica) for which the vibration direction of the slower component is known may also serve to identify the vibration direction of the slower component in any specimen by finding whether superimposition with vibration directions parallel gives compensation or increases the optical thickness. The wedge is generally applicable; the usefulness of other compensators depends upon the thickness (interference color) of the specimen.

If necessary, restore the combination of wedge and gypsum plate showing the black band of complete compensation. Then rotate the analyzer into the position parallel and study the effect.

Repeat the experiment by using the combination of quarter wave length mica plate and quartz wedge; use first crossed, then parallel polars. Note the shift of colors in the wedge when the mica plate is added in positions of addition and subtraction (compensation). Confirm that the mica plate gives a gray of first order.

Using the analyzer in position crossed, superimpose the gypsum and mica plates. If possible, use one upon the stage and the other in the slot of the tube or in the substage; also try to arrange that the plate upon the stage occupies only half of the field of vision. Focus upon the preparation upon the stage, and rotate the stage so that the (standard) retardation plate upon it obtains the 45-degree position of maximum brightness. Adjust its position so that it occupies half the field, and then superimpose the other plate so that the vibration directions of the slower components are parallel. Observe that the combination produces blue (of second order = first order red + first order gray). Rotate the analyzer to parallel, note the effect, and return it to crossed. Rotate the stage through 90 degrees to obtain compensation. Observe that the combination gives yellow (first order = first order red − first order gray). Rotate the analyzer to parallel.

From the experiment recognize the truth of the following statements. If the specimen is thin, i. e., gives by itself between crossed polars a gray of first order, bring it into the 45-degree position of maximum brightness and superimpose a first order red plate (gypsum, also called selenite). It will, because of its small optical effect, change the interference color of the thick plate only slightly, from first order red to either second order blue (addition) or to first order yellow-orange (compensation). There cannot be any doubt concerning the meaning of the color change, and the directions of the slow and fast components in the specimen follow from the known directions in the first order red plate.

If the specimen is thick, i. e., gives by itself between crossed nicols a color, use a quarter wave length (mica) plate. The very thin (standard) retardation plate can change the interference color of the specimen only little, just to the adjacent colors, and the decision whether addition or compensation has taken place will offer no difficulty.

Obviously, use of a compensating wedge of widely varying thickness is most promising if the specimen is quite thick and exhibits the white of higher order.

Practice the identification of interference colors and the establishment of the vibration directions of slow and fast components with crystals showing between crossed nicols the gray of first order, such showing colors, and possibly some showing the white of higher order. Use crystals of NH_4ClO_4 (orthorhombic plates, for D line: $\alpha = 1.4818$, $\beta = 1.4833$, $\gamma = 1.4881$), $NH_4H_2PO_4$ (tetragonal combinations of prism and short bipyramid which latter may show the color fringes of a wedge; for D line: $\varepsilon = 1.4792$, $\omega = 1.5246$), $Na_2S_2O_3 \cdot 5\ H_2O$ (monoclinic prisms, for D line: $\alpha = 1.4886$, $\beta = 1.5079$, $\gamma = 1.5360$), $BaCl_2 \cdot 2\ H_2O$ (monoclinic octagonal tablets, $\alpha = 1.63$, $\beta = 1.65$, $\gamma = 1.66$), and $CuSO_4 \cdot 5\ H_2O$ (triclinic, short prismatic, $\alpha = 1.51$, $\beta = 1.537$, $\gamma = 1.543$). Prepare drawings and indicate the directions of fast and slow components. Save all preparations containing well developed crystals for Expt. 13.

Place the quartz wedge upon the stage of the microscope and inspect the band of first order red with crossed polars and the wedge in the 45-degree position. Observe that the characteristic purple band is quite narrow so that a slight variation in optical thickness suffices to change the color to violet-blue or orange-yellow. For this reason, the red of first order is also known as the sensitive tint, and the first order red plate is used for the confirmation of birefringency (anisotropism) in instances where the faint brightening of the object might be due to stray light.

Focus a preparation (NH_4ClO_4) that contains crystals which appear very dark gray between crossed polars. Cup the hands around the stage or use a black tent around it, and move a crystal into the center of the field, that barely brightens up when the stage is rotated. Insert the gypsum plate in the 45-degree position, and again rotate the stage.

Obviously, the first order red plate may also be used in demonstrations by microprojection to enliven the appearance of specimens showing mostly grays of first order or to obtain light for observation of manipulations in the isotropic parts of specimens.

Experiment 10

On the Determination of the Refractive Indices of Anisotropic Materials (96, 110)

Use the slide prepared in Expt. 6: orthorhombic needles of anthraquinone in nitrobenzene, n_D at $20°$ C $= 1.5520$.

Remove analyzer and polarizer, and focus upon the edge of the cover slip with an objective of magnification 10 or higher. Select a long needle with light outlines, and move the slide so that one of its long edges goes through the intersection of the cross hairs. Rotate the stage and observe that the outlines of the needle do not change their distinctness. Try the Becke test as outlined in Expt. 4.

Insert the polarizer or the analyzer (or both in parallel orientation) so that the vibration direction of the transmitted light is 6 to 12 o'clock. Illuminate the field brightly, rotate the stage, and observe that the outlines of the needle nearly disappear when its long edge is parallel to the cross hair 6 to 12 o'clock and become strongest in a position vertical to this. Explain the phenomenon and use crossed polarizers to confirm that these two positions of the needle correspond to extinction positions. Why is it sufficient to use either the polarizer or the analyzer?

Return to the use of one polar giving light of the vibration direction 6 to 12 o'clock. Apply the Becke test in both positions of extinction; use monochromatic light if color fringes should be observed. Make a drawing of the needle, and provide the vibration directions with estimates of the corresponding indices of refraction. Try to confirm the finding with the use of a suitable compensator (preceding Expt.).

If a preparation with well developed crystals of $NH_4H_2PO_4$ is available, use it to repeat the experiment after applying benzene as immersion liquid and a cover slip.

Experiment 11

Observation of Pleochroism (96, 110)

Benzoquinone, solid; 5% solution of hydroquinone in ethanol; 5% solution of resorcinol in ethanol.

Take two slides and deposit upon the center of each about 0.5 mg (0.5 µl) benzoquinone in the form of several small particles. Treat one deposit with 1 drop (10 µl) of the hydroquinone solution and the other with a like volume of the resorcinol solution. Allow the solutions to spread and evaporate.

First investigate the residue containing resorcinol quinhydrone. Use one polar to give white light of vibration direction 6 to 12 o'clock. Use a bright field and low magnification to find crystals and aggregates (feathery or reminiscent of oriental rug designs) that exhibit a striking color change when the stage is rotated.

For study, select a well developed single crystal. Bring it into the center of the field and change the objective according to need for higher magnification. Be certain to obtain a brightly illuminated field (top lens of condenser swung in). Prepare a drawing of the crystal; (using crossed polars) determine the vibration directions and then (with one polar) the colors belonging to them, and enter the information into the drawing. If a spectroscopic eyepiece is available, use it to get the spectra of the two colors. For an alternative, the image of the crystal may be projected onto the slit of a standard spectroscope (1283).

Repeat the experiment with the hydroquinone quinhydrone. The preparations evaporate within a few days.

Experiment 12

Observation of Axial Figures (96, 110)

Objective, 4 mm; 2.5 cm × 2.5 cm square of mica (muscovite) sheet, more than 0.5 mm thick; iodoform, solid; xylene.

In a test tube, prepare about 2 ml of a saturated solution of iodoform in xylene. Place a large drop, 0.5 ml, of the solution on a clean 3-inch watch glass and put aside for slow evaporation. If necessary, retard the evaporation by partly covering with a second watch glass. Inspect from time to time without disturbing the solution. Try to get the hexagonal plates as large as possible, but remove the mother liquor by taking it up with filter paper before parts of the plates project above the liquid. Leave the crystals upon the watch glass.

Place the mica sheet (monoclinic, for D line: $\alpha = 1.561$; $\beta = 1.590$, $\gamma = 1.594$) upon the stage, remove the analyzer, and focus upon the top surface with an objective of magnification 20. Use white light, the plane mirror, and focus with the condenser (top lens in place) to fill the aperture of the objective. Cross the polars and rotate the stage to a (45-degree) position of maximum brightness. Try the three methods for the observation of the axial figure:

a) LASSAULX. Remove the eyepiece and look down through the body tube at the back aperture of the objective. If the aperture is not filled with light, try opening the substage diaphragm, using the concave mirror, focusing the condenser, and using the condenser without top lens. When best conditions are obtained, rotate the stage and observe that a black cross opens to form two hyperbolas.

b) BERTRAND. Replace the eyepiece, insert the Bertrand lens into the tube, and focus upon the back focal plane of the objective by moving the draw tube up and down. Rotate the stage.

c) KLEIN. Remove the Bertrand lens. Examine the Ramsden disk with unaided eye or (better) with a magnifying lens. Rotate the stage.

Repeat the experiment first using a magnification 40 and then a magnification 10 objective. Each time, adjust the illumination accordingly.

Repeat the experiment with the microscope focused upon the surface of a hexagonal tablet of iodoform (hexagonal; for D line: $\varepsilon = 1.750$, $\omega = 1.800$), which should be large enough to occupy nearly the whole field of vision, better all of it. Naturally, this also depends upon the objective magnification. Observe that one is viewing along the principal axis (and axis of isotropism) of the crystal which remains black during the rotation of the stage when crossed nicols are used. Note that the black cross of the axial figure does not open up when the stage is rotated. Determine the **sign of double refraction** by inserting a first order red plate. Note the appearance of yellow and blue patches in the four quadrants close to the center of the figure (near the intersection of the purple cross).

Imagine a line drawn through the two yellow patches. If this line is vertical to the vibration direction of the slower component in the compensator, the specimen is positively birefringent (+); if the directions are parallel, negatively birefringent.

Flash figures of a uniaxial substance may be observed with $NH_4H_2PO_4$ if large crystals have been obtained. If desired, further experiments may be found described in the literature (88). The interpretation of biaxial interference figures requires much practice and study (96, 110). The observation of interference figures on small crystals or grains requires careful centering of the whole optical equipment and of the specimen as well as additional field diaphragms to narrow the field of vision so that it is completely occupied by the specimen (88).

Experiment 13

Transition Phenomena in Polarized Light (159, 160, 163)

Narrow slides which may be obtained by lengthwise cutting a standard microscope slide to get two or three strips of 8- to 12-mm width. — Solids: NH_4NO_3, $Na_2S_2O_3 \cdot 5\,H_2O$, thymol, ethyl ester of p-azoxybenzoic acid (or some other substance that gives a crystalline liquid).

Narrow slides are preferable when strong heating is required since they are less liable to crack. Put into the center of such a slide about 0.5 mg (0.5 μl) of thymol ($\varepsilon = 1.609$, $\omega = 1.525$ for line D; m. pt. 49.6° C), and place a half of a cover slip over it. Heat by moving the slide back-and-forth about 5 mm above a small microflame or pilotflame of a Bunsen burner until the thymol melts. The melt fills the space between slide and cover slip and, as a rule, remains supercooled and does not crystallize. Remove any excess of melt by taking it up with the edge of absorbent filter paper; to obtain a very thin film of liquid, place the slide on paper and press down upon the cover slip while touching the edge of the absorbent paper to the edge of the cover glass.

Proceed in the same manner to obtain preparations with NH_4NO_3 and the benzoic ester. These substances crystallize when the slide cools. Depending upon the volatility of the substance, preparations of this type will keep for months or years and may be used for many repetitions of the experiments.

A. Place the melt of thymol upon the stage, remove the analyzer, and with objective magnification 10 focus upon a gas bubble in the melt or on the edge of the cover slip. Move into the field a spot where the melt seems accessible at the edge of the cover slip. Make the field medium bright and use somewhat oblique illumination. Touch the point of a needle to a crystal of thymol, and then–while observing through the microscope–touch the melt with the needle. Observe that crystal plates begin to form and to grow slowly. Move the slide and observe the front of the crystal mass as it advances through the melt until it reaches the opposite edge of the cover glass. The plates are colorless like the melt and are not too distinctly visible if the illumination is not cleverly selected. Test this by using strong axial illumination.

Heat the preparation over the microflame to melt the crystals, and then return it to the stage for a repetition of the experiment. This time, use axial illumination with strong light and crossed polars. Observe that polarized light provides a sensitive test for the formation of an anisotropic phase. When crystallization is complete, rotate the stage and observe the positions of extinction for groups of crystal plates. Air bubbles remain dark when the stage is rotated. Large plates or areas of like color may be used for the observation of axial figures (preceding Expt.); plates that remain dark during the whole revolution of the stage should be tried first; in addition, one might try to make a new preparation with a much thicker layer of thymol.

Repeat the experiment once more, especially if many of the plates show the gray of first order. This time use crossed polars and the first order red plate in the 45-degree position. Note that this renders the field bright so that the seeding is easily observed without hindering the ready recognition of the separating crystals. The colors will be more brilliant than without the gypsum plate if the preparation is very thin; they may be less impressive if the preparation is thick. Rotate the stage, and observe that the plates change their color when they enter a new quadrant (provided that the gypsum plate is stationary). Try the polars in the position parallel.

B. Use the $Na_2S_2O_3 \cdot 5 H_2O$ preparation made for Expt. 7. First focus the microscope upon the crystals. Place the preparation upon the steam bath to liquify it. Allow it to cool, and then place it on the microscope stage so that the edge of the cover slip appears in the field. Do not change the focus of the instrument. Cross the polars. While looking into the eyepiece, seed at the edge of the cover glass with a needle that has been touched to a crystal of the salt. Move the slide to keep the advancing crystal front in the field until the crystallization is complete. The phenomenon proceeds rapidly through the melt. Finally, rotate the stage, change the polars to parallel, etc., melt the salt again and repeat the crystallization with crossed polars and the gypsum plate in place, etc.

C. Focus the NH_4NO_3 preparation with polars parallel. To prevent damage, adjust the condenser so that its front lens is at least 5 mm below the slide. Without changing the focus, remove the preparation and heat it above a microflame until it is completely liquified. Without delay, place the hot preparation upon the stage so that the axis of the microscope passes through its center. Watch through the microscope while the preparation cools. Gas bubbles, recognizable by their round outlines, may be used to correct the focus. When the temperature has dropped to 170° C, cubic NH_4NO_3 separates, and the melt solidifies. Rotate the polar to position crossed and note that the field is dark even when the stage is rotated. Keep watching through the eyepiece. Below 125° C, anisotropic rhombohedral crystals form, and the field becomes bright. Continue watching the preparation. Further transitions occur at 84° C (rhombic I), at 32° C (rhombic II), and − 17° C (tetragonal); the first two may be observed and are recognized by a motion in the preparation.

Repeat the experiment with crossed polars and the gypsum plate inserted in position of maximum brightness. Try it also with a 5 x objective.

D. Focus the crystals of the benzoic ester without the analyzer. Use medium brightness and oblique illumination to obtain good contrast. Make certain that the front lens of the condenser remains 5 mm below the slide. Then heat the preparation over a microflame until the cloudy melt, which formes first, becomes clear. Without delay, place the hot preparation upon the stage and watch it through the eyepiece. Note that striations appear in the clear melt at a certain temperature and persist until crystallization takes place.

Repeat the experiment with strong axial illumination and crossed polars (without and with addition of gypsum plate). Note that the first anisotropic phase is liquid; this is indicated by the motion in the preparation and the round outlines of gas bubbles, and it may be demonstrated by the flow produced by tapping the cover glass with a rod.

Save the preparations used.

Experiment 14

Preserving Microscopical Preparations

Round cover slips; rings, 2 to 3 mm wide, cut of strong writing paper to fit the cover slips; small brushes. — Canada balsam, benzene, xylene, chloroform, gelatin, glycerol, adhesive (Ad-A-Grip glue), Duco cement, Glyptal 1201, red.

Whether the preparations are intended for evidence or for demonstrations, one will have the desire to preserve confirmatory tests without change just as they were obtained. This is reasonably simple if only solids are involved. Frequently, however, the preparation consists of some solid that has separated from a solution which will evaporate and leave a copious residue that may completely hide the objects of interest. There are two ways to avoid this. Either evaporation is prevented, or the liquid is removed before evaporation is able to mar the preparation.

Evaporation is safely *prevented* by sealing in glass, and this is easily done if the test or preparation has been obtained in a narrow tubing or capillary. The tubing is sealed at both ends and kept together with a description inside a stoppered test tube in a dark place (locker) with little variation of temperature. In this way, the specimen may be centrifuged, warmed, or cooled each time before mounting it for microscopic inspection.

A *liquid preparation on a slide* may be preserved for a limited time by covering it and sealing the edge of the cover. A small drop may be saved, without disturbing it at all, by inverting over it a small watch glass so that it does not touch the drop and sealing the edge by applying melted wax, Duco cement, or Glyptal with a brush.

Putting a cover slip upon a liquid preparation has very often the effect of completely destroying its characteristic features. Solids that do not adhere to the slide, will shift their positions and frequently float to the edges of the cover slip so that they get lost or can no longer be observed because of the refraction phenomenon at the edge. Before applying this procedure to an important preparation, it will be wise to practice it with duplicates obtained for this end. The size of the solid particles must be considered since it may determine the thickness of the liquid layer. The area of the square or round cover slip should be adjusted to the volume of the liquid (a layer of 0.2-mm thickness below a cover of 1-cm edge obviously has a volume of $20\,\mu l = 0.02$ ml). In general, it may be preferable to reduce the volume of the liquid by means of touching it with filter paper (Expt. 40) or a capillary pipet (Expt. 43) or to use a cover slip of such size that the liquid is unable to fill the whole space. When the cover slip is in place, its edge is sealed by applying melted wax, Duco cement, or Glyptal with a brush. Instead of sealing in this manner, one may place a drop of Canada balsam upon the cover slip and then superimpose a second, larger cover glass; this may be quite practical if the first cover slip is small, possibly only a fragment of a cover slip. Enough Canada balsam must be taken to fill the whole space between the large cover and the slide.

As an alternative, one may consider first painting around the drop on the slide a barrier of repellent wax or cement and then dropping upon this ring the cover slip, the surface of which may have been treated to make it repellent (Dri-Film, Desicote).

The *removal of the liquid* will give more permanent mounts. One will use the techniques described in Expts. 40 and 43. As a rule, washing will be necessary to prevent the marring of the preparation by residues from small amounts of liquid left behind. Obviously, these operations will change the appearance of the preparation if the solid particles to not cling to the slide, and again it will be necessary to practice on expendable duplicates before proceeding with a specimen that shall be preserved as evidence. The resulting dry preparation is then treated like a solid.

The *mounting of solids* requires careful consideration of the nature of the solid. Preliminary experiments may be needed to make certain that the adding of a mounting medium will not destroy the appearance of the preparation and that the mounting medium has no effect upon the specimen. The color and the refractive index of the mounting medium should be considered. The shape of transparent objects is best visible if the refractive index of the surrounding medium differs by about 0.1; the color of the object and structures in its interior are best seen when it is immersed in a medium of the same refractive index. The mounting medium must not be anisotropic if observation in polarized light is intended.

Mounting in Canada Balsam or equivalent plastic preparations requires that the specimen is dry. Drying may be accomplished by heating or by washing with alcohol.

To exclude air bubbles, first wet the dry preparation with a small volume of benzene or other suitable solvent. Without delay, add a drop of the balsam not larger than just sufficient for filling the space below the cover slip. Place the latter upon the balsam drop, and allow it to settle by gravity into the intended position. If it seems desirable, finally press down upon the cover slip with the eraser end of a pencil to extrude excess mounting medium which is removed with filter paper that may be moistened with solvent.

If permissible, heat the preparation several hours at 70 to 80° C. Otherwise allow it to stand for several days at room temperature; cover it with a watch glass or bell jar so that solvent vapors may dissipate. When the preparation has hardened, remove any excess mounting medium showing at the edge of the cover slip by scraping and finally wiping with filter paper moistened with solvent. Finally seal by painting the edge of the cover glass with Duco varnish or Glyptal.

Mounting in glycerol jelly does not require dry preparations and may also be found suitable for organic crystals which are attacked or dissolved by Canada balsam.

Proceed in principle as with Canada balsam. A solution with 10% gelatine, 30% glycerol, and 60% water must be applied quite hot. If this is not permissible, add more water; a solution with 1% to 2% gelatine remains liquid down to room temperature and may require some evaporation for setting to a gel.

Dry mounting has been recommended by EMICH for its wide applicability and for the fact that it does not affect the arrangement of objects in the preparation. To-day, it appears especially attractive because of its simplicity and the availability of excellent waterproof adhesives. The disadvantage that many transparent objects give poor images by transmitted light when surrounded by air may often be overcome by changing to observation by reflected light.

Rings of writing paper, the outer diameter equal to that of the round cover slips, may be obtained with the use of dies. Make certain that the surface of the slide around the specimen is clean. Holding the paper ring with forceps, paint both sides of it with adhesive (Ad-A-Grip or similar). Place the ring upon the slide so that it surrounds the specimens, and place the clean cover slip on top of it. Place a 200-g weight upon the cover slip and allow to stand until the adhesive has set. After 24 hours, paint the edge of the cover slip with Duco varnish or Glyptal.

It is obvious that all permanent mounts should be labelled with special care. This may be done with special ink for writing on glass, or by attaching a label to the top of the slide.

Technique of Experimentation and Observation

Basic Rules

To have a good probability of success beware of haste and **make it a habit to let reasoning precede action.** Before starting any experiment, make certain: (1) that you know its purpose, (2) that you **understand the**

principles involved in its performance, (3) that you know **how** to perform it without danger to yourself and your neighbors and without risk of losing the material under investigation, (4) that you know **what** phenomena may be expected and how they are **best** and **most sensitively** observed.

To begin with, realize that a difficult task is not made easier by increasing its complexity. Contaminating the material under investigation by the use of apparatus which is not perfectly clean or reagents that contain impurities renders the task of identification more complex and may make futile the interpretation of the phenomena which may be caused by the introduced foreign matter and not by anything contained in the original object. Observation is made difficult and sometimes impossible by dirt on the outside of transparent containers.

These considerations lead to the conclusion that **cleanliness** must be practiced **to the utmost**. Not only all apparatus containing the material under investigation, but also everything that gets into contact with it, including the hands, must be clean. Bench tops and shelves should be clean at all times; this will prevent that utensils and hands become soiled. For the same reason do never permit dirty apparatus to accumulate, but clean it immediately after use. The success of cleaning is also better assured when it is still known what the apparatus contained and the proper solvent for the removal of adhering matter is obvious.

Use water and **reagents** of good quality, and maintain their quality most conscientiously as follows: (1) **Before** removing a stopper make certain that the outside of the container is clean. If necessary, remove dust or crusts with a cloth (if the contents are hygroscopic or react with water) or by rinsing under the tap and then carefully drying with a cloth. (2) **Clean spoons, spatulas, pipets before and after** use; do not use utensils that are attacked by the reagent. (3) Handle the stopper so that it cannot become contaminated; if this should happen by accident, wash and dry the stopper. (4) Do not pour from a container directly into apparatus from which gases (NH_3, HCl, NO_2, Cl_2, Br_2) or vapors (H_2O, I_2, solvents) may be taken up. These would be drawn into the reagent container. (5) Close the reagent container after removal of the needed amount. (6) Take care that a reagent which does not appear reliable for any reason is tested and, if necessary, discarded.

Work on the Gram Scale

A triple-beam balance responding to 0.01 g is suited for the weighing of samples; a completely enclosed torsion balance with a capacity of 500 g to 1 Kg is recommended for the weighing of reagents. Suitable measuring spoons may be used for adding approximately known amounts of powdered or finely granular reagents. Graduated cylinders of 10-ml and 50-ml capacity suffice for the measuring of solutions.

Most of the *chemical treatment* is performed in low-form beakers of 100-ml to 400-ml capacity and in standard test tubes.

Solids are handled with a spoon made of horn, plastic, porcelain, or glass. Reagents are often added directly from the reagent bottle, but this should not be done if vapors or gases are contained or may form in the receiving vessel since these would be drawn with the air into the bottle and taken up by the reagent. In such instances, pour a small amount of the reagent into a clean test tube, beaker, dish, etc. and add it from there to the reaction mixture; discard whatever is left and not needed of this contaminated portion.

Whenever this is required, add the reagent dropwise by cautious pouring. The task is facilitated by touching the rim of the reagent container to a stirring rod that guides the liquid into the receiving vessel; the use of medicine droppers is not recommended since it increases the danger of introducing contamination.

It is understood that mixing is required after addition of a reagent or each portion of a reagent. Do this by stirring or swirling; to mix the contents of a test tube, grasp it with three fingers just below the opening and impart jerky motions through the wrist so that the closed end moves sideways through short arcs.

To adjust the pH of a solution, add small portions, finally drops, of the required reagent. Mix after each addition and, using a stirring rod, transfer a small amount of the mixture to the indicator paper. If the desired pH is overstepped, go back by adding still smaller portions of the suitable neutralizing agent.

For *saturating with hydrogen sulfide* (or any other gas), transfer the solution to an Erlenmeyer flask of suitable capacity, 250 ml. Close the flask with a one-hole rubber stopper provided with a glass tubing, fire-polished at both ends, that reaches to the bottom of the flask. Use a clean length of flexible rubber tubing for connecting to the gas line. The latter should be supplied with a pressure head of about 50 cm water column (4 cm mercury); if the line contains a simple gas washing bottle, this will also serve for a convenient bubble counter. To start with, lift the stopper in the neck of the Erlenmeyer and use a rapid stream of gas to displace the air above the liquid. Then, without shutting off the gas supply, tighten the stopper and shake the flask to get intimate mixing of the liquid and the gas in it. Continue until no more bubbles rise in the washing bottle (Erlenmeyer flask), which indicates that saturation must have been reached since no more gas is consumed. Turn off the gas supply and lift the stopper before disconnecting. Remember, there is pressure in the flask and its contents will be forced out through the line if the latter is broken or disconnected. To safeguard against such loss due to some accident (washing bottle pops open or is broken, rubber tubing slipping off glass connector, failure of generator) raise the inlet tubing in the Erlenmeyer so that it ends in the gas space above the liquid; of course, this makes it desirable to have a bubble counter in the line.

For *stirring rods*, cut clean Pyrex rod of 3- to 4-mm diameter into lengths of 8, 12, and 16 cm and carefully firepolish **both** ends.

For *heating*, place beaker and flasks upon a wire gauze. Heat in test tubes by letting the gas flame play directly upon the glass. A steam bath for test tubes, watch glass, or evaporating dishes is easily improvised by using a Pyrex beaker with boiling water; a test tube is simply placed into the water; a watch glass or dish is placed upon the opening of the beaker.

Evaporation of solutions perform properly by heating on the steam bath in open, flat, thin-walled dishes of suitable material (Pyrex, porcelain, vitreous silica, platinum). The process may be considerably hastened by blowing a stream of filtered air from a capillary upon the surface of the liquid. This also aids in keeping dust away. Use a dish of 3- to 4-inch (7 to 10 cm) diameter, and add the solution in several portions, but never fill the dish close to the rim. (Rapidly boiling in an open container causes loss by spray and spattering; if it is done in covered containers, flasks, or test tubes, condensate dripping or flowing back to the hot part of the apparatus may cause breakage and loss.)

Fig. 8. Filtration; about $1/_3$ nat. size

Fusions perform in crucibles of 5- to 10-ml capacity of suitable material (porcelain, vitreous silica, iron, nickel, platinum). Rest the crucible on a clean clay or silica triangle supported by a ring stand or tripod. Ashing and some fusions may be advantageously performed in a small electric muffle furnace.

For *bead tests* use a straight platinum wire, 6 cm long and 0.3 to 0.5 mm in diameter, one end of which is sealed into a length of glass tubing that serves for handle. Use the technique described in Expt. 59 and P. 39. Mount the handle in a cork which fits a standard test tube, and keep the wire suspended in the stoppered test tube when it it not in use.

When *transferring a liquid* from one container to another, render the outside of the pouring edge repellent. With the finger, get some oil from the scalp or the side of the nose and rub it on the glass surface. Either rest the pouring edge on the rim of the receiving vessel, which may be held inclined, and allow the liquid to flow down along the inside wall of it, or guide the stream of the liquid with the stirring rod, Fig. 8.

Use *filtration* for the separation of solid from liquid. To this end, keep ready at least two porosities of filter paper, a very slow paper for solids consisting of minute particles (precipitate of $BaSO_4$, etc.) and a very fast paper for gelatinous precipitates (hydrous oxides, sulfides). Use fluted filters when only the filtrate is wanted.

Fold the disk of paper exactly in half and once again to the outline of the quadrant. If the amount of solid is very small, now trim the size of the filter with shears so that the solid will fill about one-fourth of the capacity (cone of paper). Open the cone and note that one side is formed by a triple layer of paper. *Tear* off the corner of the outside double layer, and then tear off some more of the outermost sheet. Fit the dry cone into the glass funnel, best a 58-degree funnel with long stem. The rim of the cone must fit snugly to the glass, but the apex of the cone need not touch the funnel. Moisten the filter and, with clean fingers, press its edge to the glass to obtain a tight fit all around. Fill the filter with water and keep on repairing the fit of the edge until the stem remains completely filled with water even after the filter has been drained empty. Leave this water in the stem; it will provide a slight amount of suction and speed the filtration.

Place the funnel into the filter stand and put the receiving vessel underneath so that the stem touches its side and the filtrate will flow down without splashing. With the stirring rod, guide the liquid into the filter, Fig. 8. Fill about three-fourth of the filter and allow it to drain completely before replenishing. If there is much liquid and little solid, first decant most of the clear liquid through the filter and save this portion of the filtrate before starting the transfer of the solid. For the transfer of the solid, obtain a slurry by mixing with the stirring rod or swirling, and guide the slurry into the filter by means of the stirring rod. Save the filtrate before starting with the washing of the solid. The washings usually may be rejected since they would only dilute the filtrate.

Washing may be combined with the finishing of the transfer of the solid. The wash liquid supplied from a wash bottle (preferably a plastic squeeze bottle) may be used to rinse solid into the filter. For washing, fill the filter nearly full with the wash liquid and allow it to drain completely before adding the next portion of wash liquid. The latter may be added from a beaker by guiding the stream with a stirring rod. To avoid spattering, use caution when adding wash liquid from a wash bottle, especially when using the force of the jet for stirring up the solid in the paper cone or washing the solid down into the tip of the cone. In many instances, it will suffice to wash a grainy or crystalline precipitate twice or a gelatinous precipitate four or five times. If the substances contained in the filtrate interfere with subsequent operations, however, it may become necessary to continue washing to the practically complete elimination of the objectionable substances.

Separation by Decantation is preferable and saves time if the solid phase is quite insoluble and settles reasonably fast to occupy a space less than one tenth of the capacity of the container. In such instances, prepare a filter as usual. Allow the precipitate to settle in the container which is already inclined nearly to pouring position (by suitably supporting it in a beaker, cardboard box, mortar, etc.) and then pour off the nearly clear solution into the filter. Thoroughly mix the remaining thick slurry with ten times its volume of wash liquid, allow to settle, and again decant through the filter. If the washing is repeated twice in this manner, only 0.0001 of the solutes contained in the "original" liquid phase will remain with the "final" slurry. One may decide to remove most of them by transferring the slurry to the filter, or one may prefer to treat the slurry as it is; the reagent may be added by percolating it first through the filter that has collected part of the solid.

Removing the Solid from the Filter. Simply use a spatula or spoon on the moist cake if much solid has been collected and not all of it is needed. Moderate and small amounts of collected solid, which are needed in their entirety, treat depending upon their solubility behavior.

If the solid is readily soluble and is to be dissolved for further treatment, dissolve it on the filter cone by adding the solvent drop by drop. As an alternative, close the stem of the funnel by applying a rubber or plastic cap or stopper to its opening and then fill the filter cone with the solvent. In either instance, allow the filter to drain completely when dissolution has become complete. If the solution should be turbid or opalescent, you may try to clear it by passing it repeatedly through the filter, possibly after stirring into it a small amount of paper pulp ("filter aid"). If the volume of solution is small, it may become desirable to improve the yield by finally folding the filter and pressing it into the funnel (finger cots may be used). If diluting is permissible, the yield may be improved by washing the filter previous to pressing it out; diluting may be avoided by using to this end a reagent solution which would have to be added anyway.

If the solid is only partly soluble and requires thorough digesting, its treatment in the filter cone is not recommended. Either punch a hole into the apex of the filter cone with the stirring rod and use the jet of the wash bottle to wash the solid through the stem of the funnel into a suitable receiving vessel, or remove the filter from the funnel, unfold it, tear off the parts holding none of the solid, and place the rest of the filter with the solid adhering to it into the beaker or dish selected for treatment. Undesirable diluting may be avoided in the first procedure by using, instead of water, a measured volume of reagent solution from a small washbottle that may be made of a test tube. The second procedure must be avoided if the digestion with the reagent puts products of the breakdown of cellulose into the solution, which could interfere in the subsequent steps. Glass fiber paper could be used in such instances. Obviously, the material of the broken-up filter will be collected together with the insoluble part of the solid; this may be inconvenient at some times and desirable at others when the disintegrated tissue acts as filter aid.

Small amounts of insoluble solid may be isolated by ashing the filter following the procedure of quantitative analysis (13). This may cause changes and losses because of reduction, oxidation, and volatilization.

Extraction of Solids. Use the technique for the washing of solids by filtration, decantation, or a combination of the two.

Extraction of Liquids. Perform the extraction in a Squibb type separatory funnel of pear shape, in a test tube with the use of a siphon as described below for the centigram scale, or in a syringe (45) of 20- to 100-ml capacity. When using solvents with high vapor tension, release the pressure in the separatory funnel from time to time by opening the stopcock for a brief time while holding the funnel in an inverted position (thumb upon the stopper) so that the liquid does not reach the outlet to the stopcock (13).

Beware of the formation of more or less stable emulsions. Of course, vigorous shaking with intimate intermingling of the two liquids quickly establishes equilibrium, but do not use this expedient unless you have assured yourself by a side test with a small amount that the emulsion will break up readily even after vigorous shaking or stirring. If such assurance has not been obtained, reach equilibrium by gently rocking the container or by slowly stirring to keep the liquids in motion without mixing them.

If a stable emulsion has once formed, use small portions of it for finding a means to break it and then treat the bulk accordingly. Various ways are suggested in the literature (13, 58). Exposure to vibration of very high pitch, but below the ultrasonic range, may be tried. For small apparatus, it will suffice to fasten down the interrupter of an electromagnetic door buzzer, operating via transformer on 110 v a. c., so that it cannot make and break and vibrates with an almost imperceptible motion (413). The apparatus is touched to the oscillating arm.

Adding a small amount of alcohol or a large excess of organic solvent may break up an emulsion, and one may try to change the pH or to precipitate responsible emulsifying agents. The emulsion may be removed and washed by filtration. As a last resort, evaporate the whole mixture with the proper precautions until the emulsion disappears. Small amounts of solid often act as stabilizer; thus be certain to obtain a perfectly clear solution before making another attempt at extraction.

For methods of continuous extraction, which may avoid the formation of emulsions, see the literature (13, 45, 147).

Distillation. A distilling flask of 25- to 250-ml capacity with the side arm placed at the center of the neck is generally useful. If the side arm is made 30 cm long, it may serve as condenser tube, and it will accomodate a short cooling jacket. If a ground glass stopper is not provided, corrosive liquids may be distilled by placing a very short cold-finger condenser (short test tube with cooling liquid) into the opening of the neck and regulating the rate of distillation so that the ring of condensate does not climb much above the opening of the side arm.

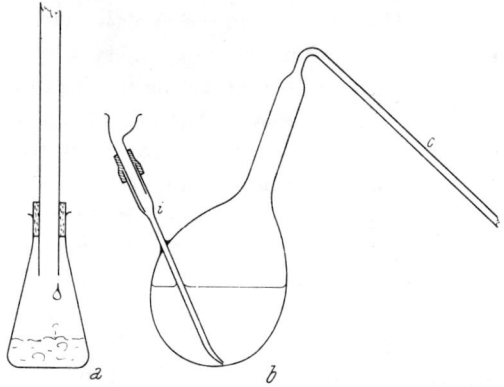

Fig. 9. Refluxing and Distillation; about $1/3$ nat. size

If loss of volatile components must be carefully avoided, the distilling flask shown in Fig. 9b is recommended, which is designed to avoid dead spaces and contact of the vapors with joints of any kind. The condenser tube c may receive a water jacket; the inlet tube i may be closed during distillation or connected to a supply of suitable inert gas. The residue in the flask may be siphoned off through i (861).

Sublimation. Place the substance into a boat of porcelain, vitreous silica, or platinum and introduce the boat into a glass tube of 8- to 12-mm bore and 25- to 40-cm length, which has been clamped in a horizontal position. At the location of the boat, place a roll of wire gauze around the tube to obtain uniform heating. Connect one side of the tube to a supply of suitable gas and the other side to absorption apparatus that may seem indicated. Heat the material in a very slow stream of gas. The sublimate will collect downstream from the boat which will finally hold the non-volatile matter.

The sublimate may often be very conveniently recovered by mounting the tube, Fig. 9a, in the neck of a flask by means of a rubber stopper, cork, or ring of asbestos paper and using it as a reflux condenser with a suitable solvent boiling in the flask.

Confirmatory Tests. Crystal precipitations upon the microscope slide and spot tests are recommended for use, Expts. 24 to 42. The traditional confirmatory tests of the gram scale are performed in test tubes. From a few drops to about 2 ml of the solution to be tested are placed into a test tube and treated with the reagent. Effervescence, the separation of a precipitate, and the appearance of a strong coloration in the liquid are readily perceived. A small amount of precipitate, turbidity, opalescence, and fluorescence are best seen before a black background and with a strong beam of light more or less vertical to the line of viewing. Opalescence and turbidity indicate the formation of a new phase, provided that the reagents have been clear. Weak colorations may become visible when looking axially through the test tube which is held over a well illuminated, white paper.

Effervescence indicates the liberation of a gas. Especially gases that are not lighter than air will collect above the liquid contents of the test tube and may be identified by means of reagents (test paper, reagent drop in platinum loop or hanging on stirring rod) inserted into the gas space. These tests are far more sensitive when using special apparatus recommended for the centigram- and milligram scales, *see* below.

Storing Material under Investigation. Solids that have been collected upon filter paper are best immediately transferred to the apparatus in which they will undergo the next treatment. Beakers or dishes containing material under investigation are covered with watch glasses; the contents of flasks or test tubes are best protected by inverting over the opening a small beaker or wide vial; stoppers are not recommended.

Experiment 15

Phenomena Observed Upon Heating in an Inert Atmosphere

Tubing of 3- to 6-mm bore of glass with high softening point, 21-cm lengths. — Brass, filings or turnings; barium chloride dihydrate or clear cleavage plates of gypsum; sucrose; benzoic acid of tested purity; pyrite; ammonium sulfate; cinnabar, HgS; anhydrous sodium carbonate; pH test paper; Nessler reagent (dissolve 11.5 g HgI_2 and 8 g KI in water to give 50 ml, add a like volume 6-F NaOH, mix, and decant from any precipitate that may form on standing); mercurous test solution (1.4 g $Hg_2(NO_3)_2 \cdot 2\ H_2O$ in 100 ml 1-F HNO_3); 1-F $BaCl_2$; 0.02-F $KMnO_4$.

The material is heated at the bottom of a narrow test tube so that air has little access to it and oxidation will not take place to an extent that would obscure other phenomena.

Starting with lengths of 21 cm, prepare test tubes of 3- to 6-mm bore and 10-cm length. Using an oxygen blast flame, obtain two test tubes from each length by first drawing a capillary at the middle of the tubing, and then sealing at the tapers. The sealed ends may be rounded off by reheating and blowing to obtain hemispherical shape; if this is done, place the finished tubing into a drying oven of 130° C for the removal of moisture that may have collected in the tubing.

To perform an experiment, bring about 2 to 10 mg (10 μl, i. e. a cube of about 2-mm edge) of the solid to the bottom of the tube. Grasp the tube near the opening with the fingers or with a narrow strip of paper that has been folded two to three times (form the strip into a V and hold the tube in the apex of the V). Holding the tube at an angle of 45 degrees to the horizontal, slowly heat the bottom with the non-luminous flame of the Tirrill burner. Whenever necessary, interrupt heating for the observation of the phenomena: change in appearance, melting, sublimation, distillation, liberation of vapors or gases. If this promises additional information, finally raise the temperature to the softening point of the glass, about 800 to 900° C (faint red incandescence visible only in a dark place: 550° C; dark red glow: 700° C; bright red glow: 950° C). Try the experiment with the substances listed below.

a) Brass filings or turning. Heat to a high temperature. If necessary, use a magnifying glass to inspect the color changes on the residue and the distillate of zinc and zinc oxide.

b) Clear crystals of $BaCl_2 \cdot 2\ H_2O$ or clear cleavage piece of gypsum. Note the change in the appearance of the residue. Test the condensate with pH test paper.

c) Sugar or filter paper or sliver of wood. Note the odor and test the pH of a distillate.

d) Small amount of pure benzoic acid. Finally try to obtain pyrolysis by heating the tube above the sublimate.

e) Pyrite, FeS_2. Observe the changes on the residue and the appearance and volatility of the distillate. Note the odor.

f) Ammonium sulfate. Observe odor and pH of the gases (liquids) obtained at different stages of the decomposition. Test the fumes with (1) a solution being 0.5 F in both $BaCl_2$ and HCl, (2) a solution 0.01 F in $KMnO_4$ and 1 F in H_2SO_4; take up a drop of the reagent with a platinum or glass loop, P. 36–or with the end of a stirring rod–and hold it in this manner in the path of the fumes.

Repeat the experiment, but this time mix the ammonium sulfate with an equal volume of anhydrous sodium carbonate. Test the liberated gas with pH paper, a droplet of 12 F HCl at the end of a stirring rod, and a narrow strip of filter paper which has been treated near the end with a tiny droplet of Nessler reagent (or strong mercurous nitrate solution and then bathed in a larger volume of NaCl solution).

g) Cinnabar, HgS. Scratch the sublimate with a glass needle and observe a color change which may take place on standing at room temperature. Repeat the experiment, but mix the sulfide with an equal volume of anhydrous Na_2CO_3.

It is obvious that the sublimates, distillates, and residues obtained in these experiments could be tested and identified by the microtechniques described later.

Experiment 16

Phenomena Observed Upon Heating in a Stream of Air

Supply of glass tubing, 3- to 6-mm bore, of high softening point. — Cinnabar (HgS); galena (PbS); sulfur; selenium; tellurium; graphite or carbon; pH test paper; 0.1% fuchsine solution; 0.02-F $KMnO_4$; lime water.

Cut the tubing into lengths of 12 cm. Using a spatula (aluminum, nickel, or copper wire hammered flat at one end), deposit 2 to 10 mg (10 μl, i. e., cube of about 2-mm edge) of the substance to be tested inside the tube and approximately 3 cm from the opening. Grasp the tube near the empty end as directed in Expt. 15 and hold it inclined at an angle of 30 degrees to the horizontal so that the sample is near the lower end. Use a small (2-cm height), just non-luminous Bunsen flame to heat the part of the tube containing the sample. The hot air rises in the tube to escape at the upper end, while fresh air is drawn in below the heated sample.

Gradually raise the temperature while observing the sample. From time to time interrupt the heating and test the escaping gases at the upper opening of the tube. Try the experiment with the substances listed below.

a) Cinnabar. Heat very slowly. Test the escaping gas with pH test paper, litmus paper, a narrow strip of filter paper which has been treated near the end with a tiny droplet of fuchsine solution. Inspect the condensate with a magnifying glass, and wipe it together with a glass thread or a steel needle.

b) Galena, PbS. Note the changes on the solid. Test the escaping gas with the fuchsine spot on filter paper, with litmus paper, with a droplet of lime water, and with a droplet of solution 0.01 F in $KMnO_4$ and 1 F in H_2SO_4. Observe the odor.

c) Sulfur. Test the escaping gas as under (b). Observe any condensate with the magnifying glass.

d) Selenium. Observe the odor and the phenomena in the tube.

e) Tellurium. As under (d).

f) Graphite. Observe the sample during heating. Test the escaping gas with pH paper, odor, fuchsin spot, and a droplet of lime water.

Experiment 17

Heating Upon the Charcoal Block

Small rectangular block of dense charcoal as obtainable from laboratory supply houses or block of graphite; 20-cm length of 4- to 6-mm bore glass tubing of high softening point; small mortar; strong permanent magnet; supply of hydrogen gas at low pressure. — Brass, filings or turnings; $CdCO_3$; KNO_3; $SrSO_4$; Fe_2O_3; As_2O_3; anhydrous Na_2CO_3.

Reductions take place when material is heated in contact with carbon. Many metals separate as such either as a powder or collected into a bead depending upon whether or not the melting point is reached. Volatile metals may evaporate completely or partially and combine with the oxygen of the air surrounding the reaction mixture to deposit as oxides upon the charcoal around the heated material. In presence of sodium carbonate, compounds of halogens, sulfur, and phosphorus are converted to halides, sulfide, and phosphide.

BELCHER, HARRISON, and STEPHEN (942) recommend the following procedure without specifying the nature of the gas. It is obvious that the needed high temperatures may be reached only if the temperature of the flame is raised as its size diminishes and the amount of reaction mixture is increased. A small hydrogen flame gives the required temperature and provides a reducing atmosphere. As an alternative, a micro oxygen torch (142, 170, 888) might be used and could be adjusted to give an oxidizing flame.

Start with a 20-cm length of glass tubing of 6-mm bore and high softening point, and use an oxygen blast lamp. Heat the tubing 5 cm from one end; slowly rotate in the flame without exerting a pull so that glass collects and the tubing contracts to a diameter of about 3 mm. Then remove from the flame, allow to cool somewhat while continuously rotating the tube, and then slowly draw a short, thick-walled capillary of about 0.5-mm bore. Cut the capillary near the short end of the tubing, and remove the latter. Then heat the capillary in a Tirrill flame about 2 cm from the wide tubing, and draw it out to a fine point. Cut the taper to get an orifice of 0.1- to 0.2-mm diameter. Finally make a right-angle bend in the middle of the remaining short piece of capillary, and connect the wide end of the tube by means of a long piece of flexible rubber tubing to the supply of hydrogen.

With some suitable tool (cork borer, drill, end of file) cut into the surface of the charcoal block a very shallow circular depression (1 mm deep, 10 mm in diameter), the rim of which is just high enough to keep beads in place without preventing full access of the flame to the material. Place the block upon the bench and prop up one end of it so that the surface of the block is inclined at about 30 degrees to the horizontal. Place 5 mg to

50 mg of the material to be tested into the depression so that it is piled together at the lower half of the rim.

Regulate the pressure to get a soft hydrogen flame of 2- to 5-mm length and, by hand, direct this flame so that it plays upon the sample from above. According to need, raise the temperature by increasing the pressure which will give the flame a length of 10 mm to 30 mm. Do not direct a stiff hydrogen flame on powdery material before it has sintered, for it would blow the powder out of the depression. Judge the temperature by the color of the light emitted by the incandescent material. Try the experiment with the substances listed below.

a) *Brass.* Note the color of the flame at different times. Observe the changes in color and shape, which the metal undergoes. Raise the temperature to about 900° C and observe the deposit of ZnO around the depression, which is yellow when hot and white when cold. The deposit does not volatilize but may show a green luminescence when touched with the *oxidizing seam* of the flame. Moisten the copper bead with 12-F HCl and observe the color of the flame when it is again heated.

b) *Cadmium Carbonate.* Heat to about 800° C. The brown deposit may be surrounded by a thin film of the oxide showing blue iridescence similar to the "eyes" in the plumage of peacocks.

c) *Potassium Nitrate.* Note the deflagration upon the coal and the flame coloration.

d) *Strontium Sulfate.* Observe the strong incandescence of the heated sample. After cooling, transfer the residue to a bright silver coin and moisten it there with a small droplet of water. After a few minutes, rinse and observe the brown spot of AgS on the surface of the coin, *Hepar test.* Mix $SrSO_4$ with an equal volume of Na_2CO_3 and repeat the test with the mixture.

e) *Red Ferric Oxide.* Mix with a like volume of Na_2CO_3 and heat on the charcoal to the highest attainable temperature. Allow to cool, scrape the residue into a mortar, and grind it to a powder. Transfer the latter to a sheet of paper. Move a magnet along the underside of the paper to pull the magnetic particles out of the mixture and to collect them in a pile.

f) *Arsenic Trioxide.* Heat with a very small flame and note the odor and the coloration of the flame. White fumes may be seen and a white deposit may be obtained at some distance from the flame.

Experiment 18

Performance of Flame Tests

Nickel wire, 0.5- to 1-mm diameter, 20 cm long; copper wire of like dimensions; 15-cm length of glass tubing, 2 to 6 mm in diameter; small square of blue cobalt glass, 5 cm ×5 cm; test tube filled with powdered $K_2Cr_2O_7$. — NH_4Cl; KCl; LiCl; $SrSO_4$; $CaCO_3$; $MnCl_2 \cdot 4 H_2O$; $TlNO_3$; $CHCl_3$; CHI_3.

The emission of colored light by an otherwise non-luminous flame is caused by the excitation of certain atoms or molecules which then radiate with frequencies characteristic of the species and also determined by the temperature of the flame. Volatilization is the prerequisite, and differences in volatility may be used to resolve interferences. All alkali salts are volatile and may be driven off at relatively low temperature before converting refractory compounds of other metals to the volatile halides.

Flame tests are very useful for the detection of lithium, strontium, barium, thallium, and the halogens. The value of the sodium test is greatly diminished by its extraordinary sensitivity and the wide distribution of the element. The coloration given by calcium is difficult to identify without the use of a spectroscope. Violet, pale blue, and green colorations must be interpreted with careful consideration of the conditions of the test and the treatment of the sample. Obviously, spectroscopic observation of the colorations will, under all conditions, enhance the reliability of the conclusions.

The traditional use of a platinum wire for introducing the sample into the flame is convenient if the absence of heavy metals which could be reduced and alloy the wire is assured. If this assurance is lacking, a fiber of vitreous silica, drawn out from the end of a rod or cemented into the end of a glass handle, is preferable; the used end may be broken off and discarded whenever cleaning proves difficult.

Nickel and steel wire are practical substitutes. Use thick-walled glass tubing of 2-mm bore and 15-cm length. If the bore of the tubing is larger, heat the ends of the tubing so that they contract to give orifices of little more than 1-mm diameter. Draw a 20-cm length of wire through the tubing, which serves for handle, so that 5 cm of the wire is exposed at one end.

Pour some 6-F HCl into a watch glass, and clean a nickel or steel wire (just like a platinum wire or a silica fiber) by dipping the end into the acid and then igniting in the hottest part of the roaring Bunsen flame. Repeat this until there is no more coloration obtained in the flame before or after the wire becomes incandescent. Use only very small amounts of sample for each test.

a) Sodium. Touch the end of the clean wire with the fingers, and then insert it into the cool outer mantle of the roaring flame about 1 cm above the barrel of the burner. Observe how long the yellow coloration persists. Repeat the experiment, but look at the flame through the cobalt glass. Repeat the experiment in a dark room and observe that the dichromate in the test tube looks yellow when illuminated by the sodium flame.

b) Place some KCl upon a watch glass. Dip the end of the clean wire in distilled water and then pick up with it a small kernel of KCl and bring it into the seam of the roaring flame just above the barrel of the burner.

Note that the coloration is visible through the cobalt glass. Repeat the test with LiCl and with TlNO$_3$. Finally clean the wire by treating it with HCl and igniting.

c) Take up with the clean end of the wire some SrSO$_4$ and heat first in the coolest part of the flame (seam above the barrel) and finally in the hottest part above the tip of the inner blue cone. Note the strong incandescence of the sample. Try to reduce sulfate to sulfide by slowly moving the sample at the end of the wire through the tip of the inner cone down to the center of the orifice of the barrel. Allow to cool in the stream of emerging gas, and then quickly withdraw through the seam of the flame. Hold the end of the wire into the gas space of a bottle containing 12-F HCl to convert sulfide into chloride, and then again heat the end of the wire as at the beginning.

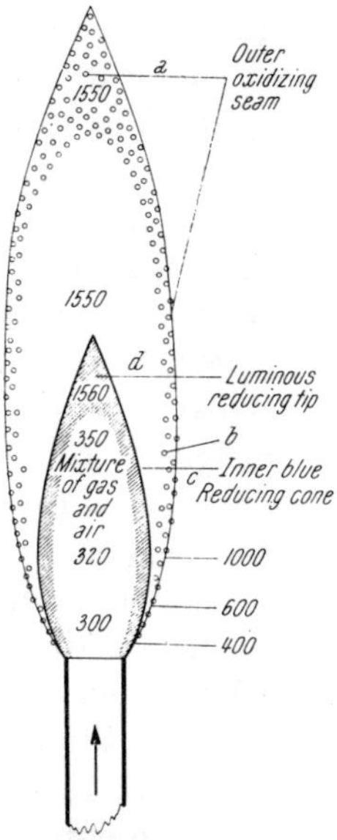

Repeat the experiment with CaCO$_3$. There is no need for reducing the sample, but it should be treated with HCl fumes.

d) For the following Beilstein tests use the just non-luminous flame; the copper wire melts in the roaring flame of the Bunsen burner. Clean the wire after each experiment by just heating it in the flame until the green coloration does no longer appear. Spread the various solids on a watch glass, and pick up small kernels with the hot end of the wire. Observe the blue to green flame obtained with NH$_4$Cl, MnCl$_2 \cdot$ 4 H$_2$O, and CHI$_3$. In all instances, quickly introduce the substance into the hot part of the flame to avoid the possibility of its evaporation before it can react with the copper oxide.

Fig. 10. Non-Luminous Bunsen Flame. Temperatures are given in degrees Celsius; the reducing tip becomes luminous when the air ports are nearly closed; about $^1/_2$ nat. size

Try chloroform by touching the liquid with the hot wire and then inserting it into the hot part of the flame. — As an alternative, heat the clean end of the copper wire in the flame and introduce chloroform vapor at the air port of the burner; to this end, take into the lower end of a capillary about 1 mg (1 μl) of chloroform and insert this end into the air port of the burner.

Try the chlorine test with kernels of KCl.

Work on the Centigram Scale

On the centigram scale it begins to become practical supplying the individual work bench with a complete set of reagents. Short, wide screwcap vials are recommended for solids; liquid reagents may be kept in bottles of 15- to 30-ml capacity and, if this seems desirable, equipped with medicine dropper and rubber bulb. Plastic vials should be used for fluorides, and plastic bottles are recommended for distilled water and alkaline solutions.

The Weighing of Samples may be performed with a pulp balance, a Sauter balance, a "Quartz Spiral Spring Balance" (Microchemical Specialties Co., Berkeley, Cal.), or some simplified Salvioni balance (142).

The Estimation of Volume requires measuring pipets of 0.1- to 0.2-ml capacity, calibrated to indicate 0.01 ml = 10 μl. Transfer pipets of 0.5-, 1-, and 2-ml capacity may be obtained by crudely calibrating a dropper obtained by drawing a capillary of 0.5-mm bore and 8-cm length at one end of a glass tube of 6-mm bore and 12-cm length; the capillary tip of such a pipet may be calibrated as is done with capillary pipets (Expt. 22) and then used for measuring volumes from 1 to 20 μl.

Centrifuge Tubes. The most useful type of apparatus for general work is a centrifuge tube of 16-mm bore in the upper cylindrical part of 2- to 3-cm length and a conical taper of 4- to 5-cm length, which should be formed to end with a hemispherical cup of 2- to 3-mm radius at the apex.

Reagents are Added by transferring them with microspatula, small measuring cup (for finely powdered solids), pipet, or dropper to the inside surface just below the opening of the centrifuge cone or test tube (microbeaker). Brief swirling in the centrifuge is sufficient to collect all material in the tip of the tube.

Adding portions of 1-μl volume is approximately equivalent to the dropwise addition of reagents on the gram scale. For the adjustment of pH (neutralizing), it will thus be necessary to add acid and base in portions of 1 μl. Ammonia and hydrogen chloride may be added by blowing upon the surface of the solution a current of air which has been passed through a gas washing bottle containing solutions of suitable strength of these reagents. In many instances, it may be preferable to evaporate the solution to dryness for the removal of ammonia or volatile acids and to dissolve the residue in a solution of the desired pH.

Mixing. To obtain a stirring rod, take a 7-cm length of glass rod of 3-mm diameter and draw out one end to a thread of 1-mm diameter and 8-cm length. Firepolish both ends. Fuse a bead of 2-mm diameter at the end of the thread; hold the rod at an angle of 45 degrees to the horizontal while touching the end of the thread to the seam of the Bunsen flame. This will place the bead at an angle to the axis of the rod and give better stirring action when twirling the thick end of the rod between the fingers. Either rinse the stirring rod back into the tube when withdrawing it or, and this does not dilute the contents, touch it repeatedly to the inside surface of the tube to strip it of liquid adhering to it.

Mixing is also obtained by holding a centrifuge tube or microbeaker below the rim between thumb and index finger of the left hand so that it hangs vertically and striking the bottom end with the fingers of the right.

Treating with Gas. The gas may be bubbled through the liquid as described on p. 107. As a convenient alternative, provide the centrifuge tube containing the liquid with a one-hole rubber stopper carrying a delivery tube of about 15-cm length (2- to 5-mm bore), one end of which is drawn to a quick taper and cut to give an orifice of about 0.5-mm diameter. Obtain a slight constriction 5 cm from the other end by allowing the glass to gather and collapse while being rotated in the flame. Cut the tube at a distance of 1 cm from the constriction, and firepolish both ends. After cooling, insert above the constriction a loose wad of cotton which will hold back any dust particles or spray carried along by the gas. Insert the tubing into the stopper so that the orifice will remain at least 3 to 4 cm above the surface of the liquid. Use flexible rubber tubing of about 30-cm length for connecting to the tap which should supply the gas at a reasonably constant, low pressure of 10- to 80-cm water column above the atmospheric pressure. Insertion of a bubble counter into the supply line is desirable.

For saturating a liquid, first hold the stopper in the opening of the centrifuge tube to leave a narrow gap for the escape and open wide the gas tap. When the air has been displaced from the tube after a few seconds, apply the stopper tightly. To promote rapid absorption of the gas, agitate the contents by striking the tip of the tube as suggested in the preceding section on mixing or incline the tube at an angle of twenty degrees to the horizontal and rotate it around its axis so that the liquid is spread over the inside surface of the tube. Shaking the contents is not recommended since this would get them into contact with the stopper and the delivery tube. For heating, the centrifuge tube may be inserted into a block, water bath, or steam bath without interrupting the connection to the gas supply. Saturation is indicated by the absence of bubbles in the counter.

Heating. For heating, immerse centrifuge tubes in water of the desired temperature. A steam bath may be readily improvised by placing the tube into a beaker with boiling water. The escape of gases or vapors may be prevented by closing the tube with a suitable stopper and tying it down with copper or aluminum wire (13). Elaborate heating blocks may be obtained from laboratory supply houses (155, 162, 904).

Fusions may be carried out in crucibles of 1- to 2-ml capacity and made of porcelain, vitreous silica, nickel, silver, or platinum. In the place of a platinum microcrucible, the cover of a large platinum crucible or a square of platinum foil may be used. If the mass of sample and flux does not exceed 150 mg, a bead may be fused at the end of a suitable wire of 0.5- to 1-mm diameter, P. 39. A small test tube, microbeaker, or centrifuge tube of vitreous silica is well suited for pyrosulfate fusions if care is taken to spread the melt in a thin layer over the inside surface by inclining and rotating the tube while solidification is taking place.

In all instances, the material should be finely ground and intimately mixed with the powdered flux before it is placed into the crucible or taken up with a wire. The apparatus is best supported by a small triangle made of 2- to 3-mm tubing of clay, ceramic, or vitreous silica (Vitreosil) which need not be clear, or of platinum wire or silica rod of 1- to 2-mm diameter. A microburner that gives a small, blue, oxidizing flame should be used for heating and may be shielded with a Pyrex tube of 20-mm height and 25-mm o. d. (625); a small electric muffle oven may prove convenient.

Centrifuges. The scale of work is already small enough to make efficient use of centrifuges for the separation of phases and the collecting of matter in the point of tubes. To-day, electric centrifuges are widely used and generally available. Hand driven centrifuges (155, 162), however, may be obtained in good quality from abroad. They are recommended for field work on the centigram scale, since it is felt that centrifuging and decantation surpass all methods of filtration in convenience and efficiency and should not be abandoned merely for lack of suitable current. On the milligram scale and smaller scales of work, hand centrifuges are convenient even in the well-equipped laboratory and save time since one turn of the crank frequently suffices to obtain the desired effect.

With the customary, gear-driven hand centrifuges, up to 2000 revolutions per minute may be obtained; with friction couplings, Gorbach obtains up to 40 000 revolutions per minute for the centrifuging of capillaries. Comparable speed is obtained by KIRK (157) with the use of compressed air for the centrifuging of small tubes of up to 0.1-ml capacity. Electric centrifuges of standard design for tubes up to 15-ml capacity give from 1700 to 6000 revolutions per minute; higher speeds than these are very rarely required, but may be obtained for work in small tubes by using "multispeed" attachments that give up to 18 000 r. p. m.

Sturdy electric centrifuges are preferable for work on the centigram scale. Hand-driven centrifuges are recommended for work on the milligram- and somewhat smaller scales. Their life will be lengthened if care is taken to start them gradually. The use of shear pins of suitably soft material in the crankshaft will save the crank which usually breaks at the notch engaging the pin.

The *relative centrifugal force* is the ratio of the centrifugal force over the gravitational force acting on the same mass. It may be estimated with the use of a nomogram (406). It is,

$$relative\ centrifugal\ force = 1.2 \cdot 10^{-5}\ r\ n^2$$

when r is the radius of the orbit of revolution in centimeters and n is the number of revolutions per minute.

The centrifuge should be provided with a *shield* having a large diameter in the plane of revolution. Exchangeable *heads* for different types of centrifuge tubes are very useful. The microhead suggested by EMICH, Fig. 11b, has long slots in the arm of aluminum sheet, which facilitate the centrifuging of long capillaries. The relatively long arms increase the centrifugal force. The makeshift arrangement of Fig. 11a, which uses the aluminum shell of a clinical hand centrifuge, is less satisfactory because of the short radius r. Heads that hold the whirling tubes in an inclined position are not recommended since difficulties may arise when long tubes (capillaries) have to be centrifuged.

The *metal shells* for microcones of 0.7-ml capacity may be either open or closed at the bottom as shown in Figs. 11b and c. The open shells are

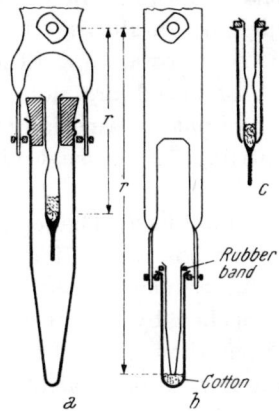

Fig. 11. Centrifuging of Microcones; about $^1/_2$ nat. size

very practical since they allow placing the tip of tubes at a maximum distance from the center of revolution so that the length of the glass apparatus is mainly limited by the radius of the shield of the centrifuge. The apparatus may be supported by the rim of the shell or at the bottom opening of it; rubber rings may serve for cushions in either instance and, to this end, are cut from rubber tubing of suitable dimensions. Closed shells may be used for floating the glass cones in water during swirling to definitely exclude the possibility of breakage. Microcones of heavy Pyrex tubing break rarely, however, and the danger may be further reduced by the use of cushions below the sturdy flanged rim and (or) underneath the tip.

Centrifuge tubes of 3-ml and larger capacity are usually centrifuged in closed shells. If they are of soft glass or thin glass, it is best to fill the space between tube and shell with water so that they float during whirling. Should a tube collapse, the shell should be immediately and thoroughly cleaned to rinse out corrosive substances and to remove glass splinters that might otherwise cause the destruction of some more centrifuge tubes.

Use of Centrifuges. The momenta acting in diametrically opposite directions of the centrifuge head must be approximately equal to prevent the appearance of a destructive force. The requirement is most conveniently met by a similar distribution of equivalent masses on opposite arms.

When working with centrifuge tubes of 3-ml or larger capacity, place the shells containing the tubes on opposite pans of a trip scale and add water to the lighter combination until equilibrium is established; depending upon the requirements of the task, place the water into the centrifuge tube or into the space between tube and metal shell. Less care is needed with the light microcones of 0.7-ml capacity and smaller apparatus; as a rule, it is sufficient to place tubes of equal size into the opposing shells.

Close the cover of the shield before starting the centrifuge and **do not open** it before the centrifuge has come to a stop. Start and stop the centrifuge **gradually;** sudden changes of speed are not only destructive but also prevent the clean separation of phases, especially in apparatus of large bore. Immediately stop the centrifuge if unusual vibration indicates lack of balance.

Separation of Solid and Liquid Phase. The separation is best accomplished by centrifuging and decanting. As a rule, it will suffice to centrifuge for about one minute. The solid frequently becomes so tightly packed that decantation may be accomplished by simply pouring off the supernatant liquid. If some solid floats upon the surface of the liquid, it may be combined with the bulk by adding a few drops of alcohol and whirling again. The alcohol is added at the rim of the tube so that it rinses down the inner surface of it.

If clean separation of the phases is needed and cannot be obtained by pouring, lift off the liquid by means of a capillary siphon, Fig. 31, a suction-operated siphon, Fig. 32, a suitable pipet, or the capillary siphon pipet, Fig. 12. In all instances, solid floating on the surface of the liquid usually collects upon the interior surface of the centrifuge tube and upon the outside of the inserted tube. If difficulties should arise, one may try to wet the tip of the siphon or pipet with alcohol before inserting it into an aqueous liquid contained in the centrifuge tube. The change in surface tension when the tip of the pipet makes contact with the liquid in the centrifuge tube may sink some of the solid particles floating on the surface and drive the rest to the wall of the tube.

The *Capillary Siphon Pipet* (868), Fig. 12, is shown in *A* in its simplest form which may be readily made in the laboratory; *B* may be further modified by placing a filter mat into the somewhat widened orifice of the intake tube and doubling the height of the heavy-walled arm of the siphon to obtain stronger suction. The apparatus is best clamped to a stand that permits raising and lowering by mechanical means, and also the centrifuge tube may be held with a clamp below the rim during decantation. Suction with the mouth is only needed to start the operation of the siphon; the collected liquid is finally transferred by either expelling it through the siphon or draining it off the bulb, whereupon the device is ready for the collection of washings.

Pipets are convenient for the removal of aliquot parts of a liquid above a precipitate, but they are not the most practical tool if a neat separation by decantation is required. For the latter purpose, the pipet should have a capacity sufficient to take the whole of the liquid at one time. Sucking

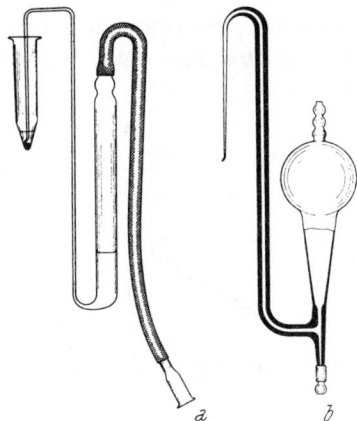

Fig. 12. Capillary Siphon Pipet; $1/3$ nat. size. CLARKE, B. L., and H. W. HERMANCE, Mikrochemie 10, 289 (1935)

with the mouth is not practical, but the pipet could be connected to an aspirator bottle. Pipet controls of the syringe type and made of glass or metal with capacities up to 10 ml may be obtained from supply houses and may permit more convenient regulation than control bulbs of rubber. Pipets may be prepared from tubing of 6-mm to 12-mm bore, which is simply drawn out to thick-walled capillaries to furnish the tip and the stem on both sides of the bulb consisting of a suitable length of the original tubing.

Clean separation requires that the solid is thoroughly compacted so that it may be closely approached with the orifice of the siphon or pipet. When the bulk of the liquid has been removed, the centrifuge tube is spun again to collect liquid adhering to the walls which is then also drawn off and combined with the bulk.

Washing the Solid. Thoroughly mix the solid with the wash liquid, centrifuge, and decant.

Transfer of Solids. Precipitates and residues are usually treated in the centrifuge tube in which they have been collected. If transfer to another apparatus is required, stir to a slurry with a liquid (reagent to be added next or solvent that does not interfere or may be readily removed) and transfer the slurry with a rapidly operating pipet or siphon to prevent the settling out of the solid. At times, it may be found possible and convenient to get rid of most of the transfer liquid by allowing the solid to settle into the tip of a pipet which may then deliver all the solid with the first drop(s) of discharge.

Obviously, if a large amount of solid has been collected, portions of the moist cake may be transferred with a small spatula.

Extraction of Solids may be performed in a centrifuge tube with the procedure for washing solids. One may heat the solid with each portion

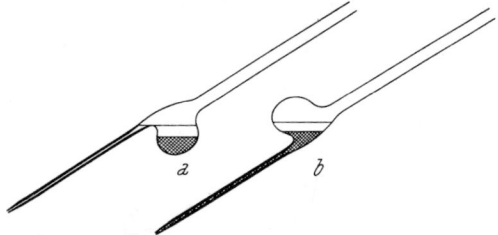

Fig. 13. Stork's Bill Designed by G. GORBACH; about $1/_3$ nat. size

of solvent and decant the hot solution. Apparatus for the automatic extraction of solids may be readily improvised with the use of either the percolating or the siphoning (Soxhlet) principle. Many such devices have been described (147, 155, 162), but special attention should be given to the comparative study of BATT and ALBER (428).

Liquid-Liquid Extraction may be performed in the centrifuge tube as described for the milligram scale, p. 116. Separatory funnels have been designed by ALBER (431) and CHERONIS (147). KIRK (157, 443) describes extractors for heavier-than-water and lighter-than-water solvents; a syringe extractor (451) may be adapted to this scale and would serve for both types of solvent. The separation obtained with the simple storkbill of GORBACH (155), which may be blown from tubing of 6-mm to 8-mm bore, will often suffice. The solution and the solvent are taken up with the aid of an aspirator and the apparatus in position of Fig. 13a. In the same position, the lighter liquid may be decanted by suitable tilting of the tube. Better control is had when draining the heavier liquid with the apparatus in position Fig. 13b. While mixing the phases, the tip should be closed with the finger or a cap. Suitable apparatus for continuous extraction have been described (45, 147), but will rarely be needed.

The simple apparatus of KIRK and CHARLOTTE L. BROWN (937, 952) for solvents heavier-than-water gives a double transfer in one operation. A sealed tube st with iron core is placed into the bottle of the apparatus shown in Fig. 14a, and 15 to 40 ml of heavier-than-water solvent or solvent mixture is added. About 22 ml of the aqueous solution to be extracted is introduced into the outer space o through the side arm, and 5 ml aqueous solution is placed into the inner tube i, which opens 6 to 7 mm above the bottom of the bottle. Obviously, the solution in the inner compartment

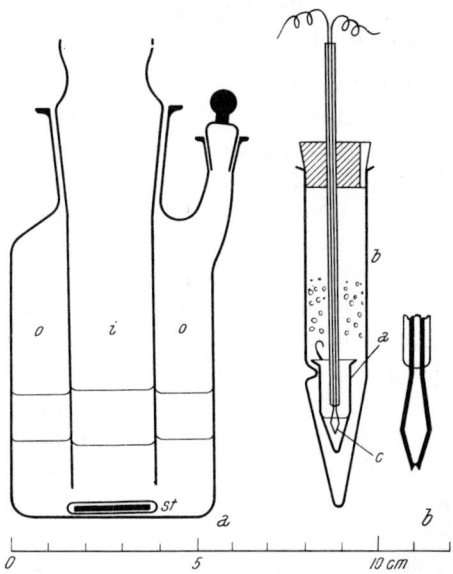

Fig. 14. a Extractor for Heavier-than-Water Solvents; KIRK, P. L., and CHARLOTTE L. BROWN, Mikrochimica Acta **1957**, 715. — b Distillation in the Centrifuge Tube

must contain a reagent which should be insoluble in the organic solvent and which strips the solvent of the extracted matter. The apparatus is placed upon the top surface of a "magnetic stirrer" and the stirring bar on the bottom of the bottle is kept in rotating motion for 3 hours. If the distribution ratios are suitable, the substance of interest will collect in the aqueous phase of the inner compartment.

The apparatus was used to isolate small amounts of basic or acidic compounds in a state of reasonable purity from complex biological materials. The pH of the test mixture was adjusted to liberate the acids or bases so that they could be extracted by chloroform, carbon tetrachloride, or mixtures of the two. The inner compartment held 1-F ammonia or 0.1- to 1-F hydrochloric acid. Thus, barbiturates and alkaloids could be isolated with yields of usually better than 50%.

Evaporation may be performed in a centrifuge tube of chemically resistant glass by heating in a steam bath or block and blowing a stream of filtered air from a capillary on the surface of the liquid, Fig. 28. Also suitable are crucibles and small dishes of porcelain, vitreous silica, and platinum, which may be heated on some bath, block, or hot plate or by means of an infrared lamp (13, 147). A jet of clean air blown across the surface of the liquid will in all instances materially speed evaporation and keep away dust.

Distillation. A variety of distilling apparatus have been described (146, 147). For analytical tasks, the simplest designs are preferable.

If the volatile matter of interest is carried over by a relatively large amount of other vapors, apparatus of the type shown in Fig. 9b will be suitable if made in the proper dimensions. Distillation in the centrifuge tube as described by Boos (141, 1251) allows simple and complete recovery of distillate and residue. Place the solution to be distilled into the small tube a, Fig. 14b, which may be graduated and has a handle attached to its rim. By means of a hook at the end of a glass rod, lower it into the centrifuge tube b so that the rim comes to rest upon the three indentations projecting into the interior of the large tube. Insert the internal heater c into the solution in the small tube. The illustration shows an enlarged view of its essential part, a platinum wire of 0.6-mm diameter which is filed down to 0.2 mm at the point of the V. Connect through a 1- to 5-volt variable step-down transformer to the 110-volt a–c line. Immerse the large centrifuge tube into a suitable cooling bath (ice water, ice and NaCl, etc.) up to the cork, and give enough current to produce slow boiling. Always keep the point of the heating wire immersed in the liquid. To finish the distillation, shut off the current and leave the apparatus for two minutes in the cooling bath. Then remove the heater and centrifuge the large tube with the small one in place; finally lift out the latter by means of the hooked rod.

Apparatus shown in Fig. 15 may be used when the residue of the distillation contains solutes that must be collected for investigation. Ground glass stoppers may replace the indicated corks. For good separation, it is essential to heat so that the ring of condensate rises very slowly through the vertical or nearly vertical tubing and finally remains stationary at the level of the side arm (Fig. 15a and b) or at the level where the fraction is collected with a capillary pipet (c). A tiny microflame may be used for heating and may be shielded with a small chimney made of glass tubing and partly closed at the bottom with a disk of cardboard or metal foil. A boiling capillary, d, is placed into the liquid to obtain even boiling, or a platinum wire of 0.05- to 0.1-mm diameter is inserted through the bottom, Fig. 15c.

The simple apparatus (1015) *a* may be held in hand during distillation; the ring of condensate is kept at the height of the side arm until the desired amount of condensate has collected in the knee of the latter, from where it is taken up into a capillary pipet. Several fractions may be obtained. The residue of the distillation is also collected with a pipet. Apparatus *b* (872)

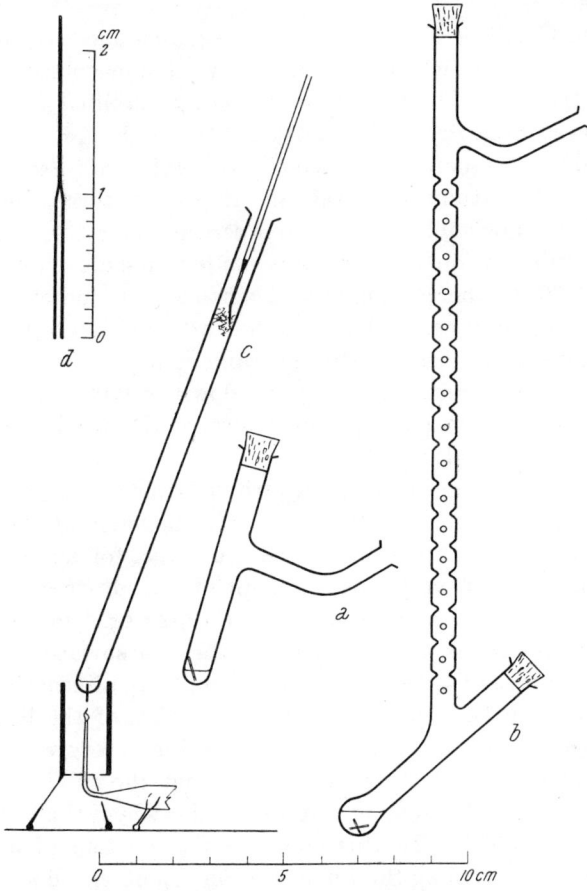

Fig. 15. Distillation on the Centigram Scale

uses an indented column which is easily made in the laboratory and gives less hold-up than packed columns. Column *b* as well as tube *c* may be surrounded by a glass mantle.

The fractionating tube *c* may be used like EMICH's (below) and whirled between fractions if a suitable centrifuge is available. During distillation, it is best clamped to a stand. The fine tips of the capillary pipets may be bent, and the fractions should be collected from the topmost part of the ring of condensate when the latter ceases rising.

A thermocouple might be used in the top of column b. As a rule, it seems more practical to determine the boiling points after collection of the fractions, Expt. 21. All apparatus of Fig. 15 could be connected to a vacuum line for distillation under reduced pressure, but the simple apparatus of BABCOCK (146, 147, 445) may be preferable for occasional analytical use.

Emich's fractionating tube, Expt. 20, is recommended for the separation of a mixture of liquids which does not leave a non-volatile residue that requires investigation. The technique is readily modified for the isolation of traces of very volatile liquids (861, 1154).

Sublimation is frequently performed to *obtain well developed crystals* for crystallographic study. Crystals developed in three dimensions are obtained by very slow sublimation at temperatures far below those usually considered as sublimation temperatures (sulfur, arsenic trioxide, vanillin, and alizarin at 50° C; mercuric chloride, benzoic acid, and caffein at 20° C; indigo, morphine, quinine, and strychnine at 100° C, etc.). Even when the condensing surface has nearly the same temperature as the heated material, very small quantities (10 μg and down to 0.001 μg) of substances may be isolated in distinctly crystalline form if the condensing surface is very close to the heated material.

R. KEMPF (1142) described an electric heating block with elaborate temperature control, which can be used also in the vacuum under a bell jar. A simple metal heating block will serve, however, for sublimation under atmospheric pressure. A very small amount of the substance (drug, textile fibers, paper) is ground up with a drop of water or some other suitable liquid and transferred as a thin slurry to the top surface of the heating block to cover an area of 5- to 10-mm diameter; as an alternative, a solution of the material may be applied. After evaporation of the liquid, a cover slip or a microscope slide is placed upon the residue to receive the sublimate. The temperature of the block is adjusted so that the sublimation requires from 4 to 48 hours. Cooling is required only when the sublimation is performed below 100° C. To this end, the flat bottom of a wide metal tube with circulating cooling fluid is placed on top of the slide or cover slip. The sublimation of indigo at 180° to 200° C from a few fibers of dyed cloth is recommended for practicing the technique.

To *obtain very thin tablets* for the measurement of profile angles, SHEAD (633, 958) uses a sublimation cell consisting of two standard microscope slides separated by a glass ring, 4 mm high and 16 mm in inner diameter, which is heated on a metal bar to a temperature giving a medium rate of sublimation (10° to 50° below the melting point of odorous solids and 80° to 150° C below the melting point of non-odorous solids). The surface of the receiving slide is slightly greased with oil from the fingers or face and then wiped to leave only an invisible film of it. When the cell and the receiving slide (which may be placed on the bar for heating it up) have attained the sublimation temperature, a small quantity of the material is placed upon the bottom of the cell which is then quickly covered with the warm receiving slide. Relatively large, thin crystal plates are obtained within two to ten minutes.

Various apparatus have been described (147), which may be used for the *purpose of analytical separation*. To gain time, vaporization at relatively high temperatures is used, and the collecting surface is usually cooled to obtain good recovery of the volatile substances. The temperature gradient becomes especially steep in those instances where the distance between charge and condensing surface is small. CHERONIS and RONZIO (147)

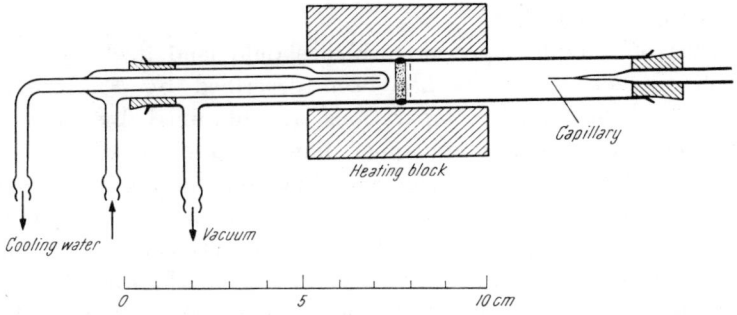

Fig. 16. Sublimation According to SOLTYS, Mikrochemie, Emich-Festschrift 275 (1930)

recommend a distance of 7 mm to 25 mm for sublimation under reduced pressure of 1 to 20 mm mercury column. The charge should be introduced as a fine powder to obtain a maximum of surface; if a non-volatile film forms on the surface and sublimation stops, the pulverizing may have to be repeated (or a drop of suitable solvent may be added and the sublimation resumed after the solvent has evaporated to leave the volatile matter behind, spread as a thin film over the heated area). Sublimation should be carried out at least 10° C below the melting point if the latter is between 40° and 100°; or 50° to 80° below, if the melting point is between 100° and 200°; and 100° to 150° below, if the melting point is above 200°. High temperatures and high rate of sublimation may, however, give poor separation of fractions and may have to be adjusted according to the needs of a given task.

Sublimation with a current of carrier gas through a straight tube of suitable dimensions, p. 74, satisfies the requirements of analytical separation. The apparatus of SOLTYS will prove quite efficient for the complete separation of one volatile fraction from a non-volatile residue. The more elaborate procedure of GETTLER, UMBERGER, and GOLDBAUM provides for the convenient separation of several volatile fractions.

SOLTYS (855) places the finely powdered substance upon a disk of coarse fritted glass, sealed into the tube, and holds it in position by pressing upon it a disk of asbestos or glass-fiber paper, which fits snugly into the tube. The tip of the cold-finger condenser is brought close to the other side of the fritted-glass disk. Air or inert gas is admitted through a capillary which is so fine that a rapid stream of bubbles of 1-mm diameter is obtained when blowing air through it into water. The tube is heated in a metal block with thermometer, and a vacuum of 20 mm is applied. The capillary may be replaced by a glass tubing with cotton filter for sublimation in a gas stream of 1 to 2 ml per second. Sublimation by atmospheric pressure requires more time and efficient cooling and gives fluffy sublimates of crystal aggregates, whereas a dense crust is obtained on the condenser when applying a vacuum.

A relatively simple procedure which should lend itself to application to much smaller quantities has been described by FLASCHENTRÄGER et al. (930). Quantitative separations are obtained by high vacuum sublimation in a tube of 8-mm inner diameter and 17- to 20-cm length, which is sealed at one end. After a preliminary heating with application of high vacuum, 10 to 20 mg of the weighed material are brought to the sealed end by means of a charging tube (13). Separation is then accomplished by suitable heating in a metal block of 17-cm length, whereby more and more of the tube is withdrawn from the block as the temperature is raised. The fractions are isolated by cutting the tube into sections.

GETTLER, UMBERGER, and GOLDBAUM (447) obtain simple recovery of sublimate fractions and residue by using a 14/20 ground joint for separating the lower part of the sublimation tube, which contains the charge, from the upper part and by lining the inside surface of the latter with a transparent sheet that catches the sublimates and may be easily withdrawn from the tube.

The lining is obtained by cutting a strip of 4-cm width and 15-cm length from a roll of cellulose sausage casing obtainable from the Visking Corporation (Chicago, Ill.). The strip is rinsed in ethyl ether, dried, rolled around a glass rod, and thus inserted through the end of the tube, which carries the male part of the ground joint, so that about 6 mm of the strip remain outside the tube. The rod is then removed, and the strip is grasped by the end projecting from the ground taper and rotated clockwise until it forms a helix inside the tube. This is followed by counter clockwise twisting so that the coil partially unwinds, expands, and wedges tightly against the wall of the tube. The sheet is then trimmed flush with the lower end of the tube.

The upper end of the sublimation tube is connected to a vacuum line able to give 1 mm pressure and less. After introducing the charge into the lower part, the ground glass joint is connected while the vacuum is turned on. Heat is applied so that the temperature of the block rises about two degrees per minute. When a sublimate appears, the temperature is held constant as long as the sublimate increases. Then the temperature is again gradually raised until another fraction sublimes, etc. In this manner, the components of a mixture may be obtained at different levels, especially when a preceding trial sublimation has shown the best rate of heating, etc.

When the sublimation is finished, the upper tube is separated from the lower, and the film is removed by hooking it with a dental probe.

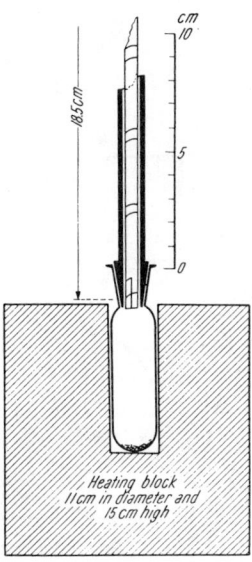

Fig. 17. Sublimation According to GETTLER, UMBERGER, and GOLDBAUM, Analyt. Chemistry 22, 600 (1950)

After withdrawal from the tube, it is unwound and fastened with thumbtacks to a cork pad for a few hours to make it lie flat. The fractions may be separated by cutting the film. Crystals may be studied on the film, but must be transferred to glass slides for inspection in polarized light because the film is anisotropic. Also chemical treatment is possible on the film which may be heated to 300° C and is resistant to most solvents, alkalies, and even cold concentrated acids. It dissolves in cellulose solvents such as acetone and ethyl acetate.

SCHMIDT (955) uses a film of polyterephthalic ester (*Hostaphan* of Kalle in Wiesbaden, Germany) rolled into a simple cylinder and a heating device (commercially available, Willi Günther, Nürnberg) producing a temperature gradient along the tube. It is pointed out that the identification of the sublimates is facilitated by the oriented growth of the crystals upon the anisotropic film. Sublimation in the horizontal tube has also the advantage that crystals of the sublimate, which separate from the supporting film as a consequence of recrystallization, cannot drop back to the sample.

Liberation of Gases for identification may be performed in one of the several *simple apparatus* developed by FEIGL (121) and his coworkers. Of these, the one shown in Fig. 18c is readily assembled in the laboratory. The sample is treated with the reagent at the bottom of the small beaker, and the small pieces of moist test paper are placed upon the outer surface of the spherical bulb. As an alternative, the latter may be moistened with

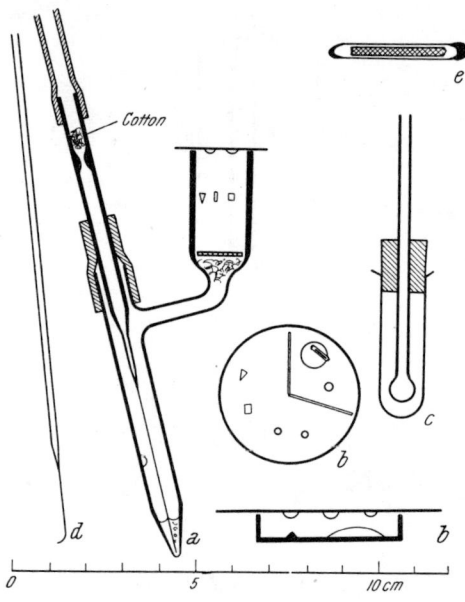

Fig. 18. Liberation and Testing of Gases; *e* is enlarged five times as compared with the other apparatus

reagent solution or absorbent solution which is later rinsed into the depression of a spot plate or taken up by spot test paper. The closed apparatus is either allowed to stand at room temperature or gently warmed.

The *gas reaction cell*, Fig. 18b, may be assembled from a dish for the cup grease test (Corning 3210) and a rectangular glass plate. The transfer of the gas is left to spontaneous action as in the apparatus of FEIGL, but spray may be prevented from getting to the reagent. The bottom of the dish is treated so that drops will not spread; ruling squares with paraffin lines on clean glass is practical, but making the surface repellent renders difficult the deposition of drops. The sample and a drop of reagent for the liberation of gas are placed upon the bottom of the dish so that they will not mix. Into the reagent drop is placed a piece of sealed capillary containing a 6-mm length of thin iron wire, shown magnified in Fig. 18e. Small pieces of reagent papers cut into triangles, squares, and rectangles of 1- to 3-mm² area are moistened and pasted to the clean underside of the glass plate forming the cover of the cell. Small reagent drops are deposited upon this surface. After the cell has been closed, a strong magnet (Alnico) is applied to its underside, and the reagent drop and sample are wiped together by suitably moving the capillary containing the wire.

A blank test is readily obtained by allowing the closed cell to stand for some time before combining the reagent with the test drop. The change taking place after adding the reagent gives convincing proof that the active agent comes from the test drop. The transfer of spray is prevented by placing into the cell a partition made from a strip of filter paper or glass fiber paper, which is as long as the diameter of the cell and nearly as wide as the cell is high. The strip is folded in the middle to give a right angle and then placed into the cell so that the legs of the angle take the positions of the hands of a clock at 3 hours. Test drop and reagent drop are deposited on one side of the partition; the reagent drops and the test papers, on the other. The strip of paper may also be treated to absorb interfering gases.

The *more elaborate apparatus*, Fig. 18a is recommended if spray must be kept away from the reagents or the gas is only slowly given off by the reaction mixture. A capillary is used to place the reagent which liberates the gas into the tip of the tube, and the short piece of capillary E with the sealed-in iron wire is added. A droplet of the solution to be tested is deposited upon the inner surface of the tube, about 15 mm above the reagent solution. If the sample is a solid, a droplet of water is deposited first, and then a particle of the solid is transferred to this droplet. Capillary pipets and fine glass rods with the tips bent at an angle, Fig. 18d, are suited for these tasks; the interior of the tube may be made liquid repellent.

Scrubbed air or inert gas is introduced through a fine capillary of 0.05- to 0.2-mm bore, which is broken off to the proper length so that it just reaches the bottom of the conical end of the tube. A very satisfactory *supply of air* is obtained by securing with wire a 2-hole rubber stopper carrying two inlet tubes in the opening of a 10- to 20-liter bottle which may contain about 100 ml of 1-F NaOH. A short piece of rubber tubing connects a rubber bulb (bellows) to the inlet, and the outlet is provided with a long flexible rubber tubing carrying a screw clamp. A supply of air pressure built up with the rubber bulb is sufficient to maintain a fine stream of air bubbles for a long time.

Into the bottom of the wide funnel of the apparatus is dropped a disk of filter paper or glass fiber paper for catching spray. A strip of paper, cotton, or glass wool treated with reagent for the removal of interfering gases may be placed below the paper disk. Small pieces of test papers may be attached moist to the inner surface of the funnel. Reagent droplets are more advantageously deposited upon the underside of a cover slip placed upon the opening of the funnel.

A blank test may be performed by bubbling gas through the apparatus for some time before using the glass-enclosed iron wire for wiping together the sample and the reagent solution. The stirrer is operated by means of an Alnico magnet applied to the outside of the tube. The apparatus may be made in the laboratory, whereas the more elaborate one of MALISSA (162, 894) requires the skill of a glass blower.

Confirmatory Tests may be carried out in small test tubes and centrifuge cones. As a rule, one prefers to use a test plate or spotting on filter paper. These procedures have the merits of simplicity and speed, but the techniques described for use on the milligram scale are preferable when reliability must be the first consideration.

The *test plates* are made of white or black, glazed porcelain or of chemically resistant glass. In the latter instance, the background may be readily changed for observation by laying the plate upon paper or tile of suitable contrasting color. The customary depressions in the top surface of the plates are convenient but not essential.

It is important to clean the test plates immediately after use. After drying, the top surface should be made liquid repellent to prevent the spreading of the drops. This may be accomplished by wiping it with a cloth or wad of cotton that has been moistened with a mixture of ethanol and diethyl ether.

Drops of the test solution and of the reagents are usually transferred to the plate by means of stirring rods or medicine droppers. To avoid contamination of these tools and to render unnecessary their cleaning after use, the drops should be deposited side by side without merging, which is easily accomplished if the drops do not spread. Use of a rod for mixing may be avoided since combining of the drops and mixing may be accomplished by blowing the air from a capillary tangentially onto the side of the drops.

Difficulties may arise when organic solvents are used and the drops spread. Thus precipitates and colorations may be carried up to and over the rim of the depression so that it becomes necessary to interpret the outcome of the test by comparison with blanks and controls (tests with known amounts of sought substance).

For the performance of spot tests on paper, it is recommended to place a strip or square of it upon the opening of a small beaker, porcelain crucible, etc. and to deposit drops of test solution or reagents upon the center of the horizontally supported sheet so that they may spread in the paper without hindrance. The drops are transferred by means of stirring rods, medicine droppers, or pipets. It is obvious that spot test paper must be handled and stored in a manner that excludes contamination.

Infrared heat lamps are recommended for the warming of spot tests on test plates (432) and paper, for the drying of test papers, etc.

Storing Material Under Investigation is simple. The centrifuge tube containing the liquid, solid, or mixture of both is closed with a suitable stopper, or a small beaker or wide vial is inverted and placed over the opening of the tube. The centrifuge tube is placed into a suitable block or beaker to keep it in upright position, possibly under a bell jar. Precipitates which oxidize in air or age and change their solubility should not be stored, but dissolved without delay. Solutions in volatile solvents may be sealed into glass containers; the tube containing the solution may be slid into a test tube which is then fused shut.

Cleaning Apparatus. Feathers and pipe cleaners will be found useful for the cleaning of centrifuge tubes and similar apparatus. A suitable suction device (p. 124) may serve for rinsing. For quick drying, apparatus may be inverted over a vertical glass tubing from which emerges hot air coming from a heated aluminum box or coil.

Experiment 19

Preparation of Capillaries, Capillary Pipets, and Microburners

Supply of glass tubing of 6- to 8-mm bore. Tubing of 12- to 18-mm bore may be used to advantage if a strong blast flame is available. The tubing should be tested by drawing capillaries of about 0.2-mm bore and bending them to get an estimate of the elasticity. For most work, one will prefer a glass that is not brittle like Pyrex. Certain kinds of soft glass are well suited.

Cut the tubing into suitable lengths, and then clean it outside and inside with soap solution and a wad of cotton tied into the middle of an aluminum wire (or suitable brush). Rinse with tap water and distilled water. Finally wrap bundles of the tubing into paper towels and store them in an upright position for draining and drying.

For *drawing capillaries*, grasp a piece of tubing at the first and third quarter of its length. While rotating it, heat its center just above the inner blue cone of the large roaring flame of a Tirrill burner (Meker or Fisher burner). If the heating is uniform, the flame will become luminous at the same time along the whole length of the heated part of the tubing. When the glass has become soft, remove the tubing from the flame; continue rotating it and leisurely draw it out to a capillary of 0.5- to 1-mm bore. The rate of drawing determines bore and length of the resulting capillary. Watch closely while drawing and regulate the rate according to requirement. Do not release the pull until the glass has hardened since otherwise bent capillaries will result.

With the sharp edge of broken porcelain, cut the capillary so that 6 cm of it remain with each piece of the original tubing. These short pieces serve as handles when drawing the next length of capillary from the portion of the tubing adjoining the taper. The procedure gives lengths of capillary separated by small bulbs. Cut the capillaries so that 6-cm lengths of capillary remain on either side of the bulbs which are saved for use as pipets. Seal the capillaries at both ends and keep them in a box or tube until needed. — The wide ends of the original tubing, each with a 6-cm capillary attached, may be fused together to get a handle on each side and then drawn into capillaries.

Several meters of capillary may be obtained by one drawing from tubing of 12- to 18-mm bore. Heat the tubing in a large blue blast flame until soft. Then remove the tubing from the flame and hold one end steady while a second person walks away with the other end.

Preparation of Capillary Pipets. Select a piece of capillary of 0.5- to 1-mm bore and 20-cm length. Heat the center of it in the seam of the

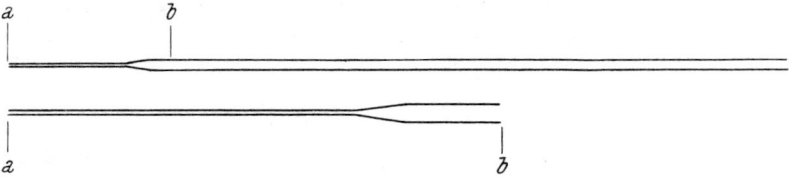

Fig. 19. Capillary Pipet; approx. nat. size. The tip is shown three times enlarged

lowest part of the Bunsen flame, just above the barrel, until the glass just softens. Then remove it from the flame and pull with a quick motion so that a fine capillary of 0.05- to 0.2-mm bore and about 10- to 20-cm length results. If the fine capillary can be bent into a loop without breaking, it probably has the desired diameter.

Break the fine capillary about 2 cm from the tapered portions to get two pipets, Fig. 19. Finally, for a crude estimation of the bore of the fine capillary, lay the pipet upon a slide and inspect under the microscope with a total magnification of about 30 diameters. Measure the bore at the fine orifice with the eyepiece micrometer.

Microburner. Draw out one end of Pyrex tubing, 6 to 8 mm in diameter and about 15 cm long, to a capillary of about 1-mm outer diameter. Cut this capillary at a point 2 to 3 cm from the taper and bend it as desired. The wide tube may be held in a clamp, or two glass prongs may be fused on near the taper to serve for legs when the tube is placed upon the bench. Flexible rubber tubing is used to connect the wide tube to the gas tap. For adjusting the size of the flame, place a loose plug of cotton inside the rubber tubing and apply a screw clamp at the location of the cotton plug.

The size of the microflame is determined by the orifice of the capillary. To obtain a very small flame, proceed as follows: light the microflame and open wide the tap to get the maximum height of the flame. Then touch the orifice of the capillary to the edge of a roaring Bunsen flame (oxygen blast flame if the glass has a high softening point) until the height of the microflame is reduced to 5 mm. Regulation with cotton plug and screw clamp will then permit to obtain a flame the size of a pinhead.

Concerning a simple shield for the microflame, made from a 20-mm length of Pyrex tubing of 25-mm outer diameter, see STOCK and FILL (625).

Experiment 20

Emich's Method of Fractional Distillation (1011)

Fractionating Tube and Handle. — The fractionating tube, Fig. 20, is made of Pyrex tubing of 4-mm bore. A slight constriction separates the two chambers, each of which is 3 cm long. The lower chamber is half filled with asbestos which is briefly ignited before it is introduced into the tube and pressed together. From the bottom of the tube is drawn a solid spike of glass, which snugly fits into the handle, a piece of Pyrex tubing, 15 cm long and 3 mm in bore.

Mixture of equal volumes of ethanol and water or of equal volumes of acetone, carbon tetrachloride, and benzene or toluene.

Drying the Fractionating Tube. Place the tube in its handle and, holding and rotating it as indicated in Fig. 20, heat it in a just non-luminous large Bunsen flame to 200° to 300° C. When dry, place the handle into the opening of a block to hold it and tube in perpendicular position for cooling.

Distillation. Place a pencil on a sheet of paper to rest upon it the tips of about ten capillary pipets intended for the collection of the fractions. To avoid confusion, write the numbers of the pipets upon the paper where they rest on it.

Take up a 4-cm length (0.03 ml = 30 μl) of the liquid to be distilled into a capillary of 1-mm bore and touch the end containing the liquid to the asbestos in the cold fractionating tube. When the liquid has been taken up by the asbestos, withdraw the capillary and centrifuge any droplets adhering to the walls of the tube into the asbestos; see Fig. 11 a and c.

To perform the distillation, insert the spike of the fractionating tube into the handle. Rotate the latter in the left hand and heat above a just non-luminous Bunsen flame of 2-cm height as shown in Fig. 20. Incline the tube at 45 degrees and hold it so that the top surface of the asbestos is about 4 cm above the tip of the flame. Watch the portion of the tube above the asbestos for the appearance of a ring of condensate. If the heating is properly done, this ring will rise slowly up the lower chamber. Remove the tube from the flame when the condensate reaches the constriction, and wait until it has passed into the upper chamber. Stop rotating the tube and hold it nearly horizontal. Insert the tip of capillary pipet No. 1 into the uppermost portion of the condensate, and allow the liquid to rise until its meniscus arrives about 1 mm above the taper of the capillary.

Storing the Fraction. Hold the capillary pipet containing the fraction slightly inclined to the horizontal so that the tip points upward, and touch the very end of the latter to the lower part of the seam of the Bunsen flame. If the tip is very fine, the liquid quietly recedes and the end of the capillary fuses to a small sphere. If the tip is too wide, there may be some sputtering before the sealing takes place. To make certain, place the pipet upon a slide and inspect the sealed tip under the microscope. There usually is a gas bubble or several of them in the fine capillary, which is essential for the determination of boiling point, Expt. 21. If the tip is properly sealed, return the pipet to its place upon the sheet of paper.

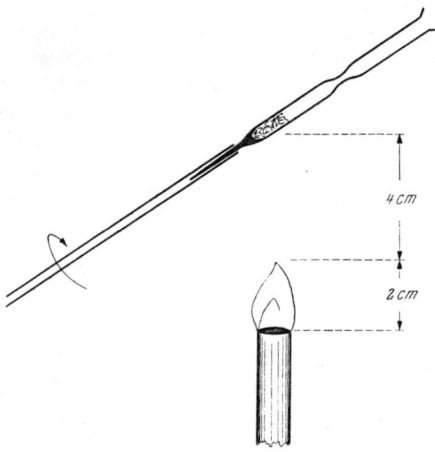

Fig. 20. Fractional Distillation According to EMICH

Repetition of Distillation. Allow the fractionating tube to stand until it has acquired room temperature. Then centrifuge it briefly to collect all liquid in the asbestos, and repeat the distillation as before to collect an additional fraction. Repeat this until no more liquid is obtained when the tube is heated. Six to twelve fractions will be obtained in this manner. These are used for the determination of their boiling points in Expt. 21 and any additional observations needed for identification.

Any non-volatile residue will remain with the asbestos which should be removed with a probing wire and replaced by a fresh batch after fractionation of an unknown mixture.

Experiment 21

Emich's Method for the Determination of Boiling Points (1007)

Pyrex beaker, tall form, 250-ml capacity; stirrer, Fig. 21; thermometer for melting point determinations; standard microscope slide; rubber band which may be cut from wide tubing; bath liquid for melting point determinations.

Expt. 21 Work on the Centigram Scale 103

The boiling point may be determined with 0.2 µl of liquid and less. As described in the preceding experiment, the liquid is taken up with the fine tip (0.05- to 0.2-mm bore and about 15 mm long) of a capillary pipet of about 0.5-mm bore in the wide part. The amount of liquid should be enough to fill a 1-mm length of the wide capillary so that the drop will not break while rising. The gas (air) bubble in the fine tip may be very small; it may fill several millimeters of the fine capillary, but it must not extend into the tapered portion of the pipet.

Set up the melting point bath of Fig. 21. Insert the top of the thermometer into a cork and, using a rubber band, attach the boiling point capillary

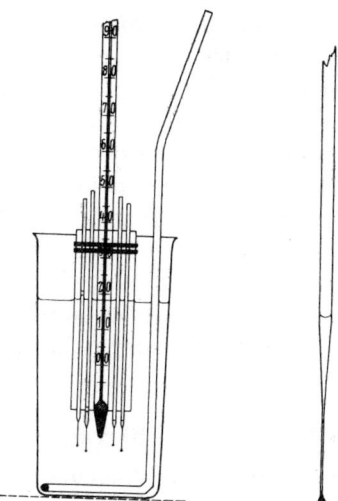

Fig. 21. Determination of Boiling Points According to EMICH. The enlarged tip of a sealed boiling point capillary is shown at the right

to the bulb end. To determine in one heating the boiling points of several liquids or fractions, attach a microscope slide to the bulb end of the thermometer as shown in Fig. 21. Arrange the boiling point capillaries on the slide in their proper order on both sides of the thermometer and at a height that the drops in the capillaries are level with the center of the bulb. When the apparatus has been set up as shown in Fig. 21, pour the bath liquid into the beaker; 30% sulfuric acid will do up to 108° C.

Adjust the illumination until the mercury in the thermometer and the droplets in the capillaries can be seen without difficulty. Then heat the bath rapidly with stirring until the boiling range of the droplets (fractions) has been approached. At this time, reduce the size of the flame to get a temperature rise of 2° C per minute and begin to stir continuously so that the loop of the stirrer moves up and down through the whole height of the bath.

Keep an eye upon the droplet representing the lowest boiling liquid (fraction). When the boiling point is approached, the droplet begins to quiver, and the rate of stirring should be increased. The droplet finally rises through the capillary, and the temperature is read when it arrives at the level of the surface of the bath. With an essentially pure liquid or an azeotropic mixture, this happens at the boiling point. A mixture may be recognized by the fact that its composition and boiling point changes as it rises in the capillary, i. e., is lifted up by the vapor given off into the gas space below the droplet. In such instances, the temperature at which the droplet starts to rise and the temperature at which it arrives at the bath surface should be recorded; the two may be many degrees apart.

Continue the experiment until all drops have arrived at the surface of the bath. Then remove the flame. When the temperature of the cooling bath approaches the highest boiling point observed, resume stirring and record the temperatures at which the droplets begin their descent. The experiment may be repeated until the droplets become so small, due to distillation to the cooler parts of the capillaries, that they break. At the conclusion of the experiment, the liquids may be saved by fusing shut the ends of the wide capillaries and collecting the liquids at these ends by whirling in a centrifuge.

GARCIA (436) has described a modification of the method, that permits determination of the boiling point under reduced pressure with 2 to 5 μl of liquid.

Work on the Milligram Scale

Reagents. A large number of solid reagents may be kept in small vials systematically arranged in a Behrens reagent box. Plastic vials are recommended for fluorides and strongly alkaline substances. It is advisable to supply the solids as granules of 0.1- to 1-mm diameter. Larger crystals should be crushed in a mortar and then classified by spreading and tapping the material on a sheet of paper and rejecting the very coarse and the fine material.

Solutions and reagents giving off fumes must not be included in a Behrens box but should be assembled on open shelves. Plastic containers are suggested for fluorides, alkaline solutions, and reagents responding to sodium, potassium, calcium, and silica. Closure with caps instead of stoppers would facilitate maintaining the purity of the reagents. Bottles or vials of 10- to 15-ml capacity will suffice for most reagents and may be systematically assembled in plastic blocks or trays designed for easy cleaning.

The Weighing of Samples may be done on an analytical balance or on a Sauter balance, which may also serve for the calibration of capillary pipets.

For **Measuring Liquid Samples,** a centrifugal pipet (Fig. 39) is recommended. Measuring pipets of 0.2-ml total delivery and graduated to read 0.01 ml together with capillary pipets (Expt. 22) suffice for the measuring of reagent solutions. In addition, calibrated loops and stirring hooks (Expt. 23) are used for dispensing small volumes of solutions.

One may prefer the use of commercially available microliter pipets with a syringe control obtainable from various supply houses. Also the simple shop-made pipet control of BACKUS (461), based upon a thumb wheel rolling over a rubber tube, should be applicable.

Microcones of about 0.7-ml capacity may be given a shape and size for convenient handling and efficient work with solution volumes from $10\ \mu l$ to 0.5 ml and precipitates from 10-μg to 0.5-mg weight. The shape shown in Figs. 22a and b is recommended for general work. Microcones of the type shown in Fig. 22c are useful for the collection and investigation of very small amounts of precipitates as well as for the estimation of the volume of small to medium quantities of precipitates; in addition, they are useful for the performance of liquid-liquid extractions with small volumes of solution and solvent. The capillary forming the lower part of this type of cone should have uniform bore and a length of 3 to 5 cm. The diameter of the bore may be varied to fit the task, and usually will be within the limits from 1 to 3 mm. They are made by fusing a thick-walled capillary of uniform bore to tubing of 6-mm i. d. As an alternative, one may start with a short length of tubing of 6-mm bore and heat it at the center until the bore has shrunk to 3 mm before slowly drawing out to a thick-walled capillary of about 6-cm length. Cutting the capillary at the center and the wide tubing at points 2 cm from the tapering portions gives two microcones.

Reagents are Added by depositing them on the inside surface of the microcone just below its opening. Brief swirling in the centrifuge then transfers them to the tip of the cone. Solids may be handled with a glass thread (Expt. 61) or a microspatula obtained by hammering flat the very end of a platinum, nickel, copper, or aluminum wire of 0.5- to 1-mm diameter. Liquids are added from measuring pipets or calibrated capillary pipets. Very small volumes of liquid reagents may be added with a platinum loop, a platinum hook, or a stirring rod. These may be inserted directly into the mixture and twirled for mixing; when withdrawing, these tools should be repeatedly touched to the inside wall of the microcone to strip them of the adhering solution or slurry. To test the pH of the mixture, the solution adhering to the tool may be transferred to test paper by drawing the tool over a point of the edge of the paper and inspecting this location with a magnifying glass. A platinum hook removes so little solution that it is often permissible to transfer the whole of the removed matter to the test paper by simply catching its edge with the hook.

Mixing in the Microcone requires the use of stirrers which may be obtained by heating one end of a capillary of about 9-cm length in the edge of a roaring Bunsen flame; first rotate the capillary until a bead of glass has formed, and then hold it quiet and horizontal with the end touching the flame until the bead has dropped into the position shown in Fig. 22 d. If the volume of liquid is small or the cone is very narrow, more delicate stirrers are obtained as follows. Fuse the center of a short piece of capillary by heating in a microflame or in the edge of a Bunsen flame while slowly rotating it around its axis, Fig. 22 f. Then remove the capillary from the flame and draw out the bead to a thread a few tenths of a millimeter in

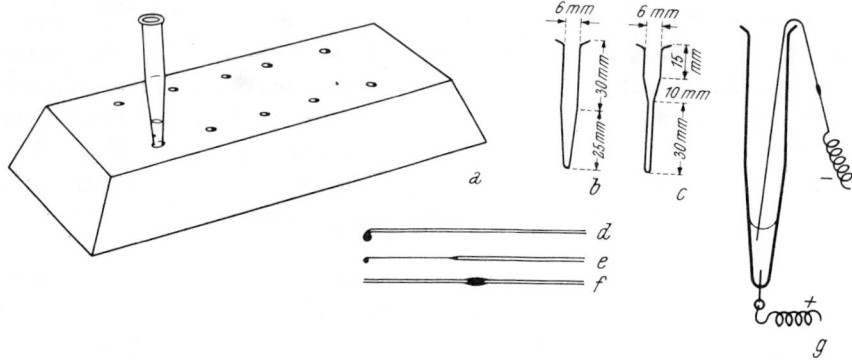

Fig. 22. Microcones, Blocks, and Stirrers. The cone g has been drawn on a larger scale to show the details of wiring

diameter. Break the thread at a point about 3 cm from the capillary handle and fuse the end of the thread into a tiny bead which is finally allowed to droop to one side, Fig. 22 e. If the stirrer is intended for a straight-tip cone, Fig. 22 c, hold in perpendicular position when fusing the bead; only very small beads are permissible in this instance and their centers must not deviate much from the axes of the rods.

For mixing, the bead is inserted into the mixture and the capillary handle is twirled between the fingers. Simultaneous motion up and down is needed when the height of the liquid exceeds a few millimeters. On withdrawing, the stirrer is stripped by touching it to the inside wall of the microcone or drawing it through a drop of solvent placed upon the inside wall just below the opening. As usual, all material is finally combined by brief swirling in the centrifuge.

Treating with Gaseous Reagents. A small volume of solution may be rapidly saturated with a gas by bubbling the latter through the liquid. To obtain suitably small bubbles of gas, proceed as follows. Draw out one end of soft-glass tubing of 6-mm bore to a capillary of 1-mm bore and 10- to 15-cm length. Insert a plug of cotton into the wide part of the tube to serve as filter for the gas. Then draw out the end of the wide capillary to a long, fine capillary of about 0.1-mm bore, which is broken off to give it a length of 10 cm. By means of a cork, clamp the tube in a perpendicular position, Fig. 23, and connect the wide tube to the source of the gas which should be supplied with a pressure of at least 30 cm of water column above the atmospheric. Start the flow of gas before inserting the orifice of the fine capillary into the liquid. A stream of very fine gas bubbles is obtained, and there is no danger that solution could be thrown out of the microcone. Finally withdraw the liquid from the capillary before shutting off the gas so that no liquid can be lost by beeing drawn into the capillary.

The technique is recommended for precipitation with hydrogen sulfide. The gas bubbles rising from the very tip of the microcone keep the solution thoroughly mixed. For testing whether or not the precipitation is complete, one may centrifuge the precipitate into the tip of the cone and treat only the supernatant solution with the gas by keeping the orifice of the capillary above the plug of precipitate.

Before shutting off the stream of gas, remove the microcone with the reaction mixture and wipe the end of the capillary, which has been inside the microcone, first with moist and then with dry filter paper. Finally break off and discard the last centimeter of the fine capillary before shutting off the gas stream.

Solutions may be neutralized by bubbling through them air which has been passed through a gas washing bottle containing ammonia or hydrochloric acid of suitable concentration.

Small volumes of liquid will soon be saturated with a gas by absorption through the relatively large surface. The apparatus (866) shown in Fig. 24 has been used for saturating the contents of microcones with highly corrosive gas, hydrogen chloride, which was generated in an apparatus similar to that of SWEENEY (699). All connections were made with ground glass joints. The inlet tube which extends almost to the bottom of the saturation chamber has four glass rings fused on, that support the microcones. The stopcocks permit to fill the chamber with gas and to remove the filled chamber from the gas line so that the action may proceed undisturbed for any desired length of time. Immersion into heating or cooling baths is possible with the saturation chamber either connected to the gas supply

Fig. 23. Passing Gas into a Solution. The bore of the fine capillary is exaggerated

or detached from it. The apparatus was used for the precipitation of $AlCl_3 \cdot 6\ H_2O$ from ether solution.

Heating in Microcones. The advantages of electrically heated metal blocks (155, 904) cannot be denied, but simple improvisations give satisfactory service.

For temperatures below 100° C, a suitable bath liquid in a 150-ml beaker will suffice, Fig. 25. The microcone may be attached to the rim of the beaker with a strong wire of aluminum, nickel, copper, or platinum. An inexpensive short thermometer may be used for stirring the bath liquid.

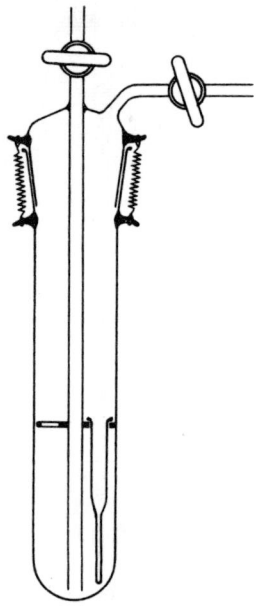

Fig. 24. Saturation Chamber; about $1/2$ nat. size

The steam bath of Fig. 26 consists of a 250-ml Pyrex Erlenmeyer flask with narrow mouth and snugly fitting inset which may be easily made of Pyrex glass tubing of 24-mm outer diameter. The cross section of the inset at the level a is shown in the side figure. The glass horns pointing toward the center of the inset support the microcones and other apparatus to be heated. If not too much water is placed into the flask, steam may be obtained in a short time with a Bunsen burner or an electric hot plate supplying the heat. A few granules of zinc added to the water will give even boiling, but this device should be avoided when searching for small amounts of zinc.

At times, solutions are heated for the expelling of dissolved gases which are rapidly carried away if steam bubbles form in and escape from the heated solution. Since boiling will not ensue when heating in a steam bath, the expelling of gases should be aided by passing a stream of gas bubbles through the heated solution. This may be done by forcing a suitable gas into the solution as suggested by Fig. 23, but frequently it will suffice to use the simple device shown in Fig. 27. A stream of fine bubbles emerges from the opening of the fine capillary as the air expands in the bulb. When the bubbling stops, one removes the gadget; if desired, one may cool the bulb in a stream of tap water and again insert the gadget into the cone.

Heating blocks of metal offer the simplest means for the maintenance of controlled temperatures above 100° C. In addition to a well for the thermometer, they are provided with holes for the insertion of apparatus. They are best made of aluminum which is not only a good conductor of heat but also has a remarkable resistance to the action of corrosive agents. Copper blocks become a nuissance because of the oxide scales which, on heating the blocks, are frequently ejected by the oxidized surface with such force that they may land in the reaction vessels and contaminate the

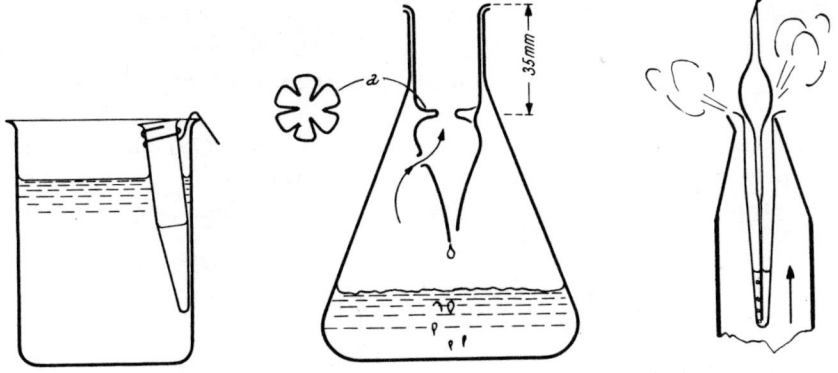

Fig. 25. Water Bath   Fig. 26. Steam Bath   Fig. 27. Expelling Gas

material under investigation. Electrically heated blocks are more easily kept clean, but heating with a gas flame is perfectly satisfactory for most purposes.

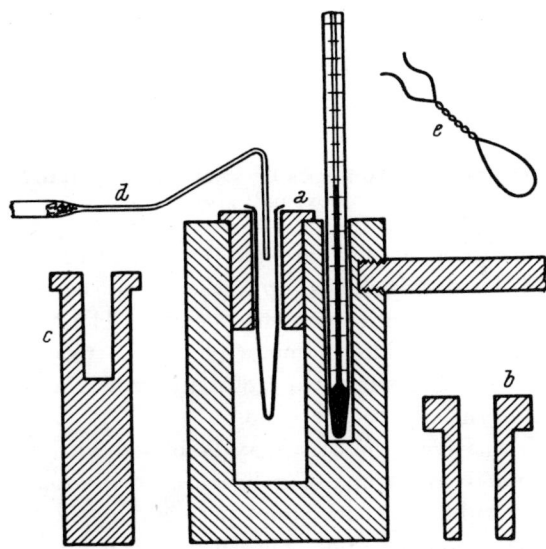

Fig. 28. Heating Block. *a* Inset for microcones; *b* inset for centrifuge tubes; *c* inset for distilling flask of Fig. 33; approx. $1/2$ nat. size

The heating block suggested by Fig. 28 permits accomodating various apparatus by exchange of insets *A*, *B*, *C*, etc. The shape of the insets shown in the figure brings about that the upper portions of the apparatus will be more rapidly heated than the bottom parts, and this will aid in the prevention of creeping. The fork E, made of aluminum wire, is used in the removal of microcones from the hot block.

For *heating under pressure*, microcones may be sealed with the use of RACHELE's pressure cap (413) shown in Fig. 29. The cone rests in a brass ring with rubber lining, to which a metal stirrup is attached. The rubber lined pressure plate is tightened on the rim of the cone by means of a screw threaded through the center of the stirrup. SCHENCK (1272) cuts two 5-mm thick slices with a very sharp knife from the thick end of a clean, new rubber stopper of about 24-mm diameter. A hole, just large enough to

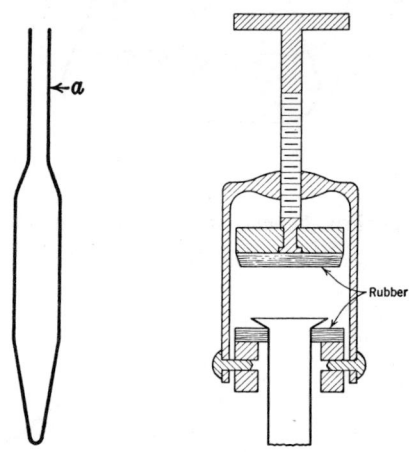

Fig. 29. Pressure Cone (left) and RACHELE's Pressure Cap Applied to Microcone (right); approx. nat. size

snugly fit the microcone, is cut through the center of the disk with the two cut faces, and this disk is shoved up to the rim of the cone. The smooth face of the other disk is placed upon the opening. The top disk is backed up by a stiff metal disk of like diameter (a 25-cent piece), and the three disks are clamped together with two Hoffman screw clamps placed along parallel chords tangentially to the cone.

For further alternatives, the cone may be closed with a rubber stopper which is then secured by means of a ligature with copper or aluminum wire (13). If use of rubber or cork is not permissible, the reaction mixture may be sealed into a pressure cone, Fig. 29, which is given approximately the dimensions of a microcone. The neck should have a bore of not less than 2 mm to facilitate the introducing of capillary pipets or siphons. Previous to heating, the cone is sealed at a by the customary procedure of heating in a small flame until soft and then drawing out without removing from the flame so that the capillary which forms is fused through and shut. After heating, the pressure cone is opened with the customary precautions (eye shield) taken when opening Carius tubes. If the absence of pressure after cooling is assured, it may be simply cut open below the seal. Finally, a standard microcone may be sealed, for heating, into a somewhat larger tube, or one may use a device like the micro autoclave of GORBACH (155), that uses the principle of the safety valve and permits adjusting the pressure by means of a weight displaceable along a lever.

Fusions may be carried out on platinum foil or in a small spoon, cup, or crucible of 0.05- to 1-ml capacity and made of porcelain, vitreous silica, platinum, tantalum, silver, nickel, or iron. Pyrosulfate fusions may be performed in a Pyrex microcone, and it is recommended to do it at the point where the taper changes to the cylindrical shape with the cone in approximately horizontal position. In the following, detailed directions are given for the performance of fusions on wires, a technique which is specially suited for the scale of work.

Straight pieces of wire of 0.5-mm diameter and 4- to 5-cm length are used. One end is either clamped into a locking forceps or sealed into the end of a capillary of 5-cm length and somewhat more than 0.5-mm bore. Assuming a density of 2.5 for the fluxes and that the volume of the wire is compensated by the elongation of the beads, one may crudely estimate that beads with 2-, 3-, and 4-mm diameter at the equator have weights of 1.5, 6, and 13 mg, respectively. These estimates will aid in getting flux in the correct proportion to sample to be treated.

The bead is formed at the end of the wire and held there by not heating the bead directly, but by heating the wire at a suitable distance from the bead which is at the end of the wire.

Sodium Carbonate Fusion. Place some anhydrous sodium carbonate upon a piece of platinum foil, and take up small portions of it by touching them with the end of the heated platinum wire and later with the molten salt on it until a bead of the desired size is obtained. Estimate the amount of flux from the equatorial diameter of the bead, or determine it by weighing the wire without and with the bead or by supplying a weighed amount of salt on the foil. Use a microflame for heating, Expt. 19, and if the sulfur content of the illuminating gas is objectionable, use a flame of purified hydrogen. To transfer the material under investigation to the bead, either make a slurry, transfer small portions of it from a capillary to the bead, and fuse after each addition–or weigh the substance on a sheet of platinum foil having a high polish and mop the substance up with the molten bead. Finally, fuse until gas bubbles no longer appear in the bead; this takes about one minute. For dissolving the melt, insert the end of the wire with the bead into the required volume of the selected solvent which may be contained in a microcone or a crucible.

Sodium Peroxide Fusion. Fuse a pellet of sodium hydroxide on a sheet of nickel. Collect a bead by dipping the end of a nickel wire repeatedly into the melt and each time allowing the hydroxide to solidify on the wire. Finally add the material under investigation as a slurry in water or mop it up from a platinum or nickel sheet with the cold bead which, to this end, is exposed briefly to air so that its surface becomes moist. After fusing briefly, take up sodium peroxide with the bead with the use of the latter procedure and fuse again. Repeat the adding of peroxide and fusion once or twice.

Pyrosulfate Fusion. On a slide or watch glass, prepare a thick paste of powdered potassium sulfate and concentrated sulfuric acid. Take the paste up with the end of a platinum wire and fuse until a bead of proper size is obtained. Heat the bead cautiously until fumes of sulfur trioxide start to appear. Then add the material under investigation as a slurry in concentrated sulfuric acid. Perform the fusion by lightly heating the wire behind the bead in the edge of the lowest part of the Bunsen flame. The bead should be 10 mm to 20 mm outside the flame. If the pyrosulfate decomposes too rapidly, there may not be time enough for getting the material completely dissolved. Thus, if undecomposed particles remain visible in the melt, the fusion may be repeated after adding to the bead some more concentrated sulfuric acid.

Treatment with Hydrofluoric Acid is sometimes used for the dissolution of silicates. The following technique considers suggestions made by VAN BRUNT (416, 1271).

Cut from platinum foil either a disk of 3-mm diameter or a rectangle of 2 mm × 6 mm and weld it to a platinum wire of 0.3-mm diameter and 4-cm length, which serves for handle. Heat the foil and the wire in the intended relative positions in a small hydrogen flame, burning at the orifice of a capillary, so that they stick together. To perform the welding proper, heat a small steel anvil to just above 100° C and mount the capillary so that the hydrogen flame burns horizontal and close above the surface of the anvil. With forceps hold the part to be welded into the flame so that it touches the anvil and, when it sends out a red glow, tap it lightly with a 4-oz. ball pein hammer. Finally clean by fusing potassium pyrosulfate upon the foil and wire and then dissolving the melt in concentrated hydrochloric acid.

Make the substance to be treated into a slurry with dilute sulfuric acid or water. Take the slurry into a capillary and transfer it to the foil at the end of the platinum wire by adding small portions at a time and evaporating after each addition. To this end, apply a microflame (burning from the orifice of a capillary) to the wire at a suitable distance from the foil. When the transfer has become complete, treat the residue upon the foil with hydrofluoric acid or with hydrofluoric-sulfuric acid mixture, a drop of which is held ready upon a platinum sheet. Add the acid by means of a platinum loop to a part of the foil, that holds little or none of the substance.

Then slowly evaporate the acid by heating the wire at a suitable distance from the foil. Repeat the treatment as often as it appears necessary. For the extraction of the residue immerse the foil in the solvent held ready in a microcone.

The technique is also suited for the removal of fluoride by evaporating with silica and sulfuric acid, and it may be used for the removal of organic matter by ashing.

Separation of the Liquid from the Solid Phase. As a rule, it is necessary to pack the solid by swirling in the centrifuge. Small amounts of liquid are then lifted off by means of a capillary pipet, whereas large volumes are removed with a siphon. Solid particles clinging to the walls of the

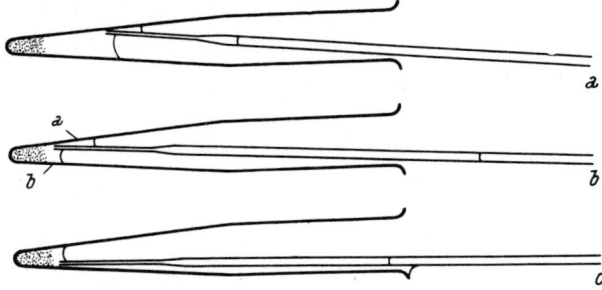

Fig. 30. Use of Capillary Pipet; *a* and *b* show the correct procedure. Approx. nat. size, but the bore of the capillary is exaggerated

microcone are stirred into the liquid and combined with the bulk of the solid by centrifuging.

To remove a small volume of supernatant liquid, select a capillary of 0.5- to 1-mm bore and 20-cm length and draw out one end to a tip of 10- to 20-mm length and not less than 0.2-mm bore. Hold the microcone almost horizontal and lay the pipet into the cone as shown in Fig. 30*a*. Regulate the rate at which the liquid enters the pipet by inclining the cone. Hold the cone in the left hand, and stepwise push the pipet into the cone with the index finger of the right hand so that the opening of the pipet always remains just below the meniscus of the liquid in the cone. Finally bring the tip of the pipet to a point 1 mm from the precipitate, Fig. 30*b*. If the cone is properly inclined, the liquid is completely taken up by the pipet. Meniscus *b* disappears first, and very soon afterwards meniscus *a*. Withdraw the pipet from the microcone and transfer its contents to another cone (or to wherever it is needed) by blowing them out with the mouth. Save the pipet for the transfer of washings, which operation will also rinse the pipet. If there is any doubt concerning the completeness of the removal of the liquid, centrifuge the microcone containing the solid and transfer any liquid collecting above the precipitate to the bulk of the centrifugate as described.

If the capillary pipet is not perfectly straight, or if a straight pipet is laid into the microcone in such a way that a narrow space results between pipet and cone, Fig. 30c, liquid is drawn into this space by capillary attraction and cannot be taken up by the pipet. When the latter is withdrawn, this liquid spreads over the walls of the microcone.

To remove a large volume of liquid, convert the capillary pipet into a siphon. Hold one end of the pipet so that the other end will give the desired angle of about 60 degrees by slowly following the pull of gravity when the middle is cautiously heated in the edge of the non-luminous Bunsen flame

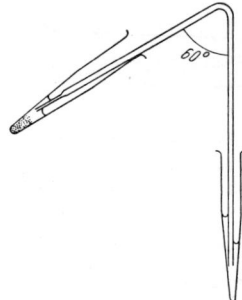

Fig. 31. Use of Capillary Siphon. Approx. $1/2$ nat. size, but the bore of the siphon is exaggerated

1 cm above the barrel of the burner. Break the fine tip close to the taper to obtain a reasonably fast flow of liquid.

The use of the capillary siphon is illustrated by Fig. 31. Place the leg with the wide opening into the microcone that is to receive the liquid. Then insert the leg with the tip into the cone containing the liquid above the solid so that the tip and the wide part of the capillary rest on opposite sides of the microcone just as in Fig. 30a indicated for the straight capillary pipet. Incline the tubes properly, and the capillary siphon will fill by itself. Proceed in principle just as described for the capillary pipet. Finally bring the tip of the siphon close to the precipitate so that practically all liquid may be removed in one operation. Withdraw the siphon so from the cone with the solid that the leg with the wide opening stays in the cone with the liquid and remains there when the cone is set aside. Proper inclining and the narrow orifice at the tip make it possible to keep the siphon continuously filled with liquid.

Whirl the cone with the solid in the centrifuge, and use the siphon for the transfer of the liquid collecting above the solid. Set siphon and centrifugate aside as before and hold them ready for the collection of washings. The first washing may serve to rinse the siphon into the centrifugate.

Washing and Extraction of Solids. Deposit the measured volume of solvent below the opening of the microcone. Without stirring up the solid in the point of the cone, spread the liquid by means of the stirrer over the inside wall of the microcone. Collect the liquid in the point of the cone by brief swirling and then mix solid and liquid thoroughly by stirring so that none of the solid is spread over the walls of the cone. At this time, heat in a suitable bath if the washing or extraction shall be performed with hot solvent. Finally, centrifuge and withdraw the liquid as described above. The microcone with the liquid will, as a rule, remain sufficiently warm to satisfy the requirement of extraction or washing with hot solvent if the work is performed without delay.

Transfer of Solids. Whenever possible, dissolve the solid and transfer the resulting solution by means of capillary pipet or siphon. If this is not feasible, add some water or some other suitable liquid and stir to obtain a slurry in the tip of the microcone while the main portion of the supernatant liquid remains clear. Obtain a capillary of 0.5- to 1-mm bore and 20-cm length and cut it off evenly at both ends. Close one end with the finger and insert the other into the microcone so that it touches the bottom of the tip. Incline cone and capillary 45 degrees to the horizontal and then remove the finger from the upper opening of the capillary. The liquid rushes into the capillary with the slurry entering first. Again close the top opening of the capillary with the finger, and transfer the contents to the location where they are wanted.

Depending upon density and particle size of the solid, it will collect more or less rapidly in the liquid at the lower end of the capillary if the latter is held in a vertical position before being emptied. It may become possible to transfer the whole slurry to the receiver while retaining most of the transfer liquid in the capillary. Of course, the transfer is repeated until it is sufficiently complete. Finally, the transfer liquid may have to be removed, which is accomplished with the use of the technique for separating liquid and solid phases and (or) evaporation.

Electrodeposition. The needle electrode, p. 186, may be used for side tests and for the detection of very small amounts of metals. In general, much simpler apparatus will suffice. A platinum wire of 0.1-mm diameter is fused into the tip of a microcone to serve as anode (1143). A like platinum wire of 6-cm length and soldered to fine, insulated copper wire is inserted through the opening of the cone, Fig. 22g. Only a few millimeters of the end of the cathode are inserted into the electrolyte. A glass fiber of 8-cm length is placed into the cone so that it touches the wire in the tip; this makes that the individual gas bubbles rise as they form on the anode and prevents the accumulation of gas in the tip. When the deposition is considered

8*

complete, the cathode is quickly withdrawn from the microcone and inserted into the rinse solution, which may be held ready in a beaker, without breaking the connection to the source of current. To remove the deposit from the cathode for further treatment or the performance of a confirmatory test, the end of the wire is dipped into the droplet of solvent held ready in a microcone or upon a slide. Electrolysis with an e. m. f. of 10 volts and an electrolyte of 3-F KOH gave efficient separation of 10 μg zinc from solutions containing up to 10 mg manganese, 0.5 mg cobalt, 0.5 mg nickel, 0.1 mg iron, 0.1 mg aluminum, 0.1 mg chromium, and 0.1 mg cadmium.

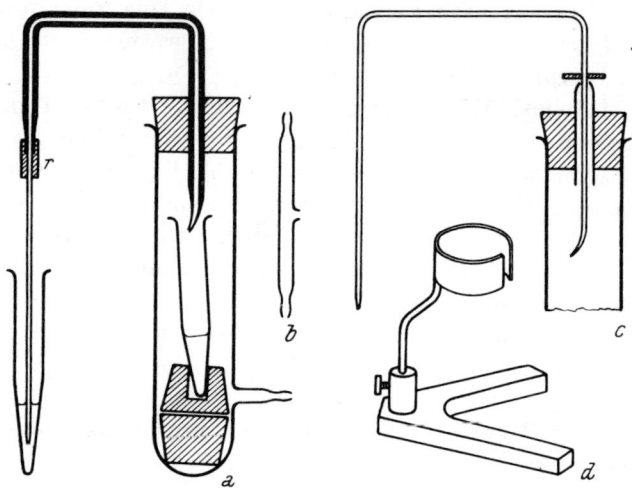

Fig. 32. Siphons Operated by Suction; a using a thermometer capillary; b tube for control of suction; c capillary siphon; d stand; about $1/2$ nat. size

Liquid-Liquid Extraction is conveniently performed in microcones. A straight-tip cone, Fig. 22c, is recommended when the volume of the bottom layer or of both layers is small. Complete separation of the layers may be obtained by use of centrifugal force; in addition, the surface of the microcone may be treated with chlorosilane (13, 90, 423) so that the liquids will not wet it, and the same treatment may be applied to the outside (or outside and inside) of the pipet or siphon used for mechanical separation of the layers.

Suction operated siphons or pipets, Fig. 32, permit removal of either top or bottom layer. The suction may be supplied by mouth or with an aspirator. In the latter instance, a tube b with side opening which is closed with the finger should be inserted into the suction line to permit instantaneous stopping of the action. Fig. 32c shows the use of a small square of thin rubber sheet for securing a tight fit of the capillary siphon in the wide tubing. As a satisfactory alternative, one may secure the siphon in the tube with sealing wax or some other suitable cement.

If a mechanical stand is available, the microcone is clamped to it in vertical position so that it may be raised and lowered with the rack-and-pinion motion. As a substitute, the cone may be attached with rubber bands to the tube of a microscope.

No difficulty should be experienced in lifting off the top layer so that only a small fraction of it is left behind. Various devices may be used to get the tip of the siphon or pipet into the bottom layer without having some of the lighter liquid entering it. The tip may be finely drawn out and sealed so that it may be broken open by pressing it against the bottom of the cone. The tip may be inserted in a drop of the heavier liquid so that it fills with it by capillary attraction before it is inserted into the cone, and (or) air pressure is maintained in the siphon or pipet to prevent the lighter liquid from entering; the last procedure offers little difficulty when a pipet is used which is operated by a syringe control obtainable from supply houses, a Pumpett (A. S. La Pine and Co., Chicago 29, Ill.), or a pressure or levelling-bulb device used on the microgram scale (90, 141). One will adjust the pressure so that an air bubble or a droplet of the heavier liquid just starts to form at the orifice of the pipet tip when it enters the heavier layer. If the outside of the pipet or siphon is not wetted by the lighter liquid, there is no objection against immediately raising the microcone until the tip of the pipet or siphon touches its bottom. Otherwise, it seems preferable to raise the cone while the lower layer is withdrawn so that only a short length of the tip of the pipet or siphon is immersed into the heavier liquid at any time.

If the volume of the layer to be removed is less than $10 \mu l$, one will prefer to use a capillary pipet. A tip of 0.2-mm bore and up to 5-cm length will give good control and will be found advantageous when working in straight-tip cones.

Evaporation may be carried out efficiently in the microcone by blowing a stream of filtered air from a capillary on the liquid, Fig. 28 d. The heating device must be selected to fit the vapor tension of the liquid to be removed; volatile solvents may be removed by simply blowing air into the cone, but creeping will be prevented if the upper part of the cone is heated in some manner (blowing steam upon the outside, heating coil, surrounding by a hot metal shell, Fig. 28 a, b, etc.). Insertion of the microcone into a steam bath suffices for the evaporation of aqueous solutions, but heating in a metal block is required for the elimination of sulfuric acid.

Use of an infrared heat lamp is recommended for evaporation in crucibles, small dishes, and watch glasses. Application of the heat from above eliminates creeping, and a stream of air blowing over the surface of the liquid gives rapid evaporation without boiling.

Description of elaborate heating devices (155, 162, 904) and apparatus for evaporation by exposure to a stream of hot air (162, 887, 891) may be found in the literature.

Distillation. Apparatus and technique have to be selected to fit the individual case.

The apparatus of Fig. 33 is useful when a substance shall be more or less completely removed from a relatively large volume of liquid (more than 10 μl) and transferred to some reagent solution as this happens in the separation of arsenic from the other metals by distillation of the trichloride. If the charge is to be concentrated to a certain volume, this volume may be indicated by a graduation mark upon the tube.

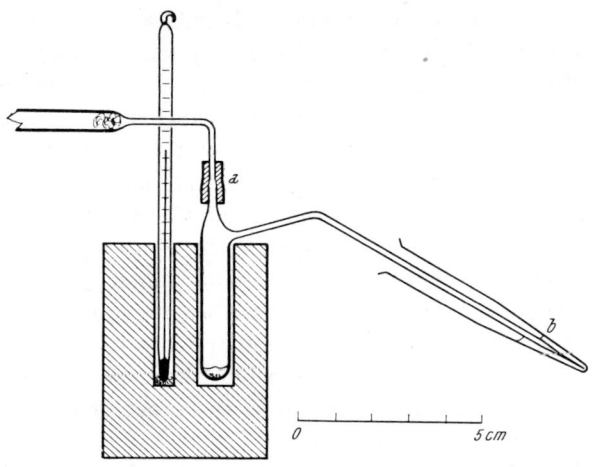

Fig. 33. Distillation Apparatus Used by ANNE G. LOSCALZO (88); about $^1/_2$ nat. size

To perform a separation, introduce the charge with a capillary, capillary pipet, or capillary siphon through the short tube a of 2-mm bore. Then place the tube into a heating block with windows for observing the progress of the distillation and connect a with a supply of inert gas (air dried by passing through a tube with Anhydrone is suitable for the distillation of $AsCl_3$). Insert the capillary b into a microcone with suitable absorbent and supply gas at a rate to obtain not more than two small bubbles per second in the microcone. For cooling, the latter may be placed into a dish with ice water, or ice water may be run over the outside of the cone to be caught in a funnel. Then heat the block to the needed temperature and hold it until the volume of the charge has been reduced to the required amount. For stopping or interrupting the distillation, remove the tube from the heating block and withdraw the microcone b before shutting off the stream of gas. At the conclusion of the distillation, rinse the capillary b

before stopping the gas stream by drawing it through a drop of suitable reagent suspended in a platinum loop.

Analytical distillations with a few microliters of liquid may be carried out quite efficiently by closely approaching the surface of the liquid which has been heated nearly to the boiling point with an efficiently cooled condenser. The procedure which uses the principle of molecular distillation has the advantages that bumping cannot occur and that spray resulting from a boiling charge will not contaminate the condensate.

The apparatus (869) may be assembled in the laboratory. Fig. 34 shows the Pyrex distilling tube of 4-mm bore and 20- to 30-mm length in the heating block. The aluminum block of 2 cm $\times$ $\times$ 3 cm $\times$ 5 cm had the four sides covered with thin sheet of mica to serve as electrical insulator. Over this mica was wound a 60-cm length of No. 27 nichrome wire (0.35-mm diameter) to serve as heating coil which was covered with asbestos board held together by two bands of copper wire. The insulating asbestos mantle extended above and below the metal block, and also the top surface of the block was covered by asbestos board. The heating coil was connected through a 64-ohms Biddle resistance to the 120-volt d. c. line. The cold-finger condenser b, attached with De Khotinsky cement to a rack-and-pinion movement, consisted of a thin-walled capillary of 6-cm length and tapering from 1.5-mm outer diameter at the lower end to 3-mm o. d. at the upper end. The inlet tubing for the ice water used in cooling opened close to the very tip of the condenser and was a capillary of less than 0.5-mm o. d. The outlet capillary of like diameter opened about 10 mm away from the tip of the condenser so that only its lower end was cooled. The outlet was connected to a water pump which sucked the ice water through the condenser.

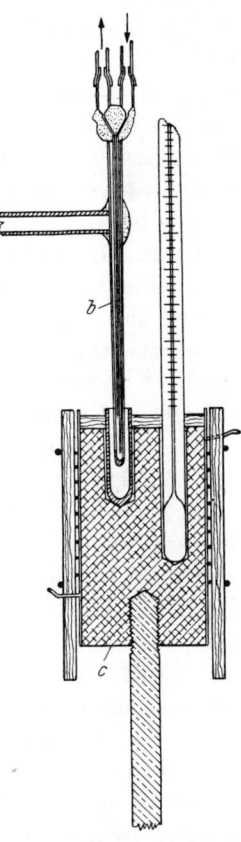

Fig. 34. Analytical Distillation with a Few Microliters of Liquid; J. R. Rachele (410); about $^2/_3$ nat. size

For efficient use provide the distilling tube with graduation marks indicating the critical volumes to be considered and indicate the corresponding positions of the condenser.

Heat the block to the proper temperature. Place the charge in the distilling tube, and centrifuge it for collection at the bottom of the tube. Start the circulation of ice water through the condenser. Place the distilling tube into the block and immediately lower the condenser into it to bring its tip close to the surface of the liquid. Hold the temperature constant, and at suitable intervals (2 to 10 minutes) raise the condenser and collect accumulated distillate by touching a capillary pipet to the tip of the condenser.

Emich's method of fractional distillation was modified by MORTON and MAHONEY (429) to give about two to three fractions for each microliter of liquid available. This is accomplished by using quite small drops for the determination of boiling points and by reducing evaporation losses during fractionation by attention to various detail such as cooling the condensate, use of a shielded block for heating, and proper timing.

The fractionating is performed with the use of a copper block, 38 mm × ×38 mm × 15-cm high, which is heated with either a gas flame or a coil of 12 meter of No. 30 (0.25-mm diameter) oxidized resistance wire wrapped around the lower 4 cm of the block. A sheet of asbestos, 1 mm thick, covers the top surface of the block, which contains two holes bored to a depth of 95 mm. The one with 6-mm diameter receives a thermometer; the other of 8-mm diameter is for the fractionating tube.

The fractionating tube is made of thin-walled tubing of 2-mm bore and 15-cm length. The lower end is sealed and blown up to form a bulb, 6- to 7-mm outer diameter. Glass wool is ground in a mortar until it is fine enough to pass into the tube, and the fractionating tube is packed with it so that the packing will not quite reach up to the top surface of the metal block. Immediately above the packing, the tube is made to form a short (3 to 4 mm) constriction reducing the bore to one half or one third. The constriction should appear just above the level of the asbestos sheet covering the top of the block, and the tube is cut off 20 mm above the constriction.

The fractionating column is shielded to reduce the rate of the transfer of heat from the metal block by slipping over it a loosely fitting tubing of glass or asbestos paper of 7-mm outer diameter. The tubing should rest upon the bulb and should reach up to nearly the surface of the copper block. Finally, the fractionating tubing is pushed through the center of a circular disk of asbestos paper, 10-mm in diameter and the underside lined with aluminum foil, which provides a heat shield that covers the opening of the well during distillation.

To perform a distillation, first prepare a suitable number of boiling point capillaries (Expt. 21) with a bore from 0.5 to 0.2 mm in the wide part of the tubing and very fine tips of 10- to 20-mm length. The longer and finer the tips, the less liquid is lost when fusing shut their orifices. Before applying the technique to an unknown, practice it first with 10 μl of benzene and then with a like volume of a mixture of equal volumes of benzene and xylene.

To start with, weigh the fractionating tube on an analytical balance. Introduce the liquid to be distilled, centrifuge it to the bottom of the tubing, and again weigh the latter to obtain the weight of the sample. Slip the insulating jacket (glass or asbestos tubing) over the packed column so that it rests on the bulb and top it off with the circular heat shield. Then wind a strip of wet filter paper, 15 mm wide and about 4 cm long, around the end of the capillary to serve as condenser; the ring of wet paper must not cover the first three millimeters above the constriction and the heat shield, where the condensate will collect. Insert the fractionating tube into the block, push the heat shield down to cover the well, turn the heat on, and direct a fine stream of air from a capillary upon the condenser, i. e., roll of wet filter paper.

If the approximate boiling point is known, heat first rapidly and then slowly until a tiny droplet forms above the constriction. Quickly turn off the heat and read the thermometer. The task is now to find the lowest temperature that will give, within 60 to 90 seconds of heating the fractionating tube in the well of the block, a fraction of not more than the size required for the determination of boiling point. To this end, withdraw the fractionating tube from the well, remove heat shield and insulating jacket, and whirl the tube in the centrifuge to return all liquid to its bulb. Wait until the temperature of the block has dropped 4 to 5 degrees, and then return the fully armed fractionating tube into the well. First hold the temperature for 90 seconds; if no condensate forms, slowly raise the temperature. Repeat the cycle of heating until a droplet forms, centrifuging, and again heating until the proper temperature has been found. Then collect the first fraction by touching the droplet of condensate with the tip of the capillary pipet intended for the boiling point determination. Wait until the wide part of the capillary is filled to a length of 1 mm; then withdraw the pipet, seal its tip as described in Expt. 20, and remove the fractionating tube for cooling and centrifuging.

The time may be used for determining the boiling point of the fraction. The size of the fraction may be determined by weighing the pipet before and after collecting the fraction, or it may be estimated from the bore of the pipet and the height of the liquid column.

Regulate the heating of the block so that its temperature will not drop more than two degrees. For the collection of the second fraction and following fractions just return the fractionating tube armed with the cold insulating jacket and heat shield to the well and repeat the procedure used in collecting the first fraction.

Up to 70 fractions may be obtained with relatively large volumes of liquid, 25 μl, and boiling point graphs may be plotted with little effort if fractions of like volume are collected, which is simple when using pipets of the same bore. If the liquids have high vapor tension, evaporation losses may be reduced by substituting for the paper condenser a coil of No. 13 copper wire (1.8-mm diameter) forming one half of the wire and the other half shaped to give a wire basket into which are placed a few lumps of dry ice or solid carbon dioxide.

Overheating the insulating jacket must be avoided by holding the time for collecting a fraction down to the 60 to 90 seconds specified. If the heating is prolonged much beyond these limits, the jacket has time to acquire the temperature of the block, about 10 degrees above the boiling point of the fraction to be collected, and the separation of the components may completely fail.

The work cannot be hurried, and the collection of a large number of fractions does require three to four hours. Obviously, when the boiling point rise to a higher boiling fraction occurs, it is advisable to again determine

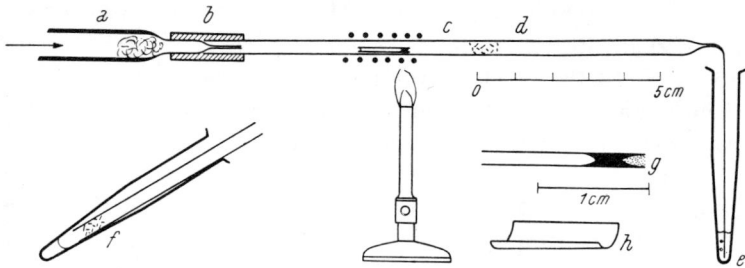

Fig. 35. Sublimation on the Milligram Scale

the proper minimum temperature of the block before collecting the next fraction. MORTON and MAHONEY (430) describe a copper block for the determination of the boiling points.

Sublimation. The techniques described in connection with work upon the centigram scale for obtaining crystalline sublimates are suited also for the milligram and submilligram scales. A modification of procedure is suggested for sublimation in a stream of gas.

Sublime in thin-walled tubing of 3- to 4-mm outside diameter and 15-cm length, which is drawn out to a capillary of 0.1- to 0.2-mm bore at one end, Fig. 35. For introducing the charge, use a technique suitable to its nature. If it is a small object, place it into the tube and push it into proper position by means of a rod. If it is a dry powder, place it into a trough made of platinum foil, 5 mm × 10 mm, which is then pushed into the tube, or introduce it scooped up into a thin-walled capillary of 1- to 2-mm bore and 2-cm length. Place a solution or slurry into the trough of platinum foil and evaporate to dryness before introducing the latter into the tube. As an alternative, take up a slurry into a wide capillary and collect it at the point of the capillary with the procedure described in

Expt. 47. Finally dry the capillary g, Fig. 35, before pushing it into the tube for sublimation.

Slip over the tube a helix of thick copper wire, which is wound loosely enough to permit observation of the charge during heating, and then connect with a supply of suitable gas by means of the wide tubing a, containing a loose plug of cotton, which is drawn out first to the diameter of the sublimation tube and then to a short capillary of 0.1-mm bore. Turn on the stream of carrier gas, insert the fine capillary e into a drop of water or suitable absorbent, and adjust the gas stream to get about two tiny bubbles per second. Push the wire helix over the charge and heat with a small non-luminous flame. If only a very small amount of sublimate is obtained, draw out the tube at d to a suitable small bore (0.5 to 1 mm) and drive the sublimate into the resulting capillary by slowly advancing the heated zone. For collecting sublimate and residue, cut the tube at c and (or) other suitable locations to obtain material for investigation near the end of a section of tubing of 6- to 7-cm length. For the extraction, place the piece of tubing into a microcone as shown in Fig. 35f or mount it in the opening of the cone by means of a cork with fitting hole. Use a small volume of solvent as indicated by the illustration and get it to the material by means of a stirrer or platinum loop inserted through the tubing or by simply heating the solvent to boiling. Finally centrifuge the microcone with the tubing in place. The liquid collects in the tip of the cone, and the tubing may be withdrawn without causing loss of any material.

Liberation of Gases. Apparatus and techniques described for the centigram scale, p. 96, are suited for the milligram scale also. If the amount of gas liberated is very small, a gas reaction cell as described for the submilligram scale, p. 159, is recommended. As an alternative, the gas may be liberated in the tip of a microcone; a wad of glass wool may be placed into the conical portion of the microcone, and the small fragments of test papers, test fibers, and reagent droplets may be deposited upon the inside wall of the cylindrical part. A paraffin-coated ring zone in the conical part would safely prevent diffusion along the glass; a test droplet could be deposited upon a fragment of a cover slip, that fits into the cylindrical part of the microcone. Obviously, a stopper should be applied, that does not absorb or react with the liberated gas. A microcone of thin-walled tubing and somewhat longer than usual in the cylindrical part could be sealed by fusing it shut.

Confirmatory Tests. Customarily used are the slide tests and the spot test technique described for the submilligram range. Obviously, the sensitivity of the confirmatory tests should be somewhat better than that of the general working technique to permit frequent testing for minor and trace constituents in small aliquots of centrifugates or precipitates.

The cell shown in Fig. 36 is useful for the microscopical inspection of small precipitates collected with the centrifuge in the point of the cone. To this end, the cell is filled with water. Illumination with daylight and observation of the reflected light reveal the true color of the collected solid.

The microcone is not suited for revealing light colorations occurring in small volumes of liquid. To this end, the liquid is best taken up with a capillary, Expt. 55.

Storing Material under Investigation. Without delay dissolve precipitates which age or oxidize upon standing. In general, it will suffice to stopper microcones containing liquids, slurries, or liquids containing precipitates or saturated with a gas. Liquids may also be taken up into a capillary which is then fused shut at both

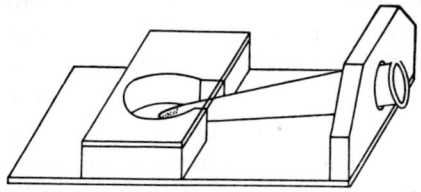
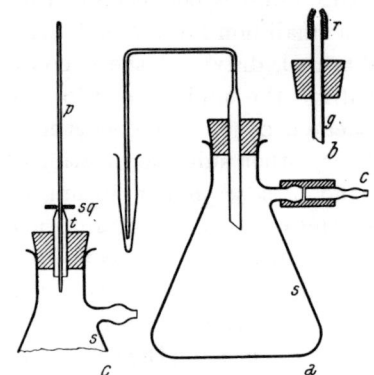

Fig. 36. Cell for the Observation of the Contents of Microcones

Fig. 37. Suction Flask for the Rinsing of Microcones and Pipets; approx. $1/3$ nat. size

ends. In this manner, the volume of gas in contact with a liquid may be reduced to a minimum.

Cleaning Apparatus. Feathers or pipe cleaners are useful for the scrubbing of microcones. Rinsing and drying are best performed with the use of suction. The suction-operated siphon, Fig. 37 a, permits efficient treatment with cleaning solution and rinsing. For quick drying, the microcone is whirled in the centrifuge, whereafter the liquid collected in the point is removed with the siphon; the remaining film of moisture is removed within one or two minutes by drawing air through the cone in position a.

The suction flask s should be connected to the line with a 30- to 50-cm length of thin-walled rubber tubing of 4-mm outer diameter; this usually requires the use of tapering connectors c at both ends. Stiff, heavy-walled suction tubing is generally undesirable, and its use should be restricted to those occasions which demand a good vacuum. The siphon a may be exchanged for a stopper b with a short length of 6-mm glass tubing g and rubber tubing r for the rinsing of thermometer capillaries, measuring pipets, microburets, etc., or it may be replaced by the gadget c for the cleaning of calibrated capillary pipets which are pushed through a pinhole in the center of the small square of thin rubber sheet, sq.

Experiment 22
Calibration of Capillary Pipets and Centrifugal Pipets

Analytical balance, millimeter rule, and microscope with calibrated micrometer scale.

To obtain a useful capillary pipet, make it from a capillary of 0.3- to 0.5-mm uniform bore and 20-cm length. TEN EYCK SCHENCK (1272) points out that for drawing such capillaries, the glass tubing must be rotated in the flame (a Meker flame is preferable to a Bunsen flame) until the whole heated section is uniformly soft, which may take twice as long as the time required for the first softening occurring in the hottest part of the flame. Proper drawing gives sharp tapers on both sides of a uniform capillary; gently sloping tapers indicate a capillary with the bore increasing from the center to both ends.

For determining the uniformity of the bore, cut from each end of the capillary a length of 1 cm. Scratch with the sharp edge of broken

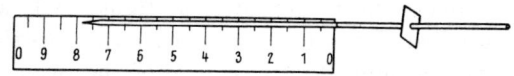

Fig. 38. Use of Calibrated Pipet

porcelain or with an ampule cutter to get a clean break vertical to the axis of the tube. Mount the short lengths on the stage of the microscope and parallel to its optic axis as outlined in Expt. 55, and focus upon the squarely cut upper ends with a $10 \times$ objective. Get the image of the opening into the center of the field, and read the diameter off the eyepiece micrometer scale; the capillary may be considered satisfactory if the bore is the same at both ends within 5%, i. e., 25 μm.

The capacity of a given length of the capillary may now be calculated from the diameter D. The distance in millimeter which the meniscus is to travel if 1 μl shall be delivered may be estimated from $4/3\ D^2$ if D is substituted in millimeters; 1 mm = 1000 μm. It will be between 3 and 8 mm. Some allowance for the wetting of the glass is made by substituting 3 for π.

Draw out one end of the capillary to a tip of 0.2-mm bore and cut it off at a distance of about 5 mm from the taper. The diameter of the orifice may be measured by attaching the capillary near the wide end to a clamp which is finally to rest upon the table top, and then inserting the tip of the pipet (after removal of mirror and condenser) from below through the central opening in the stage of the microscope; the narrow top orifice is brought into focus after moving the pipet into the optic axis of the microscope and casting light upon it.

To check the calibration, insert the tip of the pipet into a drop of water. Allow the water to enter until about 10 cm of the capillary are filled.

Then measure the length of the water column with a millimeter ruler, Fig. 38. Make allowance for the water in tip and taper by adding to the length of the cylindrical column one third of the length of the taper. Without delay, weigh the filled capillary to the nearest tenth of the milligram. Then empty the pipet slowly by holding it horizontally and touching the tip to filter paper, and weigh the empty capillary. The delivery is given by the difference of weights. Again compute the distance which the meniscus has to travel in the wide part of the pipet to deliver $1\,\mu$l. The holdup upon the walls of the capillary and consequently the amount of delivery depend much upon the rate of outflow. Thus, one should attempt to keep the latter constant within limits suggested by the precision requirements (0.1, relative, when measuring reagents).

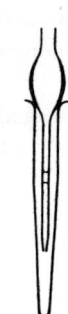

Fig. 39. Transfer from Centrifugal Pipet to Microcone

Use the pipets as in the determination of the delivery, i. e., measure the liquid column with the millimeter ruler. For cleaning, use the suction device of Fig. 37c. The rubber sheet provides a tight seal when suction is applied. Transfer cleaning solution to the upper opening by touching it with a drop of the reagent hanging at the end of a glass rod. For rinsing with water, touch the nozzle of the wash bottle to the opening. The water is sucked out of it and through the pipet. Finally, lift up the rubber square and apply some water to the tapered tube t. Again allow the rubber square to form the seal, rinse the pipet once more, and then dry it by sucking air through it for one minute.

Store the calibrated pipet between two plugs of cotton in a test tube. Indicate upon the label the length that corresponds to a delivery of $1\,\mu$l.

The *centrifugal pipet*, Fig. 39, may be calibrated for capacity since use of centrifugal force permits transferring its whole contents to the point of the microcone when the two are whirled together. It may be readily shaped from glass tubing of 6-mm o. d. by drawing a capillary of 2-mm o. d. and adjacent a capillary of 0.5- to 1-mm bore. For calibration, weigh it first empty and dry. Then add the desired weight (10 mg for $10\,\mu$l) and allow water to enter through the tip of the pipet until the weight is balanced. Mark the location of the meniscus by a line drawn with ink suited for writing on glass. Of course, a ring mark may be etched and filled with graphite (lead pencil) or ferric oxide. Keep the pipet between plugs of cotton in a test tube with label indicating the delivery and the distance of the mark from the tip of the pipet. For rinsing and drying, insert the tip of the pipet into the rubber tubing r of the device Fig. 37b and apply suction.

For adjusting the meniscus to a given mark, hold pipets horizontally and touch the tip either to the liquid to be taken in or to filter paper.

Experiment 23

Preparation and Calibration of Platinum Loops and Hooks

Platinum wire, 3- to 4-cm lengths of wire No. 31 (0.3-mm diameter); vise, steel needle or paper clip, flat-tipped forceps, and calibrated capillary pipet.

Select a stiff wire or brad of the thickness wanted for the diameter of the loop; 1 mm is suited for general work. Clamp the wire so that a 1-cm length of it is easily accessible. Use a 2- to 4-cm length of platinum wire of 0.3-mm diameter. While holding it in one hand, grasp one end with forceps and bend it around the steel wire to obtain a completely closed loop, Fig. 40a. Press the loop between two flat surfaces to get it into one plane. Bend the wire close to the loop so that its plane is inclined at an angle of 30 degrees to the rest of the wire (169), Fig. 40b. Finally, seal the other end of the wire into a capillary drawn out of a glass tube which is mounted in a test tube by means of a stopper. It has been suggested that the loop is soldered closed with gold (154), but this is not necessary if the loop is handled with care.

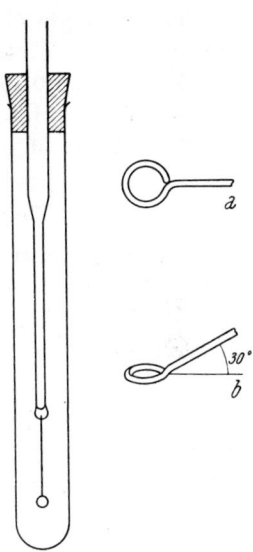

Fig. 40. Platinum Loop; *a* and *b* are approx. twice nat. size

Cleaning the Loop. Keep at hand a 30-ml glass stoppered bottle with concentrated hydrochloric acid. Immediately after each use and before use, rinse the loop in running tap water, hold for a minute into the acid, rinse it in running tap water and in distilled water, and finally ignite it in a non-luminous Bunsen flame until it glows without imparting a color to the flame. A convenient supply of running tap water is obtained by connecting one end of a U-tube to a faucet and letting the water overflow at the other opening.

Calibration. Dip the clean loop into a beaker with water and withdraw it slowly with the plane of the loop perpendicular to the surface of the water. Without delay, touch the tip of the calibrated capillary pipet to the liquid caught by the loop so that it is drawn into the pipet. Measure the length of the liquid column with the millimeter ruler, and compute the volume. Repeat the calibration five times and compute the mean volume. The procedure may be improved by having some water in the pipet from the start so that the displacement of the meniscus occurs only in the cylindrical part of the tube.

Rate of withdrawal of the loop from the liquid and its position relative to the surface of the liquid during withdrawal greatly affect the volume of liquid taken up by the loop so that it may change by a factor of two.

Use of Loop. The loop is filled as in calibration and then repeatedly touched to or set down upon the surface which is to receive the reagent (slide or wall of microcone) until the loop is empty. The loop is emptied at the first contact with a clean glass surface. If the surface is slightly oily, a succession of droplets will result, which may be wiped together with the loop held so that its plane is vertical to the surface.

Platinum Hook. Mount platinum wire of 0.3-mm diameter and 3- to 4-cm length in a glass tubing just as this is done with a loop, Fig. 40. Ignite the wire, and then give it a sharp bend 2 mm from the free end so that an angle of 30 degrees results. Clean and use the hook like the platinum loop. When slowly withdrawing it from water, it will hold a droplet of about 0.1 μl. The hook, like the loop, may be used for introducing reagents directly into solutions, for stirring by twirling, and for withdrawing small fractions of reaction mixture for side tests.

Techniques of the Submilligram Scale

Under this heading, a collection of techniques shall be presented, which may be applied to amounts of material ranging from several milligrams to fractions of micrograms and attain sensitivities from several micrograms to fractions of nanograms when applied to identification tests. The spot tests and slide tests performed with approximately 0.5 to 1 μl of solution are customarily considered adjuncts of the milligram technique of working in microcones, but a quick computation will show that, as a rule, only a few micrograms of solute derived from the sample under investigation are involved.

Spot Tests

To improve the sensitivity and sometimes also the specificity of spot tests, they are performed on paper so that the test solution is slowly added to a very small area of the paper. To this end, the solution to be tested is taken up into a capillary with fine tip which is then touched to the paper so that the solution slowly enters the paper at one point and spreads radially from this point because of the capillary action of the interstices in the paper. It seems that this procedure was first recommended by F. L. HAHN (403, 859) and then systematically investigated by CLARKE and HERMANCE (410) who also pointed out the advantages of papers impregnated with insoluble reagents.

Adsorption or precipitation of the solute in the test solution will occur in the small zone *a* where the latter enters the paper, Fig. 41. If the paper contains a soluble reagent, the spreading solution may carry it out of zone *b* and concentrate it in zone *c* before it gets a chance to react. The distribution of an insoluble reagent in the paper, however, will not be changed by the passage of fluid, and the zone *a* of precipitation or reaction will grow in proportion with the amount of reactant introduced with the test solution. Under all conditions, solutes which do not react and are not adsorbed spread in the paper with the solvent. They may be more or less completely washed into zone c and beyond by applying rinse liquid at a with a capillary pipet in the same manner in which the test solution has been added.

Spot tests on paper have the disadvantage that the outcome of a test must be judged from the observation of a mere change of color, whereby it remains undecided whether the colored substance is dissolved in the liquid, adsorbed upon the fibers of the paper, or is separated as a third phase, solid or liquid. Obviously, attention to the functions of the various zones in the paper may indicate that a color change in zones b or c does

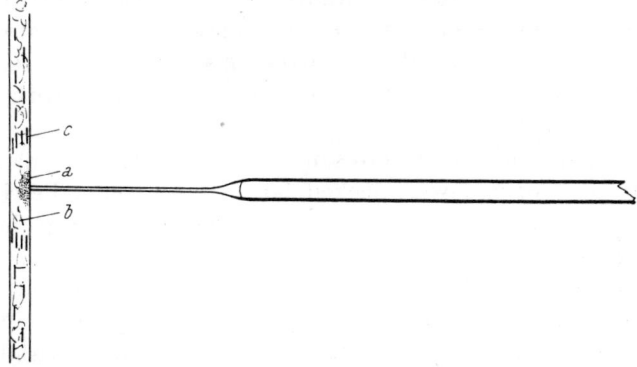

Fig. 41. Performance of Spot Tests on the Milligram Scale; approx. twice nat. size

not support the same conclusions as a color change at the spot where the liquid enters the paper. Consequently, proper performance and attention to the general appearance of the test (location of the phenomena) may prevent incorrect interpretation and improve the reliability of the tests.

The sensitivity of spot tests on paper is naturally greatly influenced by the visibility of the color phenomena. Consequently, one will prefer reagents that lead to the formation of strongly colored substances, dyes. If the reaction product is a colored solid, the perceptibility of the color will greatly depend upon the particle size. Very small particles approaching colloidal dimensions or deposition as film upon the surface of the fibers will be most desirable. It should not surprise, therefore, that the procedure of adding the reagents as well as quality of the paper may profoundly affect the sensitivity. Statements of the latter without reference to a definite procedure must be considered approximations. Strongly absorbent, thick drop test paper is suited for some tests, and thin, "ash-free" paper for others (900, 920). Provided that the test solution is first allowed to spread in the paper and that the latter is then sprayed with the reagent solution, ACKERMANN (954) has shown that the sensitivity of tests for anions and cations having little affinity to paper (alkalies) is the poorer, the larger the area is, which is wetted by the test drop; this is an obvious consequence of Beer's law. Since the wetted area will decrease with increasing thickness of the paper, thick papers should favor tests for anions and alkalies if the mentioned procedure is followed. The ion exchange capacity of the paper determines its suitability for the detection of cations with good affinity to the carboxyl groups of cellulose; thus, for uranyl, lead, stannous, cupric, hydrogen ion, etc., the suitability of the paper is determined by its action as ion exchanger, and the area which is wetted by the test drop is of no consequence.

Blank tests with the paper will show whether or not contaminants are present in objectionable concentration; the ash constituents (iron, barium, calcium, magnesium, potassium, silica, and phosphate) are rarely objectionable, but even good filter paper may contain up to 5 per cent degradation products of cellulose, that give colorations with iodine and may act reducing (1112).

The filter paper is best cut into squares of 2-cm edge or strips 2 cm $\times$ 6 cm and stored in stoppered wide-necked bottles or covered Petri dishes. It should not be touched with fingers, but handled with forceps having ivory or plastic tips.

For impregnation with reagents, strips of paper are either sprayed with the reagent solution or immersed in it for 20 to 30 minutes so that they float freely in the bath. Then they are slowly withdrawn from the bath and pressed between filter paper to remove the excess of solution.

To precipitate reagents in the paper, CLARKE and HERMANCE (418) soak the paper briefly in one solution, pass it through a wringer to remove excess liquid, dry it, insert into the precipitant, etc. To remove the excess of the last reagent, they place the paper upon an inclined glass plate and run water over it. The papers are dried in an electrically heated oven and stored in black envelopes like photographic papers.

If the substance to be identified is either strongly adsorbed by the paper fibers or reacts with a reagent or dye in the paper, that is not extracted by the test solution, the akro technique of SKALOS (881) may be used to improve the limit of identification. The test paper is cut into strips, 15 mm long and 3 mm wide, which are then cut to form a triangle of 15-mm height upon a base of 3 mm. The test solution is applied to the very point of the triangle by means of capillary pipet, loop, or hook–or by touching the droplet of the test solution with the very point of the paper. The effect may be observed with the aid of a magnifying glass. Additional improvement of the sensitivity may be obtained by concentrating the test drop before taking it up with the point of the reagent paper, by evaporating the test drop to dryness and mopping up the residue with the moist point of the paper, or by adding small fractions of the test drop to the point of the paper and drying after each addition.

Obviously, the cut surfaces constitute a major part of the area of the test, especially at the very point of the paper. Consequently, special care must be taken to avoid contaminating the paper when cutting it into shape. A sharp edge on glass or porcelain must be used for cutting if a test for iron shall be carried out.

Usually, the appearance of spot tests changes during drying. Frequently, the color change is better perceptible when the paper has become dry, but the appearance of the test may change within a few hours if it is exposed to air and light. If the test shall be preserved for evidence, one may try to improve its stability by thorough washing after development, drying, exclusion of air by spraying with suitable lacquer, impregnation with paraffin, enclosing between glass plates or sheets of cellophane, and storing protected from light in black envelopes.

A more or less crude estimation of quantity is frequently possible by comparison with tests performed under like conditions with a series of solutions containing known amounts of the identified substance.

Experiment 24

Test for Mercuric Mercury and Lead

Iodide test paper: Soak drop test paper in 1% KI solution, allow to dry, cut into strips, and store in an amber bottle.

By means of a capillary pipet add 1 μl mercuric test solution (10 mg Hg per ml) to iodide test paper. An orange-red spot forms where the solution enters the paper. Repeat the test with dilutions of the test solution containing 5 mg, 2.5 mg, 1.2 mg, 0.6 mg, and 0.3 mg Hg per milliliter, respectively, obtained by repeated dilution with an equal volume of 1-F HNO_3. Start with a drop of test solution and a drop of 1-F HNO_3 on a slide. Take into a wide capillary with narrow tip a 1-cm length of each solution and blow out the mixture upon a slide; this is the first dilution. Use 1 μl for the test, and take a 1-cm length back into the wide capillary to make the second dilution; etc. — Observe that the appearance of the test changes with the more dilute solutions: the center of the spot remains white, and the HgI_2 precipitates in a ring zone. The color is seen best when the paper has become completely dry.

Repeat the experiment with 1 μg lead test solution (10 mg Pb per ml). A yellow spot of PbI_2 appears where the solution enters the paper.

Prepare stannite reagent, just before use, by adding 2 drops of stannous chloride reagent to 3 ml 2-F NaOH in a test tube and mixing to get a clear solution.

To confirm the presence of mercury and lead, treat the colored spots by adding to each a drop of stannite reagent. The mercuric iodide is reduced to gray or black mercury, and the lead iodide dissolves to give colorless plumbite solution.

Also other metals of the hydrogen sulfide group give spots on iodide paper when present in sufficient quantity. Silver gives a pale yellow spot of AgI which turns black with stannite; bismuth gives a brown to black spot which turns black with stannite; cupric and stannic ions give blue spots (reaction of liberated iodine with the cellulose) which fade with stannite reagent. Cadmium, antimony, and arsenic leave the iodide paper unchanged.

Experiment 25

Chromate Test for Silver

Solution of ammonium chromate, 2 g in 100 ml water; sodium chloride, 1% in water.

By means of a glass rod, place drops of ammonium chromate solution upon a strip of drop reaction paper. Allow the paper to dry for five minutes at room temperature. Then, by means of a capillary pipet, add 1 μl silver test solution (10 mg Ag per ml) to the center of one of the yellow chromate spots. A brownish red spot forms where the solution enters the paper; if only little silver is present, the brown spot may appear 2 to 5 minutes after adding the test solution.

The test may be made more distinct by adding with the capillary pipet 10 to 30 μl distilled water to the center of the brown spot. The soluble yellow $(NH_4)_2CrO_4$ is carried away to leave the brown Ag_2CrO_4 in the white paper.

Confirm the presence of silver by finally adding a drop of sodium chloride solution to the brown spot. The silver chromate is converted to chloride, and the brown spot disappears immediately.

Mercuric mercury, cadmium, tin, and antimony do not give a test. Lead and bismuth produce yellow spots. The brown spot given by cupric ion after rinsing with water remains unchanged when treated with sodium chloride.

Experiment 26

Tests for Bismuth and Antimony

Quinine iodide reagent: dissolve 1 g quinine hydrochloride in 50 ml water and add 0.2 ml 6-F HCl; dissolve 2 g KI in 50 ml water; before use, mix equal small volumes of the two solutions. — Cinchonine iodide reagent: dissolve 1 g cinchonine hydrochloride in 19 ml water and add 1 ml 6-F HCl; dissolve 2 g KI in 10 ml water; before use, mix 2 small volumes of the former solution with

1 volume of the iodide solution. — Iodide-acetate reagent: dissolve 25 g ammonium acetate and 10 g KI in 50 ml water. — Stannous chloride reagent: 11 g $SnCl_2 \cdot 2 H_2O$ dissolved in ,17 ml 12-F HCl and diluted with water to 100 ml; keep in bottle containing 1 g metallic tin.

Mark six test points on a strip of drop test paper. Place the paper upon a glass plate and press down upon it with the end of a glass rod that has been drawn out to a point and cut and polished to obtain a flat circular surface of 2-mm diameter (948). Do not press hard enough to tear the paper. Using a lead pencil and writing at some distance from the indentations, label them Bi, Cu, Sb, Sb, Bi, and Pb.

With capillary pipets add to each of the marked spots about 1 μl of the indicated test solution (10 mg ion per ml) and then lay the paper aside for drying. It is not necessary to use a calibrated capillary pipet; in this instance, it will suffice to crudely estimate the diameter of the bore and from this the length of the capillary, representing 1 μl. As an alternative, the test solutions may be measured with a platinum loop and taken from the loop into the capillary pipet for transfer to the paper. The contaminated end of the pipet may be cut off before proceeding to the next solution, and a fresh tip drawn.

When the droplets have evaporated on the paper, use a glass rod or a dropper to treat the first three spots (Cu, Bi, Sb) with small drops of freshly mixed quinine iodide reagent (433, 1135) and the last three, with freshly mixed cinchonine iodide reagent (1158). Record the color of the spots. When they have become dry, treat the first three spots with a drop each of freshly prepared stannite reagent (Expt. 24) and the last three spots by adding to each 1 drop of iodide-acetate reagent.

Quinine iodobismuthite gives an orange-red precipitate (spot) which is reduced to black metallic bismuth upon adding stannite reagent; the corresponding iodoantimonite is orange-yellow and dissolves in stannite reagent to give a colorless solution. The cinchonine iodobismuthite gives a bright orange-red spot which changes little on adding iodide-acetate reagent; the corresponding antimony compound is orange-yellow, and the spot changes immediately to white when iodide-acetate reagent is added. A brown ring may form around the white central area, but it disappears when the spot dries.

Mercuric mercury, cadmium, and arsenic do not visibly react with the reagents. Lead gives yellow PbI_2 which remains unchanged with iodide-acetate reagent and dissolves in stannite reagent. The pale yellow AgI is reduced to black metallic silver by stannite reagent. Cupric ion gives a brown to black spot which remains nearly unchanged when stannite reagent is added. Stannic tin may give a blue spot of liberated iodine acting upon the cellulose.

Experiment 27

Test for Copper, Nickel, and Cobalt (1150)

Rubeanic acid, diamido-dithio-oxalic acid, 1% solution in ethanol.

With a capillary pipet transfer to different locations on a strip of drop test paper about 1 μl each of copper, nickel, and cobalt test solutions containing 10 mg ion per ml. Expose the moist spots to ammonia fumes by holding the paper over the opening of a bottle containing strong ammonia solution. Finally treat each spot with a drop of rubeanic acid solution, and record the colors of the precipitates. The tests are quite specific.

The copper salt is insoluble in acetic acid, whereas the salts of nickel and cobalt will not precipitate or precipitate only partly from acetic acid solution, depending upon the concentration of the acid. Flow through paper will produce concentration gradients aiding in the separation of the metal ions (121).

Mix small equal volumes of the copper, nickel, and cobalt test solutions, and treat the mixture with an equal volume of 4-F acetic acid. Soak a strip of drop test paper in rubeanic acid solution and let it dry. With a capillary pipet transfer about 1 μl of the acidified salt solution to the impregnated paper. The olive green or black coloration obtained where the liquid enters the paper indicates the presence of copper which may be thus detected up to the limits Cu : Co : Ni $= 1 : 2000 : 20\,000$. Try to improve the simultaneous identification of nickel and cobalt by first rinsing from the center of the spot with 10 to 20 μl of 1-F acetic acid added from a capillary pipet. Finally expose to fumes of ammonia.

Experiment 28

Test for Cadmium (590, 591)

Cadion test paper: soak ash-free filter paper or drop test paper in 0.02% solution of Cadion 3 B, benzenediazoaminobenzene-4-azo-4'-nitrobenzene, in ethanol. Press between blotting paper and allow to dry; cut the dry paper in strips. — Rochelle buffer: dissolve 10 g Rochelle salt in 100 ml water and 0.1 ml glacial acetic acid.

Mix a small volume of cadmium test solution with an equal volume of Rochelle buffer solution, and transfer about 1 μl of the mixture with a capillary pipet to the Cadion test paper. Allow the test drop to evaporate at room temperature (heating is not permissible), and then add with a glass rod a drop of a mixture of 4 volumes 2-F KOH with 1 volume ethanol. The test paper becomes purple, but the circular area where the cadmium solution entered the paper turns pink or salmon red.

The diameter of the pink spot is related to the amount of cadmium. For the detection of small amounts of cadmium, place one drop of reagent solution on drop test paper, add the test drop to the center of the spot, and then add the KOH.

Interferences by copper, nickel, cobalt, iron, chromium, and magnesium are avoided by the addition of the tartrate. Mercuric mercury gives a yellow spot which might be mistaken for cadmium, but changes to gray on adding stannite reagent, Expt. 24. Tin solutions produce a white spot, and the dye is also destroyed by heat or strongly acid solutions. Other common elements of the hydrogen sulfide group have no visible effect.

Experiment 29

Precipitation of Silver Arsenate

Buffered silver solution: dissolve 1 g $AgNO_3$ and 7.7 g ammonium acetate in a mixture of 6 ml glacial acetic acid and 200 ml water.

With a capillary pipet transfer about 1 μl slightly acid or alkaline arsenate test solution (10 mg As per ml) to ash-free filter paper. Lay the paper aside to dry, and then add 5 to 10 μl buffered silver solution by means of a capillary pipet.

Brown Ag_3AsO_4 separates. An amount of chloride ion equal to that of the arsenate does not hinder the detection of the latter, but the appearance of the brown spot is somewhat delayed. Large amounts of halide must be absent. Antimony and tin do not give colored reaction products.

Experiment 30

Molybdenum Blue Test for Tin (121)

Phosphomolybdate paper: soak strips of ash-free filter paper in 1% aqueous solution of phosphomolybdic acid; pour a few milliliters of strong ammonia solution into a 400-ml beaker, and expose each strip of paper for 10 to 20 seconds to the ammonia fumes by holding it into the gas space of the beaker immediately after removing it from the phosphomolybdic acid bath; allow the strips to dry, and store them in bottles of amber glass. — Magnesium ribbon.

Place a clean glass plate upon a sheet of black paper. Cut magnesium ribbon first into strips of about 1-mm width, and then cut the strips into 0.5-mm lengths and allow the small squares to drop on the glass plate.

Take about 1 μl of stannic test solution into a capillary pipet. Deposit a large drop of 12-F HCl upon a glass slide, and allow about 1 μl of this acid to enter the pipet containing the tin solution. Blow out the contents of the pipet upon a glass slide, and there treat the droplet with a small square of the magnesium ribbon. First dip the end of a glass thread into the tin solution; then touch the metal square with it and transfer it into the test drop.

When the evolution of hydrogen has stopped, insert the tip of the capillary pipet into the droplet and transfer the clear solution without delay to phosphomolybdate test paper. A blue spot forms, which becomes more distinct when the paper is laid aside and allowed to dry.

Arsenic, antimony, and mercury solutions do not give reduction to molybdenum blue under the conditions specified for the tin test.

Repeat the test (inclusive reduction with magnesium) with 1 μl cupric test solution, and note that a blue spot is obtained, which disappears when the paper is laid aside to dry. The presence of copper also promotes the fading of the molybdenum blue produced by stannous tin if there is not present at least four times as much tin as copper.

Slide Tests

Typical slide tests are based upon the separation from solution of more or less coarsely crystalline precipitates. They are performed upon microscope slides, and the shape of the crystals is observed under the microscope.

When comparing the different types of chemical confirmatory tests with regard to specificity and reliability, the slide tests must be given the highest rating. With spot tests and fiber tests, the outcome rests essentially upon the observation of color phenomena. At times it will be impossible to tell whether or not a new phase has formed. The test tube technique is somewhat superior, for, aside from the color effects, the separation of phases and their general appearance can be observed without difficulty. In addition to all these criteria, however, slide tests reveal also the shape of the particles of a new phase formed. The convincing finality of microscopic identification recommends slide tests for general use on any scale of work.

Since the particles of precipitates may remain so small that even the microscope is unable to reveal their shape, care is taken to favor the formation of few nuclei so that they may grow to reasonable size with the small amount of material available. The reactions are intentionally selected and conducted to favor the formation of relatively large crystals, the dimensions of which range from a few micrometers to a few millimeters. Rather soluble precipitates are preferred, and in all instances conditions are established, which give a low rate of nucleation as a consequence of the continuous maintenance of a low degree of supersaturation. In some instances, crystals of the desired size are obtained by recrystallization from suitable solvents.

Examples of the various techniques employed may be found in Expts. 31 to 41. The reagent is frequently added in the solid state so that it may spread slowly through the test drop by gradual dissolution followed by diffusion. If the reagent is a liquid, the test solution may be evaporated and the solid residue treated with the liquid reagent. The test drop and the reagent drop may be placed side by side upon the microscope slide and then connected by a narrow channel which prevents instantaneous mixing. If a gas is involved, the rate of precipitation may be controlled by the rate at which the gas is added or removed.

It is desirable to begin the microscopical observation of the test at least immediately after adding the reagent. Observation of the dissolution of the reagent, accompanied or immediately followed by the separation of the test forms, demonstrates the connection between cause and effect in a very convincing manner. If the test is set aside after adding the reagent and the crystalline precipitate is observed after lapse of some time, one cannot be certain whether the crystals have been formed by the action of the reagent or because of the evaporation of solvent. To arrive at a decision in such instances, it will be necessary to compare with a second test drop that has been standing equally long without addition of reagent, but even such comparison may fail to furnish equally convincing proof.

The time of appearance and the location of the precipitate in the test drop provide valuable clues that help avoiding misinterpretations. Crystals which are slowly growing along the edge of the drop form as a consequence of evaporation and the resulting rise of concentration, which is most pronounced along the edge of the drop. Crystallization caused by the immediate action of the reagent takes place close to the reagent, and, as a general rule, the closer the crystals are to the reagent, the smaller is their size.

The general appearance of tests makes it possible to draw conclusions concerning the conditions in the test drop, and the information may be used to recognize the changes necessary for a more successful repetition of the test. If the solid reagent does not dissolve or dissolves very slowly in the test drop and very few small crystals or none separate close to the reagent, the concentration of the sought-for substance may be so high that the reagent becomes coated with the insoluble test form so that it cannot dissolve; diluting the test drop may be followed by copious separation of characteristic crystals. Also instantaneous separation of a dense, powdery or gelatinous precipitate close to the reagent indicates that the concentration of the test drop is too high; larger crystals may form after some time at the periphery of the area of granular precipitation, but even these may remain quite small if too little reagent was added so that the granular precipitate consumed all or nearly all of it. The most favorable conditions prevail when small crystals form close to the reagent and large ones at a distance. With somewhat lower concentrations of the sought substance, the number of small crystals separating close to the reagent decreases, and there is a distinct trend toward the formation of medium-sized crystals only. With further dilution of the test solution, size and number of the crystals of the test form decrease; finally, no precipitation or crystallization follows upon adding the reagent, and crystals of the test form separate only after the concentration has been raised by evaporation. Under these conditions, the crystals are small and usually appear first at the edge of the drop. Experienced workers usually succeed in keeping near optimum conditions by suitable adjustment of the concentration of the test solution by proper choice of solvent volume, diluting the solution, or concentrating it by evaporation. Naturally, the solution volume may be adjusted at will, and the quantity of sought substance present may be estimated from the amount and nature of the sample or the volume of precipitates and other observations made in the course of the analysis.

Obviously, all dissolved substances will appear in the residue when a test drop evaporates to dryness. The soluble substances contained in the test solution usually form crystals that are much larger than those of the test form, and complete evaporation of the drop gives a crust in which the crystals of the test form become hidden. If the latter are quite insoluble, adding a drop of solvent will render them again observable by dissolving the incidental solids; if they are reasonably soluble, however, the appearance of the test must be restored by the very cautious adding of small amounts of solvent. This may be done by exposure to solvent vapors (breathing upon the residue of the test drop). When working in a very dry atmosphere, the rapid evaporation of test drops may become a nuissance; it may be prevented by frequently breathing upon the slide, covering the test drop with a small watch glass, or placing the slide into a humid atmosphere maintained inside a desiccator or under a bell jar; the slide with the test drop may be made the floor or ceiling of a gas chamber containing a drop of solvent.

The volume of the test drops is optional; naturally, the limits of identification will improve in proportion with the reduction of volume. In the following experiments, drops of 0.3- to 1-μl volume are used to get well below the milligram range. Working with much smaller drops requires use of a moist chamber to prevent their rapid evaporation, p. 198.

For work with aqueous solutions, the microscope slides should be slightly oily so that small drops do not spread without, however, assuming hemispherical or even spherical shape. This condition may be obtained by wiping the slides with tissue paper containing a trace of sebum (oil) from the face or scalp. Cover slips are rarely used; if they are placed upon the drop after adding the reagent, the characteristic appearance of the test may be completely destroyed, and crystals of the test form may be lost by being carried to the edge of the cover slip where they remain hidden to view. Omission of the cover slip also makes it possible, after washing and (or) evaporation of the drop, to treat individual crystals with reagents for further convincing confirmation of their identity.

Experiment 31

Silver Dichromate (125)

Potassium dichromate, granular, 1-mm diameter.

Upon a slide, mix 5 µl silver test solution (10 mg Ag per ml) with 1 µl 16-F HNO_3. Take up the mixture into a capillary pipet and transfer it to another slide to obtain the droplet in a suitably small area. Using transmitted light and a magnification of 20 to 80 diameters, focus the edge of this drop under the microscope. At some distance from the drop, deposit upon the slide some $K_2Cr_2O_7$.

Draw a glass thread of 0.1- to 0.2-mm diameter from the end of a capillary. Moisten the end of the thread by dipping it into the test drop, and then pick up with it a kernel $K_2Cr_2O_7$ of about 1-mm diameter (0.5-µl volume) and place it into the edge of the test drop (this may be done while observing through the microscope, compare Expt. 61). Without delay observe the test under the microscope.

Spears of $Ag_2Cr_2O_7$ grow out of the reagent kernel and form also at some distance from the reagent which spreads in the test drop as may be seen from the yellow coloration. The crystals are yellow to deep red depending upon their thickness; they are triclinic and show strong birefringency and a slight degree of pleochroism, light to dark. Prepare a sketch and collect into it all typical forms observed at this time and on later occasions. Relatively large crystals of yellow $K_2Cr_2O_7$ and colorless KNO_3 will be observed when the test drop dries out.

Distilled water or glycerol may be used for moistening the end of the needle previous to picking up the solid reagent. The use of saliva is not advisable for obvious reasons, and the removal of reagent directly from the vial with a moistened needle is likewise objectionable. Naturally, the unused reagent should not be returned to the reagent bottle or vial. The portion of the glass thread, which has come into contact with the reagent or the test solution, is broken off and discarded.

Experiment 32

Recrystallization of Silver Chloride (125)

Watch glass, 2.5-cm diameter. — Silver chloride, powder.

With a spatula, transfer some silver chloride to a slide. Select a cluster of 0.5-mm diameter (0.05-µl volume), push it to an empty area of the slide, and treat it with 5 µl 6-F NH_3 from a capillary pipet. Without delay, stir the mixture with a glass thread, take up the clear solution into a capillary pipet and transfer it to another slide. Immediately cover the droplet with a small watch glass so that it projects slightly over the edge of the slide, Fig. 42. The small opening between slide and watch glass permits slow escape of the ammonia, and the rate of its escape from the solution is sufficiently retarded to give relatively large crystals of AgCl.

Allow to stand for 10 minutes with the watch glass in place. The separation of AgCl is indicated by the appearance of a turbidity visible to the unaided eye; a clean watch glass does not interfere with the inspection of the drop under the microscope when a low magnification is used.

Finally, remove the watch glass and examine the crystals with a magnification of 80 diameters or more. When the drop has evaporated completely, place upon the residue a fragment of a cover slip with a droplet of 4-F HNO_3 hanging on its underside. The drop spreads between slide and cover slip and dissolves all solids but the AgCl. In addition, the cover slip makes possible the efficient use of medium magnifications of 100 to 300 diameters.

Colorless octahedra, tetrahedra, cubes, and combinations may be observed. When opportunities offer themselves, add to the sketch drawings

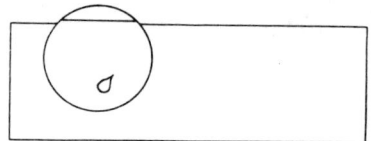

Fig. 42. Retarding Evaporation; $^2/_3$ nat. size

of skeletal forms and spherulites (168) which may be obtained if the silver chloride has been precipitated from solutions containing ions of other heavy metals.

Considering that the insolubility in acids and the solubility in ammonia is demonstrated in addition to the cubic nature of the crystals, the test appears specific for silver. In addition, the identification of AgCl is aided by its high refractive index which gives heavy outlines to the larger crystals and renders small crystals entirely black in transmitted light (reflected light shows that they are colorless).

Experiment 33

Lead Iodide

Potassium iodide, granular.

This test has found special favor with HEMMES (154).

Mix equal small volumes of lead test solution (10 mg Pb per ml) and 6-F acetic acid. With a capillary pipet, transfer 10 μl of the mixture to the center of a microscope slide. Using a magnification of about 20 diameters and transmitted light, focus the edge of the drop under the microscope. With a glass thread, place a KI grain of 0.5-mm diameter into the edge of the drop and observe that it dissolves rapidly while yellow hexagonal plates of PbI$_2$ appear in a circular zone around the reagent. As the KI diffuses through the drop, the zone of precipitation moves away from the center where the PbI$_2$ is dissolved in the excess of KI. Prepare a sketch showing the general appearance of the test and collect in it drawings of all typical shapes observed.

The plates of PbI$_2$ are usually thin enough to show interference colors. Use a dark background and strong light from the side, and observe during crystallization when the particles are kept in motion by the convection currents in the drop. When the test drop finally becomes more concentrated by evaporation, fine needles of colorless KPbI$_3 \cdot$ 2 H$_2$O appear in the center of precipitation where the reagent was introduced. Observe with crossed nicols, and rotate the stage.

Experiment 34

Potassium-Lead-Copper Nitrite (115, 116)

Potassium nitrite, 5.8-F: dissolve 5 g KNO$_2$ in water to give 10 ml solution; acetate buffer solution: treat 4.5 g sodium acetate trihydrate with 1 ml glacial acetic acid and add water to get 10 ml solution.

To prepare the reagent, transfer to a microcone 0.05 ml of 5.8-F KNO$_2$ and a like volume of sodium acetate buffer solution. Mix thoroughly. This reagent will keep for two days if the cone is stoppered and kept away from strong sources of heat or light (875). After preparation, it should be tested with a known lead residue before using it on an unknown.

By means of a platinum loop, deposit several small droplets of lead test solution near one end of a microscope slide and 5 to 10 mm apart. Hold the slide about 5 cm above a small Bunsen flame, and evaporate the droplets to dryness without overheating the residues; just heat the slide until the evaporation starts, and then remove it from the flame and finish the evaporation by blowing upon the droplets with the mouth. Quickly cool the slide by placing it upon metal, a metal block, or the base of the microscope. By means of a capillary pipet, moisten each residue with 1% (1-F) CuSO$_4$ solution (10 mg Cu per ml), and again evaporate to dryness and cool to room temperature. Finally, with a platinum loop, hook, or capillary pipet, treat one residue with a small amount of nitrite reagent. To avoid seeding, do not touch the slide with the tool, and add so little reagent that it does not completely cover the residue. Without delay, inspect with a magnification of 80 to 100 diameters and strong transmitted light. Use the front lens of the condenser.

The small squares and rectangles of K$_2$CuPb(NO$_2$)$_6$, of 10- to 25-μm edge, form either immediately or, with small quantities of lead in a relatively large drop, after a few minutes. The color varies from yellow to black, depending upon the thickness of the crystals. Bubbles of nitrogen oxide have circular outline. When evaporation nears completion, various green and colorless salts begin to crystallize and the observation of the triple nitrite becomes difficult or impossible.

The appearance of the precipitation clearly indicates the changes that should be made to improve the test.

1. If crystals of the triple nitrite are not obtained, or only a few small ones grow near the edge of the drop within 5 minutes from the addition of the reagent, either the latter is spent, or too much of it has been taken. The triple nitrite is rather soluble, and it is necessary to keep the volume of the reagent at a minimum. Repeat the test by adding a much smaller volume of reagent to the next residue. If no improvement is obtained, it is obvious that the reagent does not have the proper composition. Prepare a new batch of reagent; at times, it becomes necessary to prepare a fresh solution of potassium nitrite.

It should be understood that the slow appearance of few small crystals of triple nitrite must be expected whenever the quantity of lead approaches the limit of identification.

2. If a brown or black mass of precipitate forms immediately and the individual crystals are so small that their shape cannot be recognized, the supersaturation in the drop is too high. Repeat the test with another residue as follows: breathe upon the slide so that the residue liquifies, and then add the reagent.

3. If relatively large black crystals form within 3 minutes, the conditions in the drop are properly established.

The triple nitrite test is specific for lead since no other element is known which could substitute for it without changing the color of the crystals. Barium, strontium, and calcium may substitute for lead, but the crystals will be colorless or green. Nickel replaces copper without giving much change in the appearance of the precipitate; rubidium, cesium, and thallous ion may be substituted for the potassium, which decreases the solubility of the triple salt. Concerning the amount of copper to be added in the test for lead, the ratio $Pb : Cu = 1 : 10$ gives the best results; it need not be closely adhered to since the limiting proportions allow large deviations. The test may be directly applied to insoluble lead salts such as the sulfate, see Expt. 47.

Experiment 35

Cesium Iodobismuthite (115, 128) *and Cesium Iodoantimonite* (115, 168)

Cesium chloride, granular; potassium iodide, granular; stannite reagent, Expt. 24.

With a capillary pipet, transfer 5 μl bismuth test solution to a slide. For adjusting the acidity of the test drop, evaporate just to dryness and dissolve the residue by adding 5 μl 2-F HNO_3 when the slide has attained room temperature. Place a grain of CsCl of 1-mm diameter into the edge of the drop; it dissolves quickly, and colorless cesium chlorobismuthite separates around the dissolving reagent. Hexagonal plates are usually seen in the midst of a variety of other forms.

Without removing the slide from the stage of the microscope, introduce a grain of KI of 1-mm diameter into the edge of the drop and just outside the zone of precipitation of chlorobismuthite. Yellow, red, and nearly black hexagons and stars form where the zones of diffusion of the CsCl and the KI meet. The intensity of color is, as usual, determined by the thickness of the crystals. When the diffusing iodide reaches the colorless crystals of the chlorosbismuthite, they become gradually converted to orange iodobismuthite.

Set the slide aside until the test drop has evaporated to dryness. Then, using a magnification of 20 to 50 diameters, illumination with reflected light, and a green or blue background, focus upon a portion of the preparation, which contains well developed crystals of the iodobismuthite. While observing through the eyepiece, add a large drop of stannite reagent from a medicine dropper, but be certain that none of the reagent gets on the lenses of the microscope (it strongly acts upon optical glass, and reduction of lead glass may cause blackening).

Observe that the crystals of iodobismuthite turn black and opaque while mostly retaining their shape. They become pseudomorphs of tiny particles of metallic bismuth, which remain clustered together to keep the shape of the iodobismuthite. The shape of pseudomorphs reveals their history but bears no relation to their lattice structure.

Repeat the test with 5 μl of antimony test solution (10 mg Sb per ml). Adjustment of the acidity by evaporation is not possible because of the volatility of the chlorides of antimony, and it is not necessary since the HCl of the test solution will not oxidize the iodide. Thus add the CsCl and then the KI directly to the drop of test solution. Observe that antimony behaves much like bismuth. Finally, allow the test drop to evaporate to dryness, and then treat it with stannite reagent which dissolves all antimony compounds and thus permits distinguishing between antimony and bismuth.

It may be mentioned that solutions of tartar emetic, $K(SbO)C_4H_4O_6 \cdot \tfrac{1}{2} H_2O$, must be acidified with HCl in order to give the test with CsCl and KI.

Experiment 36

Bismuth Cobalticyanide Pentahydrate (1143)

Potassium cobalticyanide, granular (the preparation of the salt is described in the paper cited above); stannite reagent, Expt. 24; bismuth-lead test solution: dissolve 1.2 g $Bi(NO_3)_3 \cdot 5 H_2O$ and 8 g $Pb(NO_3)_2$ in 3-F HNO_3 to make 100 ml solution.

Transfer 5 μl of bismuth test solution to a slide and adjust the acidity as in Expt. 35 by evaporation and dissolution of the residue in 5 μl 2-F HNO_3. Place a grain of $K_3Co(CN)_6$ of 1-mm diameter into the edge of the drop and observe with transmitted light and a magnification of 70 to 100 diameters. Close to the reagent, the $BiCo(CN)_6 \cdot 5\, H_2O$ precipitates as a fine powder, but larger crystals grow slowly at the outermost boundary of the area of precipitation and exhibit very characteristic shapes, drawings of which should be collected. The crystals grow rather slowly, and the observation should be continued for 5 minutes or more. Remarkable twin crystals are occasionally observed.

Finally set the test aside and allow it to go to dryness. Using a magnification of 20 to 50 diameters, illumination for observation in reflected light, and a colored background, focus upon characteristic crystals and add a large drop of stannite reagent as in Expt. 35. Black pseudomorphs of metallic bismuth will be obtained.

The shape of the crystals of bismuth cobalticyanide is greatly modified by the presence of lead, stannous tin, and mercuric mercury. For a demonstration, repeat the above experiment with a solution containing 50 mg of lead and 5 mg of bismuth per milliliter. The resulting precipitate which contains about 10% Pb consists of lens-shaped forms that, seen from the front, exhibit the appearance of oil drops; side views have sharp outlines. Stannous tin behaves very much like lead, and the lens shapes are converted to black pseudomorphs by adding NaOH solution; the required stannous ion is already contained in the lenses (and in the residue of the test drop).

Experiment 37

The Mercurithiocyanates of Copper, Zinc, Cadmium, and Cobalt (115, 118, 896)

Potassium mercurithiocyanate, granular: dissolve 27 g $HgCl_2$ in 100 ml boiling water and precipitate by slowly adding a solution of 19 g KCNS in 20 ml water; cool to room temperature, collect the $Hg(CNS)_2$ in a Buchner funnel, and wash it with water and dry; dissolve 16.8 g KCNS in 15 ml water, and add 40 g $Hg(CNS)_2$; stir and add water until solution is complete; transfer to a flat dish and place for evaporation into a desiccator with 18-F H_2SO_4; grind to grains of 0.5- to 1-mm diameter. — Ammonium mercurithiocyanate solution: dissolve 5 g $HgCl_2$ and 5 g NH_4CNS in 6 ml water.

Deposit 1-μl portions of copper, zinc, cadmium, and cobalt test solutions (10 mg metal ion per ml) upon a slide so that the drops are kept 5 to 10 mm apart. Complete the collection by adding 1-μl portions of the following six mixtures of test solutions: 1 volume of Zn (test solution, 10 mg ion per ml) + 10 vols. Cu (test solution); 10 vols. Zn + 1 vol. Cu; 1 vol. Zn + + 10 vols. Co; 10 vols. Zn + 1 vol. Co; 10 vols. Cd + 1 vol. Co; equal volumes of Zn, Cu, and Co test solutions.

If necessary, warm the slide and blow air upon it until all test drops are evaporated to dryness. Treat one residue after the other as follows: dissolve it by adding to it 1 μl 2-F HNO_3, and place into the edge of the resulting drop a grain of 0.8-mm diameter of $K_2Hg(CNS)_4$; observe the resulting precipitates with suitable magnification and illumination (strong transmitted light as well as reflected light with white background). Prepare drawings.

The tests for zinc, copper, and cobalt are quite specific. In practice, one will separate these metals from other heavy metals before applying the test. To increase the certainty of identification, it is recommended to repeat the zinc test after adding to a fresh test drop a small amount of copper, to repeat the copper test with some zinc added, and to repeat the cobalt test with some cadmium added. In all instances, it will be wise to adjust the amount of added metal to that of the metal to be identified to a ratio giving a striking effect (1 copper to 5 or 10 zinc, and 1 cobalt to 10 cadmium).

Ferrous and auric ion give yellow needles and dendrites which may be mistaken for those of the copper mercurithiocyanate. Ferric salt imparts a red coloration to the solution, and dark red, almost black hexagons may separate. Manganese, lead, and silver give colorless crystals which may be mistaken for the cadmium or zinc mercurithiocyanate. Only strong solutions of nickel give small pale yellow disks and masses of radiating needles which may appear brown in transmitted light (Tyndall effect) (88). Mercuric mercury precipitates $Hg(CNS)_2$; iodide and cyanide should be absent since they bind the mercury of the reagent.

SCHOORL (168) has demonstrated in the instance of the cobalt that the identification limit may be considerably improved by decreasing the volume of the test drop. In addition, one may increase the bulk of the test form by adding a second metal ion and precipitating the mixed mercurithiocyanates copper-zinc (1 : 10), zinc-copper (1 : 3), cobalt-zinc (1 : 10), and cadmium-cobalt (1 : 10) instead of the simple mercurithiocyanates of copper, zinc, cobalt, and cadmium, respectively. Determine the resulting limits of identification of the cobalt and of the zinc tests as follows.

Prepare a small volume of a dilution of 1 volume of cobalt test solution with 9 volumes of water. Mix, and dilute 1 volume of the mixture with 4 volumes of water to obtain a solution containing 0.2 mg Co per ml. Continue diluting in the ratio 1 to 5 to obtain solutions with 0.04, 0.008, 0.0016, and 0.0003 mg Co per ml. Start with the strongest solution (1 mg Co per ml), and transfer two 1-μl portions each of every dilution to a microscope slide which is sufficiently oily to prevent the spreading of the droplets. Warm the slide and blow upon it to evaporate the droplets to dryness. After cooling to room temperature, moisten the first, largest residue with ammonium mercurithiocyanate reagent, and inspect under the microscope.

Add the reagent with a platinum loop of about 0.3-μl capacity, and avoid touching the glass with the platinum, which might give seeding and numerous small crystals. Repeat the test with residues containing less cobalt until the dilution has been found, that does no longer give a positive test in every trial. Then return to the next stronger solution and try to obtain the test with smaller volumes of the solution: 0.5, 0.3, 0.1 μl, which may be transferred to the slide with a platinum loop or hook. It is essential that the evaporation residues cover correspondingly smaller areas on the slide and that correspondingly smaller amounts of reagent are added by means of smaller loops or platinum hooks. Compute the smallest mass of cobalt, which reliably gives a positive test; it represents the limit of identification.

In a similar way determine the limit of identification for copper by starting with a mixture of 1 volume copper and 9 volumes zinc test solution.

Analogous mercuriselenocyanates are mostly of theoretical interest because of the instability of the reagent, but they may be used to test cobalt salts for presence of small amounts of iron, copper, and nickel (863).

Experiment 38

Test for Cadmium with Brucine and Bromide (136, 560)

Brucine bromide reagent: dissolve 0.65 g brucine and 1 g NaBr in 5 ml 4-F acetic acid and 15 ml water; sodium sulfide-hydroxide reagent: dissolve 24 g $Na_2S \cdot 9 H_2O$ and 2 g NaOH in water to make 50 ml solution.

Transfer 1 μl cadmium test solution (10 mg Cd per ml) to the microscope slide and deposit next to it a drop of like volume of brucine bromide reagent. With a glass needle or the platinum loop, draw a channel to connect the two drops. Prepare a drawing of the general appearance of the test showing the characteristic clusters of supposedly monoclinic plates. — It is necessary that the cadmium solution does not contain free acid; if it does, evaporate the test drop on the slide to dryness and dissolve the residue in water.

When the test has evaporated to dryness, treat the residue with a relatively large drop of sodium sulfide-hydroxide reagent, which has been diluted with 10 volumes of water, and inspect the "pseudomorphs" of CdS with reflected light and a white or purple background.

Experiment 39

Test for Mercury with Zinc, Copper, and Thiocyanate (119)

Potassium thiocyanate, 5% solution; mixture of equal volumes of zinc and copper test solutions (5 mg Zn and 5 mg Cu per ml).

This reversal of the zinc test of Expt. 37 is specific for mercury. Evaporation is carried out in presence of nitric acid to reduce volatilization of the mercuric chloride.

Deposit a series of droplets of about 0.3-μl volume of the copper-zinc solution upon a microscope slide and evaporate them just to dryness. Reserve the slide with the residues.

Transfer about 0.3 μl mercury test solution (10 mg Hg per ml) to a slide, add a like volume of 6-F nitric acid, and evaporate just to dryness without heating much above 100° C. After cooling, add from a capillary pipet 0.5 μl thiocyanate solution, and make certain that the whole residue is dissolved. Take up the solution with the capillary pipet and cover with it one of the copper-zinc residues without touching either the residue or the slide with the point of the pipet. Observe the test with suitable magnification and strong transmitted light, top lens of condenser inserted. Compare with the observations made in Expt. 37. The bundles of characteristic needles of purplish hue are replaced by tiny grains if the amount of mercury is small, and daylight illumination becomes essential for the observation of the color of the precipitate.

Experiment 40

Magnesium-Ammonium Arsenate and Silver Arsenate (115, 152, 168)

Magnesium acetate tetrahydrate, granular; or calcium chloride hexahydrate.

Transfer 0.5 μl arsenate test solution (10 mg As per ml) to the center of a microscope slide, and add 2 μl 2-F HNO$_3$. If the drop should have spread out too much, take it up into a capillary pipet and transfer it to the center of another slide. Invert the slide and place it upon the opening of a bottle containing strong ammonia so that the hanging test drop is exposed to the ammonia gas. It requires only a few minutes to render the test drop ammoniacal. Then remove the slide from the bottle, and place a grain of 0.6-mm diameter (0.1-μl volume) of magnesium acetate or calcium chloride into the edge of the test drop. Observe with a total magnification of 70 to 100 diameters and transmitted light of low intensity. Prepare a sketch of the general appearance of the test, and include drawings of characteristic forms such as X shapes, prismatic crystals, and feathery dendrites of the NH$_4$(Mg or Ca)AsO$_4 \cdot$ 6 H$_2$O.

The crystals of the corresponding phosphate have the same shape. To confirm the presence of arsenate, set the test aside and allow it to evaporate to dryness. Then place a large drop of 1-F NH$_3$ upon the residue so as to cover it completely.

If the surface of the slide does not repel the solution strongly, drag off the wash liquid as described by BEHRENS (115, 116, 118). With the slide held horizontally, touch the point of a glass needle (stirring rod of 0.5- to 1-mm diameter) to the edge of the drop and slowly draw it over the slide away from the drop so that the liquid follows the point and forms a narrow channel. The crystals of the test form cling to the slide and remain in place if the motion of the liquid is slow. While regulating the rate of flow by suitably inclining the slide, start a circular motion with the point of the needle in the channel to widen it to a pool about 1 cm from the test drop. When most of the wash liquid has been collected in this pool, remove it by taking the drop up with filter paper. Inspect the area of the test drop under the microscope to make certain that the crystals of the test form have not been disturbed. Then add another large drop of 1-F NH_3, and remove the second wash liquid like the first.

If the liquid is so strongly repelled by the surface of the slide that it is impossible to draw a channel, proceed as follows. Cut strongly absorbing filter paper (drop test paper) into squares of 1-cm edge. While inclining the slide so that the liquid must flow upward to reach the paper, insert the corner of a square into the edge of the drop, Fig. 43a. Observe the precipitate and regulate the flow by proper inclining of the slide so that the crystals remain in place. When most of the wash liquid has been taken up by the paper, treat the moist residue with the second drop of dilute ammonia, which is then removed in like manner.

Finally, set the slide aside until wash liquid left with the precipitate has completely evaporated. Then focus crystals to be tested with low magnification, 20 to 80 diameters depending upon the size of the crystals. Using reflected light and a green or blue background, treat individual crystals with silver test solution; add the latter by means of a platinum hook or platinum loop while observing through the eyepiece. The crystals become opaque and change their color, but they usually retain their shape. The resulting pseudomorphs of Ag_3AsO_4 are reddish brown; Ag_3PO_4 is yellow.

If desired, repeat the experiment with phosphate test solution and magnesium acetate.

Experiment 41

Rubidium Chlorostannate (125, 168)

Rubidium chloride, granular.

Dilute 1 volume of stannic test solution (10 mg Sn per ml) with 4 volumes 3-F HCl, and transfer 1 to 2 μl of the mixture to a microscope slide. The diluting may be done by taking into a capillary pipet 1 loop (0.3 to 0.5 μl) of test solution and 4 loops (1.2 to 2 μl) HCl, taken from a drop on a slide, and finally emptying the pipet upon a slide.

Introduce a grain of RbCl of 0.6-mm diameter (0.1-μl volume) into the edge of the drop, and observe with transmitted light and a magnification of first 20 and later about 80 diameters.

A fine powder separates close to the reagent. Isometric crystals, predominantly octahedra and tetrahedra, form outside the area of rapid precipitation or, if only little tin is present, after some time near the edge of the drop. Because of lesser solubility, the corresponding cesium compound affords a more sensitive test. The crystals of Cs_2SnCl_6, however, are smaller and interferences are more to be feared since cesium has a greater tendency than rubidium to form insoluble compounds.

Experiment 42

Estimation of the Relative Quantities of the Metals in a Slurry Containing Arsenic, Antimony, Copper, and Silver

Slurry in a microcone. The slurry is prepared by placing into the cone first a measured volume (1 to 10 μl) of silver test solution (10 mg Ag per ml) and adding 20 μl of a solution obtained by taking measured volumes (1 to 10 ml) of arsenic, antimony, and copper stock solution (50 mg metal per ml), adding 25 ml 12-F HCl, and diluting with water to 100 ml.

Centrifuge; 6 microcones; 2 straight-tip cones with 2-mm bore; equipment for working with microcones including a heating block of the type shown in Fig. 28, distilling apparatus, Fig. 33, and supply of dry air. — Reagents used in Expts. 32, 35, 37, and 40; supply of hydrogen sulfide; ammonium sulfide, 7-F: saturate 14-F NH_3 with H_2S, and then add a like volume of the NH_3; ammonium chloride, 3-F; ammonium acetate, 6-F; hydrobromic acid, 9-F; KI, KCNS, and $KBrO_3$, granular.

Treat the slurry in the microcone with H_2S and separate the copper and arsenic groups as directed in P. 64 to 66; consider that the volumes given in these directions must be divided by one thousand when working on the milligram scale.

The hydrogen sulfide precipitate will probably contain a total of 200 to 600 μg of metals, and it might be compared with a sulfide precipitate obtained from 300 μg Hg. Reject supernate 2, P. 64.

Dissolve the copper group, residue 3, in 20 μl 3-F HNO_3, p. 401, separate the clear solution from any sulfur, and then treat it with 1-μl portions of 12-F HCl until all silver is precipitated. Use the volume of the AgCl for estimation of quantity. Lift off the supernate, treat it with an excess of NH_3, and estimate the amount of copper from the intensity of the blue color of the ammonia complex.

Treat the washed sulfides of the arsenic group according to P. 68, p. 410, "Dissolution of Sample" and "Distillation of Extract A 1". Treat the distillate with 1-μl portions of 3-F $NH_2OH \cdot HCl$ until it is colorless and then with H_2S. Use the yellow sulfide for estimation of the quantity of arsenic. Treat the residue of the distillation with 1 μl 3-F $NH_2OH \cdot HCl$, 20 μl H_2O, and finally with H_2S. Use the orange sulfide to estimate the quantity of antimony. Confirm the presence of Ag, Cu, As, and Sb.

The estimation of quantity is based upon the experience that like amounts of a substance give precipitates of nearly the same volume, provided that the precipitation is carried out with the same reagent under comparable conditions and the precipitates are collected and compacted by the same centrifugal force acting for the same length of time. In practice, solutions of known content may have to be precipitated after the precipitate has been obtained in the analysis of the unknown substance. It is possible, however, to postpone the whirling of the precipitate until the precipitates of known amounts have been prepared so that all precipitates to be compared may be centrifuged simultaneously. Experience (143) has also shown that like amounts of different metals of the hydrogen sulfide group give sulfide precipitates of nearly the same volume; it is thus possible to estimate the sum total of metals in a mixture of sulfides by comparing the volume with that of some sulfide precipitate having a known metal content and to crudely estimate the quantity of one metal by comparing the volume of its sulfide precipitate with that of the sulfide precipitate of another metal. It may be expected that a like reasoning may be applied to hydroxides, sulfates, silver halides, etc. Simplifying assumptions of this kind will be frequently permissible since crude sedimetric comparison will not give a relative precision better than $\pm$ 0.1 of the estimated quantity, which is more than sufficient, however, for distinguishing between majors, minors, and traces.

Estimation of quantity by colorimetric comparison with standards may be practiced to advantage whenever a stable coloration is obtained. Titrimetry, advocated by SWIFT (53), could be performed with the use of simple capillary pipets and a millimeter rule; the meniscus at the fine tip will serve for stopcock (90, 888).

Working Upon the Surface of a Slide

The performance of chemical separations upon a microscope slide was introduced by BEHRENS (115, 116) to extend the usefulness of the slide tests by the removal of interfering substances. He succeeded in performing complete qualitative analyses with the material under investigation being continuously upon a slide. His use of decantation by "dragging off" or "drawing off", described in Expt. 39, in place of filtration requires nicety of judgement based upon wide experience and may have been responsible for the fact that the technique never found general favor and that a search for more efficient techniques started about 1900. At this time it may be stated, however, that the technique of BEHRENS is still useful in its original form and in a great variety of modifications. The principle of working with the material under investigation clinging to a flat or slightly curved surface is used in some of the most sensitive procedures (944). The limits of identification are related to the areas actually occupied by the matter under investigation; the smaller the areas used, the better is the sensitivity.

The decantation technique of BEHRENS has been modified in many ways (118, 1070) and it may be greatly simplified whenever either only the liquid or only the solid is to be collected. Whether or not a clear filtrate is obtained depends upon the extent to which the solid adheres to the slide. Any form of decantation is impossible, of course, if the solid does not settle. On the other hand, solid matter clinging tenaciously to the slide may be washed by simply running solvent (water from the wash bottle or faucet) over the slide. Aside from these extremes, most solids divide themselves, part going into the decanted liquid and the washings and some remaining on the slide. BEHRENS found, however, that many solids can be made to behave by evaporating the mixture of solid and liquid to dryness; when afterwards the residue is extracted, the insoluble solid adheres to the slide in a degree which makes it possible to obtain a clean separation by decantation.

A variation of the technique has been used by KOFLER (159, 607) for the purification of organic substances by heating and the removal of the eutectic melt. The melting point of the eutectic mixture is determined by a preliminary experiment, in which a small amount of the substance is heated on the hot stage under the microscope. Then, a square of 1-cm edge of hardened filter paper (S. & S. No. 576 which is replacing No. 575 used by KOFLER) is put upon a slide resting on the hot stage. A thin layer of the substance is spread upon the paper, and a second slide is layed on top of it. The stage is heated until the eutectic temperature is exceeded by a few degrees, whereupon a downward pressure is put on the top slide (with the raser end of a pencil). The melt is taken up by the paper which becomes translucent in places. Usually, the solid residue sticks to the top slide, and the paper square may be replaced by a fresh one after lifting the top slide. The paper square is changed several times, and after each change, the temperature is slightly raised. In this manner, a satisfactorily pure substance may be obtained in 10 to 15 minutes.

Filterpaper may be used up to 300° C if each square is heated for a short time only, which will suffice for efficient work. As an alternative, one may heat the substance upon small squares, 18 mm × 18 mm × 1.5 mm thick, of porous clay (605, 890). *See* also Expt. 54.

Experiment 43

Conversion of Silver Chloride to Silver Dichromate (152, 154)
Handling Precipitates and Solutions, Fusion, and Electrolytic Reduction

Magnesium ribbon; potassium dichromate, granular; ammonium chromate, 2 g in 100 ml water.

Obtain narrow slides, about 8 mm by 75 mm, by cutting standard slides parallel to the long edge. If a glass cutter is not available, try the points of a triangular file. With most cutting tools, it is necessary to find by trial the position that gives a barely visible trace with a humming rather than scratching sound when going with a smooth motion and moderate pressure across the surface.

If the trace is properly drawn, the slide will readily break along the desired line when it is lightly tapped along it with the metal tool. Use the file to blunt the resulting sharp edges and corners.

The silver chloride is collected, washed, dried, and fused; the solidified melt is reduced with metallic magnesium. The silver sponge is washed free of chloride and then dissolved in nitric acid for testing with chromate.

From a measuring pipet, deposit 0.05 ml of silver test solution near the end of a narrow slide. Place a large drop of 12-F HCl upon another (standard) slide, and transfer small portions of this acid with a platinum loop to the drop of silver test solution until all silver is precipitated as AgCl, i. e., until adding another portion of acid no longer causes further

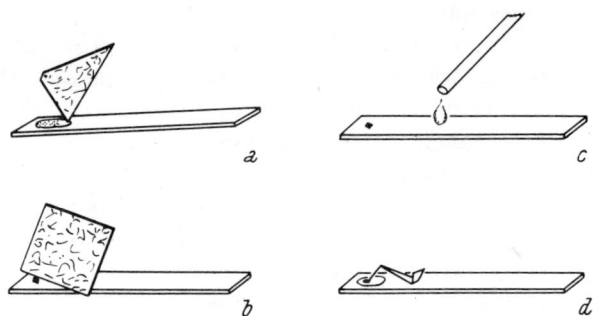

Fig. 43. Working upon the Slide; $^1/_2$ nat. size

precipitation. It is understood that the mixture is stirred with the loop after each addition of acid.

Removal of Solution and Washing of Precipitate. Use filter paper or drop test paper cut into squares of about 2-cm edge. Remove the major part of the liquid as described in Expt. 40 and shown by Fig. 43a. Then, with the dry edge of a fresh piece of paper, scrape the precipitate together into a compact little tablet, Fig. 43b. Whenever the edge of the paper becomes moist and soft, take another part of the edge, which is still dry and stiff.

For washing, place a large drop of water upon the slide near the precipitate, Fig. 43c. Then tilt the slide so that the drop flows over the precipitate. Do not stir, but allow to stand for a minute; then remove the wash liquid with filter paper and dry out the tablet of precipitate as before. Repeat the washing once, and do it so that finally the precipitate is collected into a small tablet near the end of the slide. The technique of collecting the precipitate is simple, but rather wasteful. Obviously, it is not suited for the collection of finely powdered solids.

Drying and Fusing. Hold the slide with the precipitate a few centimeters above a microflame, and perform a slight lateral oscillatory motion which ensures the heating of the entire width of the slide and prevents its cracking.

When the precipitate appears dry, lower the slide until it nearly touches the microflame. Heat until a clear drop of fused AgCl is obtained (m. pt. 450° C). Then place the slide upon an asbestos board or upon wire gauze, and allow it to cool to room temperature.

Reduction to Metallic Silver. The fused AgCl adheres tenaciously to the slide, which simplifies the following operation. Place a large drop of 1-F acetic acid upon the AgCl. Cut to a point one end of a magnesium ribbon of 3-cm length; bend the ribbon into the form of a Z, and place it upon the slide so that the pointed end touches the AgCl, Fig. 43d. The liberation of hydrogen starts immediately. Remove the magnesium ribbon when the reduction has become complete, which is usually indicated by the phenomenon that the spongy metallic silver does no longer adhere to the slide and floats to the surface of the drop.

Removal of Solution and Washing the Silver. The metallic silver adheres to paper, but there is no reason why it should adhere to the point of a pipet rather than to the surface of the slide. Consequently, remove the solution and the washings by means of a capillary pipet, but otherwise use the procedure of decanting as when working with the squares of paper. The tip of the pipet is too narrow to permit passage of the rather coarse silver particles. Regulate the intake of liquid by properly inclining the capillary pipet and the slide. For the practically complete removal of chloride, wash twice with water; use a large drop each time. Use the same capillary pipet for the removal of all liquids, and reject filtrate and washings.

Dissolving the Metal and Adjusting the Acidity. Warm the slide over a small Bunsen flame until it is completely dry. Then add to the silver from a capillary pipet 2 μl 16-F HNO$_3$, and warm the mixture to start the reaction of the acid with the metal. The dissolution requires seconds only. Warm the slide and blow upon the droplet until the solution is evaporated just to dryness. Allow the slide to cool to room temperature.

Confirming the Presence of Silver. Dissolve the residue of AgNO$_3$ in 5 to 10 μl of distilled water. Take up the solution into the capillary pipet in which the water has been measured; transfer half of the solution to a (standard) microscope slide, and use the rest for the spot test of Expt. 25. Warm the slide so that the drop of silver solution evaporates. After cooling to room temperature, dissolve the residue in 2.5 to 5 μl 2-F HNO$_3$ and test by adding K$_2$Cr$_2$O$_7$, Expt. 31.

Experiment 44
Separation of Bismuth and Lead (1143)
Evaporation, Extraction of "Invisible" Residues

Bi : Pb = 1 : 100 000; Pb : Bi = 1 : 100 or better.

Phillips beaker, 250-ml; watch glass, 9 cm. — Solution containing 0.1 mg Bi and 1 mg Pb per milliliter: dilute 1 ml of Bi-Pb test solution (5 mg Bi and

50 mg Pb per ml) with 3-F HNO$_3$ to 50 ml, and mix well. — Potassium iodide paper; quinine iodide reagent; stannite reagent; triple nitrite reagent; potassium cobalticyanide (i. e. the reagents used in Expts. 24, 26, 34, and 36).

Bismuth and lead are completely separated by repeated evaporation of their nitrates with water. Bismuth is hydrolized to the water-insoluble basic nitrate BiO · NO$_3$. Lead nitrate is extracted from the final residue with water, dilute ammonium acetate, or ammonium nitrate solution (750, 1250).

Take 1 μl of solution containing 0.1 mg Bi and 1 mg Pb per milliliter if slide tests are to be used for final confirmation; take 5 μl, if spot tests shall be used. Measure the volume in a calibrated capillary pipet, and

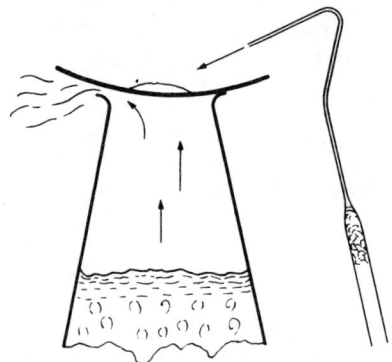

Fig. 44. Evaporation upon the Steam Bath; $^1/_2$ nat. size

transfer the solution to the center of a clean watch glass so that the residue will occupy a very small area; proceed as described below.

Evaporation Upon the Steam Bath. Place into the Phillips beaker, Fig. 44, a few granules of metallic zinc and about 100 ml water. Heat on the wire gauze to even boiling. Place the watch glass upon the beaker so that the steam escapes through the lip, and blow filtered air from a capillary toward the center of the watch glass to hasten evaporation.

Deposit upon the center of the hot watch glass about 0.5 μl solution at one time, and allow to evaporate to dryness before adding the next portion to the residue. Continue until all of the solution has been evaporated in a small spot. The residue clings to the glass, and caution may be relaxed during the following evaporations with water.

Without removing the watch glass from the steam bath, add 10 to 20 μl water to the evaporation residue without touching the latter with the tip of the pipet. As a rule, the whole residue will be covered by the large drop of water; if this should not happen by itself, spread the drop with a glass thread. When the first drop has completely evaporated, repeat the treatment four more times. This requires only a few minutes.

Finally, remove the watch glass from the steam bath, wipe its underside dry, and allow it to acquire room temperature.

Extraction of Residue. Inspection will show a relatively copious residue upon the watch glass, whereas the total quantity of lead and bismuth salts is less than 2 μg. The residue consists mainly of substances extracted from the glass apparatus and of impurities in the reagents. Exclusive use of vitreous silica and platinum in the experiment and in the preparation and storing of the reagents would be required to reduce the amount of residue to the theoretical size.

Whether or not a residue is visible, treat the area which was originally occupied by the test drop with 5 to 10 μl distilled water. If necessary, spread the water with a glass thread, but do not loosen the residue. After standing for 3 minutes, take up the clear extract into a capillary pipet.

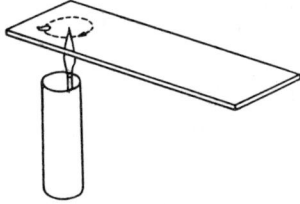

Fig. 45. Evaporation upon the Slide; $1/2$ nat. size

Fractional Evaporation. To obtain the residue in as small an area as possible, transfer 0.5-μl portions of the extract to a clean microscope slide and, each time, evaporate to dryness before adding the next portion to the residue. For evaporating, heat the slide over a microflame, Fig. 45, while moving the slide in a horizontal plane so that the drop performes a circular motion around the point of the flame. This causes the slide to be heated in a circular area around the drop, and creeping of the drop is prevented. Remove the slide from the flame when the drop begins to vanish. Complete the evaporation by blowing upon the slide, and then add the next fraction while the slide is still hot. Continue in this way until the extract has been used up.

Repeat the extraction of the residue upon the watch glass with two more 5- to 10-μl portions of distilled water. Using always the same capillary pipet, concentrate all extracts in the small area upon the microscope slide.

Confirmation of Lead. To perform the triple nitrite test, proceed with the evaporation residue upon the slide as directed in Expt. 34. If a spot test shall be used, dissolve this residue in 1 μl water, take up the solution into a capillary pipet, and transfer it to potassium iodide paper, Expt. 24.

Dissolution of Basic Salt and Test for Bismuth. Treat the residue upon the watch glass with 10 μl 6-F HNO$_3$. Use a glass thread to spread the acid over the whole area treated during the extractions with water. Finally, collect the solution in a capillary pipet, and transfer it in 0.5 μl fractions to a slide for evaporation in a small area. Evaporate just to dryness and, after cooling to room temperature, cover the small residue with 0.5 μl or less 2-F HNO$_3$. To perform a slide test, without delay add a grain of potassium cobalticyanide of 0.5-mm diameter, Expt. 36. Otherwise take up the solution into a capillary pipet and proceed according to Expt. 26.

Experiment 45

Separation of Silver, Lead, and Mercurous Mercury (168), *Sublimation, Extraction with Boiling Solutions*

Narrow slides. — Solution containing 10 mg each of silver, lead, and mercurous ions per milliliter solution: mix 2 ml each of the corresponding stock solutions (50 mg ion per ml) and dilute with 1-F HNO$_3$ to 10 ml. — Potassium iodide paper; nitrite reagent; potassium chromate, 2% solution; sodium chloride, 1% solution; potassium cyanide, granular.

Silver, lead, and mercurous mercury are precipitated as the chlorides and washed to remove the nitrate ion. Mercurous chloride is isolated by sublimation, and lead chloride is extracted from the residue with boiling dilute acid. Boiling water is not suited since it does not dissolve basic chloride which sometimes forms during the sublimation.

Precipitation of the Chlorides. With a measuring pipet transfer about 0.1 ml of silver-lead-mercurous solution (10 mg of each ion per ml) to the end of a narrow slide. Deposit 20 μl 12-F HCl close to the test drop and, with a glass thread, combine the two drops and stir until the white precipitate flocculates.

Remove the solution with filter paper as in Expt. 43. Wash the precipitate with two 0.02-ml portions of 2-F HCl, and collect it as a small tablet near the end of the slide. Reject filtrate and washings.

Sublimation from Slide to Slide. Hold the narrow slide with the washed precipitate in the left hand and a microscope slide of standard dimensions in the right. Rest the latter on the edge of the former so that the two form an angle of 5 to 10 degrees and enclose a narrow wedge-shaped air space, Fig. 46a. Briefly heat the precipitate by holding it about 2 cm above the point of a microflame. Every few seconds, remove the narrow slide from the flame and, along the edge, move the microscope slide forward over the precipitate and hold it there until the amount of condensate does no longer increase, Fig. 46b. It should be understood that the angle between the slides and the wedge-shaped air space are maintained when moving the upper slide back-and-forth. When condensation ceases, move the upper slide back into its former position, Fig. 46a, and again heat the precipitate for a short time.

A quickly evaporating condensate of water will be obtained in the first two or three tests. When no more water condenses, assume that the residue has become dry and begin the sublimation. For the brief periods of heating, decrease the distance between precipitate and tip of flame to 5 mm. The vapor usually cannot be seen. Thus, test at intervals of a few seconds if vapor is given off by moving the large slide over the heated substance. The white sublimate fills a circular area above the heated precipitate. When condensation ceases, draw the top slide back and heat the precipitate some more, etc.

Collect the whole sublimate in the same spot upon the microscope slide and in as small an area as possible. Finally prove the completeness of the separation by further heating of the precipitate and then holding over it

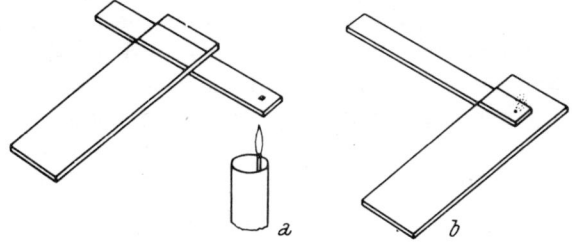

Fig. 46. Sublimation from Slide to Slide; about $1/2$ nat. size

a fresh portion of the microscope slide. If no sublimate is obtained, assume that all Hg_2Cl_2 has been removed from the precipitate.

A temperature of 300° C suffices for the volatilization of Hg_2Cl_2. Excessive heating will cause the residue to melt and will render difficult the subsequent extraction of $PbCl_2$.

Identification of Mercurous Chloride. Place the slide upon a sheet of white paper, and use a platinum loop to place a droplet of conc. NH_3 upon part of the sublimate. The moistened portion turns black. This test will not succeed if the HNO_3 is not sufficiently removed and is able to oxidize Hg_2Cl_2 to $HgCl_2$ when the precipitate is heated for sublimation.

Extraction with Hot Solvent. The heated precipitate clings to the glass, and this makes it possible to use the following simple technique. Treat the residue upon the narrow slide with 0.05 ml 2-F HCl. Hold ready in the right hand a capillary pipet with relatively wide tip for the quick removal of the hot extract. Grasp the narrow slide with the left hand and hold the residue with the drop of acid about 3 cm above the tip of the microflame. When steam bubbles just start to appear in the drop, remove from the flame and quickly insert the tip of the capillary pipet so that the whole hot solution is quickly taken into the latter.

Identification of Lead. Evaporate the HCl extract in small portions on a slide to obtain the residue in a small area, Expt. 44. Perform the triple nitrite test directly with the residue, Expt. 34, or dissolve the residue by heating close to boiling with about 20 μl water; take the hot solution into a capillary pipet, and use it for the potassium iodide test, Expt. 33.

Identification of Silver. From a capillary pipet, add 5 μl 12-F NH$_3$ to the residue from the extraction on the narrow slide. Without delay, stir with a glass thread, and then take the clear solution back into the capillary pipet and transfer it to a clean slide. Cover the test drop and continue as outlined in Expt. 32 or (and) 43.

To obtain a spot test, dry the AgCl residue upon the narrow slide by warming over a flame, and then heat until the salt melts. After cooling, add a grain of KCN about twice the volume of the AgCl, and heat over a small Bunsen flame until the KCN melts and the reduction of the AgCl is complete. The metallic silver adheres firmly to the slide. Wash it free from cyanide and chloride by running distilled water from the wash bottle over the slightly inclined slide. After drying by holding the slide over a flame and cooling to room temperature, dissolve the silver by warming it lightly with 5 μl 16-F HNO$_3$. Evaporate just to dryness, and dissolve the residue in 5 to 10 μl water. Use the solution for the test of Expt. 25.

Experiment 46

Test for Ammonium Ion, Use of the Gas Reaction Cell (149)

L. I., 0.1 μg NH$_4$.

Glass ring, 25-mm outer diameter, 5 mm high, polished at top and bottom. For substitutes may serve: crucible of 0.5- to 1-ml capacity; bottom part of wide vial, cut 5 mm above the bottom and polished flat; screw cap of plastic. — Sealed capillary, 5-mm long, containing iron wire, p. 97; strong permanent magnet (Alnico or Cunife). — Chloroplatinic acid, 5%: keep in vial of plastic or of vitreous silica and test from time to time by allowing 5 μl of it to evaporate at room temperature in a desiccator for the microscopical inspection of the residue.

The test substance is treated with sodium hydroxide. The liberated ammonia is absorbed by a solution of chloroplatinic acid and precipitates ammonium chloroplatinate. The test is performed in a completely closed gas reaction cell which is assembled from two standard microscope slides and the glass ring.

Assemble the cell of Fig. 47 and use a strip of writing paper (width nearly equal the height of the ring, and length nearly equal to the diameter of the ring), folded to give an angle of 90 degrees, to divide the chamber into two compartments. Cellophane may be used in place of paper, or the paper may be impregnated with paraffin wax if one wishes to exclude adsorption on the paper.

Deposit in the smaller compartment on the floor of the cell 1 μl of ammonium test solution (10 mg NH_4 per ml) which has been diluted with nine volumes of water and next to it 1 μl 6-F NaOH. Place the capillary with the iron core into the same compartment. Deposit 1 μl H_2PtCl_6 solution upon the top slide so that it hangs on the ceiling of the larger compartment of the cell. Place the assembled cell upon the stage of the microscope and focus through the top slide upon the drop of chloroplatinate. Observe for a period of 5 minutes and ascertain that the drop remains clear.

Without opening it, remove the cell from the stage, and by applying the magnet from below the bottom slide, wipe together the droplet of NaOH and the test drop. Return the cell to the stage of the microscope and again focus upon the drop of chloroplatinate. Use a magnification of 30 to

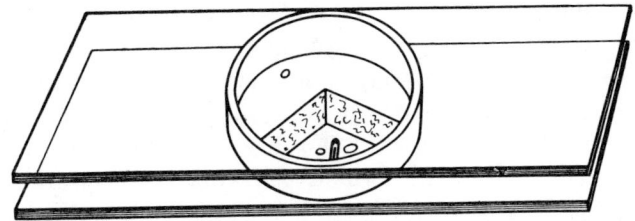

Fig. 47. Cell for Gas Reactions; about nat. size

100 diameters for observing the yellow tetrahedra and octahedra of $(NH_4)_2PtCl_6$. If a large number of very small crystals is formed, exchange the top slide for another one with a fresh droplet of reagent; repeat this until the NH_3 concentration in the chamber has become low enough to make possible the slow separation of large crystals.

Instead of chloroplatinate, one may use tiny fragments of test papers, moistened and pasted to the ceiling of the cell. Suitable are red litmus paper, pH test paper (Hydrion paper for the range 7 to 10), and filter paper dipped into Nessler reagent. The limit of identification will improve as the size of the test papers decreases; squares or triangles of 1-mm edge are convenient, but a few fibers suffice if the color may be observed with magnifiers or microscope.

Working in Capillaries

The technique of working in glass capillaries of 0.3- to 1-mm bore was developed by EMICH and his coworkers (154, 1007, 1010, 1016, 1139, 1141). It may be used for the preparation of derivatives of organic substances on a milligram scale; in inorganic analysis, it is advantageous for the performance of certain tests with about 1 to 100 μg of solid material and a few microliters of solvent. In addition, it is often useful as an adjunct to the technique of working upon the microscope slide and in the microcone. The relatively large wall surface of capillaries several centimeters long, however, presents disadvantages in the performance of a sequence of separations and, for this reason, lengthy analytical procedures have never been performed in capillaries of these dimensions.

In the performance of confirmatory tests on the submilligram scale, the capillary takes the place of the test tube of the gram scale.

Experiment 47

Lead Sulfate, Triple Nitrite, Lead Chromate
The Capillary as Adjunct to Working Upon the Microscope Slide

Nitrite reagent and 2% ammonium chromate solution as used in Expts. 25 and 34; ammonium acetate, solid.

Place 5 μl of lead test solution (10 mg Pb per ml) upon a slide, and deposit a large drop of 4-F H_2SO_4 upon another. With a platinum loop, transfer portions of the acid to the drop of lead solution until the precipitation is complete. Note that the $PbSO_4$ does not adhere to the slide and that decantation upon the slide would be difficult to perform.

Transfer of Solid as a Slurry. Cut off squarely both ends of a capillary of about 0.5-mm bore and 10-cm length. Close one of its ends with the index finger, and with the other end stir the mixture upon the slide into a slurry. When a rather uniform mixture has been obtained, incline slide and capillary, and lift the index finger off the top of the latter so that the slurry enters quickly into the tube.

Separating the Phases in the Capillary. By means of a microflame, fuse the capillary into a bead at a point 3 cm from its dry end, Fig. 48a. When the bead cools, contraction of the air inside causes the slurry to go further into the capillary; Fig. 48b.

Cut off all except 1 cm of the empty capillary beyond the glass bead, and place the capillary, bead downward, into a microcone. After swirling in the centrifuge, cut the capillary at the boundary line, Fig. 48c, so that one piece contains the clear solution and the other just the precipitate. In this instance, reject the solution which is not needed.

Transferring the Precipitate to a Slide. If the triple nitrite test is to be used, place upon a microscope slide 5 μl of copper test solution (10 mg Cu per ml). Grasp the empty part of the capillary with fingers or forceps, and dip the end containing the $PbSO_4$ into the copper solution. Without lifting it out of the drop, repeatedly tap the end of the capillary against the slide. The $PbSO_4$ usually leaves the capillary as one lump, Fig. 48d, if the bore of the tube tapers bluntly at the bead and care is taken that no air bubble forms in the capillary. Sometimes it is necessary to loosen the precipitate in the capillary by stirring with a fine glass thread. To complete the experiment, evaporate the mixture of lead sulfate and copper solution to dryness, and treat the residue with nitrite reagent as directed in Expt. 34. Note that the slow dissolution of the PbSO$_4$ obviously favors the separation of relatively large crystals of the triple nitrite.

If the experiment shall be concluded with a spot test, place 5 μl water upon a slide, and transfer the precipitate from the capillary to this drop as outlined above. Treat the mixture of PbSO$_4$ and water with a crystal of ammonium acetate about 1.5 mm in diameter (1 μl volume), and aid the dissolution of the PbSO$_4$ by stirring with a glass thread. Add to the clear solution 1 μl 6-F acetic acid and take it into a capillary pipet for performance of a chromate test as described for silver in Expt. 25.

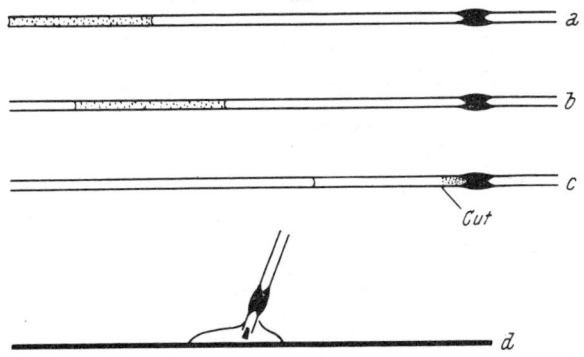

Fig. 48. Separation of Solution and Precipitate; about nat. size

Experiment 48

Recrystallization of Lead Iodide (154)

Place upon a slide so that they do not mix one large drop each of lead test solution, iodide test solution (10 mg ion per ml), and 2-F acetic acid. With a capillary pipet of 0.5-mm bore in the wide part, take up portions of the three solutions in the following order: first a 5-mm length of lead solution, then a 6-mm length of iodide solution, and finally a 20-mm length of acetic acid.

Sealing Capillaries. Use a microflame, pilot flame, or the edge of a Bunsen flame. About 6 cm from the drop, draw out a fine capillary, Fig. 49a, but do this without removing the tube from the flame so that this fine capillary is fused shut, Fig. 49b, while it is drawn. When the sealed end cools, the solution is drawn into the tube, Fig. 49c. Break the now empty tip close to the taper, and seal the opening by touching it to the edge of the flame, Fig. 49d. Inspect this second seal under the microscope.

Mixing. Cut open the end of the capillary, that has been sealed first, and thoroughly mix the contents with a fine glass thread with a bead at the end. Again draw out and seal the capillary near its opening. Repeated opening and sealing a capillary will not shorten it much if, before sealing, a thin glass rod is fused to the opening of the capillary to serve for a handle.

Heating Liquids in Capillaries. Place the capillary into a test tube, and add water to submerse the whole capillary. Heat the water in the test

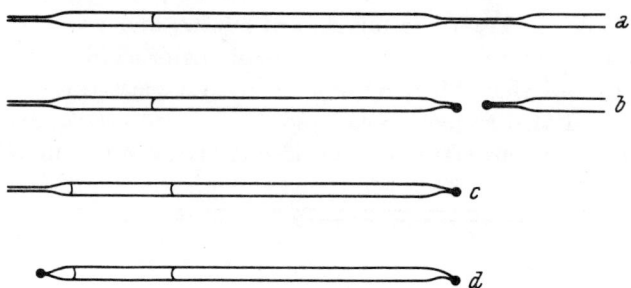

Fig. 49. Sealing a Capillary. The bore is exaggerated

tube and keep it boiling until all of the PbI_2 in the capillary has dissolved. Then set the test tube with its contents aside, and allow to cool slowly to room temperature.

Examination of the Contents of a Capillary. Pour the water out of the test tube. Then remove the capillary, place it upon a slide, and examine

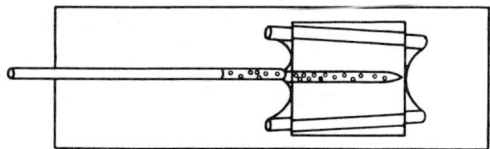

Fig. 50. Cell for Use in Examining the Contents of a Capillary. The bore of the capillary is exaggerated

with magnifying glass and under a microscope with both transmitted and reflected light. Use low magnification and improve the optical conditions by immersing the capillary in water. A suitable cell may be improvised with two glass rods of 2-mm diameter and a cover slip, Fig. 50.

Finally, cut the capillary open at both ends, and blow out its contents upon a slide for additional examination: measurement of the angles of the hexagonal plates, testing in polarized light.

Experiment 49

Isolation of Metallic Mercury, Conversion to Iodide (1140)

Copper wire, 0.1 mm diameter; iodine, powder; ammonium oxalate, saturated solution in water.

Expt. 49 Working in Capillaries 163

Metallic mercury is precipitated by inserting copper metal into the solution which has been treated with ammonium oxalate to assure a satisfactory deposit (651). The amalgamated copper is heated, and a condensate of droplets of mercury is obtained. Exposure to iodine vapors gives conversion of the metal to mercuric iodide.

Into a capillary pipet having 0.5-mm bore in the wide part, take up 1 μl mercuric test solution (10 mg per ml) and 3 μl ammonium oxalate solution. If a microscope is not available for the inspection of the test, take ten times larger amounts. Seal the capillary as outlined in the preceding experiment and illustrated by Fig. 49.

Cut open the end of the capillary, which was sealed first, and introduce with forceps a 2-mm length of clean, bright copper wire. Place the capillary, sealed end down, into a cone and centrifuge briefly; this collects wire and

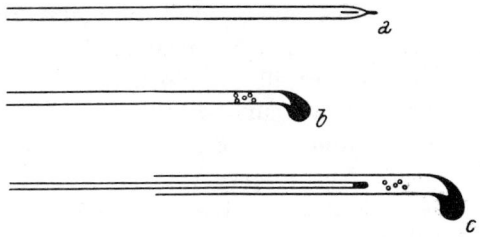

Fig. 51. Distillation of Mercury. In c the bore of the capillary is greatly exaggerated

solution at the sealed end. Draw out and fuse shut the open end, and heat the sealed capillary for at least 1 minute in a steam bath or a test tube with boiling water. Finally, lay the capillary upon filter paper, and cut it so that the copper wire becomes accessible.

Wash the wire upon the filter paper by adding a drop of water from the wash bottle. When the water has been absorbed by the paper, pick up the wire with clean forceps, and transfer it to dry filter paper. Remove all moisture by gently pressing the wire between sheets of the paper.

Distillation of Mercury. The capillary must not contain any moisture. Thus, prepare it from a freshly drawn capillary of 0.5-mm bore and 15-cm length. To prevent flame gases from entering the tube, heat and fuse shut the *center* portion. This gives two dry capillaries which are sealed at one end. With forceps, introduce the amalgamated copper wire into one of them, and bring it down to the sealed end by tapping with the capillary on the bench top or by centrifuging, Fig. 51 a.

For the distillation, heat the sealed end of the capillary in the edge of a non-luminous Bunsen flame so that the rest of the tube remains cool. Continue heating until a bead of glass has completely enclosed the copper metal, Fig. 51 b.

11*

Examination of Distillate. Place the capillary upon a microscope slide, and use reflected light and a magnification of 20 to 50 diameters. The mercury condenses in small drops near the heated end. Thus, first focus upon the glass bead, and then search the whole length of the capillary. The silvery color of the metal and the mirror action of the surface of the drops are clearly visible with a black background. When the droplets have been found, observe them also with transmitted light. They will appear as black circular disks when the hands are cupped around the stage to exclude light coming from above. The directions apply, in a general way, when a magnifying glass is used for examination. Holding the capillary in front of a lamp gives the effect of observation with transmitted light; for observation with reflected light, lay the capillary on dull, black paper or cloth.

Conversion to Iodide. Push one end of a capillary of about 0.3-mm outer diameter into the iodine contained in the vial so that some of it gets wedged into the bore of the fine tube. A very small amount does suffice (1272). Wipe clean the outside of the capillary, and then insert it into the capillary containing the distillate of mercury so that the iodine approaches the droplets within a few millimeters, Fig. 51c. Without delay, place the combination upon a slide and observe with reflected light at intervals of several minutes. Depending upon the temperature, the conversion to yellow and orange HgI_2 proceeds more or less slowly. The droplets become coated with a film of the iodide. At times, the coating bursts and liquid mercury is exuded to acquire a film of iodide that assumes the shape of strings and sausages.

Experiment 50

Bettendorff's Test for Arsenic (1009)

Stannous chloride reagent: dissolve 11.5 g $SnCl_2 \cdot 2\ H_2O$ in 17 ml 12-F HCl and dilute with water to 100 ml; place some metallic tin into each bottle containing this reagent. — Isoamyl alcohol, b. pt. 130° C.

Arsenate and arsenite are reduced in strongly acid solution to elemental arsenic; selenium, tellurium, mercury, and the noble metals interfere since they too are reduced to the elemental state.

Take up 1 μl arsenic test solution (10 mg As per ml) into a capillary pipet of 0.5-mm bore in the wide part. Deposit drops of stannous chloride reagent and 12-F HCl upon a slide, and from there take into the capillary pipet about 1 μl $SnCl_2$ reagent and 6 μl HCl. Estimate these volumes from the length of capillary filled by the 1-μl drop of arsenic solution. Take the HCl into the pipet last because it does not contain solid matter and is, therefore, suited for rinsing the tip previous to sealing.

Seal both ends of the capillary pipet by the standard procedure described in Expt. 48 and illustrated by Fig. 49.

Mixing. After inspecting the last seal under the microscope, mix the contents of the capillary by centrifuging them two times from one end of the capillary to the other.

Heating. Place the capillary into a clean, dry test tube, and add about 2 ml of amyl alcohol. Boil the amyl alcohol over a small Bunsen flame for 3 minutes so that the ring of condensate remains about 5 cm below the opening of the test tube while the capillary is entirely immersed in the boiling liquid and its vapor. After heating, set the test tube and contents aside for 2 minutes. Then pour the amyl alcohol back into the reagent bottle, and remove the capillary from the test tube.

Examination of Precipitate. With the centrifuge, collect the precipitate at one end of the capillary, preferably the one which tapers to a finer point. Then place the capillary upon a slide, and inspect its contents with reflected light before a white background and a magnification of 20 to 30 diameters. Immerse the capillary in water to better the optical conditions, Fig. 50. The arsenic separates either in the black or in the brown form. Save the capillary with the precipitate for the next experiment.

Experiment 51

Oxidation of Arsenic to Arsenic Acid
Carius' Treatment in Capillaries (1141)

Goggles or safety shield. — Buffered silver solution as for Expt. 29 or magnesium acetate tetrahydrate, granular, as for Expt. 40. — Capillary with arsenic precipitate obtained in Expt. 50.

The elemental arsenic is oxidized to H_3AsO_4 by heating with concentrated nitric acid in a sealed tube.

Separation of Precipitate and Solution. Cut open the empty end of the capillary containing the arsenic precipitate. If necessary, whirl again in the centrifuge to collect the arsenic as a compact plug in the point of the tube. For the removal of the liquid, either use a suction-operated siphon, Fig. 32a, with the intake arm drawn out to a capillary of 0.2-mm outer diameter, or use a so-called "contraction pipet", Fig 52 (401).

The latter is quickly made from one of the pipets with elongated bulbs obtained as a by-product of the drawing of capillaries from glass tubing. For lifting off the liquid, insert the fine tip a into the capillary containing the arsenic, but do not yet immerse the orifice of a into the liquid. Now, heat the bulb c of the pipet with a small Bunsen flame to 200° or 300° C. Remove the flame, and push the tip of the pipet into the liquid while the bulb is still hot. The liquid is drawn into the pipet as the air in bulb c cools. Advance the pipet into the capillary as the meniscus of the liquid recedes until the orifice of the tip gets close to the precipitate. When air begins to enter the tip of the pipet, withdraw the latter from the capillary and heat bulb c again to expel the contents of b and a. In this instance, the centrifugate is rejected, but it may be received in another capillary, in a microcone, or on a slide.

Since it cannot be avoided that a considerable amount of liquid remains behind on the surface of the capillary, centrifuge the latter and remove the collected liquid as before.

Washing Precipitates in Capillaries. Place a large drop of water upon a slide. Heat the empty part of the capillary containing the arsenic by drawing it through a Bunsen flame, and quickly insert the opening of the capillary into the drop of water. The latter enters the capillary as the glass cools. Transfer the water to the sealed end of the tube by brief swirling in the centrifuge, and then mix thoroughly by means of a glass thread with a small bead at the end. Centrifuge again, and remove the wash liquid in the same manner as the centrifugate. Repeat the washing once, and take special care to remove the second wash liquid as completely as possible since it would dilute the HNO_3 added in the next step.

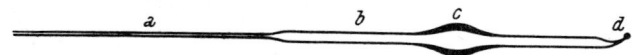

Fig. 52. Contraction Pipet. The bore of parts a and b is exaggerated

Oxidation According to Carius. Place a large drop of 16-F HNO_3 upon a slide. Briefly heat the empty part of the capillary containing the arsenic, and insert the opening of the hot capillary into the drop of HNO_3. Allow not more than 2 μl of the HNO_3 to enter the tube. Bring the acid to the arsenic at the sealed end of the tube by whirling in the centrifuge, and then seal the tube by drawing out just below the opening and fusing shut, Fig. 49. Inspect the seal under the microscope.

For heating, place the sealed capillary into a clean, dry test tube of Pyrex glass. To avoid serious consequences in the event of an explosion, use goggles or a safety shield and keep the test tube pointed toward a wall as long as the capillary remains sealed. Hold the test tube in a nearly horizontal position. Heat the part containing the capillary by moving the test tube backwards and forwards over a non-luminous Bunsen flame, 2 cm high, so that the entire length of the capillary is uniformly warmed. Do not heat too strongly; the temperature need not exceed 250° C and the HNO_3 should not completely evaporate at any time. Uniform heating keeps the HNO_3 at the end of the capillary and in contact with the arsenic.

When the solid is no longer visible, allow the test tube to cool to room temperature. Keep the eyes protected until the tube is open. First let the capillary slide into a centrifuge cone, and whirl it to collect the contents at one end. Then remove it from the cone, and cut it open in the middle.

Testing for Arsenic Acid. Take the liquid contents of the capillary into a contraction pipet, and transfer them to the center of a 9-cm watch glass for evaporation on the steam bath.

To perform a slide test, dissolve the residue in 5 μl 2-F HNO$_3$. Transfer the solution to a microscope slide and treat with magnesium acetate as directed in Expt. 40.

To get a spot test, dissolve the residue in 2 μl 0.5-F HNO$_3$. Take the solution into a capillary pipet and proceed as in Expt. 29.

Experiment 52

Conversion of Aniline to Acetanilide (1010)

Melting point apparatus consisting of beaker with bath liquid, stirrer, and thermometer, 250° C; supply of air under low pressure. — Aniline, purified; anisol; benzene, reagent grade; carbon; asbestos for Gooch crucibles.

Aniline is heated with acetic acid in a sealed tube. The resulting acetanilide is washed with water and recrystallized from benzene until the melting point remains constant. To conserve material, the procedure is arranged so that a change of container is not required. Capillaries are able to withstand high pressure; consequently, procedures involving the handling of substances with high vapor tension are advantageously performed in sealed capillaries, which simplifies heating of reaction mixtures and working with boiling solvents so that the customarily used reflux condensers are not needed. Explosions occur only rarely if unreasonably high temperatures are avoided. They are quite harmless since the exploding capillary, as a rule, is unable to destroy the surrounding test tube.

Heating the Reaction Mixture. Draw out a fine tip at one end of a capillary of 1- to 1.5-mm bore and 12-cm length. Take into the capillary 2 μl aniline and 3 μl glacial acetic acid; measure the volumes from a calibrated capillary pipet into a microcone, and then take the mixture into the wide capillary. Seal both ends of the capillary by the standard procedure, Fig. 49, inspect the seals under the microscope, and then heat the capillary 15 minutes at 150° C. To this end, either place the capillary into a drying oven, or use a bath of anisol (b. pt. 154° C) and the technique described in Expt. 50; the test tube may be provided with a cork stopper carrying a glass tube of 20-cm length, which will serve for reflux condenser. Finally, remove the capillary from the bath and, if necessary, collect its contents at one end by means of the centrifuge. Cut it open at the empty end.

Washing and Drying in the Capillary. From the bore of the capillary compute the length of it, that holds 10 μl. Introduce 40 μl water either by means of a capillary pipet with long fine tip, or by drawing the empty part of the tube through a Bunsen flame and quickly inserting the opening into a drop of water. Whirl in the centrifuge to bring the water to the sealed end. Stir and mix with a glass thread having a small bead at the end. This causes crystallization of the acetanilide. Mix, centrifuge, and remove the aqueous solution with a suction-operated siphon or a contraction pipet, Expt. 51. Centrifuge a second time, and again remove the collected liquid. Repeat the washing twice, and use each time 30 μl water. Test the removed wash liquids with pH test paper.

For drying the acetanilide in the capillary, clamp the finely drawn out tube shown in Fig. 23 to the stand holding the melting point apparatus, and connect it to a supply of clean air, p. 97. Place a wad of cotton into the wide part, and bend the fine capillary about 4 cm from the orifice through an angle of about 10 degrees, just sufficient to make it act as a spring when it is inserted into the capillary containing the acetanilide. Insert it far enough so that the orifice of the fine capillary is only a few millimeters from the wet substance. Turn on the flow of air and insert the wide capillary into the melting point bath by suitable adjustment of the position of the fine capillary which holds the wider tube by its spring action, Fig. 53 a.

Slowly raise the temperature of the bath to 100° C and hold this temperature for 5 minutes. Then shut off the flow of air, and raise the temperature for the determination of the melting point which probably will be found below 114° C.

Treatment with Carbon and Recrystallization. First, determine the suitable amount of solvent. Introduce a volume of benzene about equal to that of the acetanilide. Whirl in the centrifuge to get the solvent to the substance, Fig. 53 b. Then fuse a glass thread to the open end of the capillary, draw out close to the opening, and seal at the same time, Fig. 53 c.

Place the capillary into an empty test tube which serves as air bath and as protective tube, compare Expt. 51. Hold the test tube in a nearly horizontal position, and heat the part containing the capillary by moving the tube backwards and forwards over a non-luminous Bunsen flame, 2 cm high, so that the entire length of the capillary is uniformly warmed. If solvent distils and condenses away from the solution, it may be brought back to the solution by a flip with the hand holding the test tube or by more strongly heating that part of the capillary in which condensation occurs. If the amount of solvent is correct, the acetanilide will completely dissolve in the hot benzene and crystallize when the solution cools, which may be hastened by sliding the capillary out of the test tube. Crystallization may be delayed by supersaturation, but may be started by further cooling; to this end, wrap the capillary in some cotton, and let ether drip on it.

If necessary, adjust the amount of benzene until testing shows that recrystallization may be performed with success. If too much benzene has been added, the excess may be evaporated by the technique used for drying the substance.

Finally, cut open the empty end of the capillary and introduce a small amount of carbon, Fig. 53 d. Seal again, and place the capillary with the freshly sealed end first into the test tube. Heat to dissolve the acetanilide and transfer the solution to the end containing the carbon. Do this by either flipping or heating, and complete the transfer by means of the centrifuge. In this manner, it is possible to treat the solution with adsorbent without getting the latter all over the tube (1270).

Again cut open the empty end of the capillary, and heat with a micro-flame 10 mm from the solution to obtain a constriction. With forceps,

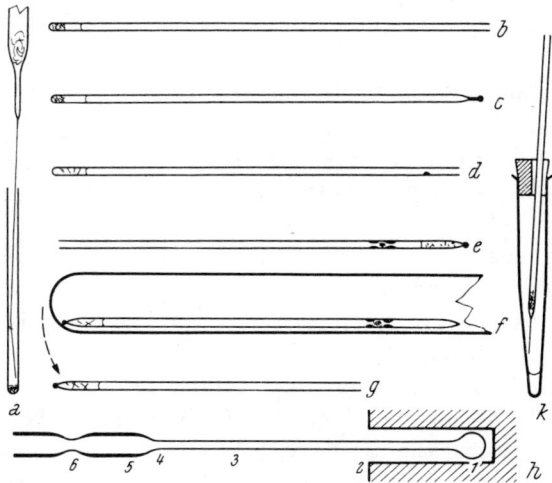

Fig. 53. Preparation of Derivatives in the Capillary

introduce some freshly ignited asbestos fibers into the capillary and, with a thin rod, push them against the constriction to form a filter mat of not more than 1- to 2-mm length. To hold the mat in place, make a constriction also on the other side of it, Fig. 53 e.

Seal the open end of the capillary, and place it with the freshly sealed end last into the test tube. Heat the whole length of the capillary to get complete dissolution and mixing of the solution with the coal. Use flipping and suitable heating to keep all solvent with the solution in the smaller compartment of the tube. Finally, allow to cool, reverse the position of the capillary in the test tube, warm to assure complete dissolution, and then heat mostly the sealed end of the smaller compartment to force the hot solution through the filter. Collect the filtrate at the end of the larger compartment by the jerky motion, the direction of which is indicated by the arrow of Fig. 53 f. (As an alternative, the filtration may be brought about by whirling in the centrifuge with the capillary immersed in a hot bath liquid.)

Allow the capillary to cool, and then cut it open so that the filter is removed, Fig. 53g. For separating the mother liquor from the crystals, press the latter somewhat together by means of a glass rod of 1-mm diameter, centrifuge, and lift off the liquid. Repeat the pressing, centrifuging, and decanting several times. Then dry the crystal mass as before by heating while blowing air through the capillary, and determine the melting point. Omit the adding of charcoal, but repeat the recrystallization until the melting point remains constant, 115° C.

The removed mother liquors may be evaporated, and the residues used for slide tests, etc.

Experiment 53

Conversion of Urea to Symmetrical Diphenyl Urea (152)

Heating block, 3 cm ×3 cm ×5 cm long, with horizontal wells for a short thermometer and for the bulb (25 mm deep, 6-mm bore); short thermometer; melting point bath as in preceding experiment. — Urea; aniline, purified; nitrobenzene (b. pt. 211° C).

The urea is heated with aniline in a sealed tube. Distillation and sublimation under reduced pressure are used for the removal of unreacted aniline and the isolation of the diphenyl urea. The latter is then purified by recrystallization from ethanol.

Heating with Aniline. From glass tubing of 6-mm outer diameter, draw out the tube h shown in Fig. 53. The bulb should be somewhat less than 6 mm in diameter; the capillary, 1 to 1.5 mm in bore and 8 to 10 cm long. Introduce 3 mg urea into the bulb 1, and, with a capillary pipet, add 9 μl aniline. Fuse the tube shut at 6, and inspect the seal under the microscope. Place it into a test tube, add nitrobenzene, and boil the latter for 15 minutes so that the capillary is completely immersed in the hot vapor; compare Expt. 50. Finally, allow to cool, return the warm nitrobenzene to its storage bottle, remove the capillary from the test tube, and wipe dry its outside. Cut open the seal at 6.

Distillation of Aniline. Insert the bulb 1 into the well of the heating block so that not more than 20 mm of the capillary are inside the well. Connect 5 to the vacuum line (water pump giving about 60 mm Hg), and begin heating the block. When its temperature reaches about 120° C, droplets of aniline condens at 2; fan the capillary with a small Bunsen flame, and drive the droplets into the wide part 5. When all aniline has been driven off, cool at 3 by hanging a strip of filter paper over the tube and keeping it wet with water.

Sublimation of Diphenyl Urea. Raise the temperature of the block. The sublimate will start collecting when a temperature of approximately 180° C is reached. Hold the temperature until the amount of sublimate does no longer increase. Then remove the tube from the heating block and, without breaking the vacuum, draw it out and seal it shut at *2*; finally cut it at *4*, Fig. 53*h*.

Recrystallization from Ethanol. The sublimate, which usually is slightly discolored, is now contained in a capillary of about 7- to 8-cm length. Recrystallize twice from ethanol and determine the melting point by using the technique outlined in the preceding experiment. The pure substance is supposed to melt at 240° C. It has been suggested (1260) to get complete removal of the last mother liquor by use of centrifugal force as advocated by RICHARDS (147, 149, 696, 698). To this end, follow the directions of the next section.

Spindrying in the Capillary. The crystals of diphenyl urea get large enough that they cannot pass through a fine capillary; thus it is not necessary to use a filter mat.

When sealing the capillary for heating with the solvent, draw out the end to be sealed to a capillary of 0.1- to 0.2-mm bore, and fuse shut the fine capillary at a distance of 10 mm from the taper to the wide part. Place the capillary with the fine end first into the test tube, and heat to obtain dissolution. Then, by one-sided heating and flipping, get the solution to the end which has been drawn out to the fine capillary, but do not force it into the latter. Thus, crystallization will take place in the wide part when the tube is allowed to cool. For separating the mother liquor from the crystals, first cut open at the wide end, and then break the seal of the narrow capillary. By means of a cork or rubber stopper with a lengthwise slit, mount the capillary in the opening of a microcone as indicated by Fig. 53*k*. When the cone is whirled in the centrifuge, the mother liquor collects in the tip of the microcone, and the melting point of the crystals may be determined in the usual manner after sealing the end of the fine capillary.

If recrystallization and spindrying shall be repeated after determination of the melting point, first seal the end of the wide part of the capillary. Then place it with the wide end first into a centrifuge cone, and heat by fanning with a flame or by placing into a drying oven to melt the substance. Without delay, whirl in the centrifuge to get the melt to the wide end of the capillary. Open the end of the fine capillary for taking in the solvent, etc.

Experiment 54

Purification of Benzene; Separation by Partial Melting in the Capillary (755, 1016)

Apparatus for the observation of *schlieren* as described in Expt. 5; V-shaped cell, not more than 0.5 mm thick, to permit observation of *schlieren* with $10\,\mu l$ in the cell. — Mixture of 98% benzene with 2% *p*-xylene; freezing mixture of sodium chloride and ice.

The liquid to be purified is sealed into a capillary with two compartments separated by a filter. The liquid is frozen in the larger compartment. The capillary is centrifuged while its temperature rises. Only part of the solid is allowed to melt during centrifuging so that the liquid containing most of the impurity passes through the filter and collects in the smaller compartment.

A rather wide capillary is used in the following experiment to get sufficient liquid for the demonstration of the change of refractive index by means of *schlieren* observation. Much smaller amounts of material

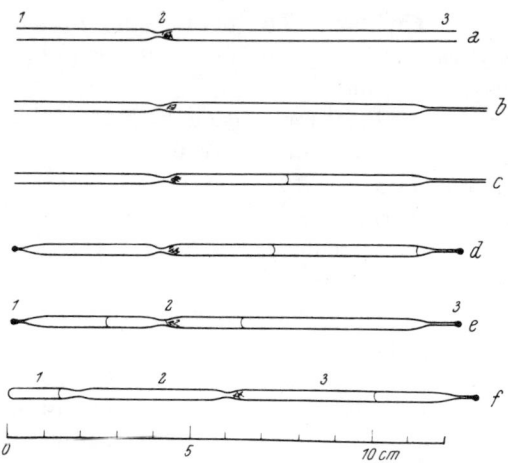

Fig. 54. Separations by Fractional Melting

will suffice if the bore of the capillary is reduced to 0.4 mm, which will not affect the ease of manipulation.

Procedure Giving Two Fractions. Get a capillary of 12-cm length and 1.2- to 1.5-mm bore with a constriction about 4 cm from one end as shown in Fig. 54a. Insert through *3* a few fibers of freshly ignited asbestos, and press them lightly together by means of a glass rod of 1-mm diameter to obtain a filter mat at *2*. By means of a microflame, attach a glass thread at *3* and draw out to obtain a fine tip of 1-cm length, Fig. 54b. Through this, take into the pipet 50 μl of the benzene-xylene mixture: c. Draw out and fuse shut the wide end, and while this end is cooling and the liquid is drawn into the capillary, fuse shut also the tip of the fine capillary, Fig. 54d.

Place the capillary with the end *1* first into a heavy-walled centrifuge tube of 10-cm length with a small plug of cotton in the tip for the protection of the sealed end of the capillary. For an alternative, a test tube may be made from glass tubing of 6-mm outer diameter by sealing one end;

it should be at least 10 cm long and by all means 1 cm longer than the metal shell of the centrifuge so that it may be easily lifted out for inspection. Balance the tube for centrifuging.

Immerse the tube, with the capillary inside, into the freezing mixture and leave it there until the contents of the capillary have solidified. Then remove the tube, wipe dry its outside, and without delay place it into the shell of the centrifuge and set the latter in motion. Stop the centrifuge at suitable time intervals (20 seconds for a first trial) to inspect the contents of the capillary. When nearly half of the crystal mass has melted, stop centrifuging, remove the capillary from the tube, and place it upon a metal block so that it may acquire room temperature: Fig. 54e. First break open the seal at *3*, and then cut apart at *2*.

To test the effect of the interrupted melting, transfer the mother liquor in *1* to the *schlieren* cell by means of a capillary pipet. Draw a long fine tip from end *3*, and allow the liquid obtained from the melting of the crystals to flow through this tip into the mother liquor contained in the cell. Note the character and estimate the strength of the *schlieren*.

Procedure Giving Three Fractions. Repeat the experiment, but use a capillary of 1.5- to 2-mm bore with three compartments, Fig. 54*f*, and a filter mat between *2* and *3*. Take 80 μl into compartment *3*, freeze, and centrifuge to collects 15 μl of mother liquor in compartment *1*. Fuse shut at the constriction between *1* and *2*, and set aside the first mother liquor which has been sealed into compartment *1*. Again freeze the contents of compartment *3*, and centrifuge to get 40 μl of mother liquor into compartment *2*. Allow the capillary to acquire room temperature. Break the seal at the tip of *3*, and cut off compartment *2* with the second mother liquor. Transfer the first mother liquor into the *schlieren* cell, and allow the melt of the crystals to flow into it; again note the character and estimate the strength of the *schlieren*.

Instead of observing the *schlieren* phenomenon, one may determine the melting point (and possibly boiling point) of the crystallized substance remaining in compartment *3*. To this end, it is only necessary to seal both sides of compartment *3* with the liquid inside and to attach it with the liquid close to the bulb of a thermometer which is mounted in a stoppered test tube. The test tube is filled with a suitable liquid (ethanol) so that an air bubble of about 3 ml is left. The test tube is immersed into the freezing mixture. When the contents of the capillary have crystallized, the test tube is removed from the bath and wiped clean (this may be followed by wiping with a glycerol-treated cloth). For the observation of the melting point, rock the test tube around a horizontal position so that the air bubble mixes the liquid in the test tube. Read the thermometer when the contents of the capillary become liquid.

By means of a centrifuge permitting controlled heating during operation, the technique could be extended for the purification of solids by removal of the eutectic melt. The use of suction in place of the centrifugal force provides a simpler solution which may, however, be less efficient and, in addition, may cause loss of substance if its vapor tension is high. FISCHER (421) first determines the melting point of the eutectic mixture by heating a small sample on the hot stage under a microscope. The purification is carried out in a capillary of 1.3- to 1.4-mm bore and 20-cm length, which has a filter mat 8 cm from one end. FISCHER prepares the filter mat by sintering glass powder, but asbestos or even paper filter may be substituted depending upon the required temperature. The substance is introduced through the opening closer to the mat and pressed against the latter by means of a glass rod which is left in the capillary. For heating, the capillary is placed upon the hot stage between two metal strips, 10 mm $\times$ 60 mm $\times$ 2.5 mm thick, which are held to the sides of the capillary by means of clamps. Two small holes must be drilled into the ring of the KOFLER hot stage so that the capillary may be placed along a diameter across the center of the stage. A microscope slide may be placed upon the two metal strips before putting the circular glass plate upon the ring. The capillary is moved so that the substance is in the field of vision and may be observed during heating; the end of the capillary which is farther away from the substance is connected to vacuum line and manometer. The stage is heated to a temperature about 3 degrees above the melting point of the eutectic mixture. When the material in the capillary has sintered, the suction is turned on and the material is pressed together with the glass rod until most of the eutectic melt has been removed.

For the final purification, FISCHER suggests three means: (a) gradual raising of the temperature close to the expected melting point of the purified substance with continuous or intermittent (if the vapor tension is high) application of suction while the substance is pressed together; (b) pressing a small wad of asbestos with the rod against the substance so that it will absorb melt left on the walls of the capillary and near the free surface of the substance; and (c) pulling the intake end of the capillary so far out of the hot stage that the substance may cool to room temperature, and applying 4 to 5 μl of wash liquid while the suction is on; the wash liquid should be a poor solvent for the substance to be purified; when the suction has removed the wash liquid, the capillary is pushed back into the hot stage for the drying of the substance.

The capillary is finally removed from the hot stage and cut at the location of the dry purified substance which may be easily removed. The procedure may be practiced with a mixture of 80% salicylic acid and 20% benzoic acid, the eutectic mixture of which melts at 110° C; the purified salicylic acid may be washed with ethanol.

Experiment 55

Cupric Ammonia Complex, Observation of Color in the Capillary (1004, 1005, 1139)

Capillary clamp or substitute.

Faint colorations in small volumes of liquid are made perceptible by taking the liquid into a long, narrow capillary and looking lengthwise through the column of liquid. Lengths of 1 cm to 10 cm of heavy-walled capillary of 0.1- to 0.5-mm bore (80 nl to 19 μl capacity) may be used and polished flat at both ends. For filling, the *clear* liquid is added at one end until convex menisci are obtained at both orifices, Fig. 55a; long capillaries

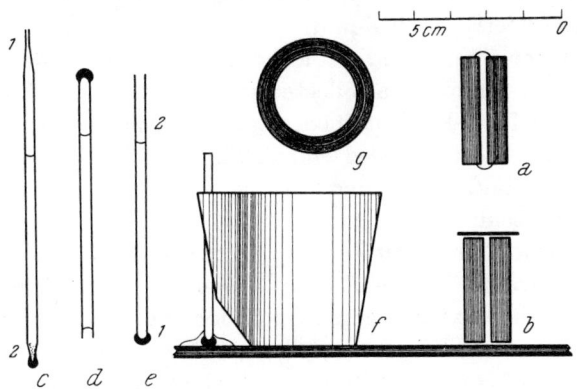

Fig. 55. Coloriscopic Capillaries. The bore of the tubes is somewhat exaggerated

are better held horizontally until the cell is completely assembled. One end is closed by placing the tube upon (against) a microscope slide. If necessary, more liquid is added to the other end to get a convex meniscus, whereupon this end is closed with a fragment of a cover slip. Air bubbles must be absent, but may be dislodged by taking off the cover slip and stirring with a fine platinum wire. The slide is placed upon the stage of the microscope, Fig. 55b; the condenser is used to concentrate the light upon the lower orifice of the tube, and the microscope is focused upon the top opening. Obviously, it is desirable that the glass of the capillary is colorless or nearly so.

Capillaries of black glass are recommended for the observation of the absorption spectrum. To this end, a spectroscopic eyepiece may be used in conjunction with the microscope (49), or the capillary is horizontally mounted in the path of the light before it enters the slit of a spectroscope; in the latter instance, the capillary is mounted in a stand for filling to avoid warming it with the hands, and both ends are closed with fragments of cover slips (1005). Cells have been designed for use with spectrophotometers (441). The technique described in the following uses simple capillary tubes.

Development of Color and Filling of Tube. With a capillary pipet of 0.5-mm bore, take up 1 μl of a mixture of 1 volume ferric test solution and 10 volumes copper test solution (1 mg Fe and 10 mg Cu per ml). Place a drop of 6-F NH_3 on a slide, and insert the tip of the pipet containing the test solution. Allow ammonia to enter until it fills 6 cm of the pipet. Then draw out and seal the wide part of the pipet about 2 cm from the meniscus of the solution, and fuse shut also the tip of the pipet, *compare* Fig. 49. Mix the contents of the capillary by briefly whirling the liquid first to one end of the capillary and then to the other. Finally collect the precipitate at one end by somewhat prolonged centrifuging which must give a perfectly clear liquid.

First open the empty end *1* of the capillary, Fig. 55*c*, and then cut and break at *2* to remove the collected precipitate. Fuse shut the end *1* so that a rounded-off seal is obtained, Fig. 55*d*. Whirl in the centrifuge to transfer the liquid to the sealed end *1*, Fig. 55*e*; lay the capillary upon a slide and inspect under the microscope to make certain that no solid has collected at the seal *1*.

Observation of Color. Cut the capillary just below the meniscus *2* so that it breaks evenly. Without delay, mount the capillary in a vertical position upon a slide so that there is room for applying a drop of immersion liquid (cedar wood oil, lubricating oil, or water) where the sealed end rest upon the slide. The commercially available clamps are most convenient (417, 862), but many substitutes are available: a stopper with a slit for inserting the capillary, Fig. 55*f*; a metal block with vertical wells of different bore to fit the capillaries; etc.

Concentrate the light with the condenser upon the sealed end of the capillary and, first using a $2 \times$ to $5 \times$ objective, focus upon the top rim of the capillary. If the liquid does not absorb strongly, the bore of the capillary showing the color of the liquid will be brighter than the ring representing the cross section of the tube, Fig. 55*g*. From time to time and by means of a platinum loop, add water to the top opening of the capillary to maintain a plane or convex meniscus. Otherwise, the image of the bore will turn gray or black depending upon the curvature, and the observation of the coloration is rendered impossible when the meniscus becomes concave and recedes into the capillary because of evaporation. The evaporation may be delayed by placing a tiny fragment of a cover slip upon the convex meniscus projecting above the top opening. The immersion liquid applied to the base of the capillary improves the optical path.

The recognition of light coloration, especially if it has a hue similar to the color of the glass of the capillary, requires comparison with blanks performed in capillaries of the same kind. To this end, suitable capillary clamps should permit getting the bore of two or more capillaries into one microscopic image (focusing the ends of a bundle or row of capillaries). The absorption spectrum may be observed with the use of a spectroscopic eyepiece (88); for an alternative, the image of the colored circle may be projected onto the slit of a standard spectroscope (1283).

The microscope provides a convincing demonstration, but the color of the liquid may be seen with unaided eye or a magnifying lens when viewing the cut end while holding the capillary over a brightly illuminated sheet of white paper. Even when the light is directed to the side of the capillary, enough of it is reflected in an axial direction to show the color of the liquid when looking vertically toward the cut end.

Working in Filter Paper

Only such procedures shall be considered in this connection, in which the paper is an essentially inert medium supporting and retaining solids and permitting the flow of liquids. The separations may be considered filtration, washing, and extraction in the plane of the paper, whereby only one liquid phase is involved and the paper acts neither as adsorbent nor as ion exchanger as it does in chromatography.

Whereas it is unavoidable that separations of this type occasionally occur in the performance of spot tests on paper, it probably was the sulfide fiber of EMICH (550) rather than spot analysis, that lead CLARKE and HERMANCE (410, 418) to the performance of chemical separations in filter paper. The separation is brought about by a process of extraction. Consequently, a solid sample may be deposited upon the paper and washed by flowing a solvent upon the sample, which dissolves only certain substances that are then carried away by the spreading of the liquid in the paper. Solutions are deposited in a small area of the paper and sometimes allowed to evaporate; the moist spot or the residue is treated with some precipitating reagent, and this is followed by extraction of the solubles. Finally, the paper may be impregnated with a reagent that immediately precipitates certain substances while the liquid sample spreads through the paper. In all instances, the insoluble part of the material is held in the paper at the point where the sample is added, and the soluble substances are carried away and diffused through the paper. If the soluble substances shall be collected, it is necessary to regulate the process in such manner that the solvent flows to a certain, well-defined, and small area, where it evaporates to leave the solutes behind in the paper. So far, actual practice has followed two ways.

CLARKE and HERMANCE (874) added the sample and the solvent to the lower end of a narrow strip of paper which was suspended inside a vertical glass tube. The solvent rose to a level at which a current of air was passed across the strip of paper, which was entering and leaving through side arms on opposite sides of the vertical tube. The zone of evaporation was thus determined by the level of the side arms and could be moved at will by raising or lowering the strip of paper in the tube. The arrangement gave the additional benefit that it must have eliminated the effect of increased evaporation along the edge of the strip (947). The sensitivity could have been increased by trimming the strip to a narrow waist at the height where the solute is collected by evaporation.

The substance on corroded electrical contacts was collected about 1 cm from the lower end of the paper strip which was then immersed in organic solvent. Lubricating oil and tar were caught in an evaporation zone 12 cm above the sample; after changing the solvent, soaps were collected in a band 10 cm above the sample. After this followed extraction with water, which was repeated after digesting the water-insoluble residue with nitric acid. The water-soluble substances were collected 8 cm above the sample, and the acid-soluble, 6 cm above it. The acid-insoluble residue and the above listed fractions were cut out of the strip and individually subjected to analysis.

For the separation of traces of nickel from the solution of a steel, the lower portion of the paper strip was impregnated with barium carbonate which precipitated the iron, but allowed washing the nickel ions into the zone of evaporation, where they accumulated and could be detected with dimethylglyoxime. In a somewhat similar manner, but using a sewing thread in place of the strip of paper, MAHON (779, 963) was able to separate and identify down to 0.015 μg Co or Fe, 0.1 μg Cu, and 1 μg Zn in solutions containing these metals (limiting proportions of 1 : 5 and better).

Paper may be impregnated with "insoluble" reagent like dimethylglyoxime by simply soaking it in the alcoholic solution and drying. The precipitation of reagents in paper has been described on p. 130, and papers loaded with precipitates of silver chromate, zinc sulfide, cadmium sulfide, antimony sulfide, zinc ferrocyanide, cadmium ferrocyanide, and cadmium xanthate have been used.

WEISZ (172, 916, 917, 922) escapes the effect of the edge of the paper by allowing the solvent to flow from the center of a disk of paper toward the circumference. This would give a very undesirable dissipation of the dissolved solutes, but resting the paper upon a heated ring and supplying the solvent at a suitable rate permit concentrating the solutes in a circular zone of only 0.3 to 0.1 mm and less in width so that the amount of substance per unit area may become larger than it was in the original central spot. Thus, this ring-oven technique collects the solutes with high areal concentration into a long (about 7 cm) but very narrow zone which may be cut into ten or more sections for the performance of various tests. If the tests are performed so that the solutes remain in the ring zone, a reasonably precise estimation of quantity is possible (918, 931), and comparison with a standard silver sulfide scale suffices for a whole series of different sulfides (951). Autoradiography may be used for locating and estimating substances in the evaporation zone (928).

The concentrating effect of the narrow evaporation zone is good enough to invite use of the ring oven for evaporating solutions previous to the performance of spot tests. Thus WEST and MUKHERJI (473) separate the metal ions by liquid-liquid extraction into five groups. The solution of each group is then fed to the center of a paper disk resting upon the ring oven, and the solutes are washed into the evaporation zone. The circle containing the solutes is cut into sections which are used for the confirmatory tests. In this manner, the presence or absence of 35 metallic ions may be determined within one hour.

Quite lately, WEISZ (973) has demonstrated that a narrow paper strip may be used with the ring oven in place of the circular disk. In this manner, the length of the narrow zone of evaporation may be decreased from 7 cm to 6 mm, which should improve the sensitivity of the technique by about ten fold. To the same end, the disk could be replaced by a 5- or 6-pointed star, possibly with the tips ending in 1-mm wide ribbons that pass through the evaporation zone.

For work with disks, WEISZ uses "ash-free", fast filter paper, Schleicher & Schuell No. 589, black ribbon[1], and transfers $1.5\,\mu l$ of the solution to be treated with a capillary pipet to the center of the disk[2]. A reagent is added, which precipitates at least some of the substances present, and gaseous reagents are preferred for this purpose since they do not cause spreading of the initial spot. The disk is laid upon the ring oven, and solvent is added to the center of the spot to rinse the soluble substances through the paper to the inner edge of the heated ring. Portions of $10\,\mu l$ of wash liquid are added with a thick-walled capillary with a conical tip that is polished flat at the orifice to provide good contact with the paper. The amount of liquid added at one time must be adjusted to the capacity of the paper within the ring zone; too much would cause flooding and overrunning the heated surface of the ring.

The flow of wash liquid regulates itself since it is drawn out of the capillary at the rate dictated by the spreading in the paper due to capillarity. When the capillary runs empty, it is again filled and applied to the spot;

---

[1] Glass fiber paper does not seem to have been tried, but it might be useful for working with strongly acid or alkaline solutions.

[2] Minute amounts of insoluble substances are fused with a suitable reagent, P. 42, on a square of platinum foil, $5\,mm \times 5\,mm \times 0.03\,mm$ thick. The foil with the melt is placed upon the center of the (55-mm diameter) disk of paper and covered with a disk of 10-mm diameter, which is provided along its circumference with 5 or 6 tiny droplets of a suitable cement. The large disk is placed on the ring oven, and the melt is extracted by adding solvent to the center of the sandwich, etc. (974).

this has to be repeated five to ten times for washing the solutes completely into the zone of evaporation.

The paper disk is finally dried in an oven. The initial spot is isolated by punching it out of the dry disk with a circular die of 10-mm diameter. The circle of evaporation of 22-mm diameter is cut into as many sectors as the problem requires. The small circle with the initial spot may be placed upon the center of a fresh disk of paper, treated with reagent, and again washed to collect another group of soluble substances in the evaporation circle of the new disk.

The ring oven consists of a circular cylinder of metal, 35 mm high and 55 mm in diameter, which has been drilled along its axis to provide a central hole of 22-mm diameter. It may be made of aluminum or of copper and plated with gold. It is supported by a tripod and heated electrically to a temperature 5 to 10 degrees above the boiling point of the solvent. The disk of paper is placed upon the top surface and secured there by laying on it a ring of glass or porcelain with an inner diameter of 25 mm. The market also offers a ring oven made of glass (Paul Haack, Garnisongasse 3, Vienna IX, Austria), which is heated by the vapor of boiling liquids (953). Tetrachlorethylene is used when washing with water since the boiling point of the liquid must be taken about 10 degrees higher than the temperature required at the surface of the oven. This oven is also provided with a thin glass plate with a central circular opening of 12-mm diameter; by placing this plate upon the paper disk, a second circle of evaporation with a diameter of 12-mm may be obtained and later isolated by using a die of 15-mm diameter in addition to the one with 10-mm diameter.

The technique of working in paper has the attractions and disadvantages associated with spot testing. The technique is simple and time saving. On the other hand, the substances must be finally extracted from the paper if they shall be obtained or observed in the pure state. The paper is attacked by strongly acid and alkaline solutions, but this difficulty may be overcome by using paper made of glass fibers and other suitable materials. Even high-grade paper contains impurities (Fe, Se) which become concentrated in the evaporation zones, may interfere with the tests, and may require washing of the paper previous to use (975).

The special literature should be consulted concerning the various techniques of chromatography (63–76), which offer unique possibilities for the sensitive separation of closely related substances.

Experiment 56

Ring-Oven Technique for Extraction and Evaporation in Paper

L. I., usually fractions of a microgram.

Ring oven with capillary for adding wash liquid and circular die of 10-mm diameter; apparatus for treating paper with gaseous reagents (916): a glass

Expt. 56 Working in Filter Paper

tripod for supporting the paper disk in a covered beaker filled with the gas may be substituted; test tube and atomizer head for spraying reagents; "ash-free" filter paper, S. & S., No. 589, Black Ribbon, disks of 55-mm diameter. — Ferric chloride solution, 0.1 mg Fe/ml; copper-iron-nickel solution (0.3 mg of each ion per ml): a mixture of 3 ml each of the test solutions (10 mg ion per ml) diluted with 0.1-F HCl to 100 ml; potassium ferrocyanide, 2% solution; rubeanic acid, 1% solution in ethanol; dimethylglyoxime, 1% in ethanol.

The following practice experiments have been suggested by WEISZ (916); the second has been slightly modified.

Concentrating into the Ring Zone for Testing. With a capillary pipet, transfer 1.5 μg of ferric chloride solution (0.1 mg Fe/ml) to the center of a disk of paper. Place the disk upon the ring oven so that the moist spot is located above the center of the oven which is heated to 105 to 110° C. Place upon the paper the glass or porcelain ring of the oven with 25-mm inner diameter. Take 10 μl 0.1-F HCl into the pipet. Insert the pipet into the holder of the oven and lower it until the tip touches the paper in the center of the moist spot. The contents of the capillary are drawn into the paper. When the capillary is empty, fill it again and apply it as before. Repeat this 5 to 10 times to wash the ferric chloride completely into the zone of evaporation. This will require about two minutes. Finally, remove the paper disk from the ring oven and place it into a drying oven. Support the dry disk horizontally and spray it with a mixture of equal volumes of ferrocyanide solution and 0.1-F HCl. A sharp circular blue line, not more than 0.3 mm wide, should be obtained, and the central portion of the disk should not show a blue coloration.

Separation by Solid-Liquid Extraction. With a capillary pipet, transfer 1.5 μl of the copper-iron-nickel solution (0.3 mg of each ion per ml) to the center of a paper disk. Place the disk with the wet spot into an atmosphere of H_2S and leave it there for 3 minutes. Test for complete precipitation of the copper by adding to the center of the dark spot from a capillary pipet just enough of a mixture of equal volumes of ethanol and 0.1-F HCl to get a wet zone, about 1 mm wide around the spot. Return the disk into the H_2S atmosphere; the precipitation of the copper was complete if the color of the moist zone around the sulfide spot does not change.

To extract iron and nickel, place the paper disk upon the ring oven which has been heated to 105 to 110° C. Treat the central spot first with 2 μl ethanol, which seems to change the nature or distribution of the precipitate so that the extraction may be efficiently performed. Then use 10-μl portions of 0.1-F HCl to wash iron and nickel into the zone of evaporation; this requires 1 to 2 minutes. Finally dry the paper in the drying oven, and then cut out the central sulfide spot with the 10-mm die.

For the identification of iron and nickel, cut the resulting paper ring into three sectors. Spray the first with the mixture of equal volumes of ferrocyanide solution and 0.1-F HCl; the second, with rubeanic acid;

and the third with dimethylglyoxime solution. Expose the last two sectors to fumes of ammonia.

For the identification of the copper, moisten the small disk of paper carrying the sulfide spot with 0.1-F HCl and expose it briefly to bromine vapors. When the dark color of the sulfide has disappeared, place the small disk upon the center of a fresh disk of paper mounted upon the ring oven. Lower the capillary to the center of the small disk, and wash the copper with 10-μl portions of 0.1-F HCl into the evaporation zone of the large disk. Finally dry the large disk in the drying oven, and cut the dry circle into four quadrants. Expose one quandrant to fumes of ammonia; spray the second with rubeanic acid; the third, with ferrocyanide; and expose the latter two also to fumes of ammonia.

Particles of Ion Exchange Resins as Reaction Media

A review of this technique, developed for the performance of confirmatory tests since 1952, has been written by FUJIMOTO (622). In all instances, a colored compound is formed on the resin, but fluorescence and luminescence might provide even higher sensitivities. In spite of the fact that the smallest particles have a diameter of 0.25 mm, 0.05 mm^2 cross-section area, and several particles are used, the limits of identification compare favorably with those obtained when collecting the matter upon the tip of textile fibers (p. 183), which permits concentrating it into an area of only 0.0004 mm^2: 0.02-mm length of fiber of 0.02-mm diameter. This is not strange when considering that resins may be chosen for their special affinity to the substance to be detected, whereas the collection of test substance on textile fibers occasionally relies upon adhesion during evaporation (Expt. 58) and may be far from complete. If either reagent or tested substance has an affinity to the fiber as in the instance of dyes, the resulting limit of identification is quite satisfactory (Expt. 57).

FUJIMOTO, who traces the history back to the experiments of GÜNTHER-SCHULZE (1210) with natural zeolite, lists as advantages offered by the synthetic resins their uniform reproducibility, their selectivity, and the improved stability of the obtained colorations.

The performance is as simple as that of spot tests. "On a white (or black) spot plate, mix a few grains of a colorless or lightly colored ion exchange resin (cation or anion exchange type, usually of low cross-linkage) with a drop of the test solution (or of the reagent solution). Adjust the conditions in the reaction medium and allow to stand until the ions under test (or the ions of the reagent) are strongly adsorbed in the resin phase. Now add a drop of reagent solution (or test solution) and, if necessary, adjust the pH and the ionic strength of the medium. Allow to stand appropriately and observe the coloration developed in the resin phase under magnification.

(In some cases the resin grains may be previously impregnated with the reagent and stored.)" (622)

It is obvious that the limits of identification listed by FUJIMOTO (622) may be improved in several ways: (a) by the use of small drops and a single grain or a fraction of it; (b) by concentrating the test solution before introducing the grain; (c) possibly by removing the exhausted test solution before adding the reagent or *vice versa*; and (d) by removing the grain from the colored reagent solution for inspection. In addition, the grain might be placed into the taper of a capillary pipet with fine tip through which it cannot pass so that test and reagent solutions may be added and removed by the customary pipetting technique. Finally, the grain may be taken up (speared) with the point of a glass needle and thus transferred to test drops and reagent drops contained in capillary cones (p. 198). Both techniques would permit multiple tests (Expt. 26) and use of the exhausted test solution for testing for an ion of the opposite charge.

As a rule, strongly acidic cation exchange resins and strongly basic anion exchange resins of low cross-linkage are used, which are colorless or lightly colored to assure high sensitivity. Before use, the commercial resins must be purified regardless of claims concerning absence of metal contamination. They are packed into a tiny column, washed with dilute hydrochloric acid, converted to the desired form in the customary manner, washed with metal-free water, and dried in air. For examples of tests see FUJIMOTO (622).

Working on Textile Fibers and Wires

EMICH conceived the idea that the limits of identification of colorations could be vastly improved by using tiny objects such as yeast cells or even bacteria for carriers of the colored matter. To-day, one might profitably generalize by including all tests using a distinctive property: radiation, fluorescence, absorbtivity for any type of radiation, electric potential, magnetic susceptibility, catalytic action. Textile fibers were chosen for the practical utilization of the principle since they may be efficiently handled without resorting to manipulation under the microscope. Using fibers dyed with indicators (1000, 1001), adsorption of colloidal matter (1002), and the deposition of colored precipitates (550), the limits of identification were pushed into the nanogram range. Logical extension of the reasoning led to the collecting of matter upon the cross section of thin wires by electrodeposition, whereby like sensitivities were obtained (858).

Systematic investigation (680, 681) lead CHAMOT (118) to the conclusion that the type of fiber should be selected so that it adsorbs the reagent strongly in relatively high concentration and does not "bleed", i. e., give it off too readily when immersed into the test solution. Obviously, the fiber must not react with the reagent in any manner which would decrease its efficiency. Thus, silk fibers (9 μm to 21 μm diameter) are recommended for red and blue litmus, Congo red, and the adsorption of colloidal gold; viscose-rayon (30 μm) for turmeric, potassium ferrocyanide, potassium thiocyanate, and gold; and wool (12 μm to 60 μm) or gun cotton (10 μm to 40 μm) for carrier of zinc sulfide. Flax (about 20 μm) is a close second for use with turmeric.

Very important is that fibers intended for use with aqueous solutions are prepared and handled so that they do not become water repellent. The suitability of reactions for fiber tests depends, of course, upon the possibility of concentrating the characteristic product on or in the fiber. If the reaction product is a colored precipitate, one might expect that a test that is succesful when performed as drop reaction on paper should also be suited for the fiber technique. This need not happen, however, since single fibers do not possess the same adsorptive power as paper. Also the particle size may very with the technique and with it, the coloring power. The coloring power increases with growing size for some substances, and with others, it decreases (670, 682, 740).

Aside from the performance of confirmatory tests, the fiber technique may be used for side tests when working upon the submilligram and microgram scales. Only a permissible amount of solution will be lost when the end of some textile fiber carrying a pH indicator, a redox indicator, or a precipitating agent is briefly inserted into a minute droplet. Fibers treated with reagent may also be used to greatly increase the sensitivity of tests for gases, P. 36. If the material to be investigated may be made to cling to a fiber, it becomes possible to perform separations upon a very small scale or to determine the presence or absence of groups of substances (414).

So far, only simple procedures were tried. To improve the sensitivity for the detection of very small amounts of matter, EMICH concentrated it by allowing the droplet of test solution to evaporate while only the very tip of the fiber is immersed. A relatively small number of reagents have been tested, and nothing is known about attempts to get separations by having test solutions rising on fibers treated with selective precipitating agents. Migration in the electric field, however, has been used for the separation of nucleic acids (1052, 1053) and proteins (766) on fibers.

Litmus Fiber. L. I., 30 pg hydrogen ion; 200 pg hydroxyl ion. CHAMOT and COLE (680, 681) recommend the following procedure. Litmus is purified according to WARTHA (600) by extracting the commercial "cubes" with 95% ethanol until the extract has no longer a reddish hue. The dye remaining in the residue is then extracted with water, whereby air is blown through the mixture to prevent reduction of the dye. The aqueous extract is filtered and then concentrated at steam bath temperature to the consistency of a syrup. The syrup is repeatedly treated with acetic acid and ethanol and evaporated to dryness for the destruction of carbonates. The residue is extracted with portions of absolute alcohol (ethanol) until the light reflected from the extract no longer shows a reddish hue. The thus purified residue is dissolved in water and evaporated to a thick syrup which is then treated with absolute alcohol. The paste obtained is stirred with fresh portions of absolute alcohol until the latter does no longer extract red coloring matter. The final residue is dissolved in water. The solution is evaporated to a thick syrup which is poured into absolute alcohol. The gummy precipitate is spread upon a clay tablet and dried at about 75° C to a hard mass which is readily soluble in water.

Raw silk of good quality is boiled in a very weak soap solution, rinsed thoroughly, and then digested at room temperature for 2 hours in a solution of 10 g NaOH in 100 ml water. The silk is then washed thoroughly with distilled water and transferred into a 10% solution of the purified litmus dye, which has been acidified by adding 3 drops 3-F H_2SO_4 per 100 ml. The solution with the silk in it is evaporated to a thick syrup, whereupon the silk is removed and washed in running water. To give the fibers the middle tint, the silk is stirred with water to which very dilute NaOH is added until the silk becomes violet. The silk is again washed in running water and then dried. To get blue litmus fibers, the adding of NaOH is simply continued until the color changes to blue; in a similar manner, red litmus silk is obtained by stirring silk of the middle tint with water and adding dilute acetic acid.

Red Congo Fibers. Silk is purified as described above and then dyed in a 0.5% solution of Congo red, made alkaline with NaOH. — Viscose rayon is boiled with a 2% (alkaline) solution of the dye for 15 minutes, washed, and dried by pressing between filter paper.

It is obvious that the test solution must not be allowed to evaporate upon a fiber if the latter shall indicate the pH of the solution as it is given.

Sulfide Fiber. White wool is washed with warm soap solution, rinsed in water, and dried. The wool (or gun cotton with long fibers) is soaked in acetone, dried by pressing between filter paper, and then digested over night in 1% NaOH of room temperature (118). After brief rinsing in water, the fibers are dipped 5 to 6 times alternately into 10% zinc acetate and Na_2S. Between dippings, the fibers are not washed, but pressed to remove excess solution. The Na_2S is prepared by passing H_2S into 10% NaOH until a drop of the solution does no longer give a precipitate when mixed with a drop of $MgCl_2$ solution. After the final dipping, the impregnated fibers are washed in water and dried by pressing between filter paper.

Sulfide fibers which have been stored for any length of time should be tested with a known mercuric chloride solution before using them on an unknown. When testing for small amounts of heavy metals, the test drop is acidified with HCl, and the fiber is inserted as directed in Expt. 57, *below*. If the tip of the fiber turns black, Hg, Ag, Bi, Pb, Cu, or Pt may be present. If the tip turns yellow, the fiber should be transferred into a drop of 6-*F* NaOH; the following may happen: (a) the tip turns black: Hg, Ag, Bi, Pb, Cu, Pt; (b) the tip remains yellow: Cd; (c) the tip turns white or colorless: As, Sn. An orange coloration that disappears with NaOH indicates Sb. The interpretation must take into consideration that several of these metals may be present, that also some other elements may react, and that the outcome of the test depends very much upon the acidity of the solution and the concentration of the elements concerned.

The textile fibers must be replaced by filament of heat resistant material for tests which require the application of high temperatures. Thus, wires of suitable metals are used for the performance of fusions and bead tests; they are the obvious choice also for electrodeposition.

Electrodes. EMICH's idea of using the cross sections of very thin wires for carriers of deposits has been systematically investigated by BRENNEIS (154, 162, 858). The so-called *"rod electrode"* was obtained by inserting one half of a thin wire of 11-cm length into a narrow, thin-walled capillary of 5-cm length and bending the other half of the wire back so that it runs along the outside of the capillary. The combination was then pushed, with the bend first, 4 cm deep into the bore of a thick-walled capillary of 0.5-mm i. d., 5-mm o. d., and 10-cm length. At the point where the wire inside turned back, the thick capillary was then heated so that its bore collapsed and the wire became imbedded in the glass. After slow cooling, the resulting short piece of glass rod was cut below the bend of the wire, and the circular cross section was polished to get the ends of the wires in the level of the glass surface. Using wires of platinum or platinum-iridium of 0.025-mm to 0.1-mm diameter, circular electrode areas of $5 \cdot 10^{-4}$ to $8 \cdot 10^{-3}$ mm^2 could be obtained. Since the distance between the wires need not exceed 0.1 mm, hemispherical drops of electrolyte of 0.15- to 0.3-mm diameter and 0.8 to 7 nl volume could be used in the performance of electrolyses. To prevent evaporation of the test drop, a test tube with an inner lining of moist filter paper was inverted over the rod electrode.

A *"needle electrode"* of not more than 1-mm outer diameter was obtained by using a thin-walled capillary for the outer glass tube. The needle electrode is intended for insertion into a small amount of electrolyte which may be on a slide or in a tube or crucible. For the performance of a side test, it may be inserted into a solution which is treated in a centrifuge tube or microcone. The needle electrode as well as the rod electrode is easily mounted in a plastic handle provided with binding posts for connecting to the supply of current.

The *electrolytic slide*, which is commercially available, is obtained by fusing three insulated wires into a capillary of 5-mm o. d., cutting it to a length of 15 mm, and mounting it on a plate of hard rubber, which also carries three binding posts. The short piece of glass rod containing the three wires is surrounded by a plastic ring of 12-mm i. d., which can be raised or lowered by turning it. A closed cell and retarding of evaporation are obtained by placing a cover slip upon the ring. The area around the three circular electrodes may be made water repellent by applying paraffin, and the droplet of electrolyte may be applied to the underside of the cover slip and then brought into contact with the electrode surfaces by lowering the ring. The slide has been given standard dimensions so that it may be readily mounted under the microscope and also fits a mechanical stage. Focusing through the cover slip permits microscopic examination during electrolysis. The third electrode is not connected to a source of current and serves as a blank for comparison. The electrodes are so small and so close together that all three are included in the field even when high magnifications are used. A vertical illuminator or equivalent is naturally required for observation by reflected light with the use of high magnification.

Electrolyses are performed by the standard procedures. If these require it or if it is desirable for the better visibility of the deposit, the cathode (and the blank electrode) may first be coated with a suitable metal; coatings of gold, silver, and copper have given good results. The needle electrode is dipped into the properly prepared electrolyte, or the latter is deposited by means of a loop, hook, capillary pipet, or micropipet upon the electrode area of the rod electrode or the electrolytic slide. The connections are made in the customary manner. If much of the sought ion is present, the deposit may be visible after a few seconds. Near the limit of identification (about 1 ng in the instance of Cu, Ag, Au, and Pb), the electrolysis may have to be continued for 15 minutes to 2 hours; 24 hours in the instance of blanks. If the electrolyte contains sulfuric acid, this suffices to prevent evaporation. With other electrolytes, a moist atmosphere must be provided. The deposit is finally washed without interrupting the flow of current by adding a droplet of water to the electrolyte, removing it with a capillary pipet or a piece of filter paper, and repeating the adding and removal of water as often as seems advisable.

The appearance of the deposit suffices for the recognition of copper, gold, mercury, lead dioxide, and manganese dioxide. Obviously, the deposit may be dissolved and the solution used for further tests. A deposit of mercury may be exposed to vapors of iodine; the resulting HgI_2 should be soluble in KI solution. A silver deposit turns black when moistened with Na_2S solution or exposed to H_2S, etc. It also should be possible to obtain contact prints of the coated electrodes (162, 434). *Vice versa*, the tiny electrodes coated with suitable deposits may be used as reagents. A silver deposit should provide a sensitive reagent for sulfide (Hepar test);

an electrode covered with tin and connected to a clean platinum electrode could be used to test for antimony.

For practicing, one may use an experiment described by BRENNEIS. A few milligrams of fresh beef liver (containing about 40 p. p. m. of copper) are impaled upon the end of a platinum wire and ashed in the flame of a Bunsen burner constructed of glass or porcelain. The residual black, crusty mass is placed into a microcone containing 20 to 30 μl 16-F HNO_3, broken up with a glass rod, and heated on the steam bath. After whirling in the centrifuge, the clear liquid is taken up into a capillary pipet and in small portions evaporated upon a narrow slide, Expt. 44. The dry residue is dissolved in 0.2 μl 2-F H_2SO_4, and the solution is electrolyzed with an e. m. f. of 2 volts across the electrodes. The washed deposit may be transferred to the other electrode by reversing the current after adding a droplet of fresh 2-F H_2SO_4. A blank test may be performed with a small tuft of purified asbestos and should give no copper deposit.

Experiment 57

Turmeric Test for Boric Acid (550)

Turmeric linen: dip unbleached linen fibers into acetone, and dry them by pressing between filter paper. Reflux 5 g turmeric powder for 5 minutes with 10 g 95% ethanol; filter the solution, and evaporate it to dryness. Dissolve the residue in 4 ml 50% ethanol, and treat the solution with small portions of solid Na_2CO_3 until it becomes clear. Introduce the linen fibers, and heat just to boiling. With a glass hook, remove the fibers from the bath, and dry them by pressing between filter paper. Without handling them with the fingers, immerse the fibers for 30 seconds in 1-F H_2SO_4, rinse them thoroughly in running water, press them between filter paper, and then allow them to dry between two sheets of paper. With forceps, transfer the fibers which should have the color of egg yolk to a suitable container, glass vial. — Canada balsam or other cement suitable for attaching the fiber to glass; Plasticine or beeswax for temporarily fastening glass to metal or glass to glass.

Selection of Fiber. It is essential to handle the fibers only with forceps and needles so that they will be readily wetted by aqueous solutions. Above a sheet of paper, cut the fibers with scissors into lengths of 10 to 15 mm. Transfer several of them to a slide for inspection under the microscope with medium magnification and strong transmitted light. Select an individual fiber (not a yarn or twist of several) which is reasonably straight, not pointed but cut squarely at the end, and has a distinct yellow color over its entire length.

Draw a fine thread from one end of a capillary of 5-cm length, and break the thread 2 cm from the capillary. Moisten the end of the thread with a trace of Canada balsam or some other suitable cement, and then touch with it the lesser end of the selected fiber. Finally, attach the capillary to the body tube of a microscope with Plasticine or beeswax as indicated by Fig. 56.

Concentrating the Test Substance at the End of the Fiber. Take into a capillary pipet about 0.2 µl of borate test solution (1 mg BO_3 per ml)

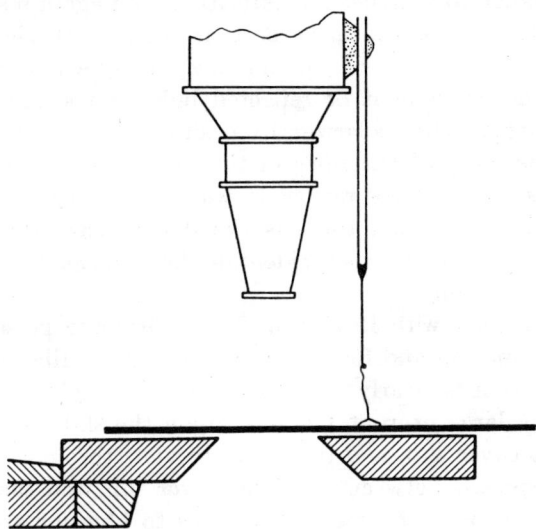

Fig. 56. Concentrating Solutes upon the End of a Fiber. The thickness of the fiber is greatly exaggerated

and an equal volume of 6-F HCl. Transfer the liquid to the center of a slide. Place this slide upon the stage of the microscope, and insert the end of the fiber into the droplet as shown in Fig. 56. Observe the evaporation of the liquid with the aid of a magnifying glass; the end of the fiber should remain in contact with the liquid until the latter has completely evaporated.

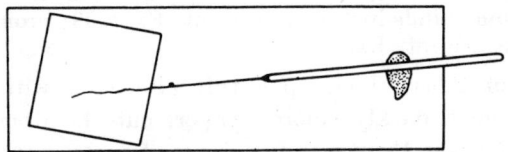

Fig. 57. Fiber Mounted for Examination. Its thickness is greatly exaggerated

If the evaporation is nearly complete before the fiber has been properly inserted, treat the residue on the slide with 0.5 µl 6-F HCl, and again insert the end of the fiber.

Observation of Test. When the test droplet has completely evaporated, raise the tube of the microscope, remove the capillary with the fiber on the end, and place it upon a clean slide, Fig. 57. If desirable, attach the capillary to the slide with Plasticine or beeswax. Place a dry cover slip or a fragment of a cover slip on the fiber to hold it close to the surface of the slide.

With low magnification, focus first upon the point of the glass needle, and then follow the fiber through its entire length until its free end is found. If desirable, change to a higher magnification, and again inspect the entire length of the fiber. For the most part, it should still show the original yellow color, but the free end should now be brown or reddish brown. Use strong transmitted light or reflected light and a white background.

If a microscope with polarizer in position is used, rotate the stage and observe the color of the fiber in the various positions of the latter. On linen fibers, the red coloration shows very strong pleochroism from red to nearly colorless, and there is the danger that the test could be completely missed if the fiber is accidentally kept in the "colorless" position during the observation.

Treating the Fiber with Reagents. To confirm the presence of borate, adjust the illumination and focus the free end of the fiber so that the red or brown coloration is clearly seen. By means of a glass rod or medicine dropper, place a large drop of 1-F NH_3 upon the slide so that it touches the edge of the cover slip, and immediately view through the microscope. The ammonia spreads between slide and cover slip, and the fiber becomes immersed. The color of its free end changes to blue (or green with small amounts of boric acid), whereas the rest of the fiber becomes purplish red. The blue color of the tip fades quickly, which makes it advisable to observe while the NH_3 is being added. Also the blue coloration shows pleochroism to colorless.

Experiment 58

Test for Bismuth, Precipitation of the Sulfide Upon the Fiber and Conversion to Sulfate, Chromate, and Elemental Bismuth

Surgical cotton with long fibers; acetone, pure; Canada balsam; Plasticine or beeswax; sodium sulfide-hydroxide reagent, Expt. 38; bromine water; 0.1-F $K_2Cr_2O_7$; stannite reagent, Expt. 24.

Preparation of Fiber. Pick up a tuft of cotton with forceps. With scissors and above a darkly colored paper, cut the fibers to lengths of 10 to 15 mm and bathe them in a few drops of acetone for removal of fat. Transfer several fibers to a slide for inspection under the microscope. Select a fiber which is straight and does not have pointed ends, but is squarely cut. As described in the preceding experiment, attach the lesser end of the fiber to a glass needle, and attach the latter to the body tube of the microscope.

Concentrating the Test Substance on the End of the Fiber and Precipitating the Sulfide. Transfer about 0.2 μl of bismuth test solution (10 mg Bi per ml) to the center of a slide, and insert the end of the cotton fiber as outlined in the preceding experiment. Make certain that not more than the very end of the fiber is in continuous contact with the evaporating drop. When evaporation is complete, raise the fiber with the body tube, exchange the slide for another one carrying a large drop of sodium sulfide-hydroxide reagent diluted with 25 volumes of water, and lower the end of the fiber into this solution.

Rinsing the Fiber and Microscopic Examination. After a few seconds, lift the fiber out of the sulfide solution. Remove the capillary carrying the fiber from the tube of the microscope, and place it upon a slide, Fig. 57. Place upon the fiber a cover slip with a large drop of water hanging on the underside of it. Examine the fiber under the microscope. The black Bi_2S_3 should be clearly visible at the end of the fiber when using reflected light and a white background.

Oxidation of the Sulfide and Precipitation of Chromate. Grasp the capillary and pull the fiber out from under the cover slip; this will not be difficult if there is enough water between the two glass plates. Insert the end of the glass needle with the fiber attached into the gas space of a bromine water bottle, and keep it there for about 3 minutes. Do not touch the neck of the bottle with the fiber when inserting and withdrawing it; if the neck is dangerously narrow, pour some bromine vapor into a small bottle with wide mouth.

Again place the capillary with the fiber upon a clean, dry slide, and cover the fiber with a dry cover slip. Inspect it under the microscope with reflected light and a dark background. The end of the fiber should be white or colorless if the sulfide has been completely oxidized to sulfate.

Deposit a large drop (0.05 ml) of 0.1-F $K_2Cr_2O_7$ on the slide so that it touches the cover slip and is drawn into the space between slide and cover glass. When the fiber has become immersed in the solution, withdraw it by means of the capillary to which it is attached, and transfer it to a fresh slide. Place a cover slip upon the fiber, and add a large drop of water so that the fiber becomes immersed. The yellow $(BiO)_2Cr_2O_7$ is very difficult to recognize. Try reflected light with white, black, and violet background.

Reduction to Metallic Bismuth. Focus the end of the fiber, and use reflected light and a white background. Add a large drop of freshly prepared stannite reagent to one edge of the cover slip, and remove water at the opposite edge by touching to it a piece of filter paper. When the stannite solution reaches the fiber, black metallic bismuth becomes visible at its end.

Experiment 59

Bead Test for Cobalt (152)

L. I., 30 ng Co.

Platinum wire, 0.05- to 0.1-mm diameter, 25 mm long: fuse one end of the straight wire into a wide glass capillary of 4-cm length, which serves for handle. — Borax, granular; xylene.

Scatter some borax upon a microscope slide, and select a kernel of about 1-mm diameter (0.5-μl volume) for the experiment. By means of a platinum loop or a capillary pipet, treat it with 0.2 μl of cobalt test solution diluted with 9 volumes of water to contain 1 mg Co/ml. Then heat the platinum wire in the edge of a non-luminous Bunsen flame to incandescence, and take up the treated kernel of borax by touching it with the end of the hot wire. Adjust the Bunsen flame to a height of about 2 cm, and fuse the borax by heating the center portion of the wire in the edge of the flame and about 5 mm above the orifice of the barrel. The bead tends to move to the cooler parts of the wire. Thus, it may be kept at the point of the wire or driven back-and-forth over the length of the wire by proper application of heat.

When the borax has been fused to a clear glass, drive the bead to the end of the wire, and then allow it to cool. Put a large drop of xylene or water near one end of a microscope slide, and lay the capillary with the wire so upon the slide that the bead becomes immersed in the liquid. Hold the slide over a brightly illuminated sheet of white paper; the blue coloration of the bead may be observed with a magnifying glass.

For microscopic inspection, focus first upon the wire with a low magnification. Then follow the wire to its end to bring the bead into the field of vision. Use the front lens of the condenser to concentrate the light on the bead, and change to a stronger objective if necessary. If the bead should be too small, heat it again, and add more borax by touching a kernel with the hot bead.

A blank test may be carried out on a second wire, and the two beads may be immersed into the same drop and viewed side by side in the field of the microscope. Very faint colorations may be made visible by drawing the bead into a fiber (767). The wire with the bead is introduced into a capillary of soft glass which is then cautiously heated at the point where the bead is located. When the borax glass melts, the wire is withdrawn. The capillary is heated, and when it has collapsed somewhat so that the borax glass fills its bore, it is drawn out to a length of a few centimeters of narrow capillary with a borax glass core. This fine capillary is cut evenly at both ends and mounted as a coloriscopic capillary, Expt. 55. Immersion oil should be added at the base, and a fragment of a cover slip should be placed with immersion liquid upon the top of the capillary. In this manner, the blue color should become visible with 3 ng Co.

For cleaning the wire, again fuse the bead, and throw the molten bead off the wire by a flip with the hand. Then take up a relatively large amount of borax, fuse it to a bead, drive it back-and-forth over the wire, and flip it off. Repeat this once. If this is desired, the borax glass may be completely removed from the wire by finally placing it for 30 minutes into dilute HCl.

Experiment 60

Luminescence Test for Bismuth, Antimony, and Manganese (604, 1006)

Platinum loop or microspatula: hammer flat the end of a wire of about 0.3-mm diameter; two such wires may be fused side by side into the end of one glass handle. — Hydrogen, pure: if prepared from zinc and acid, pass it through a cotton filter; $CaCO_3$, powder, of highest purity.

Lime containing a trace of bismuth, antimony, or manganese gives a luminescence of characteristic color when touched by a hydrogen flame so that it does not get hot enough to become incandescent. The test should be performed in a dark room if small traces of the metals are to be discovered. Furthermore, the tube supplying the hydrogen should have a vitreous silica or porcelain tip so that the hydrogen flame does not become luminous.

Performance of Test. In a dark corner, obtain a small hydrogen flame, not more than 5 mm high, burning from the orifice of a capillary which may be drawn out from glass tubing of high softening point. Clean the platinum loop or spatula by dipping into HCl and ignition until it does no longer color a Bunsen flame. Make a thin paste of $CaCO_3$ and water, and take a small portion of it upon the loop or spatula. First dry and then ignite in the hydrogen flame. By means of a capillary pipet or a platinum hook or loop, add about 0.3 μl of diluted bismuth test solution (1 volume of 10 mg Bi per ml diluted with 10 volumes of 3-F HNO_3) to the white crust of calcined lime upon the tool. Dry and ignite lightly. Allow to cool, and then approach the lower edge of the flame with the preparation. A deep sky-blue luminescence will emanate from the parts of the lime which have received bismuth solution when they first meet the flame. When the chalk begins to glow, the luminescence is displaced by the yellow incandescence.

If two spatulas are mounted close together in one handle, one is made to carry the blank test which may then be carried out and repeated side by side with the test. If the handle is mounted in a test tube, Fig. 40, the preparation may be kept for demonstration purposes.

The thermoluminescence is a light greenish blue with antimony and a deep yellow with manganese. The tests seem to be specific.

Work on the Microgram Scale

The technique described in the following has been developed for work with solid samples of 0.5 to 1 μg (139, 141, 330, 411, 433, 435, 770, 888). It should be applicable without much change to work on the submicrogram and possibly the nanogram scale; it has been used on the submilligram scale, i. e., for the investigation of solid samples of 0.1 mg to several micrograms weight. In the latter instance, the apparatus is simplified (770, 772). The capacity of the capillary cones is increased, and since evaporation is to be less feared with somewhat larger volumes of solutions, the capillary cones may be mounted upon simple manipulators without the use of a moist chamber. The pipet may be directly inserted into a syringe control, and the latter may be mounted on a second simple manipulator facing the first. The elaborate microscope with rotating mechanical stage may be replaced by a binocular microscope of the Greenough type or by a binocular magnifier (90).

The essential feature of the technique is the use of mechanical and optical aids for the handling of substances. These aids are a great convenience on the submilligram scale, but they become a necessity on the microgram scale. Consequently, it seems proper to describe the technique as it has been developed for work on the microgram scale and to leave the simplifications to the ingenuity of the experimenter who desires to apply it to somewhat larger amounts of material. Even on the microgram scale, apparatus and technique permit various modifications for which the reader may be referred to the literature (90, 903, 770, 772).

Most of the work is performed in tiny centrifuge cones of 1-μl capacity, whereby the technique is essentially the same as in working with the microcone. Because of the small volumes which must be handled, it is necessary to perform transfers with the aid of mechanical devices while observing through a low-power microscope. Furthermore, solutions of considerably less than $1 \mu l$ volume evaporate quickly when exposed to air which is not saturated with water vapor. Thus, most of the work is performed inside a moist chamber mounted upon the stage of the microscope. Small volumes of reagents and wash liquids are held ready within this chamber, and a mechanically operated pipet which is inserted through the open side of the chamber serves for the transfer of liquids.

By taking one million times smaller masses and volumes of sample, reagents, and solutions than on the gram scale, one may follow any tested procedure or scheme of separation with complete assurance of success. Because of the mechanical and optical aids, all operations may be performed with ease and confidence. By projecting the microscopic images upon a screen, Dr. CEFOLA has repeatedly demonstrated precipitations, filtrations, and distillations to audiences of sixty to several hundred people on occasion of meetings of the Metropolitan Microchemical Society of New York (1280), the New York Section of the American Chemical Society (1281), and the In-Service Training Course of the City of New York (1282). The usefulness has been demonstrated in the investigation of transuranium elements (772); the scale of work renders radiation hazards bearable; explosions become harmless.

Apparatus

Microscope. The stand should be that of microscopes used by biologists for micromanipulation under high magnification. It should be equipped with a built-in revolving and centerable mechanical stage without excessive superstructure and mechanical motions permitting displacements of 6 to 9 cm in two directions. The fine adjustment coming with such stands is rarely needed; the rack-and-pinion device for the coarse adjustment should permit tightening so that the body tube will keep its position when it carries heavy auxiliary apparatus. A revolving nosepiece which provides for individual centering of the objectives is desirable. The optical equipment should consist of bright-field condenser which can be focused on a plane 10 mm above the stage, three objectives with magnifications of 5, 10, and 20 diameters, and one micrometer eyepiece with magnification $5 \times$. The rulings of the eyepiece micrometer should occupy at least two thirds of the diameter of the field. A totally reflecting prism to be used above the eyepiece is convenient for microprojection.

Manipulator (770, 903). A simple manipulator with three rack-and-pinion motions, each permitting a displacement of about 10 cm, is needed; in addition, rotation around the vertical axis of the manipulator is desirable. A very sturdy construction is essential so that vibrations will not originate in the manipulator, that would be transmitted to the pipetholder and magnified at the tip of the pipet.

Base. Microscope and manipulator are placed side by side upon a board and secured in their correct positions by means of small wooden blocks as indicated in Fig. 58. To determine the proper distance between the microscope and the manipulator, the following procedure is recommended. The left-right motion of the manipulator is operated to advance the clamp b all the way toward the microscope while the two other pinions engage at the centers of their racks. A capillary of 15-cm length is then inserted into the pipet holder which is placed into the clamp b so that it is held close to the capillary. The correct position of the manipulator is found by moving it until the free end of the capillary is about 1 cm to the left (referring to Fig. 58) of the optic axis of the microscope and about 1 cm above the stage. To this end, the height of the microscope or manipulator may have to be adjusted by means of additional boards. When the correct positions have been obtained, the positions of the apparatus are secured with the wooden blocks. A cover for the whole assembly should be made to fit the base board; it will permit to have the equipment always ready for use.

If a strong source is available for illumination with transmitted light, an image of one to several feet in diameter may be obtained at the most convenient location. Use of the screen image gives less eye strain than viewing through the eyepiece; in addition, the height of the table may be selected to give a maximum of comfort during the operation of the various contrcls. If use of a strong source of light is not practical, it may still be possible to project an image of 5 to 10 cm in diameter on a piece of white Bristol board which may be mounted inside a black box and a short distance above the eyepiece as indicated in Fig. 58. The box should be mounted in a manner that it always remains at a fixed distance above the eyepiece and may be simply swung aside whenever direct observation becomes necessary to get the image with reflected light.

The eyepiece micrometer is focused upon the screen by adjustment of the eye lens after focusing the object. If necessary, the positions of

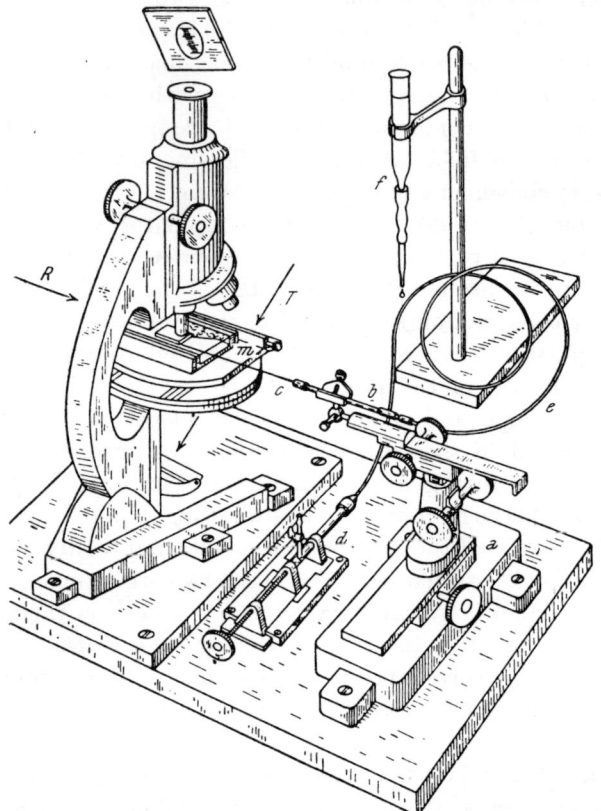

Fig. 58. Assembly for Work on the Microgram Scale

the microscope tube and of the eye lens may be alternatingly adjusted until object and scale appear simultaneously with sharp outlines. The value of the scale division of the micrometer scale will be approximately the same as when looking into the eyepiece. It is preferable, however, to calibrate the micrometer under the conditions of use. Of course, the eyepiece micrometer is not essential when projecting since the dimensions may be measured with a millimeter rule on the screen image.

Illumination. Two lamps are required. One is placed in front of the microscope to send light in the direction of arrow T to the mirror. The second lamp is placed to the left of the microscope so that it may send light in the direction of arrow R horizontally into the moist chamber. This lamp should permit collecting the light into a narrow pencil of not more than 6-mm diameter at the focus. Switches for both lamps should be mounted in a handy location so that observation with transmitted light and with reflected light may be used in quick succession.

Pipet Holder. It is now possible to obtain micrometer syringes which may be mounted in the clamp of the manipulator and into which the micropipet may be directly inserted (903). The separation of the plunger device from the manipulator will be of advantage, however, if the latter is not very rigid; in addition, the pipet holder makes also possible to use the very simple device of regulating the pressure with a levelling bulb (437).

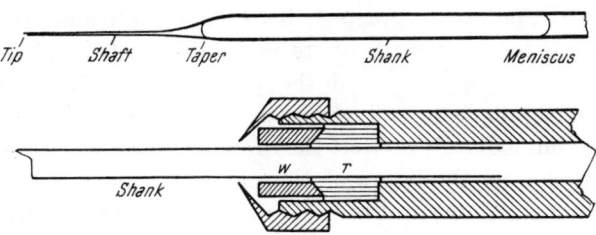

Fig. 59. Micropipet and Pipet Holder. w metal washer; r rubber gasket; bore and thickness of the tubes are exaggerated

The pipet holder shown in Fig. 59 is a metal tube of 1- to 2-mm bore, about 4 mm in outer diameter and about 12 cm long, which is fitted with rubber gaskets r, metal washers w, and screw caps at both ends. The rubber gasket consists of flexible rubber tubing which is compressed by the advancing concave face of the metal washer w. If fine-drawn copper tubing is used to make the connection to the plunger control d, Fig. 58, it may be inserted into the pipet holder just like the shank of the pipet, or it may be soldered to it. As an alternative, the connection may be made with plastic tubing which is stiff enough to maintain the capacity of its bore during manipulation. Even rubber tubing is suitable if the tube contains air connected with a larger air space the pressure of which is adjusted (437).

Obviously, the pipet holder may consist of a glass tubing of suitable dimensions, and the shaft of the pipet may be inserted with the use of sealing wax.

Plunger Control. The device indicated in Fig. 58 was fashioned after the buret control of JOHNSON and SHREWSBURY (876). Any similar device will serve, and so will a micrometer operated syringe. The control, the connecting tubing, and the pipet holder are filled with water that has been boiled to remove air. The water is used while still lukewarm, and the filling is done so that air bubbles are excluded.

Micropipet. The micropipet proper consists of the shaft and the tip, Fig. 59. The small volumes of solutions will rarely reach the taper. The shank of 6- to 10-cm length contains the air cushion which separates the solution in the pipet proper from the water used for transferring the pressure from the plunger control.

The micropipets are made from soft-glass tubing. Capillaries of 20-cm length and 0.5- to 1-mm outside diameter to fit the pipet holder are drawn out in the middle to get a quick taper and a shaft that gradually tapers to an orifice of 30- to 40-μm diameter. The drawing is best done by mechanical devices as they have been described by DU BOIS (1080) and RACHELE (424).

A microflame the size of a pinhead is needed for drawing micropipets by hand. The capillary is grasped, between thumb and index finger, at two points 3 cm left and right of its middle. The hands are steadied by resting the outer, fleshy parts of the palms on the bench top, and the middle of the capillary is brought over the flame. A steady horizontal pull is applied immediately so that the drawing starts when the glass begins to soften. By rolling the edges of the palms on the bench, the capillary is removed from the flame and at the same time symmetrically pulled (903), which takes a fraction of a second. The fine capillary may be 3 to 5 cm long, and it may snap in the center. The pipet is placed upon a slide and inspected under the microscope to learn what variations of technique will give a pipet of proper dimensions. The shaft should be 5 to 8 mm long. The tip is often too fine and fused shut; it may be snipped off with scissors to obtain the desired orifice of 30 to 40 μm.

Moist and Dry Chamber. Fig. 60 shows the top and front views of a chamber similar to that designed by CEFOLA (433). The bottom is formed by a glass plate, 60 mm $\times$ 68 mm. The two long sides are formed by bars, 6 mm $\times$ 67 mm $\times$ 11 mm high, which may be made of metal or plastic and which are cemented to the base plate. A narrow strip of thin glass plate, 11 mm $\times$ 53 mm, fits into vertical grooves of the bars and forms the back of the cell. The top of the cell is a thin glass plate, 52 mm $\times$ 63 mm, which is placed upon the bars. To obtain a humid atmosphere, the sides of the cell are lined with cotton b, Fig. 60, which is kept wet with water. The dry chamber does not receive the cotton lining.

Capillary Cones. The cone, Fig. 60a, has a capacity of approximately 0.6 μl and is made of a thin-walled capillary of about 0.8-mm bore. Gloves are worn so that the capillaries will not be touched by fingers. The capillaries are freshly drawn out from clean glass tubing, Expt. 19, and cut into pieces 6 to 10 cm long. After the bore has been checked, the middle of a piece of capillary is heated in a microflame or in the edge of a non-luminous Bunsen flame until the glass fuses together to form an elongated bead.

Apparatus

The bead is withdrawn from the flame and, after brief delay to permit some cooling of the thin glass at both ends of the bead, the bead is drawn out to a rod of about 0.3-mm diameter and 4-cm length. This should give two cones, the taper of the bores of which should be quite blunt as shown in Fig. 60 a. Obviously, the procedure may be repeated at suitable intervals to give a string of 4-cm rods separated by pieces of the capillary of about 5- to 10-mm length. Cutting at the proper places gives a number of capillary cones with handles of about 2-cm length. The cutting should be done with a sharp tool so that little pressure is needed (903). The length

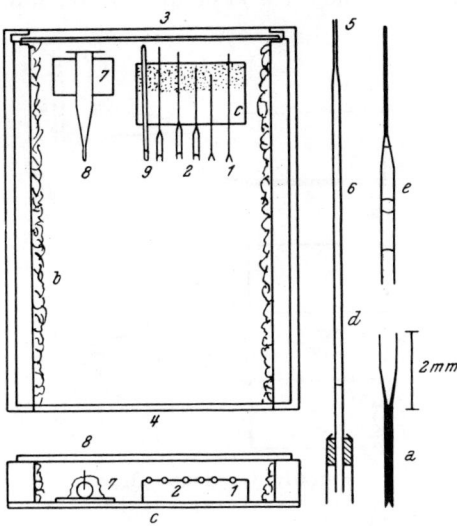

Fig. 60. Moist Chamber Ready for Chemical Work. a capillary cone; b cotton lining; c carrier slide; d micropipet; e reagent cone

of the cone proper should be 2 mm, and half of this length should be occupied by the blunt taper. The finished capillary cones are collected in a screw-cap vial. Like any other apparatus used in the moist chamber, they should never be touched with the fingers since the fingerprints would develop into a pattern of small droplets, which greatly interferes with the microscopical observation of the contents.

Reagent Containers. A large number of reagent containers is prepared from an assortment of clean capillary tubing of 0.3- to 1-mm uniform bore. The bore of each capillary is determined, Expt. 22. The outside of the capillary is wiped with a moist and then a dry cloth; hereafter it is handled with gloves. About 2 cm from one end, the capillary is fused to a bead which is then drawn out to a thin rod of about 4-cm length; no attention need be paid to the shape of the bore of the taper. The fusing and drawing is repeated so that a 2-cm length of the original capillary is left between each pair of rods. Cutting at the centers of the rods and of the capillaries gives reagent capillaries of 1-cm length with handles of 2-cm length. They are stored in a screw-cap vial, the label of which indicates the cross-section area of the bore and the number of eyepiece micrometer divisions corresponding to the length of the capillary holding 1 nl.

Measuring Capillaries, Fig. 60 9, permit a more accurate measuring of very small volumes, but are rarely needed if reagent containers of sufficiently narrow bore are used. They are thin-walled capillaries of 0.05- to 0.2-mm uniform, known bore and 2- to 3-cm length, which are sealed at one end. Like reagent containers, they are kept in labelled screw-cap vials.

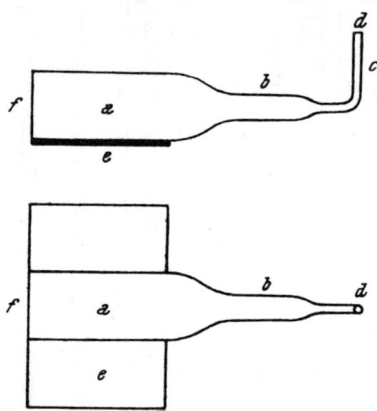

Fig. 61. Condenser Rod

Carriers. At least two should be available. The carrier c, Fig. 60, consists of a strip of plate glass, 25 mm × 35 mm × 7 mm thick. It may be made of several thin plates of glass by cementing them together with Canada balsam. One half of the top surface is coated with a layer of petrolatum, about 1 mm thick, as indicated by the dotted area in Fig. 60. To get a smooth coating, the carrier is slightly heated. EL-BADRY and WILSON (903) make the carrier of plastic; a deep groove, parallel to the front edge and 5 mm from it, is cut into the base surface so that a rubber band placed into it will not touch the surface supporting the carrier and interfere with the sliding of the latter. The light rubber band is stretched around the carrier so that the handles of the capillary cones and reagent containers may be inserted between the band and the top surface of the carrier.

Condenser Rod. Side and top views of this device are shown in Fig. 61. It consists of a short piece of glass rod which tapers to a fine thread. The latter is cut to provide a tiny platform d for the performance of tests. A strong beam of light is sent into the rod through its base f. Most of the light is collected in thread c and emerges at d to give efficient illumination for the observation of the test.

Soft glass is used, which has as little color as possible. A rod of 4- to 5-mm diameter is drawn out to a thinner rod of about 2-mm diameter so that a quick taper is obtained between a and b, Fig. 61. About 10 mm from the taper, rod b is drawn out to a thread 0.1- to 0.3-mm in diameter, which is broken 2 cm from the taper. The thick part of the rod is cut to give a a length of 10 to 12 mm; the rod is scratched with a file and then broken while part a is held with pliers. It is important that an even surface is obtained at f. The thread c is bent at a right angle by cautiously approaching it with a microflame while the rod is held horizontally; the thread bends by its own weight. Then, thread c is dipped into molten paraffin nearly up to the bend and slowly withdrawn. When the paraffin on the thread has solidified, the thread is scratched 5 mm above the bend and broken

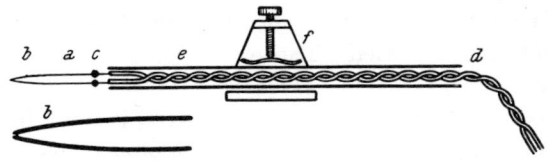

Fig. 62. Heating Element

off with the aid of forceps. The break should be clean and at a right angle to the axis of the thread. Rod a is cemented to a cover slip e so that thread c assumes a perpendicular position when e is supported horizontally. Finally, a fragment of a cover slip is attached to surface f with Canada balsam to obtain a plane surface; obviously, this is not necessary if the circular face f is given a high polish.

The film of paraffin on the cylindrical surface of the thread confines solutions to the glass surface of the cross section. The resulting circular platforms of 0.1-, 0.2-, or 0.3-mm diameter have areas of 0.008, 0.03, and 0.07 mm^2 which would support hemispherical droplets of 0.25-, 2-, and 7-nl volume.

Reservoir of Water for Cleaning Micropipets. The reservoir f, Fig. 58, has the shape of a short pinchcock (MOHR) buret. A glass bead in the rubber tubing controls the outflow. The inside of the rubber tubing is cleaned with brush and soap solution before it is attached to the tube. The reservoir is filled with distilled water and then covered with an inverted vial to keep dust out.

Heating Element. A 5-cm length of No. 24 copper-nickel alloy wire (0.5-mm diameter, 0.7 ohm) is bent in the center to give the shape of a V. The point of the V is pressed closely together by means of pliers, and it is then carefully filed down until the cross section of the wire is reduced to one-third of the original (408) as shown in the enlarged drawing of the point b, Fig. 62. The ends of the wire are soldered to ordinary insulated copper wire of the type used in radio work. The insulated wire is forced through a glass tube e of 10- to 15-cm length and of such bore that the wire fits tightly. Insulating tape is applied at d to prevent the wire from twisting around in the tube. The leads are connected to a variable transformer which is plugged into the a.-c. line; in general, not more than 5 volts need be supplied to the heating element. The glass tube may be fastened in the clamp f of the manipulator.

Forceps. Two forceps, preferably of stainless steel and with polished flat tips, are needed. One of them is provided with gripping surfaces of cork. Chips of cork, 1 mm thick, are sliced from a stopper with a razor blade. The tips of the forceps are slightly heated, and some Krönig glass cement (1 weight white beeswax and 4 weights of rosin) is applied by touching the stick of cement to the hot metal. The cement melts, and the slices of cork are placed upon the treated tips, which are squeezed together until the cement has solidified. The excess of cork is trimmed off with the razor blade.

Gloves. A pair of thin cotton or silk gloves is kept in a covered jar or in an envelope so that they remain meticulously clean.

Technique

Manipulation under the microscope does not require skill. Needed are some practice and the ability of organizing the work so that nothing will be missing when the cell has been assembled for the performance of an operation.

Mounting the Micropipet. The plunger of the pressure device is advanced until a drop of water appears at the opening of the pipet holder. The shank of the micropipet is inserted into the opening, and the screw cap is made tight. If necessary, the plunger is advanced until the meniscus of the hydraulic water in the shank is seen about 3 cm in front of the opening of the pipet holder, Fig. 59.

Filling Reagent Containers. Clear liquid reagent is taken up into a capillary pipet having a fine tip of about 1-cm length. A reagent container of a bore assuring the desired precision of measurement (*above* p. 199) is selected, and the tip of the capillary pipet is inserted so that the orifice is about 6 mm inside the container. By blowing with the mouth, liquid is gradually expelled from the pipet which is at the same time gradually withdrawn so that the container is filled close to its opening, Fig. 60e.

Assembling the Chamber. All apparatus, reagents, and wash liquids needed for an operation or a brief series of operations are assembled in the chamber which takes the rôle of the laboratory and contains bench and reagent shelf. If several reagents are needed, a pencil sketch should be prepared in which the containers are labelled; the containers are then arranged according to this plan. The required number of capillary cones *1*, reagent containers *2*, and measuring capillaries *9* are assembled upon the carrier *c* side by side. The handles of the reagent containers and capillary cones and the sealed ends of the measuring capillaries are pushed into the layer of vaseline (or under the rubber band). The openings of all tubes are brought into a straight line parallel to the edge of the carrier, which facilitates later manipulations. Without delay, the carrier is then placed into the chamber, Fig. 60, which may already hold a condenser rod or other needed devices. A droplet of water is deposited upon the bottom plate of the chamber before putting down a piece of equipment such as the carrier or the condenser rod. The water spreads between the glass surfaces, and the surface tension holds the apparatus in its assigned place. If necessary, water is added from a washbottle to the cotton lining, and the chamber is closed by putting on the cover plate.

The chamber is then clamped into the mechanical stage of the microscope so that the opening of the chamber faces the manipulator on the right of the microscope and the controls of the mechanical stage are on the left-hand side of the microscope.

Introducing the Micropipet into the Chamber. The manipulator is swung around so that the micropipet is parallel to the side walls of the chamber. It is then introduced into the chamber by operating the controls of the manipulator while watching with the unaided eye.

Handling Solid Particles. Particles of 1 μg mass and less may be weighed with suitable balances (156). To this end, the particle would be supported by a metal foil or a fragment of a cover slip. If the supporting surface is mounted inclined to the horizontal in the chamber or above the chamber, the particle may be transferred to a capillary cone by the technique of Expt. 60, or dissolved *in situ* by adding a measured amount of solvent with a micropipet which may then be used to transfer the solution to a capillary cone for further investigation.

In general, it will suffice for qualitative analysis when the mass of solid particles is computed from an estimate of its density and measurement of its dimensions under the microscope with the use of the eyepiece micrometer. The material for investigation or solid reagent is scattered upon the surface of a slide or cover slip, 2.5 cm square. By means of a short piece of glass rod (with or without the use of cement), the slide is mounted in an inclined position in or on top of the chamber, Fig. 68*c*. Obviously, solids which are not hygroscopic may be mounted inside a moist chamber. The microscope is focused upon the particles, and a particle of suitable size is selected with the aid of the eyepiece micrometer, picked up with the point of a needle or the sealed tip of a micropipet, and transferred into a capillary cone. Obviously, the needle is inserted into the pipet holder or the clamp of the manipulator. The tip is treated to receive a film of glycerol, sebum, or other suitable adhesive.

Using the lowest available magnification, the selected particle is brought into the center of the field of vision and sharply focused. The point of the needle is brought into the field of vision by using the motions of the manipulator while observing with the unaided eye from the directions 2 o'clock and 5 o'clock (the needle coming from 3 o'clock). Without changing the focus of the microscope, the point of the needle is brought into sharp focus with the vertical motion of the manipulator. After the point of the needle has been brought close to the selected particle, one may change to a higher magnification for the observation of the contact. As a rule, a loose particle will stick to the needle point that touches it and may thus be transferred to any desired location.

Fusions. One end of a straight platinum wire of not more than 0.05-mm diameter and 3-cm length is fused into the end of a glass capillary which latter is then suitably mounted on a stand (or on a manipulator) so that the wire is held horizontally and at a right angle to the axis of the microscope.

The (binocular) microscope is provided with $2.5\times$ or $5\times$ objective(s) and tilted into the horizontal position. If it must be used in the vertical position, the front lens(es) should be protected by a horizontally mounted glass pane inserted just below it (them).

The free end of the platinum wire is brought into the field of vision. A gas or hydrogen microflame, 1 mm in height, Expt. 19 (glass tubing clamped to a manipulator or to a stand that glides upon the top of the bench), is moved into position so that it heats the wire a short distance (5 to 10 mm) from the free end. The end of the wire should get hot but not incandescent. Small particles of the selected flux are placed upon the end of a microspatula (flattened end of a 0.5-mm platinum wire) or upon a narrow slide. The tool is moved by hand, while watching through the microscope, so that a particle of flux touches the hot end of the wire. The particle will adhere to the wire, and it is fused by bringing the flame gradually closer to the end of the wire. Additional particles of flux are added to the melt (or to the hot bead) until the bead has the desired size.

The dimensions of the bead are measured with the eyepiece micrometer. The bead will approach the shape of a rotation ellipsoid with radius a vertical and axis b parallel to the wire. If r is the radius of the wire and d is the density of the flux, the weight of the bead is approximately given by $2\,b\,d\,(2\,a^2 - 3\,r^2)$; this is nearly the same as $6\,a^3\,d$ which is the weight of a sphere of flux of radius a, Table IV.

When the required amount of flux has been collected in the bead, various procedures may be used for adding the sample: (a) a solid particle may be added to the molten bead by using the technique of Expt. 61 and a platinum needle operated with a manipulator; (b) the bead may be allowed to cool (possibly moistened with water or glycerol) and used to pick up the particle; (c) the particle located upon a slide or spatula may be transferred by touching it to the bead; or (d) a powder or precipitate may be made into a slurry and transferred to the cold bead by means of a micropipet with wide tip.

Table IV. *"Equatorial" Diameter of Bead as Function of Its Weight*
Assumed are: density of flux = 2.5 g/ml and diameter of supporting wire = 0.05 mm

Weight of Bead μg	Diameter $2a$ μm
1	80
2	100
4	130
6	145
8	160
10	175
20	220

Finally the bead is dried, if necessary, by approaching it with the flame which is first applied to the wire at about 20 mm from the free end with the bead. By heating the wire closer to the bead, the latter is fused while observing through the microscope. The advice on pp. 111, 322ff. should be suitably applied. The heating is continued until the bead becomes clear or the reaction stops. For dissolving the cold bead, it is transferred into the solvent held ready in a capillary cone in a moist chamber, see next section.

An electrically heated loop has been described by KOCH, MALISSA, and DITGES (570). Finally, it should be possible to perform pyrosulfate fusions in a capillary cone of vitreous silica.

Inserting a Tool into a Capillary Cone or a Capillary. Using the motions of the mechanical stage, the capillary cone (capillary) is moved into the location of the diameter 3 to 9 o'clock with the opening facing the manipulator. With transmitted light and a magnification of 40 to 60 diameters, the opening of the capillary cone is focused so that the outer contours of the walls of the capillary appear as perfectly sharp straight lines (point of the taper of the sealed end is in sharp focus). This setting of the microscope is retained during the following operations, and this assures that the tool will be located in the axial plane of the capillary in focus.

The capillary cone is withdrawn toward 9 o'clock so that it occupies only one third of the field, Fig. 63. The manipulator is rotated to get the tool (needle or micropipet) parallel to the axis of the capillary. The tool is introduced into the chamber, and its tip is brought close to the opening of the capillary by using the motions of the manipulator and watching with unaided eye from 2 o'clock and 5 o'clock. At this stage, the microscope will give a blurred image of the tool. Without touching the adjustment of the microscope, the tool is brought into focus with the vertical motion of the manipulator; this brings the tool into the mid plane of the capillary. To line it up with the axis of the capillary, the stage may have to be rotated somewhat, and the side motions of the manipulator are used for the final adjustment. Providing that stage and manipulator are set so that the side motions follow the diameters 6 to 12 o'clock and 3 to 9 o'clock, the tool may now be made to enter and leave the capillary along its axis by operating the 3 to 9 o'clock motion of either the manipulator or the mechanical stage. Fig. 63 shows capillary and tool not as they are actually

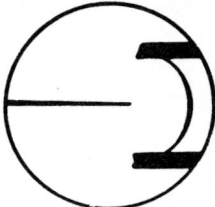

Fig. 63. Micropipet Ready to Enter a Capillary

situated in space, but as they appear when looking into a monocular microscope.

Measuring Liquids. All measuring is done with the micrometer scale. The basic measuring devices are the reagent containers and the measuring capillaries, but the measuring may also be done in the micropipet and in the capillary cone. Since the volume determinations are based upon computation of the capacity of the dry containers, the volumes actually delivered may be about 10% smaller than expected, even when the liquid is very slowly withdrawn. The fraction of the liquid remaining upon the walls seems to reach a maximum of 20% with capillaries of 0.2-mm diameter, but decreases for narrower and wider tubes (437). Obviously, these uncertainties are avoided by the Teddol treatment of EL-BADRY and WILSON (423).

Liquid reagents are measured out of the reagent containers, the cross-section areas of the bore of which are known. The length of the liquid column which represents the desired volume is readily computed in terms of the divisions of the eyepiece micrometer for the magnification in use. The eyepiece is rotated to make the scale appear in the image of the reagent container, Fig. 64a. The left hand grasps the control of the mechanical stage, and the right hand that of the plunger device. The plunger is slightly advanced to obtain some pressure in the micropipet, which will not dissipate immediately if the orifice of the pipet is fine enough. Then, without delay, the reagent container is moved forward by means of the mechanical stage until the opening of the micropipet is immersed in the liquid somewhat beyond that length of liquid which is to be taken into the pipet. As a rule, solution enters the micropipet as soon as its tip touches the liquid. The entering liquid is completely expelled from the pipet by advancing the plunger until the tiny meniscus arrives at the orifice of the tip. Then the position of the reagent container is adjusted until the meniscus in the reagent container coincides with a convenient division of the micrometer scale such as 10 in Fig. 64a. Then suction is cautiously applied by with-

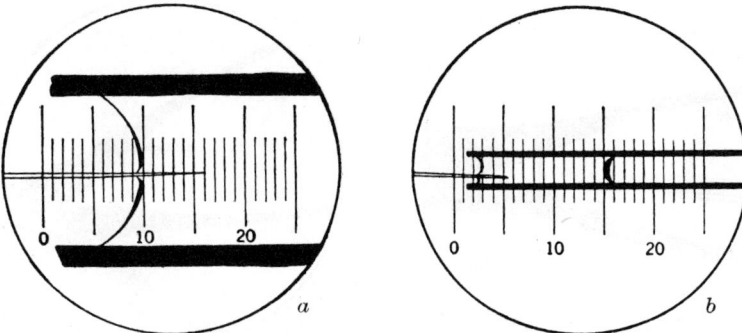

Fig. 64. Measuring Liquids. *a* in the reagent container; *b* in the measuring capillary

drawing the plunger from the pressure chamber so that the meniscus recedes slowly into the reagent container. When the meniscus in the reagent container has travelled through the desired number of scale divisions and arrived at the predetermined point of the scale, the tip of the pipet is quickly withdrawn from the liquid, which may be done by moving either the capillary or the pipet, preferably the former. The pipet now contains the desired volume of liquid.

The pipet may now be calibrated if the volume of liquid is small enough so that the meniscus in the shaft appears still on the micrometer scale when the orifice is lined up with the zero mark or the edge of the field. Obviously, use of low magnification permits extending the range of applicability. The volume and the position of the meniscus in the shaft are recorded for future use. If the bore of the shaft is uniform or tapers regularly, it will be possible to crudely estimate fractions of the determined volume.

A reasonably correct estimate of the volume *delivered* by the pipet may now be obtained by transferring the liquid to a dry measuring capillary (437). The latter is moved into the center of the field, and the micropipet is inserted so that the orifice of the tip moves into the capillary a distance equal to the estimated length of the liquid column. The plunger control is operated to transfer the liquid from the pipet to the measuring capillary, and the tip of the pipet is withdrawn as the meniscus in the capillary moves toward the opening of the latter, Fig. 64b. It does not matter whether or not the tip of the micropipet touches the wall of the measuring capillary. When the plunger is advanced, a droplet forms at the orifice of the pipet, which grows and finally fills the bore of the measuring capillary. The two menisci in the capillary move apart as liquid is added.

Calibration of the pipet would require that its whole contents are transferred to the measuring capillary. If, however, the latter is used for the direct measuring of small volumes, then the micropipet is withdrawn

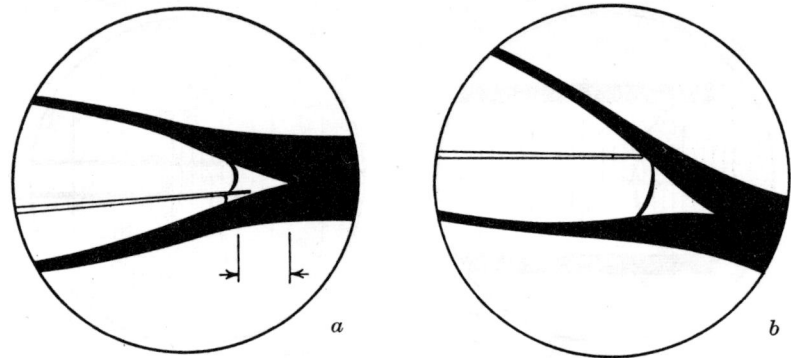

Fig. 65. Transferring Solutions to the Capillary Cone. *a* delivery of the bulk; *b* delivery of the small amount of liquid left in the tip

when a drop of predetermined length has been obtained; the micropipet is then emptied and hereafter again inserted into the drop in the measuring capillary to take it up by a reversal of the above procedure for transfer to the capillary cone in which it is needed.

When finally a measured volume of liquid has been transferred to a capillary cone, this latter may be calibrated in turn by measuring the distance from the meniscus to the point of the taper, Fig. 65a.

Transfer of Liquid to the Empty Capillary Cone. The microscope is kept focused upon the point of the taper of the capillary cone. The tip of the micropipet is advanced with the manipulator to a point just inside the capillary cone, and then the cone is moved with the mechanical stage until the orifice of the micropipet touches the wall of the cone close to the point of the taper, Fig. 65a. Slight pressure is then cautiously applied with the plunger device so that the contents of the pipet are slowly delivered to the point of the cone. The capillary cone is slowly withdrawn so as to keep the orifice of the micropipet just below the meniscus. When the meniscus in the micropipet approaches the orifice, the flow is stopped.

The cone is withdrawn until the orifice of the micropipet is above the meniscus of the liquid in the cone. Then the stage of the microscope or the manipulator is rotated so that the opening of the pipet touches the wall of the microcone close to the meniscus, Fig. 65b, and the remainder of the liquid in the pipet tip is expelled gently.

If the micropipet is emptied carelessly so that air bubbles follow the liquid, this will cause spattering, and it is then advisable to collect all liquid in the point by whirling in the centrifuge.

Emptying and Cleaning the Micropipet. The micropipet is completely withdrawn from the chamber. A strip of filter paper is grasped with the fingers by one end. The other end of it is touched to the orifice of the micropipet while the plunger is slightly advanced. When the liquid has been expelled, the meniscus of the hydraulic water in the shank of the pipet is brought back to its proper position if necessary, 3 cm from the pipet holder.

For rinsing the micropipet, the valve of the water reservoir f, Fig. 58, is operated so that a large drop of water forms at the tip of the outlet. The tip of the micropipet is inserted into this drop, and suction is applied with the plunger device. When the water has advanced to the taper of the pipet, the latter is withdrawn from the hanging drop, and its contents are expelled by touching the filter paper with the opening and advancing the plunger. The rinsing is repeated twice, each time with a fresh drop of water. Finally the position of the meniscus of the hydraulic water in the shank is again checked and corrected, if necessary.

Of course, the cleaning may require a solvent different from water, which may be supplied hanging at the end of a stirring rod.

Adding Reagents to Solutions in the Capillary Cone. The micropipet containing the reagent is brought close to the opening of the capillary cone which is then advanced so that the orifice of the micropipet is close to the surface of the liquid already in the cone. Slight pressure is now applied with the plunger device. This assures that the outflow of reagents starts as soon as the opening of the pipet makes contact with the liquid in the capillary cone.

The outflow of reagent is stopped when the meniscus in the shaft of the micropipet gets close to the orifice. This is the time for using the micropipet for stirring, *see* below. The tip of the pipet is then withdrawn from the liquid and touched to the side of the capillary cone for the delivery of the small remainder of the reagent, Fig. 65b.

The effect of the reagent upon the contents of the capillary cone is observed with reflected as well as transmitted light.

Solid reagents are ground to a fine powder and sprinkled upon a glass slide which may be mounted inside or on top of the chamber as the hygroscopicity of the reagent and the atmosphere in the chamber dictate, Fig. 68. The measuring and transfer of solid particles is described on p. 203.

Stirring in the Capillary Cone. A satisfactory stirring effect may be obtained by moving the shaft of the micropipet, which is sufficiently sealed by the presence of the last trace of reagent in the tip, through the contents of the cone by means of the mechanical stage or the manipulator. Very efficient stirring is obtained by plucking with the finger the copper tubing connecting pipet holder and plunger device while the pipet tip is immersed in the contents of the cone.

Treating the Contents of the Capillary Cone with Gaseous Reagents. A glass tubing is drawn out to a capillary of about 1-mm bore, which is

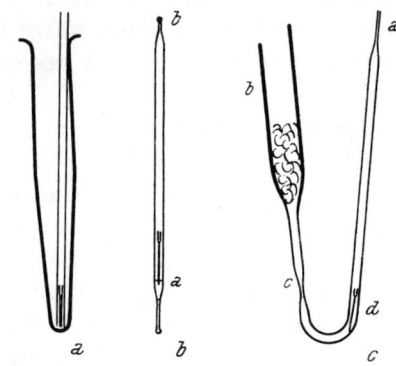

Fig. 66. Working in the Capillary Cone. *a* centrifuging; *b* heating; *c* saturating with gas

then provided with a constriction c and bent as shown in Fig. 66c. A plug of cotton is placed into the tube b. The capillary is cut squarely at a and the capillary cone with the material to be treated is introduced. To this end, a piece of hot cloth or lens paper is held ready on a heating block or in a drying oven. The chamber is opened for a moment, and the capillary cone is lifted out with cork-tipped forceps. Petrolatum adhering to the handle is removed with the warm cloth before inserting the cone into the capillary.

With the capillary cone at d, Fig. 66c, the capillary is drawn out to a fine tip at a. The wide tube b is connected to the gas supply, and the gas is allowed to flow through the capillary until the air is displaced. Then the point of tip a is sealed. The capillary may be left connected to the gas supply or it may be fused off at the constriction c. In either instance, it is possible to immerse the capillary into a suitable bath so that the gas may act at a chosen temperature. To retrieve the capillary cone with its contents, the capillary is cut at d so that the handle may be grasped and the cone returned to the carrier in the chamber.

Heating Solutions in the Capillary Cone. A capillary of 0.7- to 1-mm bore and about 10-cm length is drawn out to a fine tip at one end. A small volume of water is allowed to enter the tip which is then sealed shut. The capillary cone is removed from the carrier, and its handle is cleaned as described in the preceding section. The cone is then introduced, handle first, into the wide capillary and made to slide down to the sealed end, Fig. 66b. The open end b is sealed, and the capillary, with the cone inside, is placed into a bath or heating block having the required temperature. The presence of water in the wide capillary and the confinement of the capillary cone into a small space prevent the evaporation of the solution in the cone. To retrieve the latter, the outer capillary is allowed to cool to room temperature and then cut at a.

Collecting Precipitates in Capillary Cones. A capillary of 0.6- to 0.7-mm bore is cut to a length of 6 to 7 cm. The capillary cone containing the precipitate is grasped at the handle and introduced, opening of the cone first, into one end of the capillary which is held horizontally and then introduced without change of position into a microcone, Fig. 66a. The latter is then placed into the shell of a centrifuge. After whirling, the microcone is again held horizontally, and the capillary is withdrawn. The handle of the capillary cone usually protrudes from the opening of the capillary and may be grasped for the return of the capillary cone to the carrier.

The wide capillary facilitates the handling of the capillary cone, and it limits the air space around the capillary cone sufficiently to prevent excessive evaporation of the solution in the cone.

Separation of Solution and Precipitate. After whirling in the centrifuge, the capillary cone is returned to its former position on the carrier in the moist chamber and focused in transmitted light. The tip of the micropipet is inserted into the capillary cone and advanced until its opening touches the wall of the capillary cone at a short distance in front of the surface of the precipitate. Suction is cautiously applied by means of the plunger device so that the clear solution is taken slowly into the micropipet. Either reflected or transmitted light may be used for observation. The last portion of the solution is taken up very slowly so that the operation may be stopped as soon as air begins to enter the tip of the micropipet. The capillary cone with the precipitate is withdrawn. If needed, the contents of the micropipet are transferred to another capillary cone; if the centrifugate is to be rejected, the micropipet is withdrawn from the chamber, and its contents are discharged on a strip of filter paper. The micropipet may be rinsed once with water, but this may be omitted if it is used for washing the precipitate.

Washing the Precipitate. The required amount of the proper wash liquid is taken into the micropipet from the reagent container on the carrier. The wash liquid is delivered upon the wall of the capillary cone at a short distance from the precipitate. The liquid is slowly expelled from the micropipet so that the precipitate is not stirred up. If the washing is to be repeated with a like volume of liquid, the distance from the point of the capillary cone to the meniscus is measured and recorded for future use. The wash liquid is left in contact with the precipitate for one minute, whereafter it is again removed as described above for the centrifugate. The washing is combined with the centrifugate or rejected depending upon the requirements of the procedure.

Estimation of the Volume of the Precipitate. The reasoning from the observed cross-section area of a precipitate to its weight assumes that precipitates obtained with known and with unknown amounts of estimated substance are compacted to the same extent by means of the centrifuge. The use of a high-speed centrifuge is desirable, but whirling for 1 minute

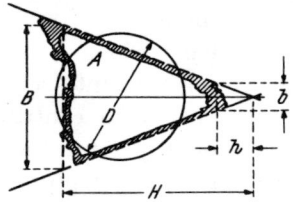

Fig. 67. Estimation of the Volume of a Precipitate

with 2000 to 3000 r. p. m. is usually satisfactory. Adherence to a standard procedure is also necessary during precipitation since a variation of the particle size affects the volume of the compacted precipitate.

Transmitted light is used for observation, and the microscope is focused on the point of the taper of the capillary cone. The precipitate often occupies an area A similar to that indicated by shading in Fig. 67. The volume of the corresponding ideal truncated cone of diameter B at the base and height $H - h$ is given by 0.26 $(B^2 H - b^2 h)$. A less accurate but simpler procedure is to imagine a sphere of the same volume as the cone of precipitate. A circle representing the cross section of this sphere is shown in Fig. 67. Its diameter D may be estimated and expressed in micrometers rather than divisions of the micrometer scale, the value of which depends upon the magnification used. The mass of the precipitate and the estimated constituent is in direct proportion to the volume 0.52 D^3 and is directly proportional to the cube of the diameter of the imagined sphere. Obviously, a mistake of 10% in selecting the diameter of the sphere will cause an error of 30% in the derived volume and mass.

Evaporation in the Capillary Cone. Placing the cone into a desiccator or under a small bell jar together with some drying agent will suffice for the evaporation of most aqueous solutions. Hydrogen chloride may be absorbed by a bead of sodium hydroxide; ammonia, by potassium acid sulfate or a drop of sulfuric acid placed near the capillary cone. Obviously, it is not permissible that liquid in the capillary cone is made to boil, but evaporation may be hastened by laying the capillary cone upon a watch glass or into a dish heated to a suitable temperature. EL-BADRY and WILSON (903) use a heating block. Creeping may be prevented by applying the "hot point" (*see* under Distillation) in front of the opening of the capillary cone. If the evaporation is performed in a dry chamber, the process may be observed with the microscope.

Expelling Dissolved Gases. Because of the small expenditure of time involved, evaporation to dryness and dissolution of the residue in solvent of the desired composition is recommended for the elimination of gases and other volatile constituents.

Distillation from Capillary Cone to Capillary Cone. A distilling capillary is obtained by giving the cylindrical part of a wide capillary cone a length

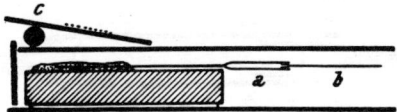

Fig. 68. Distillation from Capillary Cone to Capillary Cone; approx. nat. size

of 5 to 6 mm; it must be wide enough to permit insertion of the capillary cone containing the charge to be distilled. The capillary cone has to be calibrated (p. 208) if the distillation shall be stopped or a fraction shall be removed when the charge has evaporated down to a certain volume. Suitable allowance for the meniscus must be made when estimating the volume equal $h^3 \, \text{tg}^2 \, (\alpha/2)$ from the angle α and the height h of the cone filled by the liquid.

The charge to be distilled is placed into a capillary cone which is then spun in the centrifuge to assure that all material is collected in the point. The distilling capillary a is mounted in a dry chamber, and the capillary cone b with the charge is inserted into its opening as shown in Figs. 68 and 69, whereupon the chamber is closed and mounted on the mechanical stage of the microscope.

The heating element, Fig. 62, is inserted into the clamp of the manipulator and introduced into the dry chamber. As a rule, it is impossible to get the whole distilling capillary into the field of vision as shown in Fig. 69. Thus, the capillary cone in end a is focused in transmitted light, and the mechanical stage and the eyepiece scale are adjusted until the point of the taper of the capillary cone coincides with a convenient scale division of the micrometer. The "hot point" h of the heating element is brought into the level of the distilling capillary and moved to a point about 1 mm to one side of the capillary cone d containing the charge.

The current is turned on, and the voltage is slowly stepped up until the meniscus in the capillary cone begins to recede slowly toward the point of the taper. Before this happens, small droplets may be seen to form along the walls of the distilling capillary. The rate of movement of the meniscus in the capillary cone indicates the rate of evaporation which must be quite slow and may be controlled by either changing the position of the hot point or by regulating the voltage supplied to the heating element. It may happen that a gas bubble forms at the point of the taper, begins

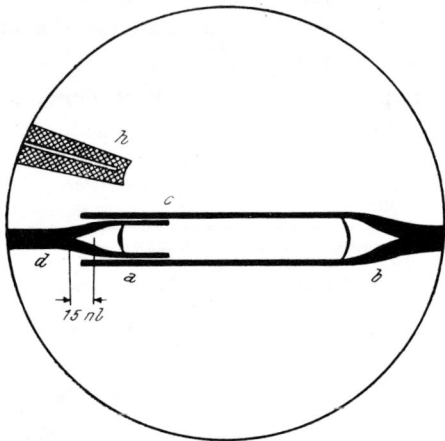

Fig. 69. Distillation from Capillary Cone to Capillary Cone as Seen with the Microscope; schematic

to grow, and pushes the liquid contents of the capillary cone toward its opening. It then becomes necessary to reduce the rate of heating to keep the bubble to a small size in the point of the taper.

When the receding meniscus reaches the predetermined mark (15 nl in Fig. 69), the hot point is quickly withdrawn from the chamber and the current is turned off. The whole length of the distilling capillary is inspected under the microscope, and it will be found that the distillate collects at the end b, Fig. 69. The capillary cone with the distillation residue may be transferred to another distilling capillary for the collecting of another fraction or it may be transferred to the carrier of a moist chamber for other treatment of its contents. The distilling capillary is spun in the centrifuge to collect all distillate at the sealed end. It is then scratched at a distance of 3 mm from the taper, and the unwanted portion of the capillary is snapped off by means of cork-tipped forceps. The part containing the distillate is mounted on the carrier of a moist chamber, and the liquid is transferred to a capillary cone of standard dimensions by means of the micropipet.

Determination of Boiling Point and Boiling Range (902). A uniformly tapering cone is drawn at the end of a wide capillary of several centimeter length. The liquid is collected in the point of the cone and confined there by a drop of mercury as shown in Fig. 70a. The capillary is then heated upon a microscope hot stage which has been calibrated by melting point (or still better, boiling point) determinations. The boiling point is taken

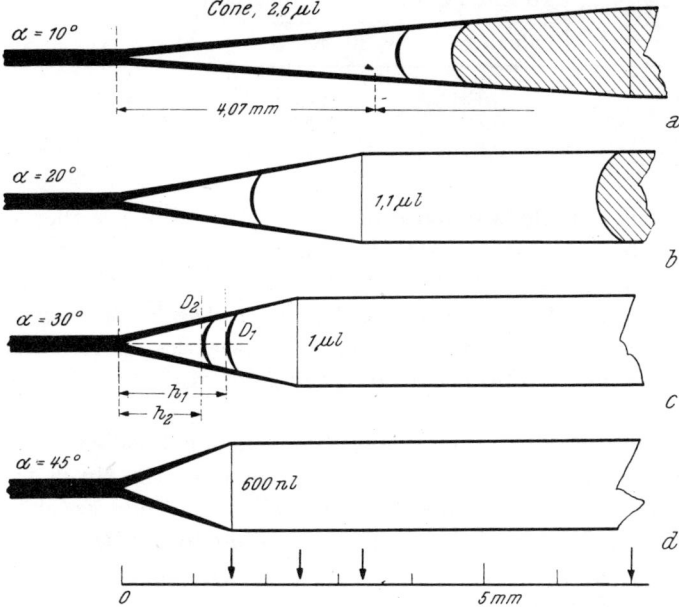

Fig. 70. Determination of Boiling Point and Boiling Range

as the temperature at which the bulk of the liquid suddenly vaporizes or condenses. Boiling ranges may be determined by plotting the temperature against the decrease of the volume of the liquid, which is computed as a fraction or percentage of the initial volume of the liquid.

Capillary tubing of soft glass, 1- to 2-mm outer diameter, and 10-cm length is heated near one end over a microflame so that the glass fuses to a solid bead which is then drawn out to a rod. The bore at the sealed end of the capillary should assume the shape of a uniform taper with an angle α of 10 to 45 degrees at the apex. It is suggested to sort the capillaries according to the angle α and to store them in stoppered test tubes. The rods are fused off about 5 mm from the taper, and the cylindrical portion of the capillaries should remain 8 cm long. Capillaries with blunt tapers will be satisfactory for boiling point determinations. A taper of

20 degrees will permit more precise measurement of the decrease of volume if a boiling range shall be determined. Lengths and capacities of the tapered portions of a capillary of 1.2-mm bore are shown and listed in Fig. 70. Table V will aid in selecting the taper according to the volume of liquid available. It should be hardly necessary to add that the capillaries must be clean and dry.

Table V. *Height of Circular Cone for a Given Volume and Angle of Taper*

Volume of Cone μl	Height h of Cone in Millimeter if the Angle α at the Apex is			
	45 degrees	30 degrees	20 degrees	10 degrees
0.001	0.18	0.24	0.32	0.51
0.01	0.39	0.52	0.69	1.1
0.1	0.84	1.1	1.5	2.4

The liquid sample is taken into a micropipet which is then completely withdrawn from the dry chamber. Without delay, the selected capillary is pushed over shaft and shank of the micropipet until the tip touches the inside wall of the taper of the capillary. The liquid is expelled from the pipet, while watching through a magnifier, and the pipet is simultaneously withdrawn. The liquid is collected in the point of the capillary by whirling in the centrifuge.

A micropipet with a relatively wide opening is filled with mercury. The capillary is pushed over the micropipet until the tip of the latter is close to the liquid in the taper of the capillary. While watching through a magnifier, a mercury column of a few millimeter length is transferred to the capillary so that only a small air space is left between the liquid and the mercury, Fig. 70a.

The capillary is bent at the middle to give it a U shape to fit the microscope hot stage. Care is taken that the capillary is not sealed during bending and that the closed end containing the sample is not heated. The capillary must be about 8 cm long so that the drop of mercury is not expelled from it when the sample vaporizes. About 6 cm of a capillary of 1-mm inner diameter will suffice to hold the vapor of 100 μg of liquid, but a capillary of 1.6-mm bore should be taken if 200 μg shall be vaporized.

The capillary is placed upon the hot stage so that the liquid sample is in the field of vision. An aluminum plate, 15 mm × 50 mm × 2 mm thick, having a 3-mm slot cut lengthwise is placed around the sealed end of the capillary. The glass baffle of the KOFLER (98) hot stage is placed crosswise over the capillary, and the top plate is put in place as for melting point determinations. The microscope is focused upon the meniscus of the liquid, which is observed while the stage is heated. The meniscus of the mercury usually moves out of the microscopic field when the boiling point is approached, Fig. 70*b*, but its position is of no consequence since the volume of the vapor is not measured. The volume of the liquid decreases slightly before the boiling point is reached. The boiling point of a pure substance is then characterized by the complete evaporation within 1 or 2 degrees and the complete disappearance of the liquid phase, Fig. 70*d*. On cooling, the boiling point is indicated by the appearance of the meniscus; in addition, the mercury drop may be seen to snap back into the field of vision.

If a boiling range is determined, the magnification should be chosen so that the meniscus of the liquid sample appears near the end of the micrometer scale when the apex of the taper is at the zero division. In the instance of fine tapers, the micrometer scale may not be able to cover the length of the liquid cone. It is then advisable to measure distances from a suitable reference point (particle of dust upon the outside of the capillary) as suggested in Fig. 70*a*. The temperature of the stage is raised very slowly, and the distance of the meniscus of the sample from the apex of the cone (or from the reference mark) is measured at the start and at frequent intervals. The volume of liquid vaporized in each given temperature interval would have to be computed as the volume of the frustum of a cone, $1.05 \, (\operatorname{tg}^2 \alpha/2) \cdot (h_1^3 - h_2^3) = 0.26 \cdot (h_1 D_1^2 - h_2 D_2^2)$, which would require also knowledge of either the angle α or measurement of the diameters of base and top of the frustum, Fig. 70*c*. The measurement of α, D_1, and D_2 is not necessary, however, since the percentage of the total volume is given by $100 \, (h_1^3 - h_2^3)/H^3$, where H is the length of the cone of liquid at the start of the distillation.

Especially when known mixtures are treated for comparison, the boiling curves obtained with 50 to 200 nl of liquid should make it possible to recognize, for example, characteristic petroleum fractions and products.

Crystal Precipitation Upon the Condenser Rod. The limits of identification of slide tests may be improved by reducing the volume of the test drops to 10 or 1 nl. Whenever, for lack of material, the crystals of the precipitate remain so small that their shape can no longer be discerned under the microscope, one is simply limited to the criteria used with test tube tests: appearance or disappearance of phases and color phenomena. Under such conditions, it will be advisable to give preference to tests giving characteristically and intensely colored precipitates which will assure a high degree of specificity and sensitivity.

A small volume of test solution, which will cover the area of the platform of the condenser rod, is measured off and taken into the micropipet. The latter is then withdrawn to a distance of 1 cm from the axis of the microscope, and the condenser rod is brought into the field while observing with the unaided eye. After focusing upon the platform, one turns off the lamp sending light to the mirror of the microscope and concentrates a strong beam of light from the lateral lamp upon the base of the condenser rod. The lamp is adjusted until the platform is brightly illuminated while the rest of the field of vision remains dark. This illumination facilitates the observation of colors and precipitates, but it is desirable to add the general illumination with transmitted light during manipulations.

By means of the manipulator and while observing with the unaided eye from the side, the tip of the micropipet is moved close to the platform of the condenser rod, which is in the focus of the microscope. While viewing through the eyepiece, the tip of the micropipet is brought into sharp focus and moved close to the platform with the manipulator. Using the controls of the manipulator only, the tip of the micropipet is slightly raised so that its image becomes somewhat blurred; it is then advanced horizontally so that the opening of the tip appears at 3 o'clock above the side of the platform. The pipet is then lowered so that the opening touches the top surface of the platform and its outlines become sharp. The plunger is slowly advanced to expel the contents of the micropipet upon the platform. When the first bubble of air appears at the opening of the pipet, the latter is first raised and then completely withdrawn from the moist chamber for cleaning.

Solid reagent may be transferred to the test drop upon the platform by the standard procedure, p. 203; a particle of about 1 μm in diameter will represent a suitable quantity. Liquid reagents are measured in and added with the micropipet. The transfer to the platform is performed as described for the test solution, but it is advisable to apply slight pressure by means of the plunger device and to cut off the general illumination with transmitted light just before lowering the tip of the micropipet into the test solution upon the platform. The pressure upon the solution in the micropipet causes it to flow out of the pipet as soon as the tip touches the test solution, and separation of precipitate inside the tip is prevented. Absence of the general illumination improves the conditions for the observation of the effect of the reagent upon the test solution.

When a bubble of air appears at the orifice of the pipet, the micropipet is removed and completely withdrawn from the moist chamber for cleaning. The test is observed with the illumination furnished by the condenser rod. The platform may finally be cleaned, after removing the condenser rod from the moist chamber, by dipping it into suitable solvents and rinsing with distilled water. If this does not give satisfactory results, a new platform is readily obtained by cutting the glass thread close below the old platform.

Performance of Spot Tests (903). EL-BADRY and WILSON provide upon the carrier a thin glass rod bent into the form of a U, 30 mm long and 6 mm wide. A cotton thread is placed across the open end of the U and attached with a suitable cement (gum or starch paste) to the glass. The thread is impregnated with reagent by applying to the middle portion of the thread the reagent solution hanging as a drop from the end of a glass rod or capillary. By means of the glass U, the thread is then mounted on the carrier, and a measured volume of solution to be tested is added from the micropipet. The latter is manipulated so that the opening of its tip is in contact with the middle of the thread. If necessary, slight pressure is then applied to transfer the solution to the thread at the rate at which it is absorbed.

Obviously, the technique might be refined by using a single fiber in place of the thread. The acro technique, too, appears promising, p. 130.

Preserving Solutions and Precipitates. Solutions and precipitates may be preserved for days by sealing the capillary cone containing them into a capillary as shown in Fig. 66b.

Starting a New Series of Procedures. It is not practical to plan too far ahead and to crowd the carrier in the chamber with a large number of apparatus and reagent containers. Thus it becomes necessary, from time to time, to clear the chamber of used apparatus and reagents no longer needed and to assemble the material needed in the next step. Obviously, it is convenient to have at least two chambers available so that the capillary cones with the material under investigation may remain undisturbed in one of them while the other is prepared for the continuation of the investigation.

Experiment 61

Mechanical Separation of the Components of a Powder (154)

Preferably a Greenough-type binocular microscope; needle mounted in a handle, 12 cm long. — Mixture of 1% bone black and 99% Al_2O_3. The particle size of the two ingredients should be checked with a total magnification of about 20 diameters before they are mixed. If the particles are too small, it will be impossible to separate them as told below. — Glycerol.

The isolation of a material by mechanical collecting assumes that the particles may be recognized by their color, shape, fluorescence, or behavior in polarized light. It is also necessary that the boundaries of the individual grains are clearly recognizable and that aggregates are either absent or readily separable into their components.

With a camel's-hair brush, dust a small amount of the carbon-alumina mixture onto a clean microscope slide. Place the slide under the microscope and focus the particles of the powder with a low-power objective which has a working distance of at least 20 mm. Use reflected light and a colored background.

For collecting the particles, place a droplet of about 0.5 μl upon the center of a clean slide, and cover it with a 1-inch (25 mm) watch glass to prevent its evaporation.

Examine the powder under the microscope and select a black particle which is not closely surrounded by white ones. By moving the slide, get this particle into the center of the field. Moisten the point of a sewing needle (glass thread, platinum wire of 0.05- to 0.1-mm diameter) mounted in a suitable long handle by rubbing some glycerol on the back of the hand and drawing the point of the needle across the treated area. With the hand resting on the stage of the microscope, hold the needle 45 degrees inclined to the horizontal and insert its point half way between the slide and the front lens of the objective. Look into the microscope and move the needle in small horizontal circles until a blurred image of its point appears in the field of vision. Then bring the point of the needle straight down upon the particle to be removed. Touch the particle and lift it out of the preparation.

While the hand which holds the needle remains resting upon the stage, use the other hand to exchange the slide with the drop of water for that with the powder mixture. Remove the watch glass and focus the edge of the drop of water. While observing through the microscope, insert the point of the needle into the space above the drop and then bring it straight down into the drop. The particle floats off when the needle touches the water.

Cover the droplet of water with the watch glass, bring the powder mixture under the microscope, and remove another black particle, etc. Repeat the procedure until 8 to 10 black particles are collected in the drop. If some white particles have been carried along, allow the drop to evaporate and remove the white particles to another drop of water. The black particles will finally be located in an area which is small enough so that one may proceed to chemical treatment without further preliminary work.

Some advice may be added. The needle should always be held as nearly vertical as possible. In this manner it is possible to avoid touching other particles in addition to the one selected. If the working distance of the objective is shorter than 15 mm, it becomes necessary to bend the end of the needle. Any visible amount of glycerol on the needle will defeat its purpose; when the particle is touched, the glycerol flows down over the particle and spreads to a drop on the slide. It is then impossible to pick up the particle floating in the glycerol. Of course, the selection of glycerol and water is arbitrary. As a general rule, the adhesive used for the treatment of the needle must be readily soluble in the liquid in which the particles are collected.

Du Fresne (978) mentions the possibility of developing on the particles an electric charge by warming them gently under an incandescent lamp for 15 minutes, whereafter they will adhere to a metallic needle. For the removal of particles from oil (immersion liquids used for the determination of the refractive index), he uses a sewing needle with a heat sink (triangular sheet of copper of about 5-g weight) soldered to it near the wooden handle. The tool is chilled by immersion in liquid nitrogen or dry ice-acetone mixture. When the point of the needle is then brought close to the immersed particle, the oil freezes around the needle, and the occluded particle may be lifted out and transferred to another location where the oil is allowed to melt. The oil may be removed with a capillary pipet, and the particle may be washed with a suitable solvent like acetone.

The practicability of the described techniques depends upon the working distance of the objective rather than upon the total magnification. Using a $10 \times$ objective and projection, one should be able to apply it with total magnifications of several hundred diameters. The use of simple manipulators for tasks of this sort (409) renders the work less tedious, removes most of the fear that some untoward accident might prevent a successful conclusion of the task, and permits the experimenter to assume a more detached attitude which is very helpful in arriving at sound decisions.

Experiment 62

Precipitation of Silver Dichromate Upon the Platform of the Condenser Rod

L. I., 1 ng Ag or less.

Equipment for working on the microgram scale. — Mixture of 8 volumes of silver test solution (10 mg Ag per ml) with 1 volume 16-F HNO_3; saturated solution of $K_2Cr_2O_7$.

Assemble in a moist chamber a condenser rod (0.2- to 0.3-mm diameter) and a carrier with a measuring capillary and two rather narrow reagent containers, one containing the acidified silver test solution and the other, the dichromate solution. With the micropipet, take up somewhat more than 1 nl of the silver solution and transfer just 1.1 nl of it to the measuring capillary; 10% of this volume will probably be left behind when the solution is then transferred to the platform of the condenser rod. For this task, empty the micropipet on a strip of filter paper; do not rinse it, but keep it wet with the solution. Take the solution from the measuring capillary into the micropipet and record the distance from the tip to the meniscus in the shaft, which corresponds to a volume of 1 nl. Transfer the solution to the platform of the condenser rod.

Rinse the micropipet, and then take into it about 1 nl dichromate solution and add it to the silver solution on the platform. As a rule, some crystals will be observed, which have the shape and color characteristic of the silver dichromate. Clean the micropipet.

The test may be repeated with a solution of 1 mg Ag per ml, which is 2-F in HNO_3. Evaporation of the test drop upon the platform may be obtained by removing the cover of the moist chamber for a short time.

Experiment 63

Estimation of the Quantities of Arsenic and Antimony in a Solution of Unknown Concentration

Sample: about $0.1\,\mu l$ of a mixture of 2 to 30 ml antimony stock solution (50 mg Sb per ml), 50 ml 12-F HCl, 2 to 20 ml arsenate stock solution (50 mg As per ml), and water to make 100 ml.

Equipment for working on the microgram scale; buzzer, p. 73; heating element. — Antimony test solution (10 mg Sb per ml); supply of hydrogen sulfide; $KBrO_3$, finely powdered solid; 9-F HBr; 3-F H_3PO_3; buffered silver nitrate (1 g $AgNO_3$ and 7.7 g ammonium acetate in 6 ml glacial acetic acid and 200 ml water); quinine-iodide reagent: before use mix equal volumes of solutions A (1 g quinine hydrochloride dissolved in 50 ml warm water; the cold solution treated with 0.2 ml 6-F HCl) and B (2 g KI in 50 ml water).

A dry chamber is needed in addition to a moist chamber. In the latter, assemble a condenser rod and a carrier with 4 capillary cones and 3 reagent containers containing the sample solution, antimony test solution, and water.

Measure out of the reagent containers 10 nl sample solution, 2 nl antimony test solution, and 10 nl antimony test solution; transfer each portion to a separate capillary cone. To correct for the liquid remaining behind, first wet the shaft of the micropipet with the solution to be taken, and then remove 10% more from the reagent container than indicated above: 11, 2.2, and 11 nl.

Add 100 nl water to the solution in each capillary cone, and then saturate each with hydrogen sulfide. Heat the mixtures of solution and precipitate for half a minute at 60° to 80° C, and then allow to stand for one hour at room temperature. Centrifuge simultaneously all three precipitates and estimate their volumes both from the dimensions of the cone or frustum occupied and the diameter of the equivalent sphere. Remove and reject the centrifugate in the instance of the sample.

In the dry chamber assemble a capillary cone which fits into a distilling capillary held ready in a vial and 3 reagent containers with 12-F HCl, 9-F HBr, and 3-F H_3PO_3. Transfer the capillary cone containing the sulfides of arsenic and antimony to the dry chamber. Treat the sulfides with 30 nl 12-F HCl; seal the capillary cone into a dry capillary and immerse for 15 seconds into a water bath of 70° C. Agitate the mixture with a buzzer and inspect under the microscope to observe that only part of the precipitate has dissolved. Return the capillary cone to the dry chamber and treat the contents with $KBrO_3$.

Grind some $KBrO_3$ to a fine powder and sprinkle it on a small slide which is mounted upon the cover of the dry chamber, Fig. 68. Transfer individual small particles of the salt to the mixture in the capillary cone, mixing after each addition, until all sulfide is dissolved and only sulfur remains behind. Centrifuge and transfer the clear solution to the capillary cone which fits into the distilling capillary. Wash the residue of sulfur with one 40-nl portion of 12-F HCl, and transfer the washing to the solution which is then treated with 10 nl HBr and 20 nl H_3PO_3. Seal the capillary cone into a dry capillary and immerse for 5 to 10 seconds into a bath of 80° to 90° C. Cool with tap water, centrifuge, remove the capillary cone from the capillary, and insert it into the distilling capillary which is then mounted in the dry chamber.

Using the graduated scale of the rotating stage, measure angle α of the taper of the bore of the capillary cone containing the solution to be distilled, and compute the lengths h of the taper corresponding to 10-nl and 15-nl volume, $h^3 = v/(\text{tg } \alpha/2)^2$. Express h in divisions of the micrometer scale. Focus the meniscus in the capillary cone and heat cautiously until the volume of the liquid has been reduced to 15 nl. Then stop heating.

Withdraw the capillary cone from the distilling capillary and place it upon the carrier. Add 10 nl 12-F HCl to the distillation residue, return the capillary cone to the distilling capillary, and continue the distillation until 10 nl liquid is left in the capillary cone. Stop the heating, and transfer the capillary cone with the distillation residue to the moist chamber. Centrifuge the distilling capillary to collect its contents at the sealed end; return it to the dry chamber and transfer its contents to a standard capillary cone, which is then transferred to the moist chamber.

Treat the residue of the distillation with 50 nl water, and dilute the distillate with water to a volume of approximately 100 nl. Precipitate the antimony with H_2S as before. Treat the distillate containing the arsenic with 40 nl 12-F HCl before saturating with H_2S. The arsenic sulfide has a tendency to become colloidal. To avoid this, heat the mixture of solution and precipitate by immersion for 30 to 45 seconds into a bath of 60° to 70° C, and then agitate by means of the buzzer. Without opening the sealed capillary, inspect the contents of the capillary cone under the microscope. If the arsenic sulfide is properly flocculated, collect it into the point of the capillary cone by centrifuging simultaneously with the antimony sulfide. Mount the capillary cones with the precipitates in the moist chamber and estimate their volumes from the dimensions of cone or frustum and equivalent sphere. Using the volumes of the known quantities of Sb_2S_3 (28 ng and 140 ng Sb_2S_3), compute the weights of the mixture of the sulfides, the antimony sulfide, and the arsenic sulfide from the sample. Finally, compute the weights of arsenic and of antimony in 10 nl of the sample solution.

The colors of the sulfide precipitates are observed with reflected light. To confirm the antimony, introduce into the moist chamber reagent containers with 12-F HCl and with quinine-iodide reagent. Remove the centrifugate and treat the remaining antimony sulfide with 10 nl

12-F HCl. Seal the capillary cone into a capillary containing some 12-F HCl, and heat at 70° C until only a white residue of sulfur remains. Spin in the centrifuge, return the capillary cone to the moist chamber, and transfer 1 to 2 nl of the clear solution to the platform of the condenser rod. Add a like volume of quinine-iodide reagent. A yellow precipitate confirms the presence of antimony; the corresponding bismuth precipitate is orange or brown.

To confirm the presence of arsenic, transfer the capillary cone with the precipitate to the dry chamber which has been cleared of other apparatus. Add a reagent container with 6-F NH$_3$. Close the chamber and transfer 10 nl of the NH$_3$ to the arsenic sulfide. Bring the hot point of the heating device in front of the opening of the capillary cone and evaporate just to dryness. Open the chamber, remove the reagent container with ammonia and replace it by containers with 16-F HNO$_3$ and 0.5-F HNO$_3$.

Treat the residue in the capillary cone with 5 nl 16-F HNO$_3$ and again evaporate. Treat the residue with 5 nl 0.5-F HNO$_3$, centrifuge, and transfer the capillary cone to the moist chamber. Clean the condenser rod, and place upon the carrier a reagent container with buffered silver solution. Transfer 1 to 2 nl of the clear arsenate solution to the platform of the condenser rod and treat with 1 nl of the buffered silver solution. A brown precipitate of Ag$_3$AsO$_4$ confirms the presence of arsenic.

Additional Practice Experiments for the Chosen Scale of Work

Experiment 64

Study of Chemical Behavior

Test solutions of silver nitrate, mercuric nitrate, mercurous nitrate, lead nitrate, bismuth nitrate, cupric nitrate, cadmium nitrate, stannous chloride, stannic chloride, antimony trichloride, and arsenate containing 1 mg of the metal per 1 ml solution; see Appendix. Reagent solutions: NaOH, Na$_2$S, KI, and (NH$_4$)$_2$S$_5$.

The purpose of the experiment is the collecting of information for the efficient identification of the cations in Expts. 65 and 66. Perform the task with the apparatus and technique chosen for confirmatory tests. Most convenient is the use of a spot plate or glass slide on darkly colored paper and stirring rods, capillaries, capillary pipets, or loops for the adding of reagent. Depending upon the amount of material available for study and upon preference, the tests may be carried out in test tubes, micro cones, or capillary cones.

a) Test a portion, droplet, of the solution of each metal ion as follows. Mixing after each addition, add small increments of 2-F NaOH until the mixture is just alkaline, test with litmus or pH paper, then add 4-F NaOH to bring the NaOH concentration of the mixture to about 1 formal. Record the observations in table form. Add to the mixture 4-F Na_2S to bring the concentration of Na_2S to about 2 formal and mix. If a clear solution results, add 4-F H_2SO_4 until the mixture is slightly acid. Record the observations and, if they do not agree with your expectations, repeat the experiments and compare with the literature.

b) Test a portion of the solution of each metal ion by adding 2-F NH_3 until just alkaline, then 6-F NH_3 to bring its concentration in the mixture to 2 formal. Add 6-F $(NH_4)_2S_5$ to bring its concentration to 1 formal, warm, and stir. If a clear solution results, add 4-F H_2SO_4 to make it slightly acid. Record as under (a).

c) Treat all but the chloride solutions with 2-F HCl until they are acid and have a chloride concentration of 0.5 formal or more. Record as under (a).

d) Treat all test solutions first with small increments of 0.1-F KI and finally with an excess bringing the iodide concentration to 0.05 formal. Acidify the arsenate solution with HCl if it is alkaline. Record as under (a).

Experiment 65

Analysis of Two Unknown Solutions

Samples: Each solution contains only one metal ion of those studied in Expt. 64 in the concentration of 1 to 10 mg per ml. Depending upon the chosen scale of work, 0.5 ml, 0.1 ml, 10 μl, or less solution is given. Recommended is work on the submilligram scale with about 2 μl of each unknown solution handed over in a capillary pipet.

Apparatus and reagents as in Expt. 64.

Determine the approximate pH of the unknown solution. Use a scheme of testing, which will permit recognition of the ion present with the use of two, at the most three, portions of the unknown solution so that enough material is left for three or four confirmatory tests. Confirm the finding with the use of at least one slide test and one spot test and try to collect some evidence (preparations, drawings, or photographs) that may be kept for the purpose of demonstration. If necessary, refresh the memory by first performing the confirmatory tests with a like volume of known test solution of similar concentration and composition before risking the loss of a portion of the unknown.

Experiment 66

Identification of Simple Compounds of the Common Metals of the Hydrogen Sulfide Group

Two different Samples: Depending upon the chosen scale of work a suitably small amount (100 mg, 10 mg, 1 mg, or less) of a solid compound of one of the metals considered in Expts. 64 and 65. Work upon the submicrogram scale is recommended, and this suggests samples of a total mass of about 0.3 to 1 mg, most conveniently about 5 or 6 particles of 0.2- to 0.5-mm diameter, which may be placed upon a microscope slide and covered with a small watch glass.

Apparatus and reagents as in Expt. 65. In addition, provision has to be made for heating in various gases or upon charcoal, pp. 78 and 276, etc. A handbook of chemistry should be readily accessible.

Inspect with magnifying glass or microscope and make use of criteria and tables given with the procedure of systematic analysis, P. 9, etc. Use color, shape, and behavior in polarized light. Depending upon amount of material available, test by heating in open and closed tube, upon "charcoal", or in a stream of air, hydrogen, etc., P. 23 to 30.

The information collected should exclude all but one or two metals and leave a choice of a small number of possible compounds. Use the residue and condensates obtained in heating tests to get a solution suited for the orientation tests used in Expt. 65 and confirmatory tests. The collected evidence should make it possible to select a suitable solvent and to decide whether it should be applied to the original or to a certain residue. Make it a habit to proceed as in Expt. 65; always collect material to be used as evidence in future demonstrations.

Using tables on the properties of compounds and minerals in a handbook, decide the identity of the sample. There should be material left for additional tests.

If time permits, it may appear desirable to insert experiments with the third analytical group (Fe—Al—Ni) corresponding to Expts. 64 to 66. In this instance, testing with a magnet should be added and, if the amount of material permits, bead tests.

Experiment 67

Identification of Simple Inorganic Compounds

Two different solid samples, each consisting of one inorganic compound, are given in the amount specified in Expt. 66.

Apparatus as in Expt. 66 with the possible addition of wire for bead tests. Reagents needed for the confirmatory tests for all common ions.

Proceed as in Expt. 66, but add testing with the magnet and the performance of bead tests if working on the gram or centigram scale.

Try to guard against being misled by the effects of incidental impurities by always making certain that the intensity of tests corresponds with the amount of unknown taken. The latter may be estimated by measuring under the microscope the dimensions of the particle taken and computing the mass with a crude approximation of the density (about 4 g/ml for compounds of heavy metals). In case of doubt, compare with a test obtained with the appropriate quantity of the substance in question.

Experiment 68

Identification of Simple Compounds

Two different solid samples, each consisting of one substance, are given in the amount specified in Expt. 66.

Equipment as for Expt. 66 with the addition of apparatus for the observation of transition temperatures and, when work is done on the gram or centigram scale, wire for flame tests and possibly a simple spectroscope.

After studying the appearance and the behavior in polarized light, heat in the melting point apparatus and determine the transition temperatures. Performance of the heating test with a **very** small amount of material and observation under the microscope is recommended. An ignition test with a **small** amount of material upon a wire must be carried out before heating any appreciable quantity of an unknown substance if there is no definite knowledge that the substance is **not** an **explosive**. The residue may be used for the following tests.

If the collected evidence does not suggest a different approach, proceed to heating in a current of air or oxygen (heating in closed and open tube) and pass the gas coming from the reaction zone through $Ba(OH)_2$. Continue as suggested in the systematic procedure. Testing the solubility in water, HCl, and organic solvents may give useful information.

Experiment 69

Identification of Simple Compounds

Two different solid samples, each consisting of one substance, are given in the amount specified in Expt. 66.

Equipment as for Expt. 68.

Proceed as in Expt. 68, but after studying the appearance and the behavior in polarized light, do not fail to test the hardness.

Experiment 70

Identification of Materials as They Occur in Nature, Industry, and Research

Any desired number of solid or liquid unknowns representing simple **or** complex materials and mixtures of such.

Proceed as suggested in the systematic procedure.

Part II

Systematic Analysis

Choice of Materials and Cleaning

It is important to use reagents and apparatus which will not introduce detectable amounts of the substances that are considered in the search. In the search for majors and minors, distilled water and reagents satisfying ACS specifications (24) will be suitable. Special precautions are required, as a rule, when testing for traces.

The label of the manufacturer guarantees the quality of the reagent only up to the time of the breaking of the original seal; from there on, it is the responsibility of the user to prevent contamination (13). Organic reagents are best kept in a dark closet since many of them deteriorate, some quite rapidly, when exposed to light.

Reagent solutions should not be kept longer than for one year; they should be discarded immediately if their appearance changes or a precipitate separates. Small glass bottles with glass stoppers serve for acids, solutions giving off corrosive vapors, strongly acid solutions, and organic solvents. Glass bottles with rubber stoppers, which have been carefully freed of coatings, are recommended for neutral and mildly acid solutions. Polyethylene containers are suggested for distilled water, ammonia, ammoniacal and strongly alkaline solutions, hydrofluoric acid and fluorides, and for metallic mercury.

The material of apparatus must be considered in every phase of the work including the preparation of reagent solutions and the storing of samples. Ordinary bottle glass is quite strongly attacked by water and harmless aqueous solutions and may introduce unpredictable impurities in addition to silica, alkalies, aluminum, iron, and calcium. The black glaze on porcelain may contain chromium, lead, cobalt, copper, manganese, and iron. Pyrex glass is recommended because of its simple composition: silica, B_2O_3, alumina, sodium, 0.4% potassium, and only traces of arsenic and antimony. The safest materials are those which contain essentially only one element: vitreous silica, platinum, silver, and nickel; the last two are used for alkaline fusions. All of them may be slightly attacked, even when properly used, and may get into the reagents or the material under investigation. New platinum ware may give off iron which should be removed by heating with strong hydrochloric acid. The surface of old platinum ware may be alloyed with other metals taken up as a consequence of improper use. The permeability of platinum at high temperature may permit flame gases (H_2, SO_2) to react with the contents of the apparatus.

Attention must also be paid to the cleaning of apparatus. Common sense suggests cleaning immediately after use. The nature of the contamination is still known, and the proper solvents may be selected for its efficient removal. The cleaning may require several steps which are in logical order: (a) disposal of dangerous contents and removal of corrosive agents by rinsing with solvent, usually water; (b) removal of greasy residues by wiping with absorbent paper or rags; (c) cleaning with brush and water; (d) use of solvents to remove residues escaping mechanical treatment: acid for metallic, sulfidic, etc. mirrors; thiosulfate for silver halides, etc.; (e) brushing with soap or detergent and water; (f) heating on the steam bath with chromic-sulfuric acid for the removal of organic residues; (g) thorough rinsing with tap water and distilled water; (h) wiping dry the outside and possibly drying the interior by either draining, or a stream of clean air and possibly heating.

Phosphates of detergent solutions and chromic ion from the chromic-sulfuric acid may be adsorbed tenaciously on glass. In addition, apparatus having been in general use may have adsorbed other substances or may be contaminated by invisible residues of insoluble substances. Cleaning may be tried by a succession of treatments: digestion with bromine water for the oxidation of sulfur and sulfides; heating with sodium carbonate solution for the conversion of insoluble compounds into carbonates; rinsing for the removal of anions; followed by digestion with 6-F nitric acid and rinsing. In the search for traces and in work with very dilute solutions, however, it may be best to use new apparatus; even this should be rinsed with 8-F nitric acid and water (13).

Especially when working on microscope slides or watch glasses, it should be remembered that fingerprints furnish sodium chloride, substances strongly absorbing in the ultraviolet region (13), oil (sebum), and cholesterin (1226).

Plastics also may contribute to contamination as has been recently summarized by DELHEZ (623). Water and solutions stored in polyethylene bottles may take up some organic substance. It has been reported that Tygon tubing introduced some plasticizer into a gas stream. Furthermore, the history of plastic apparatus should be considered; it may absorb and give off hydrocarbons and other organic substances, and it has been observed that it retains fluoride, nitrate, and sulfide ion.

Sampling for Analysis

Analysis presupposes some object of interest and a specimen of it, a sample, for investigation. It requires that some matter has been collected, the chemical identity of which is of interest *per se* or because of the deductions to be based upon it.

Sampling and analysis are related as question and answer are (463). Just as a foolish question will rarely lead to a useful answer, an analysis will not be able to solve the problem in the mind of the person suggesting it if the sample has not been selected so that knowledge of its composition will shed light upon the situation on hand. At times, a very thorough understanding of the general problem may be required to arrive at an intelligent decision concerning the objects that shall be analyzed and the questions that shall be answered by the analyses. The decision may require conferences with all persons present, who are interested in the problem or are able to contribute useful information. It is desirable that the analyst participate in the deliberations; he may gain knowledge to aid him in his task, and he may be able to suggest approaches that facilitate the analytical work and improve the probability of its success.

Qualitative micro analysis is rarely concerned with the average composition of mixtures; for the reasoning used in such instances, the reader is referred to the literature (13, 14).

As a rule, the sample for micro analysis will have to be separated and freed from extraneous material and collected in a more or less "pure" condition. Frequently, this task must be performed by the analyst himself, and it occasionally turns out the be the most difficult part of the work. Not only patience and skill, but also considerable ingenuity may be required since only very general advice can be given and the procedure must be adapted to the requirements of the special instance and the nature of the particular material. The apparatus and technique depend essentially upon the size of the specimen to be isolated and upon the amount of force required to separate it from its surroundings. Complications may arise if the specimen is either poorly defined or requires special methods of observation to render it visible. Only a few of the most obvious examples shall be mentioned in the following discussion.

In the instance of loose particles, the problem reduces to the collection of a number of them sufficient for the investigation. Considerable ingenuity has been expended upon the collection of dust particles. Filter mats made of carbon dioxide snow, potassium nitrate, benzoic acid, naphthalene, anthracene, and salicylic acid have been used since they allow isolating the collected particles free from extraneous matter by simple evaporation or dissolution of the filter (910, 1101, 1108). For the investigation of individual particles floating in air as well as for the picking up of individual particles from surfaces, it is convenient to use the impinger principle which has been widely applied in recent times. An impactor which collects the particles upon a microscope slide for investigation by variations of the microgram procedure described is used by CADLE (450). The identification may be simplified by coating the slide with a film containing the reagent (456), and even continuous counting of particles of a certain kind is possible by collecting them on a strip of cellulose acetate carrying a film with the reagent, that moves through the field of the microscope after having passed the impacting chamber (450). LODGE (455, 470, 471) collects upon Millipore filters (Lovell Chemical Co., Watertown, Mass.) which become transparent on impregnation with immersion oil so that there is no interference with the microscopic investigation of the collected particles.

Obviously, the individual particles of conglomerates may be set free by crushing and grinding until the aggregates are resolved into their components. Suitable tools are the wellknown Plattner diamond mortar and the Ellis' mortar if contamination with steel is not objectionable. Mortar and pestle for fine grinding are now available made of glass, porcelain, agate, Mullite which is synthetic $3 Al_2O_3 \cdot 2 SiO_2$, Coors U. S. A. alumina and Diamonite consisting essentially of corundum (Mohs hardness, 9), Kennametal (tungsten carbide with a binder of metallic cobalt or nickel), and pure boron carbide (468) next in hardness to diamond (13). The micromortar of ALBER (431), obtainable made of Mullite, uses interior proportions approaching the tall form in order to avoid loss of material; the inside dimensions are 15 mm high, 7 mm in diameter at the bottom, and 20 mm at the rim. Micromortar and pestle of similar shape but fashioned of a single crystal of synthetic corundum, "x-mono" (Al_2O_3), by cutting and polishing with industrial diamonds are commercially available (1290). The material is resistant to common acids and bases, has a Mohs' scratching hardness of 9, can stand pressure up to $12000 kg/cm^2$, and is more abrasion resistant than Kennametal.

Manipulators and micromanipulators (85, 86, 90, 103) may be used for dissection and the removal of particles from objects if not much force is required. Efflorescences, deposits, coatings, or pigments may be scraped loose with a steel needle and then collected by the technique of Expt. 61 (409). The simple microchemical manipulator of ALBER (862) which permits magnifications up to 200 diameters and the application of considerable force was used for the removal of small inclusions from soap bars (414). A preliminary treatment with solvents (892, 913) is applied in the investigation of paintings for loosening the protective surface layer; and solvent extraction is used to remove the binder from chips of paint (853, 860, 882). The principle may find other applications and it has been widely used for the isolation of inclusions in steel by anodic dissolution (162, 188). A microsectioner for paint films has been described by GETTENS (1092).

An interesting detailed account of the technique of dissecting metallic objects of various sizes under the microscope with the use of simple tools has been given by CHAMOT (87). The technique has been developed for the investigation of small arms' primers, but it appears generally useful. CLARKE and HERMANCE (407) used a rugged manipulator in conjunction with a dental motor with flexible shaft and a set of drills, burrs, stones, and cutting wheels for attacking complex structures in a systematic manner under the binocular microscope. Upon the object was placed a drop of oil, large enough so that the tiny tool is completely immersed; the particles removed by the burr are retained in the oil and may be collected by centrifuging and washing with a suitable solvent.

Microdrills (90) for the removal of grains and inclusions from (polished) specimens have been described by GRANIGG (768), MORITZ (1230), HAYCOCK (660), and RUSSANOW (162, 871); the rotating shaft of the drill is mounted in one tube of a binocular microscope or magnifier so that the point of the tool is in the focus of the objective carried by the other tube. Jimmying of the tool may be arrested by first coating the specimen with a 0.2-mm layer of celluloid which is applied as a solution in amyl acetate. KOCH, MALISSA, and DITGES (570) developed a simple manipulator for the isolation of tiny objects (mass of the order of 1 μg) from the surface of metallic objects. The specimen is mounted in special holders on the stage of a Zeiss microscope with binocular eyepiece attachment and special objectives giving magnifications of up to 80 diameters with working distances of 92 mm (more than 80 diameters with 27 mm). The tool is held in a clamp with ball-in-socket joint, but operated directly by the hand which rests upon a gliding support. This gives smooth motions and permits the application of considerable force.

Inclusions of 0.1- to 3-mm diameter are isolated by removing the surrounding material with suitable dental tools operated by a motor with flexible shaft. When the surrounding metal has been sufficiently removed, the inclusion is lifted out by means of a lancet or needle. Inclusions close to the surface may be squeezed out by the forces created when drilling a hole close by. Corrosion spots may be planed off for separate investigation of successive layers, and it is also possible to polish small areas for microscopic investigation. Needles, forceps, and a suction tool are described for the collection of the detritus.

Particles more than 0.2 mm in diameter are picked up and transferred by means of delicate forceps which are attached to the manipulator and may be opened and closed by turning a milled screw head at the end or the tool shaft. Smaller particles stick to surfaces by adhesion forces and may be lifted after touching them with a needle or fiber. Likewise, they are deposited by touching them to the collecting surface; this, however, becomes difficult when the diameter of the tool is large compared with that of the particle. For this reason, sharply pointed needles of steel or platinum are used for particles of about 0.1-mm diameter, but textile fibers (fine vitreous silica fibers) for smaller particles. Fibers are cemented to the end of a glass thread (Expt. 57) which is inserted into the tool holder of the manipulator. For the rapid collection of many particles of a kind from a mixture, an aspirator operated suction device may be used. It is obtained by placing a small suction tip with narrow orifice, by means of a ground joint, upon a small disk of fritted glass fused into a tiny funnel, the stem of which is connected to the glass tubing held by the clamp of the manipulator. The particles collect upon the filter disk and may be removed with suitable tools after the disk has been mounted on the stage of the microscope.

Ferromagnetic particles are collected by applying first a permanent magnet and then a small electromagnet which picks up weakly magnetic particles that have been left behind. A cover slip is attached with glycerol to the pole end of the magnet which is then moved with the manipulator over the detritus so that a gap of 1 to 2 mm remains. The magnetic particles are pulled to the surface of the cover glass and remain there when the magnet is turned upside down, whereafter the cover slip may be lifted up vertically. Most of the magnetic material may be collected by repetition of the procedure. The electromagnet is operated with alternating current and may be given the shape of a pointed needle which may be brought close to the particle that may then be dropped wherever desired by gradually reducing the current to zero. Since the wire of the coil has only 0.05-mm diameter, the current must be limited to about 20 milliamperes.

It is understood that small specimens may be imbedded in resin or plastic and thus mounted in a ring fitting into a specimen holder.

The glaze of pottery may be separated from the burned clay (141). The fragment of pottery is split through the middle by means of chisel and hammer so that only one side of each half has a glaze coating. A fragment is placed, with the glazed surface down, into molten paraffin which is allowed to solidify. While running tap water over the specimen, the clay may then be removed with a dental drill until only the glaze is left imbedded in the paraffin, which is then transferred with knife and forceps to filter paper. Pressing between filter paper while heating upon a block to 80° C, will remove most of the paraffin.

The "ultrasonic jack hammer" (90, 765) uses styli made by electrolytically pointing a drill rod of 1.1 mm in diameter. The specimen is covered with a small glass hood which is supplied with argon to prevent air oxidation as a consequence of the created frictional heat. The inclusion is focused under the microscope, whereupon the stylus is lowered to make contact with it. When this happens, the inclusion is shattered around the point of contact, and a shower of debris is scattered over the vicinity of the inclusion. The extent of destruction is readily controlled by manipulation of the micropositioner which guides the tool. The debris is finally collected in suitable manner. Inclusions of diameters down to 10 μm may be isolated.

Systematic Procedure of Analysis

It is understood that the technique of working must be adjusted to the amount of material available for investigation. It should be kept in mind that also the general approach is determined by the size of sample.

If enough material is available and more may be obtained without trouble, it is only reasonable to proceed without hesitation to destructive methods if they promise a quick solution of the problem.

On the other hand, if little material is available and cannot be replaced or cannot be increased without excessive expenditure of time and effort, it becomes imperative to derive a maximum of information from careful reasoning based upon the history of the specimen, its appearance, and non-destructive testing. Destructive testing must be delayed until careful deliberation has given a plan assuring solution of the problem with as small a part of the available material as possible. No part of the sample should be sacrificed without assurance that some vital information will be gained. If the sample cannot be replaced and the procedure is not based upon reliable routine experience, it will become necessary to first test each step with knowns representing possible compositions of the unknown and to make any changes that will improve the hope for success before proceeding with the analysis. Depending upon the nature of the specimen, work of this kind may require a considerable amount of analytical research.

In work connected with the investigation of crime and in all instances where there is no knowledge whatsoever concerning the nature of the unknown, the experimenter is advised to protect himself by first testing for the presence of dangerous radiation (P. 14) and of explosives. In addition to testing for stability upon heating (P. 21), a small sample, not more than 0.5 mg, may be treated—with the necessary precautions: goggles and heavy gloves—in a porcelain mortar to test the stability on shearing and mechanical shock.

P. 1 The History of the Sample

Extremely small quantities of sample which cannot be replaced may have to be analyzed in connection with the investigation of crime, accidents, material failures, and objects of art, archeology, basic and industrial research. In all these instances, it is desirable to sacrifice only small part of the sample for analysis and to keep most of it as evidence and for additional tests, the desire for which may occur at a later time. Under such circumstances, one should try to get so much information from inspection, nondestructive testing, and a careful study of all factors surrounding the origin of the material that a few chemical tests with a small fraction of the sample will establish the identity.

As a rule, the story of the origin is given by the agency submitting the sample, but an interview may give additional data. Inspection of the sample may reveal the need for additional information. It may become desirable to inspect the milieu from which the sample comes, the tools used in isolating it, and the wrapping materials used for shipping. It may become necessary to go through a careful study of a manufacturing process to get a list of the substances used. When dealing with objects of archeology, art, etc., the advice of experts is usually available, who should be able to point out literature and museum collections for additional information.

It is the analyst's responsibility to properly select and evaluate the information and to decide what the method of isolating and shipping the sample may have done to it.

P. 2 Description of Sample and Record of Investigation

The report should begin with the date of receipt and a description of the sample as received or of the method of its collection and the circumstances connected with it. Especially when the amount of material is small, an estimate of the quantity of sample should be given and this should be followed by statements concerning the amounts sacrificed for the various destructive tests and the quantity left for evidence and future testing. The report on the investigation should contain all facts leading to identification and consequently explained by the identification and also those observations which seem incidental and are not necessarily connected with the identity. Depending upon the importance of the problem and the scarcity of the sample material, efforts should be made to preserve fractions and tests for future reference or to obtain photographs or drawings of the specimens and significant tests. The record should finally disclose where the left-over sample is kept or for what is was used.

P. 3 Preliminary Inspection

The aid of a magnifying glass will not be needed, as a rule, in deciding whether the sample is solid or liquid. Use of centrifugal force is recommended for separating the phases of a slurry; the washing and analysis of the solid may have to be postponed until investigation of the liquid has furnished the information needed for efficient performance.

Solids should be inspected with the magnifying glass or (and) under the microscope (478). To start with, this will reveal obvious heterogeneity and may suggest mechanical separation into several fractions for separate investigation or mechanical removal of obviously incidental contaminants. Often it may be advantageous to repeat the inspection in ultraviolet light. The sample may be placed into a dark cabinet illuminated with ultraviolet radiation and observed with or without optical aid while shielding the eyes against the radiation. If available, a fluorescence microscope may be used (88). The light emitted by the sample or parts of it may reveal even very slight contamination or heterogeneity (95). Use of or contamination by lubricant will be revealed and minute stains by biological fluids will be discovered.

The presence or absence of repetitive external shape or internal structure will reveal whether or not the material is **organized,** i. e., derived from plant or animal. If not fossilized or incinerated, all organized material will later (P. 22) show the presence of organic matter. Since species and type of tissue frequently give striking differences in shape and structure, whereas the subtle differences in chemical composition are still mostly unknown, identification must rely to a large part upon the recognition of form and microscopic structure, P. 5.

Matter which is **not organized** may again be separated into three classes be mere inspection with the use of magnifying glass or microscope. The material may have a shape or structure indicating that it is the product of some human activity, an **artifact** (see P. 5); it may consist of more or less well developed **crystals** (see P. 6), and it may **not** be **distinctly crystallized** and present shapes which are more or less incidental, P. 7.

P. 4   The Sample is a Liquid

Liquids derived from organized matter are, as a rule, either oils (essential or fatty) or aqueous solutions. The surrounding tissue retards the evaporation of small droplets of these fluids. Water, aqueous solutions, and mineral oil form large bodies in nature. Outside of living matter, small samples of water and other volatile liquids will persist only if sealed into cavities; water, liquid carbon dioxide, and mineral oil may be found included in minerals and rocks. In all these natural occurrences, the history of the sample will be a good guide to recognition of the general nature of the liquid. Likewise, the circumstances connected with the origin of industrial and research samples should narrow the area of search so that identification becomes possible with a very small sample. Even when a container is found without identifying label, the location (place in a systematic collection), and the type of container should give a very good idea of the possibilities involved.

Carbon dioxide or water included in minerals is identified according to W. N. HARTLEY (320, 690, 691) by determination of the critical temperature under the microscope.

It is understood that goggles should be worn and the work performed behind a protective shield as long as there is no assurance that the unknown has not an explosive character. If sufficient sample is available, a **test for stability on heating** should be carried out before exposing any sizable amount of sample to elevated temperatures. To this end, take up about 1 mg of the liquid through the firepolished end of a thin-walled capillary. Mount the capillary in slightly inclined position behind a safety shield and push a Bunsen burner under the capillary so that the just non-luminous flame envelops the part of the capillary adjacent to its opening holding the drop, while the opening itself is in or just barely outside the seam of the flame. Observe whether the substance burns, the appearance of the flame (luminous, non-luminous, color, sooty), and the nature of decomposition products; save for analysis any ash left. If explosion or deflagration does not occur, it may be assumed that it probably is safe to heat the unknown material.

Assuming that a small sample of a liquid shall be investigated and no clues to its nature are given, it will be best to determine first its volatility so that the necessary precautions against loss may be taken in the subsequent testing. To this end, all or part of the sample is taken into a capillary pipet (p. 100 or 103) with the necessary precautions. The rate of flow through the fine intake capillary may provide a rough indication of the viscosity. The color of the liquid may become perceptible, before sealing, when viewing the opening of the fine capillary containing the liquid, p. 177. If the whole sample is taken for the test, also the appearance in ultraviolet light should be observed, P. 9, before heating. The attempt of determining the boiling point may indicate the presence of a more or less pure substance and give its boiling point; it may indicate a mixture of liquids and give its approximate boiling range; finally, it may indicate the presence of dissolved solids (or gases). By all means, the temperature should not be raised much above 300° C, and heating should be stopped at the slightest sign of decomposition. After cooling, the centrifuge will serve in collecting all material in the tip of the capillary.

If a sharp boiling point is obtained, a study of the tables in handbooks together with additional observations (color, fluorescence, fluidity, odor or fuming perceived during transfer) may suffice to get identification by means of a few additional tests. The recognition of a particular batch of the liquid may be possible by means of a *schlieren* test, Expt. 5, which will also allow to identify a particular batch of mixture.

If boiling starts above 200° C, one may assume that vaporization losses will not be excessive at room temperature. If it starts below 200° C, the liquid should be classed as volatile and special precautions should be taken when working with small samples; most tests are easily modified for performance in capillaries.

Testing the odor by P. 16 becomes practical with small samples if the volatility is low. Furthermore, a droplet may be transferred to the surface of a slide for inspection under ultraviolet light; this may be desirable since viewing the sample in glass, which absorbs most of the ultraviolet, may not reveal slight fluorescence. The refractive index too, may be determined with the glass powder scale upon the microscope slide (159). In the instance of volatile liquids, a few particles of a glass powder may be brought with the centrifuge into the taper of a capillary pipet, whereupon the tip of the pipet is inserted into the liquid and the latter permitted to enter until the particles are immersed (653). For microscopic inspection, the tip may be sealed, and the capillary may be placed into a suitable liquid, Fig. 50. After observation of the test and cleaning of the outside of the capillary, its tip may be cut open, and the contents may be transferred with the centrifuge to a pipet with another sample of glass powder or returned to the storage vessel.

The melting point may be determined in the same capillary in which the boiling point has been observed.

Nonaqueous liquids that fume in contact with air may react with water. A small sample may be treated with water, and the resulting aqueous solution may then be tested for inorganic ions; if a gas is liberated during the reaction with water, it should be collected for identification (153, 157, 1090).

If the liquid is a mixture, the procedure should be selected with due consideration of the collected evidence and the origin of the sample. Separation may be tried by distillation, gas chromatography (477), fractional crystallization (Expt. 54), or a combination; if properly performed, significant loss of material need not occur. A mixture of organic liquids may be extracted with water, dilute NaOH, or (and) dilute acid; the same treatment may be applied to solutions of solids in organic solvents, but it will be preferable to evaporate the solvent in such a manner that it and the residue may be collected for the determination of their relative masses. In the instance of an aqueous solution, it will suffice to evaporate a measured part of it so that the resulting solid residue may be weighed; neither measurement of volume nor determination of weight have to be very accurate. The residual solid is investigated as suggested in P. 6 or P. 7.

The identification of isolated liquids may be tried by infrared and ultraviolet spectroscopy (P. 34), determination of the refractive index (p. 238 *above* and P. 18), determination of the density (P. 19), of optical rotation, of surface tension, of viscosity, of melting point, of boiling point, of critical temperature, of the critical solution temperature, by ultimate analysis (P. 35), tests for functional groups, specific confirmatory tests, and methods of quantitative analysis.

Plotting one physical constant as a function of another reveals simple, nearly straight-line relationships for the points of homologous series. Charts showing the plotted data on refractive index, density, and boiling point permit classification, limit the possibilities to be considered very sharply, and indicate what tests or derivatives, if any, are needed for final identification (449).

MARION (427) has shown how to convert at little expense a polarizing microscope to a polarimeter with a cell of 0.15- to 0.18-ml capacity. For other suggestions see SMITH and EHRHARDT (439).

The surface tension may be determined with 2 μl liquid by the method of FERGUSON and KENNEDY (1068). The liquid is placed into the open end of a narrow capillary which is connected to a manometer system in which the air pressure is varied until the meniscus at the mouth of the tube is plane.

The capillary microviscosimeter of BOWMAN (423, 735) requires only 30 μl of sample and gives a precision of $\pm$ 0.001 in the range from 2 to 10 000 centistokes.

The critical temperature may be readily determined by heating, on the microscope hot stage, a sealed capillary of about 20-mm length, one third of which is filled with the liquid (653, 1019).

The critical solution temperatures have been listed (1) for 6000 pairs of liquids, about 70% of which are hydrocarbon compounds. The determination with the use of the microscope hot stage has been described by R. FISCHER (886, 889, 897). It requires from 0.2 to 2 μl of sample depending upon the bore of the capillary used, which may be varied from 0.2 to 0.9 mm. Needed is a test liquid which is pure or closely reproducible, is not completely miscible with the sample at room temperature, but becomes miscible at a higher temperature before a critical temperature is reached or decomposition starts. Suggested are: *paraffin oils* for low molecular weight alcohols, aldehydes, ketones, and phenols; *glycol* for essential oils, ethers, esters, aldehydes, ketones, and benzene derivatives; *glycerol* for aldehydes and ketones; *ethylenecyanide* for benzene, its homologous series, and halogen derivatives; *benzyl alcohol* for aliphatic hydrocarbons; *methanol* and *aniline* for aliphatic hydrocarbons and naphthenes.

Take into a capillary of 30- to 40-mm length from 5 to 7 mm of the sample and of the test liquid selected. The two liquids must be in contact. Seal the capillary at both ends to give it a length of about 30 mm.

It is assumed that the hot stage has been equipped with a metal frame which carries an aluminum sheet, 26 mm $\times$ 37 mm $\times$ 1.5 mm thick, so that the latter may be moved by means of milled heads outside the hot chamber. The aluminum sheet has a slot, 32 mm long and either 0.5 mm or 1.1 mm wide, into which the capillary fits.

Place the sealed capillary into the slot of the aluminum slide, and cover the latter with a glass slide, 26 mm × 34 mm, which is held in place in some suitable manner. Tilt the microscope to a position 45 degrees to the horizontal, and move the frame so that the meniscus appears in the field of vision with the capillary in the diameter from 6 to 12 o'clock. Raise the temperature at any desired rate and keep the interface of the two liquids in the field of vision by means of the motions of the frame. When the critical solution temperature is approached, the meniscus flattens out, and its outlines become faint and disappear when the critical temperature is reached. Allow the stage to cool very slowly (1 to 2 degrees per minute) and watch the region in the capillary, in which the meniscus was last seen. Read the temperature when the critical point is indicated by the appearance of droplets and a fine line, the meniscus.

The critical solution temperature can be reproduced within $0.2°$ C. As a rule, the effect of the pressure seems to be negligible. Because of the lack of convection currents, the two liquids mix mainly by slow diffusion which automatically produces the critical mixture somewhere in the narrow diffusion zone of about 0.5-mm height. The possibility should be kept in mind, however, that the meniscus may disappear if both liquids acquire the same refractive index at a certain temperature. Mistakes are easily avoided by the use of oblique illumination or of the Becke test; the meniscus never disappears altogether, and it becomes again more distinct if the temperature is raised further. Finally it must be pointed out that the critical solution temperature may be very strongly affected by impurities of the sample; in such instances, it may be used for a rather accurate determination of the amount of impurity.

P. 5 Identification of Organized Matter

Identification by shape and structure requires familiarity with these. The chemist with little or no training in biology and general microscopy may, with the use of introductory books and by the study of the microscopic appearance of various samples of known origin, nevertheless be able to recognize at least the class of matter to which an unknown belongs. If samples of this kind will have to be handled frequently, it becomes certainly advisable to study the specimens or materials which have to be considered, and to collect samples, drawings, photomicrographs, and microscopic mounts for comparison. Study of different samples of the same kind will show the characteristic features which are preserved in spite of incidental variations. The execution of drawings is strongly recommended; it is not only an aid to the memory, it helps developing the ability of recognizing distinctive features.

Whereas inspection under the microscope may immediately reveal the identity of a sample in some instances, it would be a mistake to

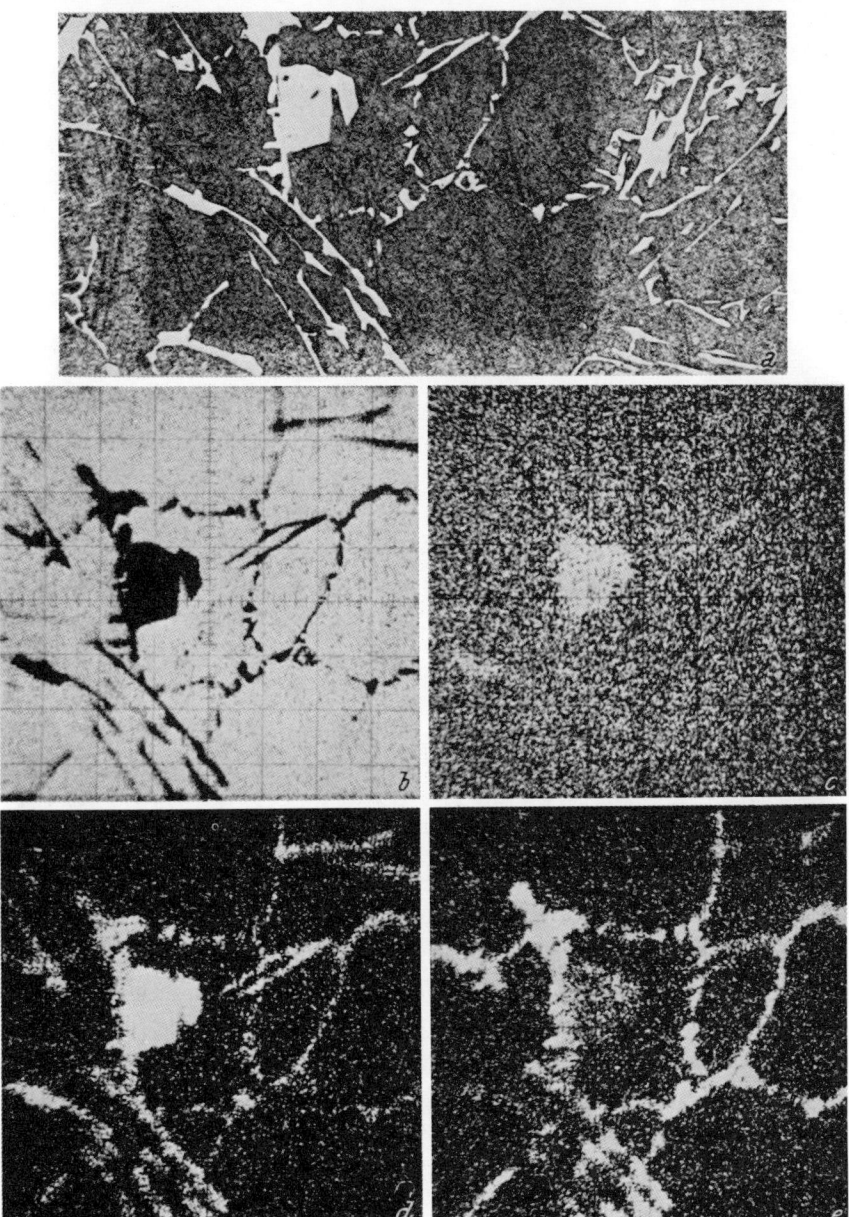

Plate I. Investigation of a Solder with an Electron Probe Microanalyzer, Magnification 400×. *a* Light photomicrograph showing the darkened area analyzed with the electron probe; the darkening is caused by contamination with oil from the diffusion pump, which polymerizes where the electron beam strikes the surface; *b* Image produced by backscattered electrons when the specimen is scanned with the electron beam; the dark areas indicate regions which are lower in average atomic number then the light areas; *c*, *d*, and *e* Images obtained with the characteristic X-rays excited by the scanning electron beam and showing the distribution of arsenic (*c*), tin (*d*), and antimony (*e*).
Micrographs by the courtesy of Research Laboratories, General Motors Corporation, Warren, Michigan

Benedetti-Pichler, Identification   Springer-Verlag in Wien

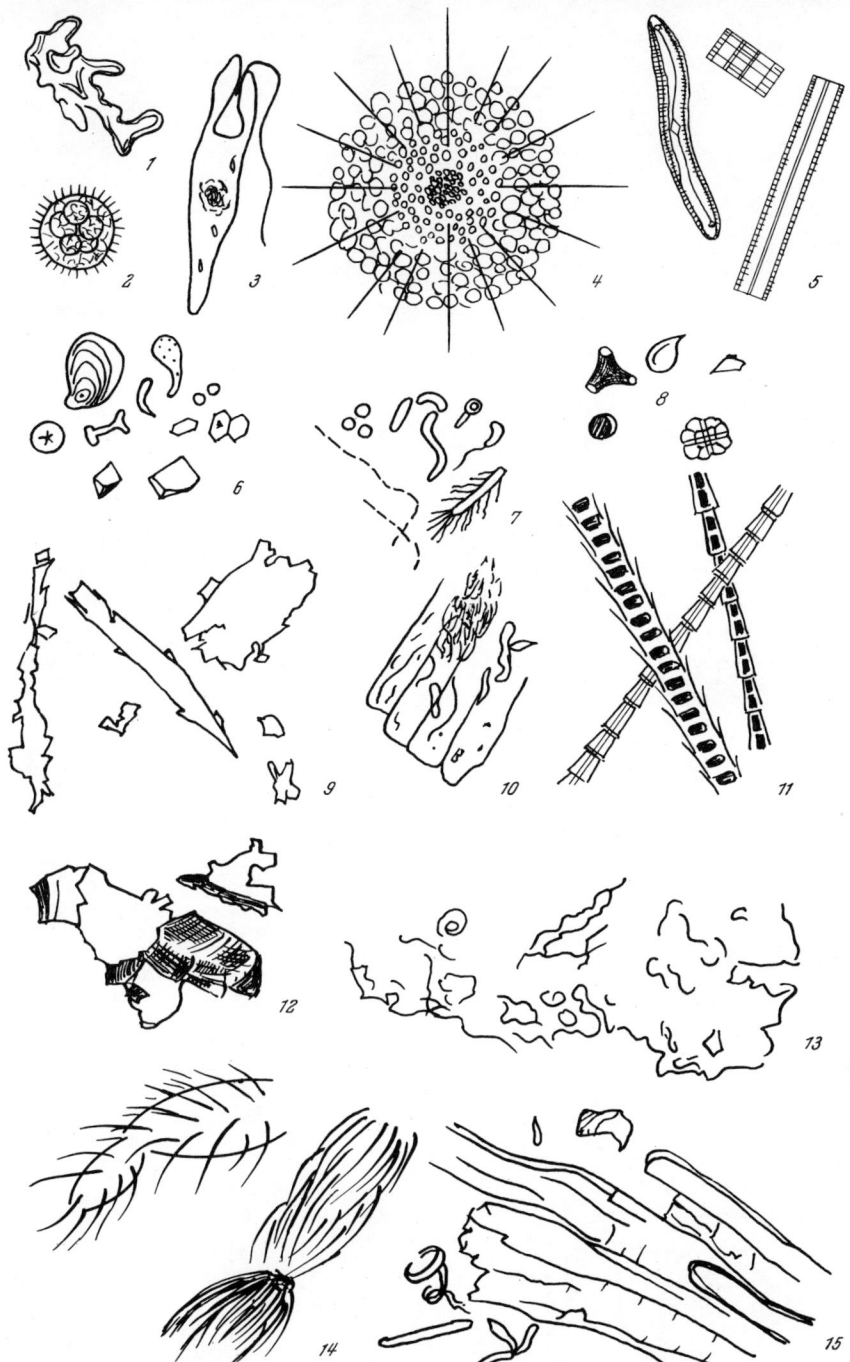

Plate II. Examples of Organized Matter. *1* Amoeba Proteus, 40×; *2* Volvox, 40×; *3* Euglena, 700×; *4* Thalassicalla Nucleata, 15×; *5* Diatoms, 450×; *6* starch grains, 80×; *7* bacteria, 1000×; *8* pollen, 120×; *9* filings from finger nails, 80×; *10* ground black pepper in cedar wood oil, 400×; *11* cat's hairs in cedar wood oil, 400×; *12* ground black pepper, dry in reflected light, 80×; *13* particle of human skin, in air and reflected light, 80×; *14* legs of insects, in air, 80×; *15* wood filings in cedar wood oil, 400×

expect this to occur as a rule. The shape of starch grains (Plate II), to mention one example, does appear also among pollen, bacteria, and occasionally even in crystalline material, Expt. 36. The very variable size may not provide a decisive clue. Soil, ground black pepper, tobacco dust, and many others may look very much alike under low magnification; filings from finger nails (Plate II) resemble chips of white paint film or plastic. In all such instances, immersion in water or cedar wood oil and change to medium or high magnification will reveal the cellular structure of organized matter. Origin and history of the sample provide most useful leads. Protozoa occur only in essentially aquatic media including living tissue and body fluids. Soil is of such variable composition that no general characteristic can be given; the presence of hard particles (quartz) is readily recognized by the scratch marks produced on moving the cover slip under slight pressure, P. 17. The recognition of the soil of a particular locality requires close comparison with a sample taken from the same spot (dirt on shoe and soil of foot print); the cooperation of a biologist may be essential.

The beginner needs advice concerning sources of materials for study and the preparation of microscopic mounts. To this end, CORRINGTON's, "Exploring with your Microscope" (175), and other elementary introductions to general microscopy (111, 190) are valuable. The following list gives an idea of the kinds of organized matter, that may be met, and supplies references to the special literature.

Micro Organisms
    Algae (175)   Molds (181, 182)
    Bacteria (189)   Protozoa (175, 181)
    Diatoms (175)   Yeasts (182)

Seeds (180)

Parts or Tissue of Plants (177, 180, 185, 194, 196, 197, 202, 205)
    Fibers, see below   Starch (180, 190, 197)
    Wood (179, 180, 190, 197, 1100) Charcoal (1100)
    Spices (195)   Vegetable Drugs (200)

Parts or Tissue of Animals (180)
    Insects (175, 181)   Skin and Leather (672)
    Excreta (181)   Hair and Feathers (181, 186, 675,
    Blood (175, 186)   1130), see also Fibers, below

Fibers (180, 183, 184, 188, 190, 191, 197, 204)
    Artificial Fibers (180, 184, 188, 204) may also be identified by the melting point (459)

Food (181, 201)

Water (207)

A few chemical tests may be mentioned, which may serve for general orientation.

1. If not completely fossilized or already ashed, the material will char and in general show the presence of carbon compounds when tested in P. 26. If the ashing of tissue is carried out with the necessary precautions, the structure is preserved in the ash and may be more clearly recognized there than in the original (194, 310, 641, 642, 852).

2. Moisten with 0.1-F I_2 in 0.4-F KI solution: starch turns blue; fat, oil, protein, wool, and silk become yellow or brown; artificial cellulose fibers become brown at first and turn blue after some time. The material may be recovered by allowing the test to stand until the iodine has evaporated.

3. Moisten with 16-F HNO_3 and heat upon the steam bath until dry. Proteins assume a yellow color which darkens upon adding (6-F) NH_3. Many resins, alkaloids, and tyrosine behave in a similar manner.

4. Moisten with a solution of 1 g phloroglucinol in 50 ml ethanol, which has been treated with 25 ml 12-F HCl: lignin containing, woody tissue assumes a bright red color which slowly changes to violet (650).

P. 6   Identification of Artifacts

Man-made materials are recognized by shapes or textures which must have been purposely brought about and cannot be imagined as the result of a chance happening. Even when the external shape is completely preserved, the criterion fails in the instance of fibers. When fragments occur, that preserve little or none of the outside shape, the structure may not afford a clue to the origin. Thus, splinters of glass and obsidian, or fragments of industrial slag and pumice cannot be differentiated by their structure. As in the instance of organized matter, a check list of the objects which may be met, appears useful.

No Outstanding Dimension, Irregular Shape

    Fragments of structural parts: metal, alloy, concrete, brick, mortar, caulking, ceramics, plastics, rubber, sponge of cellulose, plastic, glass

    Filings, shavings, chips of metal, alloy, plastic, wood (1100), cork, pulp (179)

    Beads of metal, alloy, glass, or plastics

    Powders obtained by grinding or precipitation, see P. 8

    Food Products: candy, baked and unbaked dough, bread, cake, noodles

One Outstanding Direction,

    Spun fibers, thread, wire, needles

Two Outstanding Directions,
> Metal foil, laminates, paper (179), inorganic paper, cloth, paint and lacquer films, coatings, chips of paint, enamel, glaze, plaster, and other structural materials

Since very different materials are put to identical uses and are given like shapes, chemical analysis gains in importance. As far as organized matter is involved (wood chips, paper pulp, fibers), identification by structure as outlined in P. 5 will have to be used. In general, it will be advisable to obtain good photomicrographs or (and) drawings of the sample material before it is put to destructive tests since the appearance may assume increased importance when the results of chemical analysis are known. A scale should be included into the likeness. The particle size distribution of powders may be decisive in the identification of a particular product or brand.

Inspection in ultraviolet light may produce valuable evidence concerning origin or characteristic contamination and should not be omitted with artifacts. The selection and order of the tests of the systematic procedure will be determined by the history of the sample and its appearance (shape, texture, transparency, luster, body color, surface color) in white and ultraviolet light. The use of polarized light, testing with a magnet, or a hardness test may be indicated before reviewing the evidence with the use of Table I, handbooks, and special literature. One may conclude that mechanical separations or the removal of binding matter by solvent extraction should precede the chemical investigation of parts of special interest. In general, the whole systematic testing procedure, starting with P. 9, should be considered.

In drawing the final conclusions at the close of the investigation, the question of origin or batch identity may have to be solved on the basis of appearance or characteristic contamination and comparison with samples of known provenience.

P. 7   Well-Developed Crystals

If the sample consists of more or less well-developed crystals, proceed with the systematic procedure starting with P. 9. Photomicrographs or drawings should be made. An attempt may be made to determine the crystal class, and observation of transition temperatures while heating to 300° C should not be omitted. Especially when dealing with minerals, identification by their optical properties may be considered. Under all conditions, it will pay to determine characteristic angles and to study the behavior in polarized light.

Crystals which are not clear have undergone some change, and their outward shape is no longer related to their internal structure. The outward appearance may give a clue to the chemical composition of the originally formed crystals and to the nature of the change (transition to another modification, change to a different state of solvation, dehydration, oxidation, reduction, metathesis) which has taken place.

P. 8   Solids of Random Shape and Structure

This group shall comprise crystalline and amorphous matter in the form of powder or irregular lumps or pieces; it is also assumed that the outlines of crystals cannot be clearly discerned.

Microscopic examination should be tried with transmitted and reflected light and varicolored backgrounds; observation in ultraviolet light is recommended, and the magnetic behavior should be tested, P. 15. Photomicrographs may be used to record the general appearance since type of agglomeration and luster of surface are difficult to describe. If this holds any promise, a Geiger test and an autoradiograph of the sample, P. 14, may be made before proceeding with the investigation. In general, the latter will have to follow the systematic procedure starting with P. 9; testing the solubility, P. 32, may be rewarding if recrystallization furnishes crystals suited for optical investigation and angular measurements. The hardness should be tested if abrasives may be present, and the determination of transition points below 300° C should not be omitted, if carbon compounds are not excluded.

Non-Destructive Testing

P. 9   Action Upon Light, Color

By color of an object is meant its **transmission** or **body color**, i. e., the sensory effect of the change in composition imparted to daylight which has passed through a selectively absorbing object. It is implied that illumination with artificial light may give entirely misleading color impressions, and it is assumed that the observer has "normal" color perception.

The body color is greatly affected by the thickness of the specimen. Thin layers of lightly colored substances appear colorless in transmitted light, and thin layers of strongly colored matter or highly opaque substances may show unexpected hues which do not seem related to the color of bulk. The body color of fine powders is most sensitively observed with darkfield illumination or with reflected light before a black background; a white background will serve if the particles are darkly colored. Use of all three: low-power microscope, magnifying lens, and of the unaided eye is recommended; if the particles are very fine and appear only lightly colored, one may also try to gather them into a pile for the observation of the color of the reflected light. Immersion in media of like refractive index for the elimination of disturbing refraction phenomena is not practical at this stage of the investigation if only little material is available.

Any form of darkfield illumination (reflected light and black background) may also bring forth surface and fluorescence colors. **Surface or reflection colors** are caused by selective reflection from the surface. As in the instances of crystals of $KMnO_4$ or indigo, they occur only with strongly absorbing substances and are approximately complementary to the body color.

Finally, body color should not be confused with color resulting from structural peculiarities of the object, which may produce:

a) **white** which is always caused by repeated reflection and refraction of white light from the large "internal" surface of conglomerates of small, colorless, transparent objects (flour, marble, porcelain, milk, foam, smoke); such matter prevents passage of light and, in transmitted light, may consequently show:

b) **black** which as a true body color would be the consequence of an absorption band taking in all of the visible region;

c) **interference colors** characteristic of thin plates or films or brought about by the grating effect of repetitive microscopic structures (88);

d) color resulting from difference in dispersion when the refractive indices of object and surrounding medium are approximately matched, **Christiansen effect** (88);

e) **Tyndall colors** resulting from the selective scattering of particles of about 300 nm diameter, which makes them appear blue in reflected light and orange or brown in transmitted light (88);

f) **transmitted blue** (88) and colors derived from the **chromatic aberration** of the optical system, which will disappear or change when the illuminating or (and) image-forming system is altered.

Interference colors and the Christiansen effect are characteristic of certain substances, but are not directly related to their chemical identity.

An accurate description of color requires statement of the spectral composition or reference to a chart of standard hues. No such device is being used in Table I since color and hue of many substances varies with their history. The Table must be in many ways incomplete and inaccurate; it should be taken as an aid to the memory, trying to make certain that at least the most obvious possibilities are recognized. Among others it does not mention **lakes** of organic dyes and **tungsten bronzes,** both of which may exhibit any color of the rainbow. The former char and leave an inorganic residue when ignited. Tungsten bronzes are characterized by their metallic luster, high density, and semimetallic properties. They are of variable composition and obtained by the reduction of alkali and alkaline earth tungstates. They precipitate metallic silver from ammoniacal solutions, are resistant to HCl or HNO_3 (but not always to a mixture of the two), and are oxidized to tungstates when ignited in air (16).

Luminescence is the emission of light after absorption of radiation of other frequency. In this connection, emission of visible light is wanted. If it occurs during excitation, one speaks of **fluorescence,** but if it continues after the exciting radiation has been cut off, it is called **phosphorescence.**

Ultraviolet light is most convenient for excitation of fluorescence. If it is alltogether free from visible radiation, the light emitted by the object cannot be confused with reflected or refracted visible light coming from the source.

Lenses and prisms of vitreous silica are needed for concentrating ultraviolet light into the small field of a high-power microscope (88). On the other hand, illuminating with ultraviolet for the observation of fluorescence (phosphorescence after the light has been turned off) with the unaided eye, a magnifying lens, or a low-power microscope is relatively simple. Argon glow lamps may be mounted in a black cabinet so that they irradiate objects placed into the center of the box, but cannot be seen through openings provided for observation. A cleverly designed cabinet should permit observation with the low-power microscope as well as inspection with the unaided eye of large objects held in the hand. Ultraviolet light is absorbed by the lenses of a microscope or a glass window, but stray light is difficult to avoid. Thus the observer should, by all means, wear goggles with ultraviolet absorbing glass, that enclose the eyes on all sides.

The samples to be inspected are best directly exposed to the source of ultraviolet without intervening cover glass. Containers of vitreous silica or of Corex glass are best for liquids and gases, but apparatus of thin ordinary glass may serve.

A plate of strongly fluorescing uranium glass or a thin-walled beaker with alkaline fluorescein solution is used for determining the path of the ultraviolet radiation. A smear of petrolatum or a fragment of a boric acid crystal may be used as test objects; they should be luminous when brought in the position intended for the object. In the same position, unglazed porcelain should be black if visible radiation and stray light are absent.

The outstanding feature of inspection in ultraviolet light is the extraordinary sensitivity with which fluorescent substances, contamination, and heterogeneity are shown. This recommends it for the preliminary inspection of samples and the isolation and collection of the parts of special interest. The usefulness for identification of substances is seriously impaired by the fact that trace impurities in a substance as well as its history greatly affect its behavior; they determine whether or not there is fluorescence or phosphorescence, and they determine the color of the fluorescent light. This peculiarity, however, renders observation in fluorescent light a useful tool for batch identification after the nature of the substance has been recognized by other means, and there are many applications connected with determining the origin and history of materials and with the authenticity of documents, objects of art, etc. (50).

Pure inorganic substances show little or no fluorescence, but technical grades fluoresce in various colors. Outstanding and useful for identification are the strong yellow-green fluorescence of uranium compounds, the strong phosphorescence (yellow-orange) of impure zinc sulfide and luminous paints, and of ZrO_2 (white). Strong fluorescence is also characteristic and indicative for some minerals: calcite, $CaCO_3$ (red to violet); fluorite, CaF_2 (red); celestite, $SrSO_4$; ruby, Al_2O_3; spinel, $MgO \cdot Al_2O_3$ (red); hydrozincite, $5\ ZnO \cdot 2\ CO_2 \cdot 3\ H_2O$ (blue-white); and zircon, $ZrO_2 \cdot SiO_2$ (10).

It is claimed that most organic compounds fluoresce (10, 621), but there is still doubt whether this holds for the pure compounds. Strong fluorescence is exhibited by (solutions of) derivatives of diphenylmethane, triphenylmethane, quinoline, acridine, as well as by leuco compounds of dyes (indigo), chlorophyll, porphyrins, body liquids, petroleum and its fractions.

P. 10 Investigation of Crystals and Crystal Fragments

Most of the following observations assume that material is transparent. Obviously, angular measurements may be performed on opaque objects, but the data obtained may belong to a form or compound which has undergone metamorphosis and is no longer present.

The measurement of angles requires reasonably well-developed crystals and fragments which retain characteristic facets. In the instance of isotropic fragments, the lack of crystal form makes it impossible to decide whether the material has a regular (cubic) lattice or is glassy. If a solvent which does not react with the material can be found, the result of recrystallization will readily solve this problem and will also produce crystals for measuring angles in the instance of anisotropic materials.

The determination of precise crystallographic data will assure most positive identification but requires use of special equipment; in addition, considerable difficulty may be met in the preparation of crystals suitable for microscopical measurements. The effort spent, naturally, will not lead to identification if the data are not listed in the literature or the indexing system fails, but the effort gives the data which may be listed after the material has been identified by other means.

P. 11

Interfacial Angles of pyramidal forms and of prism and dome faces of monoclinic and triclinic crystals are characteristic for each substance and permit its immediate identification. In 1912, E. S. von Fedorow submitted an index of interfacial angles of some ten thousand substances to the

Russian Academy of Sciences, but the organization of the data was not practical. Simple classification rules for selecting the key angles were developed by T. V. BARKER (83) and applied in the Barker Index of Crystals (104). The latter lists in three volumes about ten thousand of the more common inorganic and organic substances arranged in the order of their characteristic interfacial angles; in addition, other useful data are given such as melting points, refractive indices, and cross references to the X-ray powder index.

The relatively inexpensive and efficient two-circle optical goniometer of L. W. CODD and W. T. MOORE (1105) is claimed to give the interfacial angles with an accuracy of three minutes of the arc, but its usefulness in micro analysis is somewhat limited by the size required for the crystals to permit proper mounting, at least $0.1 \text{ mm} \times 0.1 \text{ mm} \times 0.2 \text{ mm} = 0.002 \text{ mm}^3$ or $6 \,\mu\text{g}$ for a density of 3 g/ml. The angular measurements, the precision of which depends mostly upon the perfection of the crystal facets, are used to construct a gnomonic projection of the faces which are identified by reference to orthogonal or (and) perspective drawings of the crystal (665). The revealed symmetry gives the crystal class, and application of the Barker rules leads to the recognition of the classification angles and identification.

If crystals are too small for mounting on the goniometer or such instrument is not available, it is possible to derive by graphic methods the interfacial angles from angles observed and measured under the microscope (438). The precision of angular measurements under the microscope, however, will rarely exceed ± 0.5 degree, but the graphically derived interfacial (Barker) angles "will be serviceable, if allowance is made for a possible error of $2°$" (438). Naturally, the interfacial angles of cubic crystals or combinations of pinacoid and prism of the hexagonal, tetragonal, and orthorhombic classes are determined by the implied symmetry and not characteristic of a particular substance.

P. 12

Profile Angles. SHEAD (412) recommends measuring the angles of the outline of very thin crystal plates which, by necessity, will always settle with the large faces parallel to the surface of the microscope slide. Thus, the profile angles observed under the microscope are readily reproducible and characteristic for certain substances which have a tendency to separate in tabular form.

The profile angles are related to the interfacial angles, and have like the latter little or no diagnostic value in the instances of regular hexagons and rectangles observed on cubic crystals or pinacoids and unmodified prisms of the hexagonal, tetragonal, and orthorhombic classes. Angles derived from pyramids and domes and their combinations with prisms and pinacoids are useful for identification purposes. But for the triclinic system and clinopinacoid faces of the monoclinic, the angles in the outline of thin tablets are related to one another in a simple manner.

An asymmetric octagon is ordinarily the most complex form met. If the characteristic profile angle of the basic parallelogram cannot be derived, it is still possible to use the angles and the order of their occurrence for comparison and identification.

SHEAD (958) suggests determining the profile angles on selected perfect thin tablets of the simplest geometric form available, preferably parallelograms, the whole outline of which is simultaneously in focus. If only one kind of parallelogram is found, the two different angles add up to 180 degrees, and either one of them may be taken as the characteristic profile angle.

If other, more complicated forms appear in addition to the simple (basic) parallelogram, the derived forms may be identified by simple reasoning which may also be used for recognizing the basic parallelogram and its characteristic profile angle (956). Five- to eight-sided forms originate when one to four corners of the basic parallelogram are cut off parallel to a diagonal of the basic parallelogram. Depending upon whether the basic parallelogram is equilateral or not, from one to two or from two to six new angles, respectively, may appear.

The equilateral square or rhombus is readily recognized by testing whether or not the cross hairs of the eyepiece can form the diagonals. Truncation of a corner or two opposite corners by a parallel to the diagonal gives a symmetrical pentagon or hexagon; there is only one new angle B which is related to the characteristic profile angle A of the basic parallelogram by $4B = 720 - 2A$. This condition holds also when the equilateral nature of the basic parallelogram is hidden by unequal growth rate which changes the square to a rectangle and the rhombus to a rhomboid.

If the basic parallelogram is a rectangle or rhomboid, truncation parallel to one diagonal gives two new angles, B and C, so that $2A + 2B + 2C = 720$, and truncation parallel to both diagonals of a rhomboid gives two more new angles, D and E, so that $D + E = 180 + A$ and $2B + 2C + 2D + 2E = 1080$.

If the truncation is so severe that nothing is left of the sides of the basic parallelogram, the derived form is a parallelogram that preserves the angle of the diagonals of the basic parallelogram.

The desired simple shapes may be obtained by recrystallization (412, 970) or sublimation (412, 958, 971). In general, an intermediate rate of growth seems to favor the formation of thin tablets of simple outline. If well developed, three-dimensional crystals are obtained, the rate is too slow. On the other hand, finely divided granular or formless as well as feathery and dendritic deposits indicate that they grow too fast. Shead sublimes from one slide to another which is separated from the first by a glass ring

Table VI. *List of Profile Angles* (412, 958)

Substance	Parallelogram acute angle degrees	Hexagon 2 angles degrees	Hexagon 4 angles degrees
$KClO_3$	79.8		
KNO_3	79.8		
$CaSO_4 \cdot 2\,H_2O$	52.5		
$CaC\,H\,O_6 \cdot 4\,H_2O$	57.5		
$AgCH_3COO$	—	90	
$Ag_2C_2O_4$	58		
$Ag_2Cr_2O_7$	44.5	87.4	136.3
$HgBr_2$	69		
HgI_2	64.4		
$HgCH_3COO$	83.5	97	131.5
Acetanilide	—	99.5	130.2
Anthracene	109.7	—	—
Antipyrene	—	128	116
Asparagin	50.7	—	—
Aspirin		119.7	120.3
Bromanil	109.2	—	—
Chloranil	113.3	—	—
p-Dichlorobenzene	59	—	—
Morphine	59.1	118.2	120.9
Naphthalene	108.2	—	—
o-Nitrobenzoic acid	39.5		
o-Nitrophenol	80		
m-Nitrophenol	58		
p-Nitrophenol	77		
p-Nitrotoluene	80		
Phenobarbital	56.7	113.4	123.3
		122.1	119.5
Picric acid	87	87	136.7
		108.5 (?)	126.3 (?)
Sulfonal	85	95	132.5
Tribromphenol bromide	69.7	110.3	124.5
Trional	86.5		
Urea nitrate	81.8	81.8	139.4
	49.8	99.5	130.2

of 4-mm height and 16-mm inside diameter. The condensing slide is slightly greased with sebum from the fingers or face, wiped to leave only an invisible film of oil, and then heated to the sublimation temperature before placing it on top of the ring. The material to be sublimed should be evenly distributed over the floor of the cell.

One estimate of the precision of the profile angles listed in Table VI is $\pm\ 0.25°$.

P. 13

Behavior in Polarized Light. The following data may be determined:

a) Test for anisotropism: Expts. 7, 9 and A. N. WINCHELL, p. 106 (113).

b) Determine the vibration directions and the angle of extinction: Expts. 8, 9, and WINCHELL, p. 106. Prepare drawings and enter the vibration directions and the angle of extinction.

c) Determine the relative velocities, and mark the vibration direction of the slower component with s: Expt. 9.

d) Using one nicol, determine pleochroism: Expt. 11. In the drawing, record the colors observed with the vibration directions.

e) Using one nicol prism, observe the intensity of shading (heavy or light outline) and possibly behavior of Becke line in both extinction positions. Record the difference found and the conclusion made concerning the strength of double refraction (magnitude of the difference between the refractive indices for the two components of light). The strength of birefringency may also be estimated from the interference color and the estimated thickness of the specimen (88).

f) Possibly try to observe axial figures, Expt. 12, and if such are obtained, try to determine the sign of double refraction; the location of the optic axial plane OAP; the magnitude of the optic axial angle $2\,E$ in air, $2\,V$ in the crystal, and $2\,H$ in an immersion medium of $n = 1.515$; and the direction of the acute bisectrix Bx_a.

Concerning the **interpretation** of the findings, some caution is indicated. Obviously, solid substances which are isotropic are either a glass or crystallized in the cubic system, but these materials become anisotropic when exposed to strain. Internal strain may result from twinning or external influences during growth, and diamond is usually anisotropic. Thus, whereas isotropism indicates the *cubic* or glassy states, anisotropism does not exclude them.

The **cubic** or **isometric class** is established when an obviously crystallized material is isotropic. In the instance of irregularly shaped bodies, recrystallization or X-ray diffraction may be used to differentiate between glass and cubic crystal.

Hexagonal crystals are recognized by the fact that they always show parallel or diagonal extinction and that "end views" (or cross sections) are isotropic and have the shape of regular hexagons (profile angle, 120°), equilateral triangles (60°), or dodecagons with alternating profile angles equal to $(120 + 2f)°$ and $(180 - 2f)°$, respectively. By isotropic end views is meant that these remain completely dark between crossed nicols while the stage is given a rotation through 360°.

Tetragonal crystals and **orthorhombic** crystals also have always parallel or diagonal extinction. If an end view may be obtained which is either a square (profile angle, 90°) or an octagon (angles of $90 + 2f$ alternating with such of $180 - 2f$ degrees), remains completely dark between crossed nicols during a complete revolution of the stage, and (or) gives a uniaxial interference figure, tetragonal structure may be safely assumed. As a rule, orthorhombic crystals will not present an "end view" with a simple rectangular profile when they approach the behavior of an isotropic substance between crossed nicols. In addition, even when viewed parallel to an axis of isotropism, they never appear as dark as a uniaxial (hexagonal or tetragonal) object, and the interference figure is of the biaxial type.

Theoretically, the biaxial classes could be differentiated by the fact that orthorhombic crystals show only parallel (diagonal) extinction, monoclinic crystals show parallel and oblique extinction, and triclinic crystals exhibit only oblique extinction. In reality, measurement under the polarizing microscope is not precise enough to assure recognition of a slight degree of obliqueness, and the danger exists that monoclinic crystals may be classified as orthorhombic or triclinic ones as monoclinic. This does not prevent, however, identifying **triclinic** crystals with certainty if **all** views give clearly oblique extinction; it will be necessary to "roll" a crystal under the microscope to get all possible views.

All classification based upon crystal profile is impossible with fragments. The positions which give darkness (uniaxial) or a more or less dark gray (biaxial) when rotating the stage with crossed nicols inserted into the path of light, are suited for starting the observation of interference figures (with biaxial objects, one will then try to view along the acute bisectrix) (88). Use of the interference figures will permit distinguishing between uniaxial and biaxial materials, but it will not be possible to assign a crystal class unless suitably developed forms are obtained by recrystallization.

Since cubic and hexagonal crystals are readily recognized and do not occur too frequently, lists of the more common of these substances have been abstracted from Lange's Handbook of Chemistry which in 1952 described about 800 minerals, 2600 inorganic substances, and 6800 organic compounds. The lists of Tables 2 and 3 (Appendix) are long and obviously incomplete, but they may be helpful since circumstances may rule out whole groups. Thus, metals are excluded if the particles are transparent, and rare minerals or exhibition specimens of synthetic chemicals need, as a rule, not be considered.

In the instance of organic compounds, the six cubic and 18 hexagonal substances listed seem to narrow the field of search close to final identification. It may be expected that there are far more cubic and hexagonal carbon compounds. Even in the group of the 6800 most common of them, many solids are described as crystalline, needles, plates, leaves, powder, prisms, scales, rhombs without specifying the crystal class. Lattices of low symmetry seem to predominate, however, and the lack of definite statements indicates reluctance to form well-developed crystals which could be recognized under the microscope. If this interpretation is correct, the following lists are fit for practical use since they give the substances which may be recognized as cubic or hexagonal with the use of the microscope.

Inorganic substances are listed by formula, and names or remarks are placed in parentheses. Minerals are indicated by giving first the name. Abundant and common minerals are emphasized by the use of **bold face**; radioactive minerals by *italics*. Most of the minerals are either uncommon, or rare, or very rare. Carbon compounds are listed by name and (in some instances) formula. The temperature of melting or decomposition is added, if known.

P. 14   Testing for Radioactive Decay (770)

A test for the radioactivity of the sample is generally advisable in legal investigations. Otherwise, if induced radioactivity or the presence of artificial radioactive isotopes need not be considered, a test for radioactivity is indicated if the substance is a mineral or an industrial product which may contain uranium (yellow-orange glass or glaze, luminous dials) or thorium (gas mantles); it is not necessarily without purpose if the sample is a carbon compound.

Natural radioactivity is shown by the elements belonging to the three series of radioactive decay: U 238 — Th 234, 230 — Ra 226 — Rn *222* — Po *218*, *214*, 210 — Bi *214*, 210 — Tl *210* — Pb *214*, 210; **Th** 232, 228 — Ra 228, *224* — Ac *228* — Rn *220* — Po *216*, *212* — Pb *212* — Bi *212* — Tl *208*; and U 235 — Th *231*, 227 — Pa 231 — Ac 227 — Fr *223* — Ra 223 — Rn *219* — Po *215*, *211* — Pb *211* — Bi *211* — Tl *207* and by the elements K 40, Rb 87, In 115, La 138, Nd 150, Sm 147, Lu 176, and Re 187. (Italics indicate a half-life of less than 10 days.) The decay of the last-named eight elements is very slow so that the radioactivity is low even with Rb ($\beta-$), In ($\beta-$), Nd (β), Sm (α), and Re (β), which contain a high ratio of radioactive isotope. The naturally occurring elements K ($\beta^+ \gamma$) and La (γ) contain only a minute fraction of radioactive isotope. Lutetium (β) need not be considered since it occurs in such small quantity that it will not be detected unless a special search is undertaken. Concerning the series of natural radioactive decay, the presence of any one element of long life will assure the presence of an equilibrium mixture of daughter elements and strong radioactivity (probably α-, β-, and γ-radiation).

Any available apparatus may be used for the detection of radioactivity. Alpha radiation is readily recognized in a spinthariscope or with an electroscope; the Geiger-Müller counter is most convenient for beta radiation but must have a very thin window for recording low-energy beta rays or must be charged with heavy gas to make it sensitive for gamma rays; oscillation counters are able to handle all kinds of radiation including X-rays. Shields are generally useful in classifying radiation according to penetration power, and they are used in autoradiography also.

Autoradiography. Place the sample or part of it upon a thin sheet of black paper or upon a perfectly flat sheet of the thinnest aluminum foil available. In the darkroom, lay the sheet with the sample upon the emulsion side of X-ray film as used by dentists, which rests on the bottom of a light-tight can. Close the can, and set it aside for at least 12 hours, best at a temperature of 0° to 5° C. Finally save the sample and develop the film in a solution consisting of 1 volume Eastman print developer D-72 and two volumes of water (434). The film will show a dark area where the radioactive matter was located; inspection under the microscope may reveal radiation tracks if the radioactivity was too low to produce general grayness.

The sensitivity of the detection of alpha and weak beta radiation is improved by placing the sample directly upon the photographic emulsion, but this may produce darkening also because of phosphorescence or chemical action. Use of special emulsions (Eastman Kodak NTB 2 for alpha particles, NTB 3 for beta radiation) is needed to get records of the radiation tracks. Prints showing the location of radioactive matter may be obtained by placing thin sections or the polished faces of rock specimens upon the emulsion. The autoradiographs may be magnified by the customary photographic enlarging procedure and compared with corresponding photomicrographs of the object.

P. 15   Testing for Ferromagnetism

Testing for ferromagnetism may be useful if metallic or darkly colored particles are present. It must not be forgotten that, to-day, organic materials also may appear magnetic as a consequence of incorporation of magnetic oxides or metals (rubber, recording tape).

Spread granular material upon a sheet of glazed paper of contrasting color, and while watching the material, move the pole of a permanent magnet along the underside of the paper. Ferromagnetic matter is recognized by its motion, orientation along the lines of magnetic force, and attraction to the pole of the magnet. If the sample is very small, observe it through a magnifying glass or microscope while approaching it with the magnet. It may be left in a container of thin glass; if it is transferred to a paper or thin glass plate, which is placed under the microscope, it is practical to use a strong Alnico magnet in the shape of a knitting needle (about 1.2-mm diameter and 12 cm long), which is easily used below the stage. A slurry is tested by moving the pole of the magnet close to the surface of the liquid; a drop of it may be placed upon a slide or watch cover of thin glass, which is placed upon the stage of the microscope and approached from below with one end of the Alnico needle.

If ferromagnetic particles are present, they are best separated from the unmagnetic material. The procedure has been described on p. 233. Individual particles may be lifted out of the microscopic field with the point of a magnetized needle and collected upon a square of glazed paper or upon a slide, which has been placed over the poles of a strong magnet, Fig. 71 d; glycerol may be used to prevent the glass plate from sliding off the metal support. For alternatives, a fine wire of soft iron (reagent wire meeting A. C. S. specifications, 0.23-mm diameter) may be used and temporarily magnetized with an electric coil or by contact with a permanent magnet. To the latter end, prepare a capillary pipet of the shape indicated by Fig. 71 a and dimensions permitting insertion of an Alnico needle into

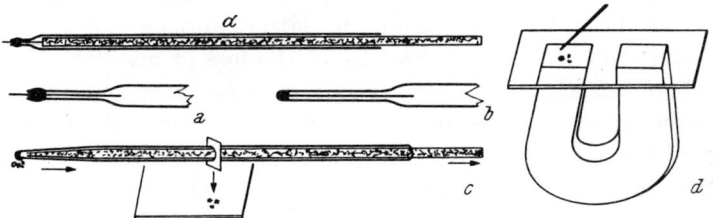

Fig. 71. Use of Needle-Shaped Permanent Magnet for the Transfer of Particles

the wide tube. Insert a 0.5- to 0.7-mm length of the soft iron wire through the tip which must be short so that the wire may contact the Alnico needle. Using a flame of about 1-mm diameter, heat a portion of the tip so that the glass tube shrinks on the wire and holds it in place. Depending upon choice, either seal the end of the tip to enclose the metal, Fig. 71 b, or leave the end of the wire exposed, Fig. 71 a. For the transfer of particles, insert the Alnico needle so that it touches the wire in the tube. Make certain that the tip of the tool is thoroughly clean and free from matter that may act as adhesive. Pick up the particle as directed in Expt. 61, and then move the tip of the tool close above the spot where the magnetic particles are to be collected. On withdrawing the Alnico needle from the tube, the particle should drop off the point of the tool. If this does not happen, try collecting with the use of a strong magnet, Fig. 71 d. Obviously, the technique may also be used for removing individual particles from a slurry and depositing them in a drop of liquid for collection.

A more crude technique may be used if sufficient material is available. One end of the Alnico needle may be pointed by dipping it into acid until its diameter is reduced to the desired extent. The needle is then, pointed end first, inserted into a glass capillary which has been sealed at the finely drawn out end. A square, 4 mm × 4 mm, of glazed paper, plastic, or thin sheet rubber is punctured at the center and pushed up upon the capillary to a point about 2 cm from the sealed end, Fig. 71c. After collecting the particle at the sealed point of the capillary, the latter is then moved so that the lower edge of the (paper) square is above the point of collection. On withdrawing the Alnico needle from the capillary, the collected magnetic material follows the Alnico magnet up to the square and then drops off.

The list of ferromagnetic substances is small: the metals iron, cobalt, and nickel and some of their compounds; alloys of Fe, Co, Ni, Mn with Al, Cu, Ag, Cr, Ti, W; the oxides Fe_3O_4, Fe_2O_3, Co_2O_3 and their mixtures; finally any other substances suitably compounded with ferromagnetic ones to produce magnetic materials.

P. 16 Odor

The sense of smell is about 25000 times as sensitive as that of taste. One may be hesitant about using the latter, and tasting small samples is completely out of question because of the loss of material. One should, however, not fail to note the odor of the sample for investigation, which may indicate the presence of a substance that other methods cannot detect because of lack of sensitivity.

All identification by smell is naturally based upon familiarity with the odor of the substance concerned or upon comparison with known samples. This fact is clearly indicated by the adjectives describing odors, which nearly always refer to things or conditions: fresh, sweet, fragrant, balmy, spicy, aromatic, flowery, sour, acrid, burnt, oily, rancid, earthy, stale, moldy, musty, foul, fetid, putrid, etc. as well as by the more close descriptions with reference to the origin: odor of rose, geranium, mint, pine, bitter almonds, onions, garlic, vinegar, old sherry, stale beer, rotten egg, burnt rubber, burnt flesh, decay, and untold other things more or less closely specified.

Some obvious disadvantages of identification by odor are derived from the high sensitivity of the olfactory organs. One may be mislead by the odor of incidentally present trace impurities which may have been introduced by the handling of the sample. The reasoning may be speciously based upon the perception of trace impurities responsible for the "odor of illuminating gas" (odor of intentionally added additive), "odor of acetylene" obtained from calcium carbide and water (odor of phosphine), or "odor of coumarin" (odor of unknown contaminant?). Furthermore, the response of the sensing organ shows individual differences and depends greatly upon the health and condition of the mucous membranes. Fatigue may be explained by a lack of response of the olfactory nerves when the membranes have become loaded with absorbed vapors. It may be connected with the well-known facts that the odor of some substances is not perceived when they are given in high concentration, that the odor changes with concentration, and that the odor is modified by the presence of other vapors as well as by preceding exposure of the membranes. To properly function, the olfactory organs must be used with caution and be given a chance to recover after each severe assault with strong odors by a rest in fresh air until the mucuos membranes have lost the absorbed vapors. It should hardly be necessary to add that testing for odor should be done in clean air (and in the absence of deodorants).

As a rule, the observation of odor will be a casual one and happen during the first inspection of the material under investigation. Any odor of the packaging material, paper or plastic sheet containing the sample should be noted. When opening a container, the odor of the interior should be observed before its atmosphere is replaced by fresh air. A stopper or cap may have absorbed vapors, and also its odor should be noted.

A deliberate test may not be practical when only a small amount of substance is available and its odor is not very strong so that its perception requires an atmosphere saturated with the vapor.

A test may be carried out by placing some of the material into a perfectly clean and odorless vial which is closed with a glass stopper. After the lapse of time sufficient to allow saturation of the gas space with the vapor of the substance (standing over night at room temperature), the vial is then opened in clean air for smelling its contents. A simple calculation shows that the procedure may not be advisable for the investigation of very small samples.

If the vial is given a capacity of only 2 ml, the mass m of evaporated substance will approach
$$m = 10^{-7}\, p\, W \text{ gram},$$
where W is the molecular weight of the substance and p its vapor pressure at room temperature in millimeter mercury.

In the instances of chloroform, acetone, and esters ($p \geq 100$ mm), this would cause evaporation of more than 1 mg, a loss which would not be tolerable on the milligram scale. When working with microgram samples, the procedure would not be permissible with substances like naphthalene and aniline ($p \approx 0.1$ mm, loss of 1 to 2 μg), or iodine, phenol, cresol, benzoyl chloride, and camphor ($p \approx 1$ mm and evaporation of about 12 μg) or acetic acid ($p \approx 20$ mm and evaporation of about 0.1 mg).

These limitations apply also to chemical tests based upon the appearance of characteristic odors. The identification of acids or alcohols by the odor of their esters does not appear inviting for the small scale. On the other hand, not more than a few milligrams of the ester should be prepared since the characteristic odor may not be recognized if too much of the vapor is produced. The fragrance of acacia or orange blossom will not be perceived in the laboratory where a milliliter of the oil is prepared, but it will be noticed (by other people) on the aired clothing of the personnel.

P. 17   Hardness

Whereas the cohesion force in a properly annealed glass is the same in all directions, the resistance offered by crystallized matter to mechanical penetration by a foreign body depends significantly upon the direction of the applied force. Crystals of various substances (mica, gypsum, etc.) are quite readily split with a knife along certain cleavage planes which are always parallel to a possible face which may or may not be shown by the particular crystal (usually prism, pinacoid, dome, cube or octahedron). The perfection, smoothness, and gloss of the resulting facet is determined by the ease of cleaving and also characteristic for the substance.

Hardness may be variously interpreted to mean ability to withstand pressure, shear, wear, etc. and the data obtained depend upon the choice of the test method. Good reproducibility is given by methods which employ a standard point attached to a lever so that it may rest under a known (and adjustable) load upon a smooth, horizontal surface of the test object. The depth and width of the scratch obtained when the latter is moved sideways are measured under the microscope. The test is performed with a very fine point under the microscope to show the hardness of the individual grains visible in polished surfaces of rocks, ores, alloys, etc. (90). UYTENBOGAARDT (109) lists the hardness obtained in this manner, crystal system, reflectivity, color, reflection pleochroism, behavior on etching, and various other characteristics for the microscopic identification of ore minerals, and BOWIE and TAYLOR (1054) claim that the task may be accomplished in minutes. Experience must be assumed since observations with polarized light on opaque materials do require it (169).

MURDOCH (101) recommends performance of scratch tests on polished surfaces (thin sections of minerals) by means of the point of a No. 10 Sharp's needle which is attached at an angle of 30 degrees at one end of a handle that is 12.5 cm long and weighs 7 g. Materials are classified as soft if the mere weight of the handle, held near the middle, suffices to scratch them. If pressure has to be added to obtain scratching, the material is called medium hard; hard materials are not scratched by the steel needle. In his tables, the minerals are classified first according to color (colored, white, and gray) and secondly according to hardness; reagents applied with a platinum loop to individual grains (169) lead finally to identification. For classification tests serve 8-F HNO_3, 6-F HCl, 20% KCN, 1.5-F $FeCl_3$, 10-F KOH, 0.2-F $HgCl_2$, aqua regia, and 3% H_2O_2.

SHORT (169) finds too many overlaps between Murdoch's medium and soft grades of hardness; a needle point loses its sharpness very quickly, and some minerals show considerable variation of hardness depending upon the crystallographic direction of the scratch. Consequently, he divides the minerals into soft ones, which are readily scratched by the needle, and hard ones, which are not scratched or only with difficulty. His determinative tables use the classes "soft" and "hard" and a subdivision into isotropic and anisotropic; etch tests lead to the final decision as in Murdoch's scheme.

If a large specimen is available for investigation, one may use the hardness scale of MOHS: (1) talc, soft to the finger nail and greasy to the touch; (2) gypsum, readily scratched with the finger nail; (3) calcite, scratched by a brass pin; (4) fluorite, easily scratched by a knife; (5) apatite, scratched with difficulty by a knife; (6) orthoclase, easily scratched by a file; (7) quartz, with difficulty scratched by a file; Nos. (7), (8) topaz, (9) corundum, and (10) diamond scratch "window" glass.

Especially when the type minerals are being used, one will try to scratch the mineral with the type and vice versa. A scratch must not be confused with a "chalk line" or streak produced by material rubbed off the scratching tool; the latter may be removed by wiping without leaving a mark. The color of the streak, which is the color of a fine powder of the softer material, is also used in the identification of minerals.

Obviously, the distinction between soft and hard, below or above Mohs' hardness number 5, is not sufficiently helpful if the material under investigation may be any inorganic or organic substance. In addition, there is no method of testing without risk of loss or contamination of a small specimen. Even with very small samples, a test for high hardness is advisable, however, if the material has a dark color, might be an abrasive, might come from some grinding or cutting tool, or has a glassy appearance combined with a high refractive index. The test is performed as follows.

Transfer a particle that may be spared to the center of a perfectly clean microscope slide and place upon it a clean square of glass (25 mm × ×25 mm) cut from a second slide. Pressing down with the raser end of a pencil, move the top square about 5 mm along the surface of the supporting slide. Inspection under the microscope will show a scratch on the glass surface which was sliding over the particle if the latter has a hardness of 7 or more. The scratch may be surrounded by glass splinters. If the material is soft, it may crumble under the pressure; if it is plastic, it may show deformation. For recovery of the specimen, lift off the glass square and place it, upside down, upon one end of the microscope slide. Inspect the slide as well as the square, and when the specimen is found, collect it with a needle as told in Expt. 61. If the specimen is transparent and colorless, it may not be recognizable in the debris of glass splinters; the latter are isotropic, however, and an anisotropic specimen will be recognized between crossed nicols. After removal of the test material, wipe the glass surfaces and confirm the presence of the scratch by inspection under the microscope.

If the material under investigation scratches glass, it must have a hardness of 7 or more, and the choice is reduced to the small list of Table VII, rocks containing some of the listed substances, and certain of the newer structural materials for high-temperature use such as Pyroceram (microcrystalline borosilicate glass of Corning containing Na, K, Mg, Zn, and Pb) and cermets (metals in ceramic matrix and *vice versa*).

Table VII. *List of Very Hard Materials*

Hardness Number of Mohs	Material
10	Borazon, cubic boron nitride, hardest material known; diamond
above 9	WC, B C, SiC
9	Bromellite, BeO; corundum, Al_2O_3; TiB; ZrB; WC
8.5	Chrysoberyl, $BeAlO_4$; Burundum, nonporous high-alumina ceramics containing 85% and more Al_2O_3; TiC; ZrC; TiN
8	Picotite, $(Fe, Mg) \cdot (Al, Cr, Fe)_2O_4$; pleonast, $(Mg, Fe) \cdot (Al, Fe)_2O_4$; spinel, $MgAl_2O_4$; topaz, $Al_2(F, OH)SiO_4$; ZrN
7.5 to 8	Beryl, $3 BeO \cdot Al_2O_3 \cdot 6 SiO_2$; gahnite, $Zn(Fe)AlO_4$; hercynite, $FeO \cdot Al_2O_3$; phenakite, Be_2SiO_4
7.5	Andalusite, $Al_2O_3 \cdot SiO_2$; laurite, $Ru(Os)S_2$; zircon, $ZrO_2 \cdot SiO_2$; Mullite, synthetic $3 Al_2O_3 \cdot 2 SiO_2$
7.0 to 7.5	Various rare silicate and borate minerals
above 7	Si, ferro-silicon, tool steels containing Si, Ti, Mo, W, Cr, Ni, Co
7	Quartz; vitreous silica; "hard" glass

For the detection of very soft materials, the test might be repeated with two polished plates or sheets of brass or copper, which are not scratched by materials of hardness less than 4. Finally, the unknown might be tested between two glass plates that have been coated with a suitable, soft (and colored) film such as collodion or gelatine.

P. 18   Refractive Index

The ability to match refraction within 0.00002 by the use of phase contrast or interference microscopy (476) may serve for batch identification, but such precision is useless for the recognition of substances, where the presence of incidental impurities will give far greater variations of refractive index. In general, the determination of refractive index under the microscope requires availability of an Abbe or a dipping refractometer for the checking of liquid standards and is only conditionally recommended to the chemist.

If a liquid substance is given, which may be expected to be reasonably pure and which is not noticeably volatile, the determination of refractive index may be performed with a purpose and little loss. If 10 to 25 μl are available, the method of the Duc de Chaulnes (88, 422) may be used with cells of 3- to 4-mm diameter and 1- to 2-mm height and give the refractive index within $\pm\ 0.005$ after calibration of the cell with standard liquids as suggested by F. E. Wright and described by Chamot and Mason. The general idea applied in the cells of A. Möhring (153) and L. Nichols (425) should permit further refinement to reduce the required volume of liquid substantially below $10\ \mu l$. Both cells have been commercially available.

The immersion method, too, should allow to work with less than $10\ \mu l$ of sample, but will not always give three decimals for n as the cell of Nichols is able to do; since variation of temperature is not advisable, as a rule, the availability of a solid of closely similar refraction determines the attainable accuracy. If necessary, Kofler's scale of glass powders may be extended by adding $MnSiO_3$ glass ($n = 1.700$), spinel (1.718), periclase, MgO (1.736), garnet (1.735), and arsenolite, As_2O_3 (1.755). The final matching may be done with the use of a microscope hot stage (P. 21) with the liquid containing a fragment of the solid of slightly lower refraction confined in a sealed capillary. The refractive indices of the more common organic liquids may be found collected in table form (10).

For the determination of the refractive index of very small solid objects, only the immersion method is available. For lack of suitable immersion liquids, this excludes the determination of refractive indices above about 1.8, but still permits the study of organic solids and of approximately 70 per cent of the minerals and inorganic solids, the refractive indices of which range from 1.3 to 3.6. If need arises, a liquid with a refraction from 1.74 to 1.83 may be obtained by dissolving sulfur in methylene iodide. It is understood that the solid must be transparent.

The mixture which perfectly matches the refraction of the solid should be prepared upon a not too small scale so that evaporation of some of the more volatile constituent cannot significantly affect the refractive index. The solid particle should be washed with this mixture before it is immersed in it for the observation in monochromatic light. Obviously, the immersion media must not dissolve the solid or react with it in any way, and this fact may have to be established by preliminary trials (P. 31, 32).

The application of the immersion method is relatively simple if the solid is either isotropic or uniaxially birefringent (hexagonal or tetragonal) so that a view vertical to the principal axis is readily obtained. Systematic compilations of refractive index data are available for minerals (99), inorganic substances (113), and some organic substances (112). In the instance of organic solids, the effort is least rewarding, and it seems preferable to determine the melting point and the refractive index of the melt as outlined in P. 21.

The determination of the outstanding values for the refractive index in the instance of anisotropic solids which are not obviously hexagonal or tetragonal involves a study of the orientation of the optic axes and is best undertaken by a crystallographer. The effort may be worth while if the test material is of mineral origin, if well developed crystals are available, or if destructive testing must be avoided by all means. Very small crystals are best studied with the use of a universal stage. The axial rotation stage (464) made by Kenneth A. Dawson Co. (Belmont, Mass.) may serve as a substitute, and minute crystals might be mounted in a droplet of Canada balsam at the end of a glass needle.

If a large number of approximately equidimensional crystals or crystal fragments may be mounted in a suitable medium between slide and cover slip, the axial figure may be studied on several specimens to find whether the solid is uniaxial or biaxial, positive or negative. After suitable change of immersion medium, the axial figures may again be used for determining the optical orientation previous to use of the immersion method. With uniaxial solids, n_0 may be determined on specimens that remain dark between crossed nicols when the stage is rotated, and n_e is then that refraction observed, which differs most from n_0. Similarly, β of biaxial solids is exhibited by specimens that remain gray during the rotation of the stage, and α as well as γ are the extreme (lowest and highest) values that may be found. If the crystals or fragments are small, use of an elaborate petrographic stand and availability of special high-power objectives for the observation of axial figures become desirable. Additional advice may be found in the literature (88, 93, 96, 97, 110).

P. 19 Density

Liquids. Provided that suitable balances are available, accurate determination of the density of small amounts of liquid may be performed with pycnometers or specific gravity pipets, and no loss of material need be involved. Data on the density of liquids may be found in handbooks and systematic compilations (54).

ANDERSON (444) described semi-self-filling Ostwald pycnometers of 1-ml capacity giving precisions from $\pm$ 0.0001 to 0.000025. The commercially available specific gravity pipets of ALBER (426) weigh about 5 g and are provided with tight-fitting ground caps to prevent evaporation or uptake of moisture or carbon dioxide during weighing. The decigram pipet has a capacity of 0.1 ml and gives a relative precision of $\pm$ 0.0005. The centigram and milligram pipets are made of heavy-walled capillary tubing of uniform (1-mm and 0.5-mm) bore and are provided with a millimeter scale to allow use of amounts of liquid varying from 20 to 80 μl and from 6 to 16 μl, respectively; a relative precision of $\pm$ 0.005 may be attained.

The principle of the specific gravity pipet may be used with very small volumes of liquid. The success will depend upon vapor tension, stability, and viscosity of the liquid. The pipet of von WARTENBERG (153, 602) will require less than 1 µl if made from capillary tubing of 0.6-mm diameter of bore so that the length of the bulb does not exceed 1 to 2 mm. The fine capillaries, Fig. 72a, may be 15 to 20 mm long and have a bore of about 0.03 mm. The microbalance may be provided with a wire rack for holding the pipet. Essential is that liquid and pipet have, at all times, the temperature of the room; to avoid warming the apparatus, it should be handled with forceps having flat tips covered with a soft sheet that does not leave material on the glass surface. Small pieces of soft rubber or plush are cemented upon the tips and then trimmed with scissors; a quickly drying cement

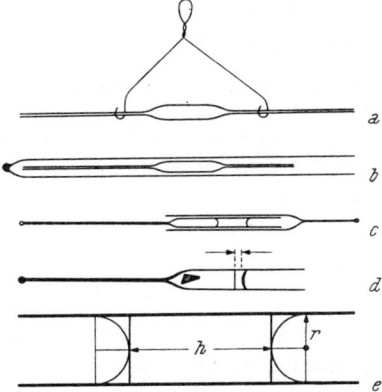

Fig. 72. Determination of Density. The bore is greatly exaggerated in e

which does not penetrate the sheet should be used. To hasten the intake of liquid, the pipet may be connected through a drying tube to the suction line so that the bulb rests on the opening of a stopper or rubber tubing mounted on the end of a glass tube. When the liquid reaches the upper capillary, the connection to the suction is broken so that the tube may fill by capillary attraction. The adjustment of the volume is automatical, but the liquid adhering to the outside of the intake capillary must be removed by touching to the edge of spot test paper. Even this task may be avoided or simplified by treating the outside of the apparatus with Dri-Film or Desicote before cutting the fine capillaries to proper lengths. Evaporation during weighing may be suppressed by inserting the pipet into a somewhat wider capillary, Fig. 72b. Water may serve for calibration. Obviously, all rinsing and drying must be done with use of suction. Von Wartenberg used a Nernst balance for a pipet of 4-µl capacity and obtained a relative precision of ± 0.003. The arbitrary scale of the balance was calibrated by weighing the pipet filled with several different liquids of known density, but calibration by customary procedure (142) should suffice.

The density of down to 0.02 μl of liquid could be estimated by introducing drops of about 1-mm length into capillary cones of 0.2-mm bore by the standard procedure, p. 208, and by taking special care that the tip of the micropipet will not touch the cone anywhere but at the point where the drop is to be deposited. The fine capillary from which the cone is to be made should be inspected for uniformity and roundness of bore by cutting out sections of 20-mm length and measuring the diameter of bore at both ends, Expt. 22. To check on the circular shape of the bore, four diameters are measured by rotating the stage through 45 degrees after each measurement. A satisfactory section of capillary is used for the preparation of the capillary cone. After introducing the liquid, the capillary cone may be placed inside a somewhat wider cone, Fig. 72c, to reduce evaporation during measuring the length h of the liquid column and weighing. If the menisci are spherical or nearly so, the volume, $v = \pi r^2 (h + \frac{2}{3} r)$, may be computed as that of the cylinder of height h increased by that of a cylinder of height $2r$ minus that of a sphere of radius r; Fig. 72e. The radius may be checked at the conclusion of the experiment by cutting the capillary cone where the drop of the liquid had been located. A relative error of 0.01 in measuring the diameter of bore will give an error of 0.02 in volume and density.

Solids. Density data are available for minerals and common inorganic solids, and the larger number of these substances has densities between 1 and 4 g/ml so that the available heavy liquids permit accurate determinations. In addition, the densities range from about 0.5 (Li) to about 22 g/ml (platinum metals) so that even crude estimations will permit to distinguish between light metals (Be 1.8, Mg 1.74, Al 2.7), the common metals (Fe 7.86, Co 8.9, Ni 8.9, Cu 8.9, Ag 10.5, Zn 7.14, Cd 8.6, Sn 7.3, Pb 11.3, Sb 6.68, Bi 9.8), and the heavy metals (Ta 16.6, Au 19.3, Pt 21.4, Ir 22.4 — W 19.3, U 18.9). On the other hand, the density of organic solids varies only from about 0.8 to 4.3 g/ml, and few data are listed so that knowledge of density will rarely permit identification.

The density of large objects is conveniently and accurately determined by weighing in air and in a suitable liquid of known density. Determination of volume by this method is not applicable to small objects, however, because the buoyant effect becomes too small as compared with the forces developing where the suspending filament enters the surface of the liquid.

The density may be found by weighing the specimen and computing its volume from its linear dimensions. BRILL and EVANS (720) obtained a relative precision from ± 0.015 (bead of tin) to ± 0.001 (crystals) with specimens of 0.06 to 1.5 μl volume (lineal dimensions of 0.5 to 2 mm). Far smaller specimens will suffice if a suitable microbalance is available and magnifications are used permitting a relative precision of ± 0.01 or better in the lineal measurements which may be carried out on screen images or enlarged photomicrographs. The precision of volume and consequently density will greatly depend upon the shape of the specimen. If it has a simple form with known exact relation between dimensions and volume, the latter may be accurately computed from a few simple measurements of length; this holds for sphere, ellipsoid of rotation, cube, prism, pyramid, tetrahedron, octahedron, cylinder, and cone. Well developed crystals may represent a simple geometry. Noble metals may be fused to a spherical bead by fusing them suspended in a bead of boric acid, carried on the end of a fiber of vitreous silica (1200). A flux of KCN or NaCN or heating in hydrogen may be tried with metals that readily oxidize. The diameter of metal wire may be measured in the side view, but as in the instance of fibers and transparent filaments, the cross section should be inspected to be certain of its nature. Deviation from sphericity is serious with beads since the relative error committed in estimating the diameter appears tripled in the volume of the sphere.

HABER and JAENICKE (1200) tried to obtain the density of small metal beads from the rate of fall through paraffin oil, but difficulties arose when the diameter was less than 30 μm, volume less than 1 pl or 10^{-12} liter.

The rate of fall through liquid as well as the buoyant effect or the displacement of liquid (volumenometric method) are affected by adsorption of gas on the surface of particles or its presence in cavities or cracks. Consequently in all these methods, care must be taken to eliminate gas adhering to the specimens by the use of wetting agents or application of the vacuum impregnating technique. The surface tension of water may be reduced by adding some ethanol, or about three drops of Triton 100 per 50 ml fluid (300), or "Anti-Creep" according to the directions of the Schleicher and Schuell Co.

Volumenometric methods require a balance having sufficient precision for the weighing of the specimen which is then immersed in a fluid contained in a calibrated tube; the volume indicated by the displacement of the meniscus is the volume of the specimen. CALEY (400) describes a calibrated cylindrical centrifuge tube which permits attaining a relative precision of about $\pm$ 0.005 with 25 to 100 mg of powdered solid. Coarse powders are preferable since fine particles tend to float on the surface. Caley recommends using diethylether and to keep the tube stoppered to prevent evaporation. It is understood that a liquid must be chosen, which does not dissolve the solid or react with it, and this may require a preliminary study of the solubility, P. 31, 32. KIRK (157) has applied the method to small particles which could have a diameter as small as 0.1 mm (0.5 μg weight) and got a relative precision of $\pm$ 0.1 or better. A 15- to 20-mm section of thin-walled capillary of uniform circular bore (*see* above p. 264), slightly larger than the diameter of the particle, is selected for preparing a capillary cone, Fig. 72d. To obtain a fixed mark, the outside of the capillary cone may receive a very fine scratch with a diamond pencil, or a short length of a straight textile fiber may be cemented to the outside so that it lies parallel to the axis of the tube and either end may serve as reference. Using the standard procedure with the micropipet, p. 208, water, butyl phthalate, or any other suitable liquid is introduced into the capillary cone until the meniscus arrives close below the mark. If it seems desirable, the tube is centrifuged before determining the distance between meniscus and reference mark with the eyepiece micrometer. The air of the "moist" chamber may be saturated with the vapor of the liquid used, and the customary precaution may be taken to prevent evaporation during centrifuging, p. 211. The weighed solid is introduced by means of a glass fiber or microforceps, p. 232, operated by a mechanical manipulator, while the capillary cone is resting upon the carrier slide in the chamber. The particle is deposited in the opening of the capillary cone and then centrifuged into the liquid, whereafter the position of the meniscus is again determined against the reference mark. The volume is computed from the displacement of the meniscus and the known diameter of the bore. The accuracy may be improved by multiple performance of all measurements and use of the averages.

In the buoyancy method, the composition of the fluid is changed until the solid specimen remains suspended without rising or falling; hereafter, the density of fluid is determined by a standard method. Since a large volume of liquid may be used, its density and thus the density of the floating particle, no matter how small, may be determined with high accuracy. Obviously, the temperature must be controlled within the limits required by the accuracy of the determination of density. The range of applicability, however, is limited by the density of the available heavy liquids, a good list of which may be found in Lange's Handbook of Chemistry.

Mixtures of benzene (d, 0.88), nitrobenzene (1.20), carbon tetrachloride (1.59), bromoform (2.89), and methylene iodide (3.33) cover a range from 0.88 to 3.33 g/ml, and this may be extended to 3.65 g/ml by dissolving iodine and iodoform in methylene iodide. The solution of ROHRBACH is prepared by dissolving 20 g BaI_2 and 26 g HgI_2 in less than 6 ml water; it has a density of 3.58 and is diluted first with a 20% solution of BaI_2. When an equal volume of the BaI_2 solution has been added to prevent separation of HgI_2, further diluting may be done with water.

For accurate determination of density, a coarse powder should be used in preference to large pieces. ANDRAE (153, 1215) selected under the microscope perfect crystals of 1-mm diameter and somewhat less. These were transferred into a test tube where the mixture of methylene iodide and benzene was prepared in which the crystals floated at room temperature. The floating of the heaviest particles should be taken as criterion since the lighter ones may be carried up by adhering air; at this stage, the mixture also might be exposed to reduced pressure in the hope of displacing adsorbed gas. ANDRAE transferred the crystals and liquid of equal density to a dilatometer with a calibrated capillary of 2-mm bore. When the dilatometer was nearly full, the small remaining air bubble was used for thoroughly mixing the contents, and then the filling of the apparatus was completed. The dilatometer was mounted in a large beaker with water, the temperature of which was regulated until the crystals remained stationary in the liquid as judged by the immobility with respect to reference marks on the outside surface of the beaker. When this condition was reached, the temperature was read and the volume of the contents was determined by reading the position of the meniscus in the capillary. For getting the weight of the contents, the capillary may be capped, but the apparatus must be wiped dry on the outside and allowed to acquire the temperature of the balance room. To obtain accurate results, it is necessary to mount the dilatometer in the water bath so that it may be rotated around its axis and that the temperature may be changed so slowly that gradients will not be established. Obviously, the weight must be corrected for the buoyant effect, and the volume of the dilatometer for the expansion of the glass.

KIRK (157) prepares the mixture of like density in a test tube of about 6-mm bore and 5 cm length. If the particle to be tested is somewhat porous, only little fluid is added and the pressure above the liquid is reduced so that boiling occurs which will aid in eliminating adsorbed gas. After each addition of heavier or lighter liquid, the contents of the tube are mixed by twirling in it a glass "zigzag stirring thread containing a number of successive sharp bends". When the specimen, which may also be a drop of immiscible liquid, remains suspended without motion, some of the mixture is taken into a specific gravity pipet for the determination of the density, p. 262—3.

HUTCHISON and JOHNSTON (701) refined the buoyancy method so that the density could be determined within $\pm$ 0.000005 g/ml with fragments of 3- to 4-mm edge. The precision was determined by the temperature control. Other factors–particle size, viscosity of liquid, length of observation, heat diffusibility of system–become dominant with smaller particles, and PRIMAK and DAY (454) accepted relative precisions from $\pm$ 0.0001 to $\pm$ 0.001 with crystal fragments 0.25 mm on edge, i. e., about 4000 times smaller than those used by HUTCHISON and JOHNSTON.

Because of the high toxicity, special precautions should be taken for the prevention and collection of spillage when working with "heavy liquids" and especially when using thallium salts.

HENDRICKS and JEFFERSON (745) as well as BERNAL and CROWFOOT (1050, 1051) centrifuge the fluid with the immersed specimen to find the mixture of like density, and it is claimed that a particle of 50 μg will suffice for the determination. Finally, it should be mentioned that separation of the constituents of a mixture (rock) may be carried out by sedimentation from a liquid. The mixture is ground to a size so that individual particles, for the most part, contain one component only. The powder, which should be as coarse as possible, is then transferred to a centrifuge tube and treated with a liquid that is so dense that only one component may settle out. After centrifuging, the liquid with the light components is transferred to another tube and mixed with a lighter fluid until the next component drops out, etc. Use of acetylene tetrabromide (d, 2.96) is recommended for the separation of ore and gangue; among others, it will float quartz, d = 2.65.

P. 20 Classification Tests

As indicated by the name, the following procedures will permit recognizing the general class or group of materials to which the unknown belongs; they also allow recognizing a number of the more common substances.

Some of the tests are destructive, and the others may lead to destruction in certain instances. Thus, if very little material is available for investigation and cannot be replaced, it becomes advisable to consider once more whether or not identification is possible by non-destructive means. A review of the collected evidence may indicate a way. If X-ray diffraction and electron probe are either not available or do not give the needed information, one may decide to try identification via the optical constants of the crystal. This is promising in the instances of minerals and inorganic materials (83, 94, 99, 113, 114, 921) and possible with organic substances (112). One may decide on first determining the solubility (P. 32) on a very small fragment of the specimen and to use the knowledge gained for the preparation of a medium for spectrophotometric testing (compare P. 34) or for recrystallizing (448) the specimen for crystallographic investigation. Slow evaporation of the solution from a flat dish, 30-mm diameter at the bottom, 40-mm diameter at the top, and 20 mm high, may give well-developed crystals since the undesirable concentration gradient toward the edge of drops on a slide is missing (924). The systematic procedure for identifying mineral particles is outlined also by FRY (93) and by KERR (97).

A technique for the performance of a series of orientation tests on a particle, 0.1 mm or less in diameter, is described by PRAZAK (969). The substance may be heated in various gases (P. 27 to 30) and finally tested for its solubility (P. 32).

The apparatus, Fig. 73, consists of a short length of glass tubing, about 6 mm in bore, which is drawn out so that a bulb is left between two short pieces of capillary. If ignition is contemplated, the apparatus may be made of combustion tubing or vitreous silica. Capillary *a* may be 0.5 mm in bore or less; capillary *c* should have about 1-mm inner diameter or more.

The particle to be investigated is introduced into bulb *b* through the wide tube *d* by means of a glass thread or a fine capillary.

"For a simple ignition test, the tube is tilted so that capillary *a* points up and a microflame is applied to the bulb. If water or another volatile substance is liberated, it will condense near the outlet *a* of the bulb and, by cautious application of heat, it may be driven into capillary *a* for collection and (or) testing. Any desired gas may be supplied through *d* and gaseous products of the reaction may be collected by inserting capillary *a* into a droplet of absorbent."

"For solubility tests, a small amount of solvent is allowed to enter capillary *a*, whereafter it is sucked into the bulb and brought into contact with the particle to be tested. The action may be observed under the

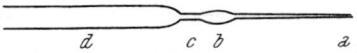

Fig. 73. Tube for Classification Tests and Dissolution of Sample; approx. nat. size.
PRAZAK, G., in Mikrochimica Acta **1961**, 899

microscope. If the particle dissolves, the solution may be blown out through capillary *a* and analyzed. If dissolution does not seem to occur, the solvent is evaporated by heating the bulb which should be inclined so that *d* points up. A suitable gas may be supplied through *a* by connecting *d* to a suction line. Thus it is possible to try a series of solvents: water, dilute and concentrated hydrochloric acid, dilute and concentrated nitric acid, aqua regia, etc. Agitation may be obtained by tapping the bulb."

"An insoluble residue may finally be heated in a current of oxygen (formation of CO_2 indicating carbon, organic substance, or carbide) or in a current of hydrogen (reduction of insoluble sulfates and of insoluble compounds of heavy metals: $AgCl$, SnO_2, etc.). Formation of water may be simply demonstrated by directing the gas escaping from *a* onto a small crystal of copper sulfate which has been heated until it turned white (anhydrous salt)."

If material is available for destructive testing, the following "chemical" procedure, which does not require specialized instrumentation, will be found quite efficient. It may be used with very little material without becoming cumbersome.

P. 21 Observation of Transition Points Below 350° C

Test for Stability Upon Heating. The possibility of the presence of explosive substances in the unknown must be excluded with certainty before an amount of the unknown exceeding a few tenths of a milligram may be heated without fear of disastrous consequences.

To test for stability, place not more than one milligram of the unknown substance into a thin-walled test tube. Mount the tube behind a pane of safety glass, and shove a lighted Bunsen burner under it so that the substance is rapidly heated by a just non-luminous flame of about 8-cm height. The unknown may be considered safe if neither explosion nor deflagration occurs.

Observation of the Behavior on Heating. If reviewing of the already established knowledge concerning the unknown does not suggest otherwise, the testing is best continued by observing the behavior of solid material under investigation while the temperature is raised until melting occurs or decomposition sets in. If a cooling stage is available, liquids may be frozen, and the behavior of the resulting solid studied in a like manner. Solids too, are best studied with the use of a heating device mounted on the stage of a microscope. The apparatus must be constructed so that a thermometer or thermocouple indicates the temperature of the specimen (98, 159). A heating block and observation with a magnifying glass may be used as a substitute, but they do not permit getting all of the information which is obtained with the microscope hot stage. Finally, a melting point determination may be carried out by the classical method with capillary attached to a thermometer and inserted into a well-stirred bath liquid.

Heating to 350° C will not give any information if the material is very hard, Mohs No. 7 or above. Useful information may be expected when the material is relatively soft and distinctly crystalline; in such instances, the experiment should never be omitted if an organic substance or an inorganic hydrate may be present.

Microscope hot stages and their use have been described by LUDWIG and ADELHEID KOFLER (159, 160) and by McCRONE (163). Apparatus is commercially available, and instructions added by the manufacturer may contain important information. Special attention must be paid to the rate of heating since the data given in the literature assume adherence to a standard procedure. It is general practice to reduce the rate of heating when approaching the melting point; a rise of 4 degrees per minute is permitted by the KOFLERS. In addition, it must be kept in mind that the rate of heating will greatly affect the temperatures at which loss of water of hydration, distillation, sublimation, and other transitions are observed. The amount of loss by vaporization and the amount of decomposition, and consequently the melting point, are also determined by the rate of heating. Thus, if nothing is known concerning the behavior of the material upon heating, it may be best to sacrifice a very small amount of sample for an exploratory experiment in which the heating is done quite rapidly.

For heating on the microscope hot stage, use less than 0.1 mg of material and crush larger particles or crystals by pressing between two glass slides. Place the material so upon the slide that the individual particles do not touch one another. Use a slide of the thickness suggested by the manufacturer of the heating device and cover the material with a clean glass slip. Slightly press upon the cover slip and impart a small rotary motion to distribute the solid between the glass surfaces and to break up larger particles so that the cover slip lies close to the surface of the slide. Place the slide upon the microscope hot stage and move it until a suitable portion of the preparation is in the field of vision. If several types of particles are present—different color or transparency, crystals of different shape—make certain that all types are included so in the field that they are not in contact. Proper selection of the field is important at this time since most of the apparatus do not make provisions for moving the slide when the heating is once started. On the use of dark-field illumination *see* FELTON (488).

Finally place a glass bridge over the preparation if the directions call for it, cover the hot stage, and adjust the heating rate in accordance with instructions. Continuously observe the preparation until either all particles have melted, or decomposition sets in, or the temperature limit of the apparatus has been reached. Whenever a change occurs in the preparation, record it together with the temperature. Use of polarizing equipment may facilitate to observe transitions from the appearance, change, and disappearance of interference colors; a compensator ($\lambda/4$-mica or first-order red plate) used together with crossed polars gives sufficient brightness to permit simultaneous observation of isotropic matter. If projection is used for observation, be certain to insert infrared-absorbing cells between the source of light and the preparation; otherwise transition temperatures will be found too low since strongly absorbing particles may acquire a temperature noticeably higher than that indicated by the thermometer.

The following phenomena may be observed: (a) sublimation or distillation with the growing of new crystals or the appearance of droplets; (b) crystallization of the distillate to give a phase stable at high temperature; (c) color changes; (d) clear particles becoming opaque because of dehydration, decomposition, or transition to another modification; (e) melting; (f) crystallization of a new compound (anhydrous substance) or of a high-temperature modification from a melt; (g) several sharp melting points if either different substances are present, or the melting points of the hydrate, the anhydrous substance, or of several modifications of one substance; (h) decomposition, sudden or gradual, possibly accompanied by melting, boiling, discoloration, charring, evolution of a gas, or separation of a solid.

Record the temperature interval in which the melting occurs since it is a criterion of the purity of the substance. Usually, the smallest fragments liquify first; then the larger particles show a rounding-off and become surrounded by melt. The temperature, at which the last trace of solid dissolves in the melt, is closest to the melting point of the pure substance. To observe the equilibrium between liquid and solid phase, stop the heating when the particles begin to melt. After a short time, the remaining crystal fragments will begin to grow in the melt, and by turning the heat on and off, the temperature corresponding to equilibrium between the last trace of solid and the liquid may be determined quite accurately.

In the evaluation of data, it is useful to know whether the unknown substance is organic or inorganic in nature. This information, however, may not be available at this time, and the evaluation may have to be postponed (P. 34) until the behavior on heating above 300° C has been tested (P. 22 to 28).

Inorganic Substances. The more common inorganic substances which melt below 900° C are listed in Table 4. For convenience, they are arranged in the order of their melting points; in some instances, transition temperatures other than melting points have been used. The data are taken from the literature, mostly from Lange's Handbook of Chemistry. The procedure of arriving at the table implies that there are about 2000 of the more common inorganic compounds which do not melt below 900° C.

When using Table 4, proper allowance (98, 159, 160, 163) has to be made for the fact that, as a rule, the listed transition temperatures have not been determined under the microscope but either in the capillary or with large amounts. Whenever doubt arises, an apparent identification may be confirmed by observing the behavior of the known substance having comparable purity and by performing a mixed melting point. This procedure will also resolve doubts concerning the accuracy of the listed data. Obviously, the significance of the zeros is doubtful for most transition temperatures given as 50, 80, 100, etc.; the mere fact that relatively large numbers of substances are reported to melt at 30, 40, 50, 100, 110, 120, etc. calls for caution. The desirability of a revision and expansion of the table after an experimental study of the behavior of the substances on the microscope hot stage is obvious. It need hardly be added that the author would be grateful for any information, corrections, and references to published data.

As further aid in identification, Table 5 lists the more common inorganic substances that sublime, and Table 6 those which burst into flame when heated in air. It is generally known that sublimation temperatures vary widely with the conditions of the experiment; tables usually give the approximate temperature at which rapid sublimation starts when the substance is heated without any special precautions.

Explosives. Table 7 gives a list of the explosive solids, extracted from tabulations of the more common 2600 inorganic and about 6800 organic compounds (10). Potential explosives are furthermore all mixtures of oxidizable matter (S, P, P_4S_3, P_2S_3, P_3S_6, P_2S_5, As_2S_2, Sb_2S_3, FeS_2, Mg, Al, C, organic substances) with oxidants such as liquid air, peroxides, chlorites, chlorates, perchlorates, bromates, iodates, periodates, permanganates, nitrites, nitrates, and others. Intimate mixing is frequently an important factor; on the other hand, the components may be solid, liquid, and gaseous to give explosive emulsions, suspensions, sprays, and dusts. Explosions of air-born combustible dust occur in coal mines and flour mills.

Black powder is an intimate mixture of C, S, and KNO_3 which is usually formed into definite shapes varying from small grain to large cubes. It is very sensitive to heat, friction, and mechanical shock. The chemical analysis of explosive mixtures usually presents little danger after the oxidant has been separated from the fuel, which may be frequently accomplished by extraction with water.

Small amounts of cellulose nitrate (nitrocellulose, gun cotton, collodion) and smokeless powder deflagrate, i. e., decompose suddenly with a flash of flame, when heated to 100° to 200° C, but do not explode. Smokeless powder is obtained by treating nitrocellulose with nitroglycerin and giving the resulting jelly the shape of flakes or grains.

Dynamite is obtained by absorbing nitroglycerin in suitable solids (wood flour, diatomaceous earth, etc.) to obtain a plastic mass. It is safer to handle than nitroglycerin itself, but will explode when heated.

Quite complex mixtures have been prepared in adapting explosives to special uses. They may contain explosives (gun cotton, nitroglycerin, nitroglycerin incorporated in a glue jelly, TNT, nitronaphthalene, dinitrobenzenes, NH_4NO_3), oxidant [KNO_3, $NaNO_3$, NH_4NO_3, $Ba(NO_3)_2$, $Sr(NO_3)_2$, $KMnO_4$, $K_2Cr_2O_7$, $K_3Fe(CN)_6$], fuel (S, C, Al, naphthalene, chloronaphthalene, kerosene, fatty oil, sugar, starch, wood flour, cellulose, resin), moderator [NH_4Cl, $(NH_4)_2SO_4$, aniline hydrochloride, NaCl, $Na_2CO_3 \cdot 10\,H_2O$, magnesium sulfate], binder (kerosene, soap, glue, dextrine, resin, starch), and absorbent (diatomaceous earth, flour, wood flour, cellulose, etc.).

Organic Substances. If the substance under investigation is an organic compound, consult suitable melting point tables and compile a list of substances that have to be considered.

The KOFLERS (159, 160) and MCCRONE (163) list the substances in the order of their melting points as observed upon the microscope hot stage. These convenient tables may also be used for a preliminary orientation if the melting point has been determined in the capillary, but in this case it will be wise to include into consideration all substances melting in the range from t observed to $t + 10$. For final identification, it may be necessary to compare with listings of melting points determined in the capillary (2, 8,

10, 54). Tables listing melting points in the order of ascending values may also be found in handbooks (10).

Residues from melting point determinations may be reserved for use in heating to higher temperatures and for chemical analysis.

The heating on the microscope hot stage may be repeated with fragments of the material under investigation immersed in paraffin oil (159, 160). This will show whether or not a gas is given off on heating. In general, however, it will be more efficient to get this information by heating in a capillary, P. 26, which will permit identifying the nature of the gas.

P. 22 Ignition Above 300° C

The material under investigation may be heated by itself in an essentially inert atmosphere or with the addition of various reagent, solid or gaseous, and by using various techniques depending upon the amount of material available.

P. 23

Heating in the Closed Tube. Customarily, a few milligrams of the substance are placed into a test tube of hard glass, which is about 4 mm in bore and 10 cm long. The procedure is described in Expt. 15.

If little substance is available, a narrower tube may be used, and a particle of only a few micrograms weight might be heated in the sealed end of a capillary tube of vitreous silica. Heating in a stream of inert gas (*see* below) will be preferable, however, when dealing with small amounts of substance. For interpretation of the observed phenomena see P. 27.

P. 24

Heating in an Open Tube. Description of the procedure is given in Expt. 16. Condensates collect in the cooler part of the tube above the substance; liquids may be removed with a capillary pipet, and solids may be tested after the tube has cooled and has been cut into suitable sections. Pointed strips of reagent papers and loops or capillaries with reagent drops are applied at the upper opening of the tube to test for escaping gases. Liquid reagents are best applied with an elliptical loop, about 10 mm long and 4 mm wide, which may be formed at the end of a glass rod of 0.5- to 1-mm diameter, P. 36. Colorations are seen by holding the loop in front of a brightly illuminated, white surface. To test for turbidity, inspect the drop in the loop before and after exposure to the gas in front of a black background with strong light coming from the side.

To test for acidity, use wide-range pH paper. Record the odor of the escaping gas, and test it with limewater (saturated solution of calcium hydroxide), fuchsin paper, 1% solution of silver nitrate, and KI-starch paper. Additional reagents may be used, and for interpretation see P. 28.

To perform the test with small quantities of substance, use the technique of P. 26 for heating in oxygen or air.

P. 25

Heating Upon Charcoal. Use the technique described in Expt. 17 if sufficient material is available. Small amounts of substance are better heated in hydrogen, P. 29.

First heat the substance by itself. Record colors appearing in the flame. Deflagration indicates the presence of an explosive material or of an oxidant such as chlorate, perchlorate, iodate, nitrite, nitrate, etc. Compounds of noble metals are converted to the metal. Oxides and compounds that are converted to oxides on heating behave as stated below for heating with sodium carbonate. Substances that do not react with either the carbon or the hydrogen behave as when heated in an inert atmosphere with some influx of oxygen from the periphery.

Germanium compounds, in the absence of an alkaline flux, are reduced to the gray metal, and a deposit of white GeO_2 may be obtained around the heated zone. Elemental selenium melts and vaporizes; the brown fumes condense around the heated area to give a deposit of gray metallic selenium which may be surrounded by a ring of red selenium. The odor of rotten radishes is characteristic for selenium. For interpretation of melting and sublimation see also Tables 4 and 5.

Phosphates, borates, and silicates may fuse to a glassy bead, the color of which may be indicative of metals present, P. 37.

If the substance is not reduced, but the presence of heavy metals is suspected, mix some of the substance with twice the amount of anhydrous sodium carbonate and again heat upon the charcoal or graphite. The results are determined by the circumstance that oxygen has access to the periphery of the reaction zone, whereas the flame and the support are reducing. Thus the vapor of a volatile metal may be oxidized outside the flame, and a deposit of oxide (incrustation) may be obtained in a ring zone surrounding the flame.

The following list may aid in the interpretation of the observed phenomena.

A white, strongly incandescent residue forms, which refuses to melt: oxides of Ba, Sr, Ca, Mg, Al, and rare earths.

Reduction to metal occurs, but no deposit forms around the heated material:
a) a malleable bead is formed, which may be flattened in the mortar: Cu, Ag, Au, Sn;
b) gray, metallic particles which may be
> malleable: Pt, Ir, Rh, Pd;
> not malleable, not magnetic: Mo, W, Re;
> not malleable, ferromagnetic: Fe, Co, Ni.

Reduction to metal occurs, and a deposit forms around the heated material:
a) malleable button and yellow deposit: Pb, In;
b) brittle metallic button that may be ground to a powder:
   white deposit, Sb;
   yellow deposit, Bi;

c) gray metallic powder and white deposit consisting of colorless crystalline scales: Mo. The formation of a condensate of MoO_3 may not occur since it requires use of an oxidizing flame. The scales of MoO_3 are yellowish when hot; when they are touched with a reducing flame, a blue oxide may form.

Reduction to metal does not occur, but a deposit forms around the heated material: This deposit is

a) white: As, the deposit is volatile and an odor of garlic is perceived;
   Zn, deposit is not volatile, but it turns yellow when heated;
   Tl, slight deposit at some distance from the hot reaction zone; the flame becomes intensely green;
   Te, the outer seam of the deposit has a brownish hue; the deposit is readily volatilized by touching it with the flame;
   Mo, colorless crystalline scales, *see* above.

b) brown: Cd, brown rings may surround a blue central area to imitate the appearance of the "eyes" in the plumage of the peacock.

c) violet: RuO_2.

It is understood that the presence of several heavy metals may lead to complications in the interpretation of the observed phenomena, and this possibility should be kept in mind. Colorations of the flame should be noted since they may provide additional clues; for their interpretation *see* P. 38. Metallic buttons are seen without difficulty, but for the detection of powdery metals it is necessary to transfer the residue of the ignition and some of the carbon supporting it to a mortar where it is ground, leached with water, and freed from the carbon by floating off the latter with a stream of water.

The Hepar test, P. 59, should be tried with a portion of the residue; blackening of the silver indicates the presence of compounds of S, Se, or (and) Te. More information and additional tests may be found in the special literature (37, 47, 55, 56, 57).

P. 26

Heating in a Current of Gas (935). The material may be placed upon a narrow slide which is then heated inside a combustion tube after the air has been displaced by the desired gas. After cooling in the chosen atmosphere, the slide is withdrawn from the tube for inspection of the residue under the microscope. Vapors and gases produced upon heating the substance will frequently escape detection, however, and the following technique is recommended for the investigation of small samples.

The substance is heated in a capillary of hard glass or vitreous silica, which has a reasonably heavy wall, a bore of 0.5 to 2 mm, and a length of 10 to 12 cm. One end is drawn out to a bore of possibly less than 0.1 mm and bent at a right angle, Fig. 74. The other end is mounted, by means of a rubber stopper or heavy-walled rubber tubing with capillary bore, in the opening of the tube which supplies the desired gas. A wad of cotton,

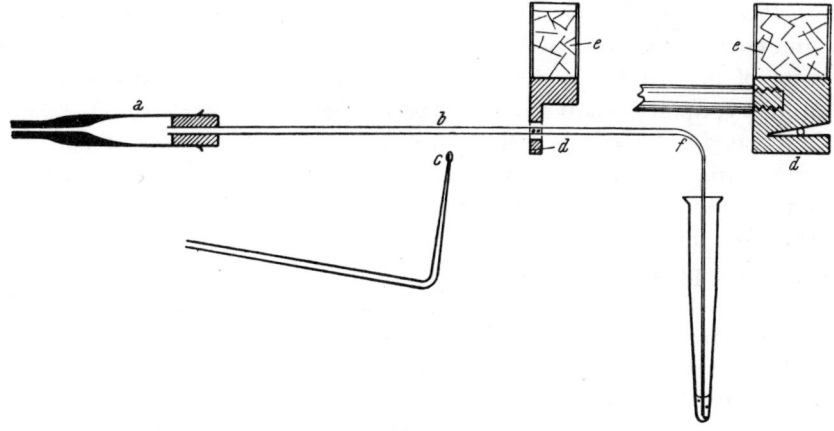

Fig. 74. Heating in a Stream of Gas

placed into space a, will retain dust and spray. The tube is held in a clamp and may be provided with a manifold stopcock arrangement admitting nitrogen, oxygen (air), or hydrogen. Gases like chlorine, hydrogen chloride, or hydrogen sulfide should be supplied through special outlets; they must not pass through the manifold so that there is no possibility for the contamination of either nitrogen, oxygen, or air. The purity of these latter gases must be assured if the gas escaping from the fine capillary shall be tested for sulfur and halogens. The presence of sulfur or halogen might also modify the behavior of the solids observed in the capillary.

The substance may be heated without and (or) with the addition of two parts of anhydrous sodium carbonate, and it may be heated first in nitrogen and then in oxygen (or air), hydrogen, chlorine (or hydrogen chloride), and finally in hydrogen sulfide. The choice of the experiments should depend upon the information available at the time. Before, during, and after heating, the material should be observed with suitable magnification. A magnifier may be used or a simple microscope with stage and substage removed and the tube in horizontal position. A simple mechanical manipulator should be available for positioning the heating device which may be a flame, an electric coil or wire, or a mirror or lens concentrating the radiation of an arc upon the substance. The following directions are based upon the simple procedure used by HERBERT E. SCHNEIDER.

Holding it at the wide end, clean the outside of the capillary by wiping with a slightly moist and then with a dry cloth to remove all fingerprints. Then introduce the material to be tested into the capillary which, to this end, is best placed upon a sheet of glazed paper of contrasting color. Transfer to the capillary a grain or small crystal upon a microspatula or upon a small piece of paper or metal foil and push it into the opening and to location b, Fig. 74, by means of a thin glass rod. If the material is a fine powder, proceed as follows. With a capillary pipet place a droplet of water of 0.5- to 1-mm diameter, too small to fill the cross section of the capillary, upon a small piece of plastic sheet which repels water strongly so that the drop becomes hemispherical. Using the flattened end of a platinum wire, add some of the powder and stir to obtain a thick paste. Cover the drop with a small watchglass, cup, or crucible and place it and the capillary upon a metal surface having a temperature considerably below the ice point (and best located within a dry box). When the drop has become solid, press on it from the side with a metal spatula to get it separated from the surface of the plastic. Then, without giving it time for melting, transfer the congealed mixture like a grain to location b in the capillary. Pass dry nitrogen through the capillary until the water has been evaporated completely and the whole interior of the capillary is again dry.

It may be possible to obtain a coherent grain by simply allowing the water to evaporate from the paste, and this procedure may be tried first. Another alternative is the use of the technique of Expt. 47 for collecting the solid of a slurry in the cut-off end of a capillary. The end of the capillary with the collected solid in it is placed into an oven for drying, and it is then transferred to location b of Fig. 74; the disadvantage of this procedure is that the observation of the heated solid is rendered difficult.

To obtain a particle representing a mixture of the substance with sodium carbonate, grind the material to be tested with an equal volume of anhydrous sodium carbonate to a very fine powder. Obtain a tiny droplet of water on the strongly repelling plastic surface, and collect next to it a pile of the powder, about twice the size of the droplet and still too small to fill the cross section of the capillary. With a plastic needle or a metal wire combine the solid and the droplet to form a tiny bead. Allow 5 minutes for the crystallization of the hydrate which will cement together the material. Then separate the bead from the plastic by pushing from the side with a spatula and transfer it to location b in the capillary,

Fig. 74, as suggested above. Place the capillary into a metal block, and heat it slowly to remove the water, while a stream of dry gas is being passed through. Raise the temperature to about 150° C.

If the available information does not suggest a different approach, it is suggested to first heat the material in nitrogen. If it does not sublime without leaving a residue and if it does not form a glassy bead when heated, it is advisable to try heating in oxygen and then hydrogen. A destillate or sublimate obtained in nitrogen should be tested by heating in oxygen. If the sublimate appears to be an inorganic halogen compound or if the substance fuses to a glassy bead, hydrogen reduction of its mixture with sodium carbonate is recommended.

The required gases are best supplied from steel cylinders by means of reducing valves, washed if necessary, and dried just before they enter the manifold. Nitrogen and oxygen may be dried with Anhydrone which, however, must **not** be used for drying **flammable** gases. Hydrogen as well as chlorine and hydrogen chloride are dried by passing them through concentrated sulfuric acid. By all means make certain that each gas has displaced the air up to the stopcock leading into the manifold. Before attaching the capillary to a, Fig. 74, test the gas escaping through a for purity. When tested with a glowing splint, nitrogen should stop the glowing, air will support it, and oxygen will cause the splint to burst into a flame. Test hydrogen by collecting a sample of the gas in a test tube (this may be done by displacing the air downward); when lighted, the contents of the test tube should burn with a quiet flame.

Assemble the apparatus in a location where it will not be exposed to draft. If tubing a, Fig. 74, is held in a clamp, the capillary will need no special support. Adjust its position and focus the horizontal microscope upon the substance which is illuminated by a strong beam of light coming from the side of the observer. First turn on the stream of gas, and then insert the orifice of the fine capillary into 5 to 10 μl 0.1-F Ba(OH)$_2$ solution contained in the tip of a microcone. Regulate to obtain a slow stream of gas so that individual bubbles of gas may be seen rising through the liquid which must remain clear.

For the collection of condensates, cool the capillary at a point about 25 mm downstream from the substance to be heated. A cooling block d, Fig. 74, made of aluminum and filled with a mixture of dry ice and acetone, is recommended, or a strip of wet filter paper may be hung over the capillary.

Start heating with a microflame of about 1-mm diameter. Have the flame burning at the orifice of a capillary drawn from one end of a tube of 15- to 25-cm length, which is—near the other end—clamped to a simple manipulator situated to the left of the microscope. By using the mechanical motions slowly and stepwise, approach the substance with the flame while watching through the microscope. Record the behavior of the substance and, each time before moving the flame closer to the material, check the appearance of the barium hydroxide solution and see whether or not a condensate has formed in the capillary.

Sublimates usually appear close to the sample. If a liquid condensate is obtained, lower the flame somewhat, stop the gas stream, and cut the capillary at f. Collect the liquid in a capillary pipet, and then again draw out a fine capillary at f and bend it as in Fig. 74. Use a small fraction of the collected condensate for determining its pH. If it is close to pH 7 and the barium hydroxide solution is still clear, continue to use the latter. Otherwise, get another microcone with a fresh portion of barium hydroxide solution for the continuation of the heating experiment, and investigate the contents of the first microcone.

Continue heating in the slow gas stream until either the whole material has been volatilized or the softening point of the capillary is reached. Remove condensing liquids and exchange the barium hydroxide solution for a fresh batch whenever this seems indicated. Finally allow the capillary to cool without stopping the gas flow in order to maintain the selected atmosphere until the sample has room temperature.

Test liquid condensates as suggested in P. 4. If the pH is close to 7, no heavy fumes were seen in the capillary, and the barium hydroxide in the microcone appears unchanged, it is probably water—and this may be confirmed without expenditure of sample by determination of the boiling and melting points, Expt. 21, and p. 215. After performance of the boiling point determination, the melting point is simply obtained according to EMICH (152) by sealing the capillary at both ends and attaching it with a rubber band to a thermometer which is then mounted by means of a cork in a test tube half-filled with ethanol. Test tube and contents are chilled to about $-10°$ C by immersion in an ice-salt mixture (if the water droplet does not solidify, the thermometer with the capillary is briefly withdrawn from the alcohol and chilled more strongly; EMICH squirted ethyl chloride upon them). When the water has frozen in the capillary, the test tube is taken from the cooling bath, wiped dry on the outside, and, by means of a cork, mounted in a wide test tube which is held in hand and continuously turned end over end to keep the contents of the inner tube thoroughly mixed. The thermometer is read when the solid in the capillary liquifies.

Test the contents of the microcone as follows. Centrifuge if a precipitate has separated. Transfer the solution to another microcone. Wash the precipitate with $5\,\mu l$ 0.1-F $Ba(NO_3)_2$, and discard the wash liquid.

Treat the precipitate with $10\,\mu l$ cold 3-F HNO_3:

a) the precipitate dissolves completely and some gas may be liberated: CO_2, SO_2; with a loop add $0.3\,\mu l$ 0.02-F $KMnO_4$ and mix: SO_2 is probably absent if the solution becomes pink; if the permanganate is reduced, add more of it until the solution remains red upon mixing; centrifuge: a white or pink precipitate of $BaSO_4$ in a still acid solution confirms SO_2.

b) The precipitate does not dissolve or does not completely dissolve: centrifuge, transfer the solution to another cone and test it according to (c); the insoluble precipitate may be $BaSO_4$, BaF_2, both finely crystalline powders, or $BaSiF_6 \cdot 2\,H_2O$, monoclinic spindle-shaped crystals, singly, in crosses or radiating masses, or (and) gelatinous silicic acid from the reaction of SiF_4 with water. The SiF_4 as well as HF may be obtained from fluorides and fluosilicates especially when heated in presence of acid salts like $KHSO_4$. The residue may be transferred as a slurry to a platinum crucible and tested for fluoride, P. 60. It may be digested with Na_2CO_3 solution, and the latter tested for sulfate with HCl and $CaCl_2$, P. 59.

c) Test the solution for SO_2 as under (a): if no reduction occurs and some of the precipitate did dissolve, CO_2 should have been present. It is suggested to use a fresh sample of the material under investigation for a confirmatory test.

Test the solution removed from the original precipitate as follows.

a) Use a small fraction of the solution to test for ammonia, P. 58.

b) Transfer another small fraction of the solution to a slide. Stir into it some potato starch. Inspecting repeatedly under the microscope, first add little 0.1-F KI solution that does not color starch, and then expose to fumes of concentrated HCl until the mixture has become acid. If the starch grains assume a blue to black color, the barium hydroxide must have absorbed some oxidant such as Cl_2, Br_2, I_2 or NO_2.

c) Treat the major portion of the solution with $2\,\mu l$ 3-F HNO_3 and then add up to $20\,\mu l$ 0.1-F $AgNO_3$ in small portions with inspecting, mixing, and centrifuging after each addition. Layers of differently colored precipitates may be obtained. Black Ag_2S would separate first and may be followed by yellow AgI, pale yellow AgBr, and white AgCl or AgCN. The last two may be readily identified by recrystallization from ammonia (P. 60) and nitric acid (P. 36), respectively. Black Ag_2S may be treated with $CaCl_2$ and oxidized by bromine, P. 36.

At the conclusion of the experiment, the capillary may be cut at various points to separate the residue from sublimates, and these may be kept for further investigation or immediately subjected to analysis.

Only a rather crude outline may be given for the interpretation of the observations, and much must be left to the experience and imagination of the observer and the information that may be obtained by a study of the literature.

P. 27

Heating in an Inert Atmosphere, Nitrogen. Heating in a stream of nitrogen is approximately equivalent to heating in a "closed tube". Aside from the fact that the alkaline earth metals combine with nitrogen to form nitrides, there is no extraneous agent to react with the sample, to oxidize it, or to reduce it. The observed phenomena may be interpreted with the use of Tables 4 and 5. The transition temperatures may be crudely estimated by touching a thermocouple to the heated capillary or by holding close to it a thin rod of vitreous silica to which are attached fragments of temperature indicating crayon or of Tempil Pellets (13).

Transition and melting points at temperatures from 350° to 900° C may be quite accurately determined under the microscope by using an electric hot stage which, according to BRADLEY (466), may be made by coating one side of a vitreous-silica microscope slide with a film of platinum and fastening it to an aluminum adapter, 7.5 cm × 2.5 cm, which fits any conventional mechanical stage. The temperature is simply determined by empirically calibrating the settings of the variable transformer while observing the melting points of known substances.

Still higher temperatures may be obtained by placing the particle upon a strip of platinum foil which is heated by an electric current (730, 1110, 1111). The particle is observed with a horizontal reading microscope or telescope, and the temperature is measured with an optical pyrometer. Use of a suitable dry box (458, 465) should permit working in nitrogen, hydrogen, or oxygen. Concerning D. M. Olson's reflecting microscope for temperatures up to 2500° C and low to 600 magnification see R. H. MÜLLER (484).

Changes will be observed on the substance if it decomposes upon heating; gases given off as well as sublimates and distillates may permit conclusions concerning the nature of the original substance. Hydrides of the alkalies and the alkaline earths will decompose with the formation of nitrides. Water of hydration is lost upon heating; water is also given off by some acids, by acid salts, as well as by hydroxides and basic salts. Salts of oxygen acids (chlorates, perchlorates, nitrates etc.) and, obviously, peroxides and salts of "per" acids may give off oxygen. HCl, HBr, etc. may occur as a consequence of hydrolysis when water is given off at elevated temperature. Cl_2, Br_2, I_2 may be derived from the decomposition of the corresponding salts of the noble metals. SiF_4 may come from the decomposition of fluosilicate or the action of acid fluoride upon SiO_2, silicate, or the glass of the tube. Sulfur (Se) is given off by some simple sulfides (FeS_2) when they are heated. But for those of the alkalies and some of those of the alkaline earths, most sulfates, sulfites, nitrates, nitrites, and carbonates decompose with the liberation of SO_2, SO_3, NO, NO_2, and

CO_2. Phosphates, borates, and some silicates may fuse to give a glassy bead.

Most organic compounds will undergo pyrolysis with the formation of a wide variety of substances; characteristic are evolution of flammable gas and the formation of tarry products and a carbonaceous residue; liberation of NH_3, HCl, SO_2, SO_3, NO_2, etc., are indicative of the particular elemental composition. The heating of acetates may give a condensate of acetone.

Salts of ammonium, hydrazine, hydroxylamine, and of ammonium or aminobases may decompose with the liberation of NH_3, HCl, SO_2, SO_3, NO_2, etc.

Complete decomposition which gives only gaseous products outside of water is observed with NH_4NO_3, oxalic acid, and various organic explosives (glyceryl nitrate, cellulose nitrate, picric acid, etc.).

Some substances show characteristic reversible color changes: ZnO, white to yellow; SnO_2, yellow to brown; HgS, red to black; HgI_2, red to yellow; Ag_2HgI_4, yellow to red at $50°$; Cu_2HgI_4, red to brown at $70°$.

The multiplicity of phenomena which are possible when **mixtures** are being heated in an inert atmosphere defies listing, and mentioning black gunpowder and Thermit may suffice to indicate extremes of unpleasantness. In general, interpretation will be simplified by separating the components and testing them separately.

Absence of chemical change upon heating to high temperature in nitrogen excludes organic and organized materials as well as compounds of noble metals and is characteristic for: (a) most salts of the alkalies and the sulfates of Ba, Sr, and Ca; (b) simple oxides, sulfides, selenides, tellurides, nitrides, and carbides; (c) phosphates, borates, silicates; and (d) anhydrous simple halides, many of which have a relatively high vapor tension and may sublime at temperatures around $500°$ C.

P. 28

Heating in a Current of Oxygen. Heating in oxygen will produce useful evidence when the presence of free elements, of organic compounds, or of sulfides, selenides, carbides, or hydrides is suspected. In general, the treatment will give oxides which then may be reduced by heating in hydrogen.

For information concerning the color of substances see P. 9; Tables 4 and 5 provide the data on melting points and volatility.

No change in appearance on heating in oxygen is characteristic for: (a) noble metals, stainless steel, and aluminum; (b) oxides; (c) most salts of the alkalies; (d) anhydrous halides; (e) phosphates, borates, and silicates.

Metals are oxidized to an oxide; some obtain tempering colors due to the formation of an oxide film; the oxides may be classified according to color, volatility, and melting point.

S, Se, Te, P, As, C, Si, and B give the oxides of which SO_2 and CO_2 are recognized by passing the combustion gases into barium hydroxide solution. SeO_2, TeO_2, P_2O_5, As_2O_3 sublime; TeO_2 melts to a yellow liquid. Diamond burns rather slowly.

Formation of water and oxide results when hydrides are heated in oxygen; carbides burn to oxide and CO_2.

Carbon dioxide and H_2O are obtained upon heating organic and organized materials in oxygen; in addition, other gases may be obtained (HCl, HBr, SO_2, SO_3, NH_3, NO, NO_2, etc.) which may be identified to reveal the specific elemental composition. If the substance burns, record the color of the flame and presence or absence of smoke. An ash may be left behind, which represents the mineral constituents.

Sulfides give SO_2 and oxides; selenides and tellurides give white sublimates of SeO_2 and TeO_2; sulfides of phosphorus, SO_2 and a sublimate of P_2O_5.

P. 29

Heating in a Current of Hydrogen. Heating in hydrogen is not recommended when the unknown substance sublimes or when it fuses to a glass. Sulfides should first be roasted to the oxides; halides, phosphates, borates, and silicates should be treated with Na_2CO_3 to obtain the corresponding carbonates or oxides. Sulfites, sulfates, nitrites, nitrates, carbonates, and salts of organic acids frequently give carbonates or oxides on heating in nitrogen or oxygen, and are then reduced by the hydrogen to the metal.

Heating in hydrogen is often useful for the investigation of the so-called insoluble residue: silver halides give metallic silver and hydrogen halide; sulfates of the alkaline earths are reduced to soluble sulfides; $PbSO_4$ and Tl_2SO_4, to the metals; SnO_2 and Sb_2O_5 are reduced to the metals, and so are Fe_2O_3, Cr_2O_3, and insoluble salts of chromium.

The condensate of neutral water on heating in hydrogen may also be used to prove the presence of an oxide.

The heating in hydrogen may follow immediately after heating in nitrogen if the presence of a reducible oxide is probable. It may follow upon the burning of sulfides, selenides, tellurides, phosphides, or carbides in oxygen. If halides, phosphates, borates, or silicates are present, it is best to first fuse a sample of the unknown with Na_2CO_3, to extract the melt with water, and to use the residue of carbonates and (or) oxides for testing in the current of hydrogen.

Heating in hydrogen produces the following effects.

The alkali and alkaline earth metals are converted to white hydrides at temperatures ranging from 150° to 700° C; the hydrides decompose when heated to higher temperatures, and they all react with water giving hydrogen and hydroxide.

Not reduced or apparently not reduced are the oxides of the alkalies; Mg, Ca, Sr, Ba; Sc, Y; Ti, Zr, Th; Nb, Ta; U; Mn; Al, Ga; Si, and Ge.

Not reduced are borates and silicates in general, and the halides, oxides, sulfides, and carbonates of the alkalies and the alkaline earths.

Reduced to sulfides and phosphides, respectively, are the sulfates and the phosphates of the alkalies and the alkaline earth; *see* confirmatory tests in P. 58 and 59.

Reduced to hydroxides and oxides are the nitrates and nitrites of the alkalies and the alkaline earths; reduced are also the peroxides of these metals.

Reduced to the corresponding halides are the alkali and alkaline earth salts of the oxygen acids of the halogens.

Reduced to the elemental state are the oxides and the salts of volatile oxygen acids of: V; Cr, Mo, W; Fe, Co, Ni; platinum metals; Cu, Ag, Au; Zn, Cd, Hg; In, Tl; Sn, Pb; As, Sb, Bi; Se, and Te. Of these, Zn, Cd, Hg, As, Se, Te distil to a colder part of the tube to give silvery droplets of Hg, silvery white metallic deposits of Zn and Cd, a brown to black mirror of As, a steel gray and red sublimate of Se, and a gray to black deposit of Te. In, Tl, Sn, Pb, Sb, and Bi distil only partially when they represent a large portion of the reduced metal. Fe, Co, and Ni are attracted by a magnet applied to the outside of the tube.

Water is formed by the reduction of oxides and the salts of oxygen acids. HCl, etc. are derived from reducible halides or halides that decompose on mere heating.

P. 30

Heating in a Current of Chlorine and of Hydrogen Sulfide. Heating in a current of chlorine or HCl is suggested when heating in hydrogen leads to a metal. Of course, the metal may be dissolved in acid for further identification; conversion to chloride by heating in a stream of gaseous reagent may be an attractive alternative. The resulting chloride may be converted to the sulfide by heating in a current of H_2S. Concerning the appearance and behavior of the resulting compounds, the experimenter will refer to P. 9 to 13 and Tables 1, 4, and 5.

Solubility

The information gained by determining the solubility may be disappointing, especially if the material is organic in nature. This is partly due to the fact that there is no clear boundary between soluble and insoluble so that arbitrary limits must be used. On the other hand, the testing for solubility does not cause loss of material, and information concerning a suitable solvent is necessary for further investigation. Provided that the amount of sample, solvent, and undissolved residue are determined so that the concentration of the resulting solution becomes known, the solution is well suited for chemical analysis and the estimation of the amount of substances discovered in it. In addition, evaporation of solutions may give well developed crystals for investigation under the microscope as outlined in P. 10 to P. 13.

The customary solvents are water, acids, bases, and diethyl ether, but any other solvent may be tried, which appears promising from the facts collected in preceding tests.

For the performance of solubility tests, the sample should be ground to a fine powder, if necessary, and weighed. The weight of small particles may be estimated from the measurement of their dimensions under the microscope; a reasonable guess concerning their density will usually be possible. The volume of the solvent is chosen depending upon the nature of the material. When equilibrium has been established, the decision is made concerning the solubility. A solvent which is not effective is removed and replaced by the next solvent to be tried—possibly after drying the solid. Evaporation of the removed ineffective solvent may furnish crystals of an impurity or of a slightly soluble major or minor constituent. If the material under investigation is a mixture, each clearly discriminating solvent should be repeatedly applied to a obtain a satisfactorily complete separation; the resulting solutions and residues are then used for further investigation.

The technique is essentially that of batch extraction. No further comment is needed if the work is carried out in a centrifuge cone, microcone, or capillary cone. In the last instance, the sample will be in the field of the microscope when the solvent is added, and evolution of a gas during dissolution will not escape detection. When working on a larger scale, there should be no difficulty in testing a liberated gas for CO_2, SO_2, H_2S, Cl_2, and NH_3, see P. 36, 37.

When performing a solubility test on the microscope slide, it is recommended to place the particle(s) of the sample next to the drop of solvent. After focusing upon the edge of the drop, one particle is picked up with a glass needle, Expt. 61, and placed into the solvent inside the field of microscopic vision. The rounding off of the edges of the particle, gas bubbles, appearance of a coloration or of a precipitate in the solution around the particle indicate action of the solvent upon the particle. If soluble, the particle may finally vanish, and crystals may appear along the edge of the drop where the solution gets concentrated as a consequence of evaporation. The use of slides with a concave depression or (and) treatment with Desicote (13) is (are) recommended for work with organic solvents that spread on the surface of the slide.

For the determination of the solubility of small particles, MONKMAN (457) and JAECKER and SCHNEIDER (959) expose it to the vapors of the solvent while observing under the microscope or with a $5 \times$ to $10 \times$ magnifying lens. The latter transfer the particle onto a narrow slide which is inserted into an absorption tube for semimicro combustion (as an alternative, a slide or cover slip with the particle attached to the underside could be made the removable cover for the circular or elliptical side opening of a glass tubing, 8- to 10-mm bore, which is mounted on the stage). By use of 3-way stopcocks, nitrogen or air is passed either directly to the tube containing the particle or via a gas washing bottle containing the liquid solvent. A manifold stopcock arrangement may be used with several gas washing bottles containing different solvents so that the sample may be subjected to the action of a series of solvents. Of course, every solvent should be completely removed by passing clean nitrogen or air through the chamber before starting a test with another solvent.

The rate of gas flow is important. If too fast, the gas does not become saturated with the solvent while passing through one simple gas washing bottle. If the rate is too slow, the solvent vapor is brought to the sample too slowly. A rate of 400 to 500 ml gas per minute was found satisfactory in experiments at room temperature with ethanol, diethyl ether, carbon tetrachloride, and benzene. Obviously, the gas washing bottles may be supplied with water, strong hydrochloric acid, or ammonia solution; in addition, the gas washing bottles may be placed into a shallow bath of warm water in order to raise the vapor tension of the solvent. The degree of solubility may be correlated to the time required for the conversion of the solid particle to a liquid droplet. Anomalous behavior has been discovered in several instances (965), where soluble substances refused to liquify, which may be explained by the formation of compounds with the solvent. Of course, withdrawal of the solvent by passing pure nitrogen or air through the cell may furnish crystals for identification according to P. 10 to P. 13.

Performance of Solubility Tests

Metallic materials are tested with acids only. Otherwise the solubility in water is determined first, regardless whether the material is inorganic or organic in nature.

P. 31

Inorganic Substances are considered soluble if a 1% or stronger solution may be obtained (1 mg/0.1 ml or higher concentration); moderately soluble, if 0.1 to 1 mg dissolve in 0.1 ml; and insoluble, if less dissolves, i. e., a residue is left when treating 1 mg material with 1 ml solvent. Most inorganic substances are either "soluble" or "insoluble". The relatively few substances which are moderately soluble in water of room temperature are listed in Table 8.

The material should be used in the form of small particles (powder) so that equilibrium is rapidly attained. Add 0.1 ml water per mg sample if only one substance seems to be present or 0.03 ml per mg sample if the latter appears to be a mixture. Agitate the mixture for several minutes. Dissolution may be hastened by heating on the steam bath, but finally the mixture must be cooled to room temperature with the proper precautions to prevent the establishment of a supersaturated solution.

The substance is classified as soluble if a clear solution is obtained and no solid residue is left behind. If the solid remains unchanged and evaporation of some of the supernatant liquid does not give a significant residue[1], the material is considered insoluble. If some residue remains, the microscopical inspection of it and of the residue of the evaporation of some of the solution may show whether it is partial solubility of one substance or the complete dissolution of one or several components of a mixture. If it is limited solubility, addition of 0.9 ml more solvent per 1 mg sample will decide whether the substance is moderately soluble or "insoluble". If a mixture of several substances is given, extraction with water should be continued to get complete separation of the soluble material from the insoluble. The amount of the insoluble part should be estimated from the volume of the residue.

Evaporation of the aqueous solution may give well developed crystals for investigation by P. 10 to P. 13. In the instance of mixtures, it is advisable to analyze the aqueous extract separately.

Materials (residues) insoluble in water are tested with 0.01 ml/mg solid of the following acids. In each instance apply the acid first at room temperature, and then heat on the steam bath. Try them in the order: 4-F HCl, 12-F HCl, 6-F HNO_3, 16-F HNO_3, and aqua regia (1 volume 16-F HNO_3 + 3 volumes 12-F HCl) until a satisfactory solvent has been found.

Tables on the solubility of common inorganic salts may be found in books on qualitative analysis (40, 55, 56). The following rules and Table 8 will serve the same purpose.

Soluble in Water Are:

most compounds of the alkalies and ammonium;

most chlorides, hypochlorites, chlorates, perchlorates, bromides, bromates, iodides, periodates, manganates, permanganates, cyanates, acetates, thiosulfates, sulfates, selenates, nitrites, nitrates, hypophosphites, and vanadates;

---

[1] The solvent itself may give a slight residue either because of impurities dissolved or because of attack of the apparatus; use of vitreous silica and comparison with a solvent blank will aid in arriving at a decision.

the alkaline earths cyanides, thiocyanates, ferrocyanides, ferricyanides, hydrosulfides, polysulfides, bisulfites, primary phosphates and arsenates, and bicarbonates;

AgF, SnF_2, SnF_4, SbF_3, BiF_3;

TlOH, TlCN, $Hg(CN)_2$;

many complex compounds of heavy metals with ammonia and amines, water, hydroxyl, cyanide, thiocyanate, thiosulfate, oxalate, tartrate, citrate, and various enolic compounds.

Insoluble in Water, but Dissolved by 6-F HNO_3 or HCl:

most metallic oxides, hydroxides, and basic salts;

most water insoluble salts of weak acids such as fluorides, iodates (those of Ag, Mg, Ca, Sr, Ba, Pb do not dissolve readily in HNO_3), cyanides, thiocyanates, ferrocyanides, ferricyanides, sulfides, sulfites, selenites (tellurites are decomposed with separation of TeO_2), chromates, tungstates, phosphates, arsenates, arsenites, borates, carbonates, oxalates, tartrates, and other salts of organic acids;

the readily hydrolysable salts of Bi, Sb, Sn(4), Fe(3), Al, etc.;

the hydrides, nitrides, phosphides, and carbides of the alkaline earths;

BeC, Al_4C_3, LaC_2, NdC_2, Fe_3C, Ni_3C, Mn_3C, UC_2, Cu_2C_2, Ag_2C_2;

AgCNO, $Pb(CNO)_2$;

vanadates of Hg(1), Pb, Fe, Cr, Al;

some silicates are decomposed by acid.

Insoluble in 6-F Acid, but (Oxidized and) Dissolved by Hot 16-F HNO_3 or Aqua Regia:

I_2, S (slowly), P, As, B;

Cu_2Cl_2, Cu_2I_2, $Cu_2(CNS)_2$, AgCN, AgCNS, AuCl, $PtCl_2$, TlCl, TlBr, TlI, Tl_2SO_4, TlCNS, Hg_2Cl_2;

HgI_2, BiI_3, HgS, As_2S_3, As_2S_5, MoS_3;

slowly attacked are CaF_2, SnO_2, Sb_2O_5, some tungsten bronzes, some silicates, Prussian blue.

Insoluble in All Solvents Tried:

S, C, Si; certain metals and alloys, P. 42;

AgCl, AgBr, AgI, AgCN, CaF_2;

SiO_2, TiO_2, ZrO_2, ThO_2, SnO_2, Nb_2O_5, Ta_2O_5, WO_3, H_2WO_4;

Fe_2O_3, Cr_2O_3, Al_2O_3 after exposure to high temperature;

$BaSO_4$, $SrSO_4$, ($CaSO_4$), $PbSO_4$, fused $PbCrO_4$;

pink anhydrous salts of chromium: $CrCl_3$, CrF_3, $Cr_2(SO_4)_3$;

carbides: B_4C, SiC, TiC, ZrC, HfC, NbC, TaC, (MoC, WC, VC);

nitrides: borazon=cubic BN, TiN, ZrN, TaN;

borides: TiB, ZrB;

silicates, chromites, aluminates formed at high temperature;

tungsten bronzes characterized by metallic luster, intense color, high density, and electric conductivity.

P. 32

Organic Substances are classified as "soluble" if 30 mg or more dissolve in 1 ml solvent of room temperature; as "insoluble" if less dissolves. Agitation for several minutes will give equilibrium if the particles of the sample are small, i. e., the sample has been ground to a powder.

If only one substance is present, use 0.033 ml solvent per mg sample; if a mixture may be present, preferably try first 0.01 ml solvent per mg sample. In the instance of a single compound, agitate until equilibrium is obtained and then decide whether or not the substance has been dissolved without leaving a residue; if there is a small residue, microscopical inspection may help to decide whether or not it is an impurity. When dealing with a mixture, inspect under magnification the residue of the solubility tests as well as the residue from the evaporation of a small portion of the solution; the appearance of the solids may indicate whether a separation has been obtained and whether it has been complete. The amount of residue should be estimated, and if a separation is obtained, it should be made complete with measured amounts of solvent to permit estimation of the degree of solubility. Any well developed crystals may be investigated according to P. 10 to P. 13.

CHERONIS and ENTRIKIN (146) determine the solubility in water, diethyl ether, $1.2\text{-}F$ HCl, $2.5\text{-}F$ NaOH, $NaHCO_3$, and in $18\text{-}F$ H_2SO_4 to recognize the following divisions:

(S_1) *Soluble in Water and Ether:*

generally monofunctional compounds with five carbons or less: alcohols, aldehydes, ketones, carboxylic acids, acetals, anhydrides, esters, ethers, lactones, some glycols, polyhydroxy phenols; amines, amides, amino heterocyclics, nitriles, nitro paraffins, oximes; halogen substituted compounds of the above list; hydroxy heterocyclic sulfur compounds, mercapto acids, thio acids; halogenated amines, amides, and nitriles; amino heterocyclic sulfur compounds.

(S_2) *Soluble in Water, Insoluble in Ether:*

compounds with moderate molecular weight, having two or more polar groups or being a sulfonic or sulfinic acid: dibasic and polybasic acids, hydroxy acids, polyhydroxy alcohols and phenols, simple carbohydrates; salts of acids and phenols, various metallic compounds; ammonium and amine salts of organic acids, amines, amino acids, amides, amino alcohols, semicarbazides, semicarbazones, ureas; halogenated acids, alcohols, and aldehydes and acyl halides; sulfonic acids, alkylsulfuric acids, sulfinic acids;

amine salts of halogenated acids; aminodisulfinic acids, bisulfates of weak bases, cyano and nitro sulfonic acids.

(B) *Insoluble in Water, Soluble in 1.2-F HCl:*
amines, amino acids, amphoteric compounds such as aminophenols, aminothiophenols, aminosulfonamides, arylsubstituted hydrazines, N-dialkylamides.

(A_1) *Insoluble in Water and in 1.2-F HCl, but Soluble in 2.5-F NaOH and 1.5-F (?) $NaHCO_3$:*
acids (usually with 10 carbons or less) and anhydrides; amino, nitro, and cyano acids, heterocyclic nitrogen carboxylic acids, polynitrophenols; halogenated acids, polyhalogenated phenols; sulfonic and sulfinic acids; aminosulfonic acids, nitrothiophenols, sulfates of weak bases; sulfonhalides.

(A_2) *Insoluble in Water, 1.2-F HCl, and 1.5-F $NaHCO_3$, but Soluble in 2.5-F NaOH:*
high molecular weight acids, anhydrides, phenols including esters of phenolic acids, enols; amino acids, nitrophenols, amides including N-monoalkyl amides, aminophenols, amphoteric compounds, cyanophenols, imides, N-monoalkyl aromatic amines, N-substituted hydroxylamines, oximes, *p*- and *s*-nitroparaffins, trinitro aromatic hydrocarbons, ureides; halogenated phenols; mercaptans, thiophenols; polynitro halogenated aromatic hydrocarbons; aminosulfonamides, aminosulfonic acids, aminothiophenols, sulfonamides, thioamides.

Insoluble in Water, in HCl, and in NaOH:

(M) *Containing Nitrogen or Sulfur:* anilides and toluidides, amides, nitro arylamines, nitro hydrocarbons, aminophenols, azo, hydrazo, and azoxy compounds, di- and triarylamines, dinitrophenylhydrazines, nitrates, nitriles; mercaptans, N-dialkylsulfonamides, sulfates, sulfonates, sulfides, disulfides, sulfones, thioesters, thiourea derivatives; sulfonamides; halogenated amines, amides, and nitriles;

(N) *Not containing Nitrogen or Sulfur and Soluble in 18-F H_2SO_4:*
alcohols, aldehydes, ketones, esters, ethers, noncyclic unsaturated hydrocarbons, unsaturated cyclics that are easily sulfonated (di- or polyalkyl-substituted benzenes), acetals, anhydrides, lactones, polysaccharides (charring noticeable);

(I) *Not Containing Nitrogen or Sulfur and Insoluble in H_2SO_4:* cyclic hydrocarbons, paraffins, halogenated hydrocarbons, and diaryl ethers.

Review: Inorganic Substances

The ignition test will have shown whether the material under investigation is an inorganic or an organic substance, or a mixture of both. The solubility test may have shown a way for their separation. For suggestions on the analysis of organic substances see P. 34.

If the material is inorganic, its history and the preceding tests (hardness) may more or less definitely indicate its nature so that only the confirmation of chemical identity remains as a final task. Analytical confirmatory tests are compiled in P. 44 to 60. Means for distinguishing similar substances (different oxides of a metal, different hydrates of a salt, primary or secondary, etc. salt of a polyprotic acid, choice between various complex compounds of like qualitative composition, distinction between synthetic and natural product, etc.) will usually have to be found from the description of the substances involved (4, 5, 8, 10) or from an inspection and study of samples of the substances to be considered. On the recognition of various types of carbon see CHAMOT (118), p. 188, and FEIGL (121), p. 308. On the identification of corundum among natural and artificial associates see CROSSMAN (442). Descriptions of the pigments used in paintings have been compiled by GETTENS (1094, 1095). Books on mineralogy may be consulted on the description and identification of minerals (169). Advice concerning the recognition of soil minerals (particles) is given by FRY (93); see also KIRK (186).

If the preceding testing has furnished no definite clues for identification, it will have produced sufficient information for arriving at a decision for the continuation of the work. At least five different approaches, outlined below, promise success. Aside from availability of apparatus, the decision will be influenced by the amount of material available and its nature: single substance or mixture, metal or non-metallic, simple or complex substance, crystalline or glassy (amorphous), soluble or insoluble, "rare" elements improbable or probable, naturally occurring or man-made, etc.

Crystallographical Optical Analysis. By the use of goniometer, elaborate polarizing microscope, universal stage, and determination of refractive indices (P. 10 to P. 13) solid particles (crystals and crystal fragments) may be identified as "molecular species" without being destroyed. The procedure is especially recommended in the instance of minerals and rock constituents (83, 94, 96, 97, 99, 101, 104, 107, 113, 114, 773).

X-Ray (or electron) Diffraction. This procedure likewise identifies "molecular species" without causing destruction. In mixtures, it will discover majors and minors. The amount of sample may be as small as $10\,\mu g$. Usually the material is used as a fine powder, but also single small crystals and thin films may be investigated (80, 467).

X-Ray Emission Spectrography (482, 770). This is a powerful and efficient non-destructive method for revealing the elemental composition if the elements below atomic number 11 (sodium) are of no interest. The sample may be very small or a thin film; majors, minors, and traces down to one part in ten thousand are discovered. Elements that are difficult to discover in the wet way (rare earth, Ta, Nb, W, etc.) present no difficulties.

A beam of X-rays or electrons is focused upon the sample to excite emission of K, L, M, N, O, P series of lines in correspondence with the atomic number of the elements present in the sample. The emitted radiation (X-rays) is passed to an X-ray spectrograph or spectrometer. The far greater simplicity of the spectra as compared with their optical counterparts, the identical appearances of the series for all elements but the lightest ones, and the systematic variation of wave length with atomic number are distinct advantages for the interpretation.

For qualitative purposes, an instrument like the Microanalyzer Camera of L. v. HAMOS (79) appears quite satisfactory. It gives a spectrogram of all characteristic rays excited by an X-ray beam incident upon a very small sample which may have a volume of 10^{-8} ml = 10 pl. Various instruments are commercially available (77).

Solids may be exposed as such or diluted (mixed) with starch, lithium carbonate, aluminum powder, alumina, or embedded in Lucite. Minerals may be dissolved in a borax flux. The sample may also be dissolved in water or suitable solvents (containing only C, H, N, and O). Spots on filter paper have been succesfully analyzed, and it seems obvious that also the circular residues of the ring-oven technique will do.

The microsonde or electron probe (1292) uses an optical microscope for selecting the field of study, whereupon a beam of electrons is focused upon a detail in the field, that may be as small as 1 μm or less in diameter: less than 0.000001 mm^2 in area and probably less than 10^{-12} ml or 0.001 pl in volume. The elements present in the bombarded detail are approximately indicated by the amount of backscattering of electrons and definitely identified by the frequency of the emitted X-rays. Accurate quantitative analyses can be performed by comparing the intensity of the X-rays emitted by the specimen with that from the pure chemical elements and applying appropriate interelement corrections (495, 1273).

At this time, the most highly developed instruments permit rapid scanning of the whole field of the optical microscope and selection of the effect which shall be observed: backscattering of electrons or emission of dial selected X-rays lines characteristic for the element sought. Integration on the screen of oscillographs then reveals those details of the field, in which the element is located, and by repeatedly resetting the selector dial the distribution of the various elements is obtained (qualitatively and at least semiquantitatively) within minutes. Photographic recording gives remarkably sharp outlines of the parts of the microstructure containing (e. g.) Ti, V, Cr, Mn, Fe, Co, Ni, Cu, Zn, Ga, etc., which match the structures visible in the optical photomicrograph (1072).

The electronprobe requires that the specimen is mounted in a high vacuum, withstands the heat produced by the electron beam, and is a conductor to prevent building up a negative charge that would deflect the stream of impinging electrons. It follows that volatile matter must be absent.

Polished metallographic specimens satisfy all requirements, but polished sections of rocks, ceramics, etc. must be coated with a suitable conducting film, e. g. vacuum coating with carbon, 10 to 20 nm (mμ) thick. Loose particles must be imbedded in a solid matrix provided with a conductive film since otherwise they would become charged and propelled into space when hit by the electron beam. For the study of the distribution of elements in biological specimens, it is possible to use the mineral skeleton left after cautious ashing of thin sections (194).

Optical Emission Spectrochemical Analysis (78, 480, 494, 1056). In spite of the fact that excitation by use of flame, arc, or spark causes complete loss of the sample, this may be justified if the metallic constituents are of principal interest and also traces (down to 1 part per million) shall be found. Only the elemental composition is revealed, but complex alloys, minerals, ceramics, and glasses containing lanthanides, Ta, Cb, W, etc. do not offer any particular difficulties.

The material under investigation may be supplied as a solid, a solution, or a vapor. The size of the sample may be reduced to a fraction of a milligram, but if no suitable precautions are being taken, one may be skeptical about the use of very small amounts of readily volatile matter.

Chemical Analysis. Chemical analysis often permits saving parts of the material of the sample in the form of precipitates or solutions; there is not necessarily a total loss of the material. In addition to the elemental composition represented by majors and minors, also some information is obtained concerning valence state and complexes present.

The continuation of the investigation is outlined in P. 36 and the following paragraphs. Samples of 1 μg and less will suffice if the technique of performance is properly selected. If very little material is available, each step of the investigation may be preceded by a pilot experiment, a control, on a known material containing small amounts of the substances to be tested for. Success of the control will show that apparatus, reagents, and technique are suitable. A blank (experiments with solvents and reagents only) should be performed simultaneously with the analysis of the sample. Negative tests in the blank will then prove that substances found have not been introduced by reagents, solvents, apparatus, or the conditions of the surroundings (CO_2, HCl, NH_3, H_2S in the air of the laboratory, copper in the distilled water, copper oxide particles ejected by the surface of a heated apparatus, etc.).

P. 34 Organic Substances

If this was not done already, carbon compounds should be heated under carefully controlled conditions for the determination of the melting point,

P. 21. This may lead to quick identification if only one substance is present and a sharp melting point is obtained. If a mixture is present, it will be preferable to separate the components for the determination of the melting point; to this end, particles of various kinds may be selected mechanically, Expt. 61, and heated separately. If this is not possible, other means for separation must be used.

If the melting is not "sharp", i. e., if it occurs during an interval of several degrees, a mixture may be present. Observation of the crystallization on cooling may show the separation of several types of crystals and of a fine-grained eutectic mixture which either remains liquid or solidifies last. In such instances, a mixture is obviously present, and the heating should be repeated to obtain a record of the melting point of the eutectic mixture and of the temperatures at which the last traces of the various types of crystals melt (melting points of the pure components). An impure substance will give one kind of crystals and only a small amount of eutectic mixture; of course, the crystals may represent a solid solution (mixed crystals). An impure solid may be purified for the determination of the melting point as directed on pp. 151, 174.

As mentioned in P. 21, the determination of the melting point and use of the tables of KOFLER (159) or McCRONE (163) will still leave a choice of three to ten compounds. As a rule, the final identification may be brought about by determining the melting points of the eutectic mixtures with the compounds listed in the table (159). A sample of the unknown substance is mixed with an approximately equal volume of one of the substances suggested in the table. The mixture is covered with a fragment of a cover slip, heated until it is completely liquid, and then allowed to solidify again. On reheating, the melting point of the eutectic mixture is found as the temperature at which melting begins (well defined temperature at which the melt is at equilibrium with the crystals of the eutectic mixture). If necessary, the experiment may be repeated with the second substance suggested by the table. Of course, the identification by melting point and eutectic melting point should be in agreement with other facts observed so far, and it should be confirmed by a "mixed melting point" (m. pt. of a mixture of the unknown and a pure sample of the substance it is supposed to be; crystallization of the melt should give only one kind of crystals separating within a temperature interval of less than one degree). Confirmatory tests may be suggested in the literature; or tests may be designed after reading the descriptions of the substance that seems to be present and closely related compounds.

If only one substance seems to be present, which decomposes on heating so that no melting point can be obtained, clues to its identification may be found by consideration of history, appearance, solubility, and elemental composition which may be found according to P. 35. The Tables for

Identification of Organic Compounds may be useful (54). If, however, these lines of attack fail, it is suggested to consider the following selection of instrumental methods in addition to the chemical procedure. If a mixture seems to be present, low temperature sublimation may be considered in addition to chromatography and liquid-liquid extraction as a means of separation previous to instrumental analysis.

Crystallographic and Crystal Optical Methods (112) as well as **X-Ray Diffraction** (82) offer non-destructive procedures for the identification of molecular species, P. 10 to 13. LINDENBERG (1065) discusses the effect of solvent upon the crystal habit of organic substances.

The collection of crystal optical data by WINCHELL (112) is being continuously augmented by publication of data in *Analytical Chemistry* and other journals. Attention may be called to the contributions of SHELL, POE, and WITT concerning chemotherapeutic drugs (933), antihistaminics (960), and steroid hormones (976, 977).

X-Ray Spectrography, P. 33, can furnish, without destruction, the elemental composition but for H, Li, Be, B, C, N, O, and F.

Absorption Spectrophotometry provides additional means for non-destructive identification of molecular species. Because of the complexity of absorption spectra, recording instruments are desirable for qualitative analysis. The visible and ultraviolet ranges (78, 81) are of obvious interest for colored substances, dyes, and pigments.

Ultraviolet Spectrophotometry (746) is much used in the investigation of fats and oils and of hydrocarbons. A systematic procedure for the identification of nearly 200 dangerous drugs, narcotics, and poisons (based upon separation by solvent extraction) has been described by BRADFORD and BRACKETT (946). Microspectrophotometry of objects as small as $1\,\mu$m in diameter is possible (479, 746).

Infrared Spectrophotometry (81 a) has been extensively used for qualitative studies on organic substances, and CONLEY (713) has written on the use of infrared spectroscopy in organic qualitative analysis; *see* also CHERONIS (146).

The sample may be used in the solid state (KBr pellet), as a film, or in a solvent which does not absorb in the regions of interest. Using the bromide disk procedure, even industrial spectrophotometers do not require more than 0.3 to 0.5 mg of sample. With a micromull technique, beam condenser, and microscope attachment, the amount of sample may be reduced to 50, 5, and even $0.5\,\mu$g (475, 486, 487, 626, 1092). Beam condensers permit also the use of microcells for work with down to $0.02\,\mu$l of 1% and weaker solutions (460, 462, 481, 486). An "Organic Microspectrophotometer" for 10 to $50\,\mu$l solution is commercially available (1291).

Chemical Analysis (771). For a continuation of the work, it will be advisable to first consult the books on qualitative organic analysis (35, 38,

39, 41, 42, 46, 51, 52, 59). Obviously it will be helpful to refer to books which have been written for work on a small scale. Of these, that of CHERONIS (146) supplies much valuable advice on reasoning and the use of the literature. SCHNEIDER (167) gives thorough consideration to the procedure to be be followed with substances which have not been described in the literature.

At this time, the task is to review all evidence collected this far and to select additional orientation tests so that the preliminary identification is accomplished with the expenditure of as little material as possible. Tests should be selected, that promise a maximum of information and permit, if possible, additional tests with the same material. Tests which may add little or no new information should be postponed until there is a good reason to try them.

With mixtures, separation must come first, of course, and preferably by purely mechanical means: lifting solids out of a mixture, Expt. 61, filtration, decantation, and separation of liquid layers. Exploratory (pilot) experiments with very small amounts of substance are recommended for finding the most efficient means of separation. The methods are discussed in the books cited above; they also have been reviewed recently by METCALFE (483). DAVIS, DUBBS, and ADAMS (485) have simplified Decker's (1060) method for the elution of paperchromatogram spots for further investigation. They cut the desired spot so from the chromatogram that a pointed tip is obtained and then wash the spot into the point of the tip. The material is collected within an area of just a few square millimeters; a recovery of 99.6% was demonstrated with 2 to 10 μg adenine by cutting off and extracting the tip of the paper. With sufficient substance, even a solid deposit may be built up at the point of the paper, which may be lifted off mechanically without need for any solvent. At times, well developed crystals may be obtained for identification according to P. 10–13. REIMERS (968) concentrates the material into a short line. On rapid separation by thin-layer chromatography see PEIFER (972) and WASICKY (493). For the concentrating of trace impurities and purification by progressive freezing (zone melting), see MATTHEWS and COGGESHALL (474) and the literature cited by them.

After separation and for the identification of the isolated components, SCHNEIDER (166, 167) follows the system of MULLIKEN (46) who divides the compounds into orders (according to elemental composition), genera by "generic" tests, etc. CHERONIS and ENTRIKIN (146, 771) use a more flexible approach in adding to the already collected evidence by testing for solubility and degree of ionization (indicator method), for functional groups, and for the "specific class". The tests with dilute and concentrated sulfuric acid recommended by McGOOKIN (42) may be tried when sufficient material is available to observe odors and liberation of heat. Whenever possible, the final identification should be based upon the preparation of several derivatives and their identification by the determination of characteristic properties (771).

The following references to the recent literature are offered as additional suggestions. Spot tests for functional groups and specific compounds may be found in the books of FEIGL (121, 122), and lists of microchemical tests for groups and compounds may be found in the reviews of CHERONIS (775, 780).

Alcohols. Melting points and eutectic melting points of the 2,4,6-trinitrobenzoates of 29 alcohols are listed by LASKOWSKI and ADAMS (469). A test for secondary alcohols and 1,2-diketones is described by FEIGL, GENTIL, and STARK-MAYER (929).

Aldehydes. Spot tests have been described by ANGER and FISCHER (961) and by MANNS and PFEIFER (for cyclic aldehydes) (949).

Fatty Acids. On chromatographic separation and recognition *see* CHURACEK (967).

Amino Acids. On the rapid identification of single crystals by determination of the eutectic melting point and the refractive indices *see* LACOURT et al. (914, 915).

Esters of Nitric and Nitrous Acids. Spot test by FEIGL et al. (939).

Nitro and Nitroso Compounds, Oximes, Hydroxylamines. Spot test by FEIGL et al. (945).

Organic Bases. Microchemical tests by SANDRI (943).

Pyrimidine Derivatives. Microchemical tests by WEISS (966).

Synthetic Drugs. Microchemical tests are listed in the *Methods of Official Agricultural Chemists*, pp. 511–515 (9).

Local Anesthetics. On identification via the melting points of salts *see* BRANDSTÄTTER-KUHNERT and GRIMM (932).

Alkaloids. Electrochromatophoretic separation is treated by BROWN and KIRK (938). Microchemical tests are described in the *Methods of Official Agricultural Chemists* (9) and by WORMLEY (138), STEPHENSON et al. (135, 1120), and WHITMORE and WOOD (879, 880). On ergot alkaloids see KOLSEK (927), and on the test of MALQUIN-DENIGES for strychnine: LUIS and CORAZZA (940).

Polymers. On characterization *see* ALLEN (33). SWANN and ADAMS (492) describe a test for epoxy coatings.

Dyes. Testing by chromatography should not be omitted. On the identification in food, *see* the *Methods of Official Agricultural Chemists* (9).

Inks. On electrochromatophoretic analysis *see* BROWN and KIRK (926).

Pigments. On classification and identification of organic pigments *see* VESCE (206).

P. 35 Elemental Analysis of Organic Substances

All of the classical tests have been transposed onto the microscale, and many of them have been modified repeatedly (146, 150, 152, 154, 161, 164, 166, 167). In this connection only the procedure of KÖRBL (925) shall be described since it permits testing for most of the common elements with the expenditure of only one sample.

Reagent. Treat a boiling solution of 19.4 g $KMnO_4$ in 400 ml water with 20.4 g solid $AgNO_3$. When the latter has dissolved, allow to cool. Collect the crystals of $AgMnO_4$ on the fritted glass plate of a Büchner funnel and wash with 150 ml cold water. Dissolve the washed crystals in 400 ml boiling water, and filter the hot solution through a fritted glass filter. Cool the filtrate and collect the crystals as before. Wash them with cold water and dry them at 60° to 70°C. The yield is about 18 g.

To obtain the oxidant, place about 2 g of the $AgMnO_4$ into a clean test tube and heat to 150°C. Sudden decomposition gives a black mass of "silver manganite" of the approximate composition $Ag_2O \cdot MnO_2 \cdot x\,O$, which is used without further ignition and kept in a stoppered test tube until needed.

Procedure of Körbl. Draw out one end of a glass tubing of 4-mm bore (1-mm bore if only 5 to 20 μg of substance may be spared) to a fine tip. Then, at a distance of 6 cm from the tip, heat the tubing and let it collapse to obtain a constriction with 0.3-mm bore. Cut the tubing 9 cm above the constriction and select a syringe or rubber bulb which will fit this wide end of the tube.

Place a wad of freshly ignited asbestos upon the constriction and press it lightly together by means of a clean glass rod or steel wire. If that much substance can be spared, mix 0.1 to 1 mg of it with about 50 mg of the silver manganite and compact this mixture in the tube above the asbestos wad. Finally introduce a second asbestos wad to wipe down the interior of the tube and to hold the mixture in place. It is assumed that even 5 μg of substance and 3 mg silver manganite will suffice if the quantities of all other reagents are reduced accordingly.

Start the reaction by cautious heating of the mixture with a microflame. It is complete within a few seconds, whereafter the search for the products of the reaction may be started without delay.

Hydrogen. Its presence is shown by droplets of water inside the upper part of the tube. These may be collected in a capillary pipet and—after testing for carbon—be used for the determination of pH, boiling point, and melting point.

Carbon. Do not delay this test. Connect the upper, wide opening of the tube to the syringe or rubber bulb. Place 0.2 to 1 ml barium hydroxide solution into an 8-mm test tube. Insert the combustion tube into the test tube so that the tip is just above the liquid and slowly force about 5 ml air through the combustion tube and onto the surface of the reagent solution. A white precipitate of $BaCO_3$ at the surface of the liquid confirms the presence of carbon.

Nitrogen. Introduce 0.1 to 0.2 ml water into the wide end of the tube so that the drop fills the whole bore. Force the drop very slowly through the tube and out the tip into a microcone. It will have dissolved any NO_2, HNO_2, or HNO_3 formed. Place a small amount of the extract upon the surface of a drop of a 1% solution of diphenylamine in 18-F H_2SO_4. A blue coloration confirms the presence of nitrogen.

Sulfur. Treat the rest of the solution obtained in the preceding paragraph with 50 μl 5% (0.2-F) solution of barium nitrate. A precipitate of $BaSO_4$ may be collected with the centrifuge and used for further tests.

Chlorine and Bromine. Introduce 0.1 to 0.2 ml 6-F NH_3 into the wide opening of the combustion tube. By means of syringe or rubber bulb, pass the drop slowly through the tube and into a microcone. Acidify the extract. A white precipitate may be AgCl or AgBr or a mixture of the two. Collect it with the centrifuge for further investigation, p. 323 (d). The solution above the precipitate, test for:

Phosphate and Arsenate. Treat the centrifugate from the silver halides with molybdate reagent and heat the mixture by inserting into water of about 70° C. If a yellow precipitate forms, collect it in the point of the cone, wash it with 2-F HNO_3, and dissolve it in 3-F NH_3. Treat the ammoniacal solution with magnesia mixture to precipitate $MgNH_4(P, As)O_4 \cdot$ $\cdot$ 6 H_2O. Wash the precipitate with 1-F NH_3. Transfer a few crystals to a microscope slide, and treat them with $AgNO_3$ solution while observing through the microscope, Expt. 40. Dissolve the main portion of the precipitate in 12-F HCl and saturate the cold solution with H_2S. If a yellow precipitate of arsenic sulfide is obtained, make certain that the precipitation of arsenic is complete. Centrifuge, transfer the clear solution to another microcone, make it ammoniacal, and treat it again with magnesia mixture to test for the presence of phosphate.

Iodine. Wash the contents of the combustion tubing with 6-F NH_3 until the last washing remains clear when acidified with nitric acid (absence of chloride and bromide). Then extract the contents of the combustion tube by passing through it 0.1 to 0.2 ml 30% solution of $AgNO_3$. In a microcone, dilute the extract with water. If iodine is present, a silver iodide-nitrate separates first and is converted to AgI on further diluting with water.

Mercury. If the presence of mercury is suspected, one may forego testing for carbon and hydrogen and pass a slow current of air through the tube already during combustion. Droplets of metallic mercury will be found at a short distance from the heated zone. KÖRBL (925) also tests with dithizone. If material can be spared, a separate test by Carius combustion (Expt. 51) and isolation of the metallic mercury (Expt. 49) seems more attractive.

For the *detection of refractory elements*, organic material is usually oxidized with or without the addition of suitable reagents. SPIALTER and BALLESTER (491) describe a simple procedure for the ashing of organosilicon and organometallic compounds. A hook of 0.8-mm radius is bent at the end of a 0.8-mm platinum or nichrome wire. A Pyrex capillary, 1.6 mm o. d. and 15 to 20 mm long, is placed over the wire so that it rests on the hook. The solid or liquid sample is collected in the hook. For ashing, the wire is held 45 degrees to the horizontal, and the hook is heated in the edge of a just non-luminous Bunsen flame, halfway between barrel and tip of flame. The material should ignite and continue to burn. When the burning stops, the wire is held horizontally, and the whole capillary is heated in the tip of the flame for about 5 seconds. Usually, a carbonaceous deposit is now visible on the glass near the hook. The glass and the hook are allowed to cool for one minute, and then they are again heated in the tip of the flame (wire horizontal) until the carbon has been oxidized. Si, Ge, Sn, and Pb give characteristic residues on the capillary near the hook. The authors list also the burning characteristics of various types of organosilicon compounds.

P. 36   **Testing with Dilute Sulfuric Acid**

If the solubility of the material in water and dilute acids has been investigated, evolution of gases will have been already observed. A repetition of the test with water or dilute sulfuric acid will have the principal purpose of identifying the liberated gas, and the experiment will have to be arranged to this end. If no effervescence was observed in solubility tests, the experiment with dilute sulfuric acid may serve to detect HCN, HN_3 and low concentrations of gases that are somewhat soluble (CO_2, SO_2, H_2S, PH_3, AsH_3, C_2H_2, Cl_2) and vapors (Br_2, I_2). To find small amounts of gases contained in a large amount of some other gas (AsH_3, PH_3, H_2S, or CH_4 in H_2, etc.), the treatment with dilute sulfuric acid must be carried out so that detection of these gases becomes assured.

The tests with sulfuric acid serve the detection of anions or acidic constituents. Obviously, one will not apply them to metals and alloys unless one is searching for the presence of carbide, phosphide, sulfide, arsenic, or antimony. If the material under investigation is a mineral or ore, the test with dilute acid will serve mainly for the detection of carbonate. Sulfides are better oxidized and recognized as sulfate, P. 59.

If the material under investigation is not metallic in appearance and the solubility has not yet been tested, it should first be tested with water.

Record all observations: solubility, rate of solution, color changes, change of the appearance of the solid phase, heat evolved or consumed, pH of the resulting solution, etc. The following gases may be liberated:

N_2 which extinguishes a glowing splint may be derived from NH_4 salts, urea, etc. reacting with oxidants like nitrite, hypochlorite, and peroxide;

H_2 from hydrides of the alkalies and alkaline earths, and from metals and alloys reacting with water: alkali metals, alkaline earth metals; Mg, Zn, Al when finely dispersed;

O_2 from hypochlorites; peroxides, percarbonates, perborates, persulfates especially in presence of a catalyst;

H_2S from hydrosulfides and the hydrolysis of sulfides;

NH_3 from the hydrolysis of ammonium salts (sulfide, carbonate, acetate);

PH_3 from phosphides;

CH_4 from carbides like Al_4C_3;

C_2H_2 from carbides of the alkalies and the alkaline earths; and

any one of the gases liberated by dilute sulfuric acid (*see* below) if the material under investigation is a mixture containing a definitely acidic constituent.

If the material reacts with water, the test with dilute sulfuric acid may be carried out with the reaction mixture. The reaction with water is allowed to come to completion, whereafter the mixture is treated with an equal volume of 8-F H_2SO_4.

As a rule, the material under investigation is treated with a small volume of 4-F H_2SO_4. The mixture is warmed if no reaction takes place at room temperature. All phenomena (dissolution, color changes, separation of new solid phases) should be recorded. The gases and vapors that may be liberated are the following:

H_2, colorless and odorless, burning with a barely visible blue flame to give water of pH 6 to 7, from hydrides, metals, alloys;

H_2S, colorless, odor of rotten eggs, burns with a blue flame to give $H_2O + SO_2$ (sulfur separates from the reaction mixture in the instance of polysulfides, or sulfide in presence of sulfite or oxidant; colored sulfides may separate from thiocomplexes);

H_2Se, colorless, strong odor of H_2S, burns with a reddish flame (odor of rotten radishes) to give $SeO_2 + H_2O \rightarrow H_2SeO_3$; H_2Se is soluble in water, and red flakes separate from the solution: $2 H_2Se + O_2$ (air) $\rightarrow 2 Se + + 2 H_2O$; from selenides;

H_2Te, colorless, odor similar to H_2S, burns with a blue flame to give TeO_2 (white) $+ H_2O \rightarrow H_2TeO_3$: H_2Te is soluble in water, and black flakes separate from the solution: $2 H_2Te + O_2 \rightarrow 2 Te + 2 H_2O$; from tellurides;

PH_3, possibly with some P_2H_4, colorless, garlic odor, burns with bright yellow flame which has a characteristic emerald green core to give $P_2O_5 +$ $+ 3 H_2O \rightarrow 2 H_3PO_4$, from phosphides or reaction of white phosphorus with water;

AsH_3, colorless, unpleasant odor, burns with a bluish white flame (garlic odor) to give $As_2O_3 + 3 H_2O$, from arsenides and all arsenic compounds when acid acts upon metal with the liberation of hydrogen;

SbH_3, colorless, odorless, burning with bluish green flame to give $Sb_2O_3 + 3 H_2O$; from antimony compounds when acid acts upon metal with the liberation of hydrogen;

CH_4, colorless, burning with a non-luminous blue flame giving $CO_2 +$ $+ 2 H_2O$; from Be_2C, Al_4C_3, Fe_3C, Ni_3C, Mn_3C;

C_2H_2, colorless, burning with a luminous yellow flame to give $2 CO_2 +$ $+ H_2O$; from the carbides of the alkalies and the alkaline earths, LaC_2, NdC_2, etc., UC_2, Cu_2C_2, Ag_2C_2;

O_2, colorless, odorless, makes glowing splint burst into flame, from percompounds especially in the presence of catalysts;

NO_2, brown, unpleasant odor, from nitrites: $3 HNO_2 \rightarrow HNO_3 + 2 NO +$ $+ H_2O$ and $2 NO + O_2$ (air) $\rightarrow 2 NO_2$;

Cl_2, greenish yellow;

Br_2, brown;

I_2, violet, from halide and oxidant, reduction of halogen-oxygen acid, slow decomposition of chloric acid;

SO_2, odor of burning sulfur, not supporting combustion, from sulfite, thiosulfate, thionate (in the instance of the last two, sulfur may separate from the solution; *see* also H_2S above);

CO_2, does not support combustion, from carbonate, percarbonate, cyanate;

HN_3, garlic odor, explosive, azides, especially above the b. pt. of HN_3, $37° C$;

HCN, odor of bitter almonds, from simple and complex cyanides.

For the performance of the test, the solid is treated with a small volume of reagent in order to reduce the solubility loss: for 1 mg substance use 0.02 ml water or 0.03 ml 4-*F* H_2SO_4. If no reaction takes place, warm the mixture to 90° C before testing for the presence of small amounts of gas. Combustibility and color of flame are useful characteristics only when working with 10 mg or more of solid sample per test. When working with very small amounts, even the color and odor of the gas may no longer be perceived, and one has to rely completely upon chemical tests.

Gram and Decigram Scale. Transfer about 10 mg of the solid to the bottom of a test tube (best 10 mm in diameter) and add 0.3 ml 4-*F* H_2SO_4 by means of a pipet so that the wall above the liquid remains dry. If there is a rapid evolution of gas, begin the testing without delay; as soon as the evolution stops. keep the tube lightly stoppered between tests. If little or no gas is liberated, lightly place a stopper into the opening of the tube and heat the contents by standing the test tube in a beaker containing some water of approximately 90° C. The absence of effervescence indicates that significant amounts of difficultly soluble gases (H_2, O_2, N_2) are not obtained; testing for these is omitted. Observe and record the following:

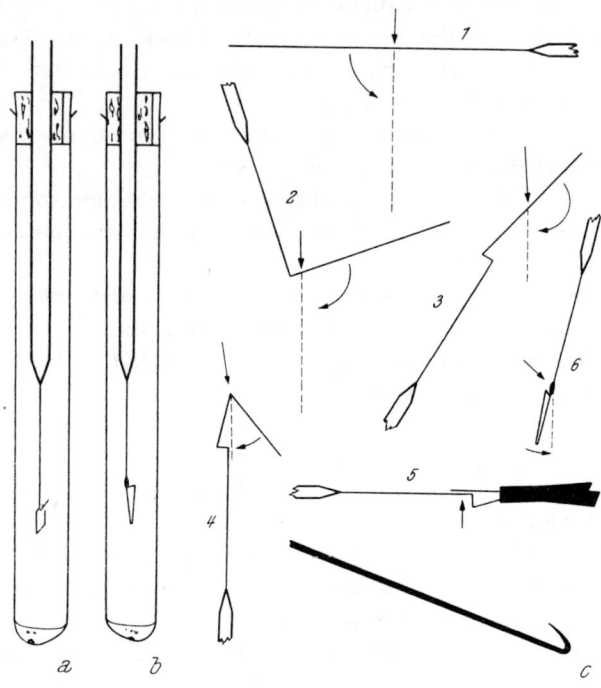

Fig. 75. Testing for Gases. *a* with reagent paper on glass hook *c*; *b* with liquid reagent; *1—6* forming of glass loop; about $^2/_3$ nat. size

1. Phenomena in the reaction mixture, color and odor of liberated gas.

2. If gas is copiously liberated, touch a small gas flame (burning at the opening of a capillary) to the mouth of the test tube. A popping sound indicates a mixture of combustible gas with air. A flame may form at the mouth of the tube or creep down the inside of the tube. Note the color of the flame and the odor of the combustion products (SO_2).

3. If a colorless gas is copiously liberated, insert a glowing wooden splint into the test tube. Oxygen is present if it bursts into flame. The other colorless gases listed will put out the glowing.

4. This and the following tests should be delayed until the rapid evolution of gas has stopped. Before testing for CO_2, it is also necessary to wait until the combustion products of the preceding test have been displaced by the liberated gas.

A loop of glass is readily obtained as shown in Fig. 75, *1–6*. A glass thread of about 0.5-mm diameter is bent by allowing it to follow the pull of gravity while it is cautiously heated by applying a pinhead flame at the points indicated by arrows. In step *5*, the rod at the left and the loop (by means of flat tipped forceps) are held while fusing the end of the thread to the handle. Finally in *6*, a small amount of bending by gravity pull establishes a slight angle between handle and plane of the loop so that the contents of the loop may be readily deposited upon a slide.

Insert the loop into a saturated solution of barium hydroxide. On withdrawing, a clear film of solution should fill the loop. Insert it quickly into the opening of the test tube. If much gas has formed, a white precipitate will separate immediately (and may even dissolve again on formation of bicarbonate or bisulfite). If there was no copious liberation of gas, insert the loop deeply into the test tube, Fig. 75*b*, and warm the reaction mixture in the test tube by standing the latter in water of 90° C. A white precipitate may be $BaCO_3$ or $BaSO_3$. Withdraw the loop from the test tube and take up its contents into a capillary which may be kept in a horizontal position until there is time for the following test: add to the contents of the capillary a like volume of a mixture of equal parts of 4-F HNO_3 and 0.02-F $KMnO_4$; seal both ends of the tube and mix by centrifuging: a **clear, red** solution proves the presence of CO_2 in the gas; a white or pink precipitate of $BaSO_4$ (and fading of the MnO_4 color) indicate SO_2 or H_2S—and the test for carbonate must be repeated with a mixture of equal volumes of 4-F HNO_3 and 0.02-F $KMnO_4$ in the place of the 4-F H_2SO_4.

If lime water (saturated solution of calcium hydroxide) is used in place of barium hydroxide, deposit the contents of the loop with the white precipitate upon a slide and expose the drop to bromine fumes by inverting the slide and placing it upon the opening of a bottle containing 12-F HCl and a few drops of bromine. After 5 minutes, remove the slide and inspect the drop under the microscope. Crystals of $CaSO_4 \cdot 2\,H_2O$ indicate SO_2 or H_2S, see P. 59.

5. If violet vapors have been noticed, suspend from a glass hook, Fig. 75*a, c*, a small square or triangle of filter paper that has been impregnated with starch. Moisten it with a droplet of water and introduce it into the test tube. Blue color indicates I_2.

6. If the test (5) was negative, test in a like manner with KI-starch paper. Blue color indicates Cl_2, Br_2, or (and) NO_2.

7. If tests (5) and (6) were negative, take up with the glass loop some 2% (0.1-F) silver nitrate solution and introduce it into the test tube, Fig. 75*b*; warm the contents of the test tube to 90° C. A precipitate may be

white: AgCN or AgN_3. On recrystallization from hot 8-F HNO_3, both compounds (and apparently also Ag_2C_2) give fine needles and bundles of such. Recrystallization from NH_3 is suggested.

Dip the loop with the white precipitate or turbidity into the gas space of a bottle with conc. NH_3 and wait until the liquid in the loop is clear. Then transfer it to a slide and observe under the microscope the crystallization taking place as the NH_3 concentration gradually decreases (*compare* Expt. 32). AgCN gives fine needles that arrange themselves to imitate the sides of polygons and of tree branches. Finally very thin plates appear with serrated outlines, which may be circular or in the shape of leaves, etc. AgN_3 gives fine needles and bundles of such.

Allow the drop to evaporate and then place upon the residue a large drop of 8-F HNO_3. AgN_3 gives a clear solution, whereas with AgCN (and Ag_2C_2) the drop of acid remains turbid.

first white, but quickly changing to yellow and light brown: Ag_2C_2. Treat with NH_3 and finally HNO_3 as directed for AgCN and AgN_3. The light brown flocks seem difficultly soluble in ammonia so that the NH_3 solution never becomes clear. On dissipation of the NH_3, extremely small yellowish brown squares, rectangles, and tiny grains separate.

yellow: $Ag_3As \cdot 3\ AgNO_3$. This may happen if the $AgNO_3$ solution has become concentrated by evaporation. Adding water gives a black precipitate: $6\ Ag + H_3AsO_3 + 3\ HNO_3$.

brown and quickly turning black: Ag_2S. Transfer the contents of the loop to a slide. Remove the liquid with filter paper (Expts. 43, 45) and wash once with water. Place upon the black precipitate a droplet of 1% $CaCl_2$ (0.1-F) solution and expose to bromine vapors for 5 minutes, P. 59. Then place the preparation under the microscope and observe the separation of crystals of $CaSO_4 \cdot 2\ H_2O$ along the edge of the drop: sheaves of fine needles, occasionally "swallow tail" twins.

black: possibly Ag_2Se or Ag_2Te

$PH_3 + 3\ AgNO_3 \rightarrow Ag_3P$ (black) $+ 3\ HNO_3$,

$AsH_3 + 6\ AgNO_3 + 3\ H_2O \rightarrow 6\ Ag$ (black) $+ H_3AsO_3 + 6\ HNO_3$,

$SbH_3 + 6\ AgNO_3 + 3\ H_2O \rightarrow 6\ Ag + Sb(OH)_3 + 6\ HNO_3$.

Transfer the contents of the loop to a microcone and add $3\ \mu l$ 16-F HNO_3 and $10\ \mu l$ 12-F HCl. Use the acids to rinse the loop into the cone. With frequent stirring heat the mixture for 15 minutes on the steam bath. Centrifuge and transfer the clear centrifugate to another cone[1]. Add an equal volume of 6-F HCl and saturate with H_2S. A yellow precipitate indicates As, and an orange precipitate, Sb. Alternately heat and saturate with H_2S to render the precipitation complete; then centrifuge and remove the clear solution. By heating on the steam bath evaporate the latter

---

[1] If Se or Te are thought to be present, the solution may be diluted with water, saturated with SO_2, and heated for the precipitation of Se and Te. The filtrate should be heated to expel SO_2 before treating it with H_2S.

inside a small area upon a microscope slide (*compare* Expt. 44). Dissolve the residue in a droplet of 3-F HNO_3, expose it to NH_3 to make it ammoniacal, and add a grain of magnesium acetate to precipitate $MgNH_4PO_4 \cdot 6\,H_2O$ (Expt. 40).

Previous to the performance of confirmatory tests (Expts. 35 and 50, 51), the sulfides of arsenic and antimony may be separated by digesting for 10 minutes at room temperature with a freshly prepared solution of ammonium bicarbonate in water.

Table VIII. *Sensitivity of Tests According to* G. C. T. CHANG (1109)

Ion	Reagent	Limit of Identification in	
		Circular Cell μg	Capillary Chamber μg
Sensitivity of Tests if the Gas is Liberated with 2-F H_2SO_4			
$CO_3^=$	$Ba(OH)_2$	—	0.1
$SO_3^=$, $S_2O_3^=$	$Ca(OH)_2$, later Br_2	0.5	0.3
Cl_2	KI-starch paper	—	0.01
$S^=$	$AgNO_3$	0.05	0.002
CN^-	$AgNO_3$	0.05	0.05
Sensitivity of Tests if the Gas is Liberated with 18-F H_2SO_4			
$CO_3^=$	$Ba(OH)_2$ and negative SO_2 test	—	0.01
$C_2O_4^=$	$Ba(OH)_2$ and negative SO_2 test	—	0.5
$SO_3^=$	bleaching of acid 0.1% $KMnO_4$	0.5	0.1
same	$Ca(OH)_2$ and then Br_2	0.5	0.2
$S_2O_3^=$	bleaching of acid 0.1% $KMnO_4$	1	0.5
same	$Ca(OH)_2$ and then Br_2	0.3	0.3
CNS^-	bleaching of acid 0.1% $KMnO_4$	0.5	0.2
same	$Ca(OH)_2$ and then Br_2	0.3	0.3
$S^=$	$AgNO_3 \rightarrow Ag_2S$	0.05	0.002
same	$AgNO_3 \rightarrow Ag_2S \rightarrow CaSO_4 \cdot 2\,H_2O$	0.3	0.3
same	lead acetate paper	0.02	0.002
Cl^-	$AgNO_3$	0.5	0.01
Br^-	$AgNO_3$	1	0.02
I^-	$AgNO_3$ giving black Ag_2S	1	0.1
ClO_3^-	KI-starch paper	0.5	0.05
$Cl^- + KMnO_4$	same	0.05	0.0005
$Br^- + KMnO_4$	same	0.3	0.0005
$I^- + KMnO_4$	same	0.5	0.1
NO_2^-	same	0.5	0.1

8. Insert into the test tube a very small triangle of moistened KI-starch paper that has been lightly blued by cautious exposure to bromine vapor.

The white color of the paper will be restored if one or several of the following gases is (are) present:

$$H_2S + I_2 = 2\,HI + S,$$

$$SO_2 + I_2 + 2\,H_2O = H_2SO_4 + 2\,HI,$$

$$HCN + I_2 = ICN + HI,$$

$$2\,HN_3 + I_2 = 3\,N_2 + 2\,HI,$$

$$PH_3,\ AsH_3,\ \text{and}\ SbH_3.$$

The last two may react with the separation of metallic arsenic or antimony.

9. A colorless and odorless gas that burns with a blue flame may be hydrogen, methane, or a mixture of the two. The presence of CH_4 may be detected by testing the combustion products for CO_2. As an alternative or confirmation, one may test the material under investigation for the presence of carbide: FEIGL, p. 281 (121).

Milligram and Submilligram Scale. The gas may be liberated in a gas reaction cell (Expt. 46 and p. 96), and the above suggested tests may be used for the detection of most of the gases. G. C. T. CHANG (778) used a circular gas reaction cell or chamber of 15-mm diameter and 5-mm height as well as a capillary chamber of only about 1.2-μl capacity and found the limits of identification listed in Table VIII.

The test for carbonate requires comparison with a blank if the laboratory air has access to the reagent solutions. The listed limit of identification for carbonate can be reached only by either working in a dry box (458, 465, 761) supplied with air free from carbon dioxide or by using the capillary chamber and jacket of CHANG (778). Also variations of Emich's technique for the sensitive detection of carbon seem feasible (150, 1141).

In all work with small amounts of gas, the volumes of aqueous solutions should be kept to a minimum since all gases are quite soluble at room temperature. Acetylene and carbon dioxide are soluble in an about equal volume of water; hydrogen sulfide and sulfur dioxide are far more soluble. One volume of NO, CH_4, O_2, CO, H_2, or N_2 (760 mm pressure) requires only from 21 to 67 volumes of water for complete dissolution. At 100° C, the solubility is usually less than one half of that at room temperature.

Considering all circumstances, the identification of small amounts of nitrogen, hydrogen, oxygen, carbon monoxide, or methane will require liberation in a closed system above mercury and analysis with a micro gas buret such as that of REEVE (721), BLACET (402, 404, 405, 408), or BURKE (893). The sample or a short piece of capillary containing the same may be treated with water or acid inside the conical top of a mercury-filled gas reservoir, which may be heated by a jet of steam before drawing the gas into the micro buret for analysis.

If the gas buret has a calibrated capillary of 0.25-mm diameter, $2\,\mu l$ gas are required to give a column of 10-mm length. On an average, this amount of gas is furnished by $10\,\mu g$ of a compound (cube of 0.2-mm edge) which reacts with water or acid to give a gas. It will thus serve little purpose to try working with solid samples much smaller than $10\,\mu g$. Using the reagents and procedure of quantitative gas analysis, the limits of identification may then be of the order of 0.2 to $1\,\mu g$.

Interesting in this connection is Chamot and Mason's test for the evolution of a gas (118). The specimen to be tested is covered with a gel or a very viscous liquid through which the reagent producing the gas diffuses. The gas bubbles liberated are trapped in the viscous medium for inspection.

In a drop of freshly boiled water dissolve one or two square millimeters of commercial sheet gelatin, just enough to get a solution that gels on cooling. It is essential that the gel shall not set too fast. Nor shall it be so thin as to require low temperature and considerable time for setting.

The specimen (particles of a powder on a slide or the evaporation residue of a solution within a very small area of the slide) is warmed and then coated with a thin film of the liquid gel, which should extend a small distance beyond the specimen in all directions. (Collodion in ether might be used if the specimen is soluble in water.) The preparation is then placed upon a cool surface until the gel has set.

For testing, the specimen is focused under the microscope, whereupon a drop of the reagent is applied to the jelly at a distance from the specimen. The reagent slowly diffuses through the jelly. When it gets to the specimen, the gas bubbles freed will persist at least long enough to be observed before they dissolve in the gel. If the gas is not too soluble in the medium, one may imagine it possible to take a sample into a micropipet, Fig. 59, operated with a droplet of mercury in the tip, and then to blow it into a reagent solution held ready in a capillary cone.

P. 37 Test with Concentrated Sulfuric Acid

If the material under investigation did not react with dilute sulfuric acid, the latter may be removed and the sample treated with the concentrated acid. If the material reacted with the dilute acid but left an insoluble residue, the latter may be treated with the concentrated acid. Since, however, dilute acid may dissolve halides, nitrates, etc. without liberating a gas or producing any other visible evidence, applying the test with concentrated acid to a fresh sample may bring evidence which could not be gotten with a residue from the treatment with dilute acid or water. (It serves little purpose to test metals or alloys with concentrated sulfuric acid.)

The following gases may be obtained with non-metallic materials.

HF, colorless, of penetrating odor, very corrosive, etching glass, is obtained from simple and complex fluorides in the absence of silicon and boron compounds (test performed with apparatus of platinum, lead, or plastic);

SiF_4, colorless gas of penetrating odor, giving a white turbidity with water ($3\ SiF_4 + 3\ H_2O \rightarrow H_2SiO_3 + 2\ H_2SiF_6$), obtained from simple and complex fluorides, fluosilicates, in presence of silicon compounds (SiO_2, silicates, glass) and in absence of more than very minor amounts of boron compounds;

BF_3, colorless gas, giving heavy white fumes with moist air, very soluble in water, imparting a yellowish green color to flames, obtained from fluorides and fluoride complexes in presence of boron and boron compounds, borates;

HCl, colorless gas of penetrating odor, giving fumes with moist air and very heavy white fumes when approached with a loop containing strong NH_3: from chlorides;

O_2, colorless, odorless gas causing a glowing splint to burst into flame: from MnO_2, persulfates;

CO, colorless, odorless gas burning with a bright blue flame to give only CO_2: derived from simple and complex cyanides, thiocyanate, formiates, oxalates, citrate, and other organic compounds;

CO_2, colorless gas with slightly acid odor, that quenches a glowing splint: from thiosulfate, oxalate, and other organic compounds;

COS, colorless, burning with a blue flame to give CO_2 and SO_2: derived from the decomposition of thiocyanate, which is accompanied by a very disagreeable smell;

SO_2, colorless, odor of burning sulfur, quenching a glowing splint: from sulfite, thiosulfate, thionate, thiocyanate, and from the reduction of sulfuric acid by iodide, H_3PO_2, H_3PO_3, and organic compounds;

Cl_2, greenish yellow gas of characteristic odor: from chlorides + oxidants or chlorates and perchlorates + reductant;

ClO_2, greenish yellow gas of chlorine-like odor, that explodes on heating or contact with organic matter (glowing splint): from chlorate, $3\ HClO_3 \rightarrow$
$\rightarrow HClO_4 + 2\ ClO_2 + H_2O$;

CrO_2Cl_2, brownish red, heavy fumes which react with water to give CrO_3 and HCl: derived from chloride in presence of chromate;

Br_2 (and some HBr), brown, heavy vapors of characteristic odor: from bromide or bromate + reductant;

I_2, violet fumes, especially upon warming, of characteristic odor: from iodide, or iodate (periodate) + reductant;

Mn_2O_7 (?), violet smoke of very characteristic odor which is obtained when some of the green or brown solution of permanganate in concentrated

H_2SO_4 decomposes (detonates) because of heat (40° C may suffice) or presence of reductant (organic matter);

NO_2 (from $NO + O_2$), brown gas of characteristic odor obtained from nitrite or nitrate + some reducing agent.

The mixture of the sample with the concentrated sulfuric acid changes to dark yellow in the presence of chlorate, to green in presence of permanganate; such mixtures may explode and should be discarded by pouring slowly into running water. The separation of sulfur may indicate the presence of thiosulfate, thionate, or sulfide. Charring, which may be accompanied by the odor of burning sugar, occurs with organic substances.

Gram and Decigram Scale. If there is any possibility of the presence of explosive mixtures (strong oxidants, such as chlorate or permanganate, and organic substances), treat first a sample of a few tenths of a milligram. If the material under investigation appears harmless, place 1 mg of it and a like amount of glass powder into a test tube and, with a pipet, add about 0.05 ml 18-F H_2SO_4. If no reaction takes place, warm slowly with a microflame. If a colored gas is obtained, test with the appropriate reagents (4, *below*). If a colorless gas is obtained, that cannot be identified satisfactorily, repeat the test with 10 mg sample, a like amount of glass powder, and 0.5 ml of the acid.

In testing the liberated gas, already available information (history, origin, nature of material) may indicate the choice of tests. If no such clues are available, use the following outline and the technique with glass loop and hook described in P. 36.

1. Observe the reaction mixture, the color and (cautiously) odor of any liberated gas as well as its effect upon a small triangle of moistened pH test paper.

2. If a colorless gas is liberated at a good rate, touch a microflame burning at the opening of a glass capillary to the mouth of the test tube:

popping sound: mixture of combustible gas and air;
gas burns with a blue flame: CO, COS;
test flame burns yellowish green: BF_3;
test flame extinguished: CO_2, SO_2, HCl.

3. If the gas is colorless, try the following tests.

a) If the gas is strongly acidic, insert a loop with water. A white turbidity indicates SiF_4. Transfer, with a capillary pipet, some of the clear solution to a microscope slide and add a small grain of NaCl. Separation of pink hexagons or six-pointed stars of Na_2SiF_6, P. 57, confirms the presence of fluoride.

b) If the gas is strongly acidic, insert a loop with 1% $AgNO_3$. If a white turbidity or precipitate is obtained, transfer the loop into the gas space of an ammonia bottle. When the contents of the loop have become clear (or distinctly ammoniacal) transfer them to a microscope slide to observe the crystallization of AgCl, Expt. 32.

c) If the gas is not strongly acidic or has a bleaching action, test with $Ba(OH)_2$ or $Ca(OH)_2$ for CO_2 and SO_2 as told in P. 36.

d) Test for small amounts of oxidizing gas by inserting a small triangle of moistened KI-starch paper. If it turns blue, it may be Cl_2, Br_2, I_2, ClO_2, CrO_2Cl_2, NO_2, Mn_2O_7.

e) If test (d) is negative, test for small amounts of reducing gas with a small triangle of filter paper that has been moistened with a mixture of equal volumes of 0.1-F freshly prepared solutions of $FeCl_3$ and $K_3Fe(CN)_6$. If the paper turns blue, it may be SO_2.

f) If (d) and (e) were negative, one may test for CO by inserting a small triangle of filter paper that has been moistened with an 0.1% solution of $PdCl_2 \cdot 2\,H_2O$ in water. The test is also given by SO_2, H_2S, NH_3, and hydrocarbons, but significant amounts of these cannot be present.

4. Colored gases have also characteristic odors so that they may be recognized by these properties. The following tests serve mainly for confirmation:

a) I_2: insert a small triangle of moistened starch paper which will turn blue. (Moisten the paper with a solution of starch in 15% NaCl brine.)

b) Br_2: insert a small triangle of filter paper which has been dipped into a saturated solution of fluorescein in 50% ethanol. A change to red proves Br_2. Test (a) must be negative.

c) Cl_2: if tests (a) and (b) have been negative, insert a small triangle of filter paper which has been treated with a droplet of 1% KBr solution, dried, and then with a droplet of the fluorescein solution. A change to red proves Cl_2.

d) CrO_2Cl_2: insert a loop containing 1% $AgNO_3$. A white precipitate forms in a yellow solution. By means of a capillary pipet transfer the clear solution to a slide. Place upon a steam bath and evaporate to dryness. Treat the residue with a 1% solution of $AgNO_3$ in 3-F HNO_3. Crystals of $Ag_2Cr_2O_7$ may appear immediately or on evaporation of the drop, Expt. 31.

e) NO_2: insert a paper which has moistened with a freshly prepared mixture of equal volumes of sulfanilic acid (1 g warmed with 100 ml 30% acetic acid) and α-naphthylamine (30 mg boiled with 70 ml water; the colorless solution decanted from the residue and treated with 30 ml glacial acetic acid). Red coloration proves NO_2 (857).

The test with conc. H_2SO_4 may be extended by adding ethanol and warming the reaction mixture. An aromatic odor of ethyl acetate and absence of carbonization indicate acetic acid. A green flame when the alcohol is boiled off and the vapors are lighted proves boric acid.

VORTMANN and LIEBER (61) finally add to the reaction mixture some metallic zinc. Liberation of NO_2 indicates presence of nitrate. A change of the color of the reaction mixture from orange or red to green suggests chromate; from colorless or yellow to blue, Mo, V, or W; from green or violet to colorless, Mn.

STEINBACH (1262) prefers the use of 15-F H_3PO_4 which, as pointed out by the above authors (61), lacks the oxidizing action of concentrated H_2SO_4 and liberates H_2S from sulfides, HCN from cyanides, HBr from bromides, and HI (and some I_2) from iodides (instead of SO_2, CO, Br_2, and I_2, respectively, obtained with H_2SO_4).

Regardless of the particular acid or concentration used (dilute or concentrated), a gas that burns in air may be efficiently identified by drying it with $CaCl_2$ and then burning it with the use of a chimney (glass tube) permitting the testing of condensates and combustion gases. For additional criteria may serve the color of the flame, the mirror obtained when holding a cold porcelain plate into the flame, the pH of the condensate, and specific tests with condensate, combustion gases, and mirror.

Milligram and Submilligram Scale. The technique and limitations have been discussed in P. 36. Some of the identification limits have been determined by CHANG (778) and are included in Table VIII.

P. 38 Flame Tests

Flame tests may be useful down to the microgram scale if some of the material can be sacrificed. If a test on charcoal has been carried out (P. 25, Expt. 17), the coloration imparted to the flame by the material may have been already observed and a repetition may be unnecessary. On the other hand, it may seem desirable to repeat the observation, possibly with the use of a spectroscope.

It is understood that any strong flame coloration may obscure any other color effect and also that the observed color may be a mixture (green = yellow + blue, etc.). The perception of several colorations is often made possible without recourse to filters or spectroscope by making use of a difference in volatility, i. e., raising the temperature of the sample in steps or addition of reagents.

The flame may be viewed through filters which absorb interfering colorations. Most widely used is blue cobalt glass which should absorb the yellow sodium light but allow passage of the radiation given by potassium.

The general technique has been demonstrated in Expt. 18. To complement it, any simple spectroscope will do, but an instrument with a wavelength scale will reduce the need for control experiments which are required for arbitrary scales. A lens may be used to give an image of the flame upon the slit of the spectroscope.

For the sake of efficiency, two people should cooperate if a spectroscope is used so that one person may concentrate upon the recording of the spectroscopic image while the other, who is feeding the sample into the flame, may give full attention to the proper procedure and also record the behavior of the material: color change, foaming, melting, sintering, volatilization, incandescence, etc. If the spectrum is somewhat complex, it will be probably necessary to repeat the test several times, and material would be saved by the use of a spectrograph.

Table IX. *Colorations Observed in Flame Tests and During Heating on Charcoal*

Coloration	Approximate Temperature	Substance	Location of Lines and Bands nm
crimson red	low	Li	lines: R 670.8, *OY 610.3*
purplish red	high	Sr	bands: RO *686, 674, 662, 649, 646, 635,* 606, B 460.7
yellowish red	high	Ca	bands: OY 620.3, 618.2, YG 554.4
orange yellow	low	Na	line: Y 589.3
greenish yellow	high	MoO_4	
pale yel.-green	—	H_3PO_4	
yellowish green	—	$MnCl_2$	bands: *two in O*, G 559.2, 539.2, 515.8
yellowish green	low	H_3BO_3	bands: *four in O*, Y 548.1, 544.0, G 519.3, 491.2, *two in B*
yellowish green	high	Ba	bands: G 513, 524, 534, *B 487*; line: G 553.5
green	high	CuO	esp. distinct on coal
green, fleeting	—	Mn	$MnCl_2$ only, spectrum between 500–600
bright green	low	Tl	line: G 535.0
green	low	$Hg(CN)_2$	with purple border
greenish white	low	Sb	around bead on coal
greenish white	—	Re	
pale green	high	Te	
pale bluish green	—	Zn	around the metal bead on coal
bluish white	—	As, Bi, Sn	
sky blue	low	$CuCl_2$	lines: G 550.7, 538.6; bands: B 443.7, 441.3, 435.4, 433.2
deep blue	low	Se	esp. distinct on coal
deep blue	low	Pb	around bead on coal
bluish violet	low	In	lines: B 451.1, V 410.1
violet	low	Ga	lines: V 417.2, 403.3
violet	low	$HgCl_2$	
violet	very low	Cs	lines: B 455.5, 459.3
violet	very low	Rb	lines: *R 795.0, 780.0,* V 421.5, 420.2
violet red	low	K	lines: R 769.9, 766.5, V 404.4

R = red, Y = yellow, O = orange, G = green, B = blue, V = violet, and the same meaning in combinations as RO, OY.

The phenomena are summarized in Table IX which also lists the positions of the characteristic spectral lines and the approximate positions of the characteristic bands. Notations in *italics* refer to bands or lines that may or may not be observed.

P. 39 Bead Tests

The sensitivity of the bead tests for the common metals is usually far inferior to that observed with the cobalt bead, and relatively large amounts of material are needed to obtain a coloration. The intensity of the color is naturally proportional to the amount of coloring matter supplied. Furthermore the colors listed in Table X may be greatly modified, changed, or completely obscured by the presence of other metals. Consequently and in general, bead tests may be recommended mainly when the elements of the middle of the periodic table are of special interest.

The material used for bead tests is not necessarily lost, however, since the bead may be dissolved for further investigation, isolation, or identification. In addition, inspection of beads in ultraviolet light provides a sensitive method for the detection of uranium and various lanthanides. Finally, use of the procedure of DONAU allows sensitive tests for some noble metals.

The classical bead tests are performed with either borax or microcosmic salt. Both fuse to a clear glass which is able to dissolve metal salts.

$$Na_2B_4O_7 \cdot 10\ H_2O \rightarrow Na_2B_4O_7,\ \text{spongy mass} + 10\ H_2O,$$

$$Na_2B_4O_7 + CoO \rightarrow 2\ NaBO_2 + Co(BO_2)_2,$$

$$NaNH_4HPO_4 \cdot 4\ H_2O \rightarrow NaPO_3 + NH_3 + 5\ H_2O,$$

$$NaPO_3 + CoO \rightarrow NaCoPO_4.$$

Sulfidic ores and compounds should be roasted for conversion to oxides before introducing them into the bead: $4\ FeS_2 + 11\ O_2 \rightarrow 2\ Fe_2O_3 + 8\ SO_2$.

The molten borax glass adheres better to the wire than the phosphate. Since both kinds of glass give about the same phenomena, the phosphate bead is recommended solely for the confirmatory test for titanium and wolfram (tungsten), for which it is better suited than the borax bead.

Even large borax beads may be fused on a straight wire with the same technique as used for microbeads in Expt. 59. The procedure may be refined for the use on a microgram scale by adopting the fusion technique described on p. 204. Electrically heated loops for fusion under the microscope have been used by EMICH (149) and by KOCH, MALISSA, and DITGES (570).

For work on the milligram scale, which is in this instance the customary classical scale, one end of a straight platinum wire of 0.3- to 0.5-mm diameter and 3-cm length is sealed into the end of a glass tube which serves for handle.

Using a non-luminous Bunsen flame, the free end of the wire is heated to redness and then quickly dipped into a small supply of borax on a watch glass. Some salt attaches itself to the wire and is fused to a glass by heating in the flame so as to hold the bead at the free end of the wire, Expt. 59. More borax is added and fused in the same manner until the bead has the desired size, 4- to 5-mm diameter at the "equator". Then some of the material to be tested is placed upon a watch glass or slide, and a small sample of it is picked up by touching it with the hot bead which is then heated in the lower oxidizing zone (b in Fig. 10) until the sample has dissolved, whereafter the color of the bead is observed while hot and after cooling to room temperature.

More sample may be added if there is no color or the coloring is too faint. More borax may be taken up and fused into the bead if the color is too dark.

To obtain reducing action, the bead is heated at c or d (Fig. 10) and then allowed to cool in the stream of gas emerging from the barrel of the burner (base of the inner blue cone of the flame). If the particular gas used should not give a suitable flame, reduction in a candle flame, near the inner dark cone, may be tried.

The colorations that may be obtained are listed in Table X. If uranium or lanthanides may be present, the bead should also be inspected under

Table X. *Interpretation of the Colors Shown by a Borax Bead*

Heated in the Oxidizing Zone		Heated in the Reducing Zone		Metal
Cold Bead	Hot Bead	Cold Bead	Hot Bead	
violet red	violet red	violet red	violet red	Nd
reddish brown	brownish violet	violet gray	gray	Ni or Fe + Mn
pale yellow	orange yellow	pale green	pale green	U
pale yellow	orange yellow	green	green	Fe
yellow	yellow brown	colorless	colorless	Ce
yellow green	yellow brown	green	pale green	V
yellow green	yellow green	yellow green	yellow green	Pr
yellow green	yellow green	green	green	Cr
greenish blue	green	ruby or opaque red	colorless	Cu
blue	blue	blue	blue	Co
violet	violet	blue	blue	Co + Mn
brown violet	violet	colorless	colorless	Mn or Ni + Co
colorless	pale yellow	dark gray and opaque	brown	Mo
colorless	pale yellow	yellow brown	yellow	W
colorless	pale yellow	pale violet	gray	Ti
colorless	pale yellow	gray	gray	Sb, Bi, (Zn)

ultraviolet light. Uranium in the amount of two millionths of the weight of the bead still gives a green fluorescence (580). HAITINGER (1087) lists the following fluorescence colors and limits of identification for the lanthanides: red, 25 µg Eu; orange, 5 µg Sm; pale yellow, Ho; yellow, 5 µg Dy or 50 µg Gd; yellowish green, 2.5 µg Tb; blue, 0.5 µg Ce; and violet blue, Tm.

For the confirmation of titanium or wolfram, fuse a bead of microcosmic salt ($NaNH_4HPO_4 \cdot 4 H_2O$) in a loop obtained by bending the free end of the platium wire into the shape of a U (8-mm height, 3-mm inner width). In the oxidizing flame, the bead is colorless, but on reducing it turns violet with Ti and blue with W. Now add a small grain of $FeSO_4$ to the bead and fuse again in the reducing flame; the bead turns blood red with both, Ti and W.

The phosphate bead may also be used to test for silica. Most, but not all, silicates leave an insoluble skeleton floating in the molten bead. On prolonged heating the skeleton slowly disappears: $SiO_2 + 2 NaPO_3 \rightarrow$ $\rightarrow Na_2SiO_3 + P_2O_5$. SnO_2 and other oxides are also "insoluble" and may give the impression of a silica skeleton.

Sensitive tests for noble metals are obtained by the procedure described by DONAU (1003). The test *solution* is added to the spongy mass obtained by just dehydrating the borax on the wire. The spongy mass is then fused to a clear glass (the proper zone of heating may have to be found by tests with controls). The color is given by the colloidal dispersion of the reduced metal and fades as the colloidal particles grow on prolonged heating. The following colors and limits of identification are listed by DONAU: red (on prolonged heating violet, blue, and colorless), 0.025 µg Au: yellow, 0.2 µg Ag, brown and turbid, 0.05 µg Pt; brown and clear, Os, Ir; brown but slate gray in reflected light, Rh; black, Pd, Ru.

The beads are removed and the wire cleaned as described in Expt. 59.

Also other substances may be used for bead tests, and oxidizing fluxes have been applied in connection with classical blowpipe analysis. According to FEIGL (340) fusion with quinolinol provides sensitive tests for ferric iron (greenish black) and vanadium (brownish black) and WEST and GRANATELLI (453) use melts of the same compound for the detection of water and the identification of chloride, nitrate, sulfate, uranyl, chromium, copper, zinc, calcium, and magnesium by the separation of characteristic crystals.

P. 40   Review of Findings

At this time the nature of the material under investigation may have been already recognized, and it is merely necessary to select confirmatory tests which prove the correctness of the identification beyond any reasonable doubt. These need not be chemical tests, and the recognition of a particular variety (modification) would even require the observation or determination of physical properties such as color, crystal form, melting point, density, electric conductivity, diffraction pattern, etc. See also P. 73.

If the identity is still in doubt, a decision concerning the general nature of the material under investigation must be made to have a guide for the continuation of the search. It should be possible to decide whether it is a simple substance (one compound), a simple mixture (two or three compounds), or a complex material (alloy, glass, silicate), or a mixture of such materials (soil, rock). Even this decision may require additional tests. It may be necessary to determine whether or not silica is present by warming a sample with sulfuric acid and ammonium fluoride as directed in P. 57; the material used for this test may be fully utilized since the residue left after evaporation of the acids may be used for the search for cations. If this is the intention, the amount of sample taken for the test should be carefully estimated.

If only little material is available, one may consider what useful tests might be performed with condensates, residues, or solutions left from preceding tests.

The presence of ammonium compounds may have been discovered by this time. If not, a fresh sample is usually needed. The test may be performed at this time and furnish additional information concerning the sample. The technique of Expt. 46 may be used or that with test tube and glass loop described in P. 36.

Use as little NaOH solution as possible, and observe the behavior of the sample: dissolution, color changes, separation of new phases. If no NH_3 is liberated and the test is negative, allow the reaction mixture to cool if it has been heated, add some Na_2S for the conversion of all metallic compounds to sulfides and again test for NH_3; it may now be liberated if it was held in a compound not decomposed by alkali alone. The reaction mixture may contain characteristic sulfide and hydroxide precipitates and thiocomplexes of various metals (As, Sb, Sn, Hg, Mo, W, V) in solution. It may be used for the detection of metals (61).

If the material under investigation appears to be a single compound, systematic testing should be able to discover the identity. If a solvent has been found, the solution obtained in P. 31 may be treated with a series of reagents to find the analytical group to which the metal(s) belong. When working in the test tube or centrifuge cone, first add a drop of $3\text{-}F$ HCl. If there is a precipitate (AgCl, Hg_2Cl_2, $PbCl_2$, TlCl, SbOCl, BiOCl; $SiO_2 \cdot x\, H_2O$, H_2WO_4, and metal hydroxides may separate if the solution was alkaline), it is a matter of choice whether or not to collect it for further investigation. The next reagent is then added to the filtrate or decantate from the HCl precipitate. As an alternative, one may allow the HCl precipitate to settle and add the next reagent so that one may observe first what it does to the clear solution and later, on mixing, its effect upon the precipitate.

Next treat with a drop of 3-F H_2SO_4. A white precipitate may be $BaSO_4$, $SrSO_4$, $CaSO_4 \cdot 2 H_2O$, or $PbSO_4$; it will hardly interfere if left in the solution which may be next treated with a little methyl red and 6-F NaOH until distinctly alkaline. Just record the appearance of any precipitates, test their solubility by adding a small excess of NaOH, and then add sodium sulfide. Record the sulfide precipitates and the color of the solution. Either treat the mixture with 3-F H_2SO_4 until acid, or separate precipitate and solution and treat them separately with the acid.

When working with a droplet on a slide, expose first to fumes of HCl and then to NH_3 vapors until alkaline. Add a grain of Na_2CO_3, expose to an atmosphere of H_2S, and finally to fumes of HCl until again acidic. Variations of the procedure and additional tests should be made as circumstances suggest.

If the material is insoluble in acids, fusing a very small sample with some Na_2O_2 (which has been tested by a control fusion with Cr_2O_3) may give valuable information, *see* also P. 42.

If everything else fails, try a method of systematic elimination. Prepare a list of all possible substances and then cross off the items that are definitely excluded by the observed facts. Recall that many things are mutually exclusive. The presence of an oxidant excludes strongly reducing substances in the same solution, and *vice versa (compare* P. 70). Solubility excludes insoluble combinations, and insolubility excludes soluble substances. If either a cation or anion is once known to be present, the solubility and color or lack of it may exclude whole series of combinations. Tables of properties of inorganic substances (8, 10) may be used to search for substances having the properties of the material under investigation and to exclude imagined possibilities; *see* also P. 73.

If the material seems to be a mixture of two or several simple solid substances, it may be best to perform a mechanical separation (Expt. 61) and to test the components separately starting with P. 5. If this procedure is not feasible, fractionation by extraction (alcohol, water, ammonia, acid) or by a combination of extraction and sublimation may be tried. If fractions containing only one or two solutes may be obtained, the task is greatly simplified. Solutions containing not more than two substances should not require separations but permit finding the cations and anions involved by systematic testing (with HCl, H_2SO_4, NaOH, Na_2S, Na_2CO_3, etc.) possibly with the aid of a few filtrations or decantations.

An elaborate procedure of systematic separations may be required to find the essential composition of complex substances such as alloys, ores, silicates, etc. Of course, analysis by systematic separation may be chosen even for simple substances when very little material is available for testing. One will try to avoid it, however, because of the effort and time involved if economy with the material under investigation is not a compelling reason. It is assumed, of course, that the problem which prompts the investigation is important enough to warrant its continuation. Depending upon the circumstances, it might suffice to know that the material is a silicate rock, a soil, some common glass, or an alloy steel.

More specific information may appear desirable when considering that there is a great variety of alloy steels, that inorganic glasses may be silicates, borates, phosphates, or even combinations of sulfur, selenium, arsenic, and thallium which become liquid above 100° C but have the chemical resistance of silicate glass. Industrial ceramics also include a wide and growing assortment of ingredients such as MgO, $Mg(AlO_2)_2$, BeO, CeO_2, ZrO_2, $ZrSiO_4$, ThO_2 as well as oxide-metal and oxide-carbide combinations (6, 12).

If it is decided to use an elaborate separation scheme, it will be well worth to keep in mind the advice given by NIEUWENBURG and LIGTEN (47) that much time and effort may be saved if the work is started only after making certain, by preliminary tests, which substances form the major constituents. During the performance of the separations, one may then focus the attention upon the minor constituents and the clarification of those details which the orientation tests failed to reveal or were unable to decide.

If very little material is available for investigation, one may also consider the advisability of a sodium carbonate fusion with the original material or that part of it which is insoluble in water; it may permit identifying the anions in the aqueous extract of the melt, while the residue from the aqueous extract is used for the separation of the cations.

P. 41 Dissolution of the Sample

The preparation of a solution of the material under investigation is necessary for a continuation of chemical testing regardless whether just a few simple confirmatory tests are needed or an involved separation has to be carried out. A suitable solvent may be known already from preceding trials, or it will have to be found now by turning to the tests of P. 31.

If a mixture of substances is present, it may not be possible to find a solvent that would dissolve all substances. The use of several solvents becomes necessary, and the outcome may depend upon the order in which they are applied. Thus a mixture of Ag_3PO_4 and Pb_3O_4 will dissolve if first extracted with dilute nitric acid and then boiled with HCl, but it will leave insoluble AgCl if the order is reversed. The proper solvents and the order of their application may be suggested by the history of the material and the findings in the orientation tests. Obviously, but not necessarily, one may try to avoid HCl if silver is present or H_2SO_4 if alkaline earths must be expected.

Metals and alloys are traditionally dissolved in HNO_3 since this will prevent the loss of S, P, As, and Sb as hydrides. Silicon, tin, and antimony are converted to insoluble oxides; the carbon of carbides is oxidized, but graphite is not attacked. The residue of the treatment with nitric acid may therefore contain graphite, SiO_2, SnO_2, Sb_2O_5; if SnO_2 is present, it carries down also PO_4 and AsO_4 and small amounts of Bi, Cu, Pb, Fe, etc. Of course, various metals (Al, Cr, W, Au, Pt, Ta, Nb) and alloys (of Au, Pt; stainless steels) are not dissolved by nitric acid. Aluminum dissolves in HCl and (readily) in strong NaOH. Gold, the platinum metals, and their alloys may be successfully treated with *aqua regia*. Powerful oxidants ($HClO_4 + HNO_3$ or fusion with Na_2O_2) are most successful with alloy steels.

For dissolution in nitric acid, cover 1 g (mg, μg) of the metallic material with 0.01 liter (ml, μl) water and, in small portions, add an equal volume of 16-F HNO_3 so that the reaction never becomes too violent. If necessary, make dissolution complete by finally heating upon the steam bath. To obtain a more complete separation of tin and silica, the mixture may be evaporated on the steam bath and the residue extracted with 3-F HNO_3, but this is not recommended if much iron (or other readily hydrolyzed ion) is in solution. Finally, separate residue and solution, preferably with the use of the centrifuge. For a simple scheme of separation *see* the literature (55, 56, 57).

Non-metallic materials (if there is no reason for a different procedure) are traditionally treated, in the order of listing, with water, dilute HCl, concentrated HCl, dilute HNO_3, concentrated HNO_3, and *aqua regia* (1 volume of concentrated $HNO_3 + 3$ volumes of concentrated HCl). The preliminary determination of the solubility will have shown which of these steps may be omitted; it may have shown that the material is insoluble in acid. For insoluble materials and the treatment of a residue insoluble in the acids listed *see* P. 42.

For obvious reasons, one tries to use as little solvent as possible. The following amounts may be profitably taken per 1 g (mg, μg) of sample or what is left of it from the preceding extraction(s):

0.01 to 0.05 liter (ml, μl) of water depending upon whether only part or the whole of the sample is dissolved;

0.005 to 0.014 liter (ml, μl) 3-F HCl;

0.003 to 0.006 liter (ml, μl) 12-F HCl;

0.002 to 0.01 liter (ml, μl) 4-F HNO_3; and

0.002 to 0.006 liter (ml, μl) 16-F HNO_3 or *aqua regia*.

The determination of the solubility will have shown whether the dissulution may be carried out at room temperature or whether it is necessary to heat the mixture. Concentrated HCl and *aqua regia* should not be heated above 50° C since this would cause a rapid loss of HCl. Strong acids are most effectively applied in small portions with removal of the solution obtained before adding the next portion of acid.

If the solvents are used sparingly, the resulting solutions will not contain an objectionably high concentration of acid. Solutions in nitric acid may be evaporated to dryness on the steam bath without any danger of losing a volatile compound of metals. The residue may then be dissolved in an acid of the desired concentration. Solutions in *aqua regia* may be evaporated after adding an equal volume of 16-F HNO_3.

The procedure of NOYES and BRAY (49, 162), P. 68, is definitely recommended if elements giving volatile oxides or halides, platinum metals, and elements of the tungsten and tantalum groups (W, Mo, V, Ti, Nb, Ta, Zr) are to be detected. It starts with the solid sample, and the dissolution is part of the procedure and is conducted in such manner that the bromides of Se, As, and Ge as well as the tetroxides of Os and Ru are simultaneously isolated in distillates. A suitable technique for the distillation is available even for the microgram scale of work, Expt. 63.

P. 42 Treatment of Substances Insoluble in Acids

The systematic treatment of insoluble substances is part of the scheme of NOYES and BRAY, P. 68, and is a matter of little concern if this scheme of separation is used. The following discussion is intended as aid in systematic testing and in the use of the classical hydrogen sulfide scheme of analysis.

A selection of procedures is given. Suitable techniques are described for the gram scale, p. 68; the centigram scale, p. 81; the milligram scale, p. 104; and the microgram scale, p. 193.

a) Heating with H_2F_2 and H_2SO_4 in platinum apparatus will bring into solution: SiO_2, ZrO_2, and silicates.

Per gram (mg, μg) of solid use 0.001 liter (ml, μl) 9-F H_2SO_4 and 0.003 liter (ml, μl) 24-F (50%) HF from a dispensing plastic bottle (as an alternative, 2 ml 18-F H_2SO_4 and 2,5 g NH_4F may be used). Use a fume hood and be certain to avoid getting any hydrofluoric acid on the skin. Allow the reaction mixture to stand for five minutes and then evaporate at low temperature so that there is no boiling at any time. Finally raise the temperature to drive off the sulfuric acid. Add to the residue a small amount of 4-F H_2SO_4 just sufficient to obtain a clear solution and again evaporate until fumes of SO_3 appear. This should remove all fluoride. The residue may be dissolved by rinsing with water into apparatus selected for the performance of the analysis.

Silicon and boron are completely removed by the treatment, and there may be significant vaporization of As, Ge, Se, Re, Sb, and Cr (as CrO_2F_2?); also some manganese may be lost (446). If fluoride is left behind, it may interfere with the precipitation of aluminum and possibly other elements by NH_3.

b) Digestion with H_2F_2 in platinum apparatus will dissolve the metals Nb and Ta as well as their oxides Nb_2O_5 and Ta_2O_5. Treating the solution with KF may precipitate the somewhat difficultly soluble K_2TaF_7; the more soluble niobium compound may hydrolyze in weakly acid solution to the readily soluble oxyfluoniobate K_2NbOF_5. Evaporation with sulfuric acid is not advisable since hydrolysis will bring back the insoluble pentoxides.

c) Fusion with Na_2CO_3 (K_2CO_3 for Ta_2O_5 and Nb_2O_5) in platinum apparatus for silicates, chromites, aluminates, Fe_2O_3, Al_2O_3, Cr_2O_3, CaF_2, $BaSO_4$, $SrSO_4$, $CaSO_4$, and insoluble phosphates and fluorides of Ti, Zr, Th, and the lanthanides.

Treat 1 part (by weight) of the finely ground sample with not more than 4 to 6 parts of anhydrous Na_2CO_3. Fuse until the melt becomes clear or until effervescence stops, 1000 to 1200° C. The treatment of the melt is simple when the work is performed on a small scale. Useful advice concerning removal of the melt from a crucible may be found on p. 846 of Applied Inorganic Analysis (7). The melt may be decomposed with dilute acid as practiced in quantitative analysis, or it may be extracted with water and the aqueous extract used for the search for anions. It should be understood that the aqueous extract may also contain antimonate, stannate, plumbate, zincate, aluminate, chromite, chromate, vanadate, wolframate (tungstate), etc. in addition to silicate and the usual anions. A green color of the melt indicates manganate and a yellow color, chromate.

d) Fusion with $KNaCO_3$ (mixture of the carbonates in the mole ratio 1 : 1) in porcelain for the halides of silver and insoluble salts of lead: $PbSO_4$ and $PbCrO_4$, fused.

Proceed as under (c). Extract the melt with water and dissolve the carefully washed residue in 3-F HNO_3. The anions will be found in the aqueous extract.

e) Fusion with $K_2S_2O_7$ in platinum, porcelain, vitreous silica, or glass apparatus for the dissolution of rhodium metal (finely divided), Fe_2O_3, Cr_2O_3, Al_2O_3, TiO_2, ZrO_2, ThO_2, and insoluble phosphates and fluorides of Ti, Zr, Th, and the lanthanides.

If the anhydrous pyrosulfate is not available, heat some $KHSO_4$ until no more steam is given off and the melt becomes quiet, $2\ KHSO_4 \rightarrow$ $\rightarrow K_2S_2O_7 + H_2O$. Pour the melt into clean porcelain dishes, break up the thin sheets of salt while still warm, and store in a stoppered bottle.

Use 3 parts (per weight) of pyrosulfate for 1 part of solid sample which should be finely powdered. Fuse at as low a temperature as possible to prevent rapid decomposition of the salt, $K_2S_2O_7 \rightarrow K_2SO_4 + SO_3$, which would not allow sufficient-time for the action of the SO_3 upon the sample.

Just maintain a liquid flux until the sample has completely dissolved. After cooling, dissolve the melt in 3-F H_2SO_4.

f) Fusion with Na_2O_2 in apparatus of nickel or iron for the dissolution of carbon, wolfram metal, finely divided ruthenium and rhodium metal, alloy steels, insoluble chromium compounds, chromite, stannic and antimonic acids, silicates, TiO_2, ZrO_2, WO_3, and acid insoluble carbides, borides, and nitrides.

When working on the gram or centigram scale, mix 1 part (by weight) of the finely powdered sample with 5 parts of dry peroxide and cover the mixture with 1 more part of the reagent. **Use eye protection.** Make certain that flame gases will not get to the mixture. Heat to first expel any water from the mixture, then raise the temperature gradually to about 700° C and hold it there for one minute. After the melt has cooled to room temperature, decompose it by the gradual addition of small amounts of water in such a manner that the violent action will not cause loss of material. With like caution, finally acidify with 3-F HCl. Since the apparatus is strongly attacked during the fusion, the solution will contain considerable amounts of Ni or Fe.

Concerning the use of zirconium for fusions with KOH *see* DODSON (489).

g) Heating in a current of hydrogen is convenient for the reduction of the halides of silver and the insoluble oxides of antimony and tin. Water may be condensed for identification, and the hydrogen halides may be absorbed to the same end.

P. 43 Confirmatory Tests

The elements are taken up in the order of the groups, starting with the alkali metals, group I A, of the long form of the periodic table.

Controls and Blanks must be carried out before applying a test to a sample which is difficult to replace. Closely imitating the conditions of actual use, try one control with a moderate amount of substance X sought and another with a small amount. Some more practicing may be desirable if the experimenter is not familiar with the test.

Controls are tests with a known amount of sought substance X.

Blanks are controls closely imitating the conditions under which the test is applied, but with the sought substance X absent.

The *Sensitivity Statements* are made according to F. L. HAHN (962):

Limit of Identification, L. I., is given in gram of X;

Limiting Concentration, L. C., is given in g X/ml;

Limiting Proportions, L. P., are given as ratio of largest mass of interfering substance per unit mass of X sought.

The statements are given in the following forms:
If a solid sample is used:

[p L. I.; symbol(s) of interfering substance(s): log L. P.]

If a sample solution is used:

[p L. C. − p L. I.; symbol(s) of interfering substance(s): log L. P.]

p L. I. = − log L. I.   p L. C. = − log L. C.

p L. C. − p L. I. = log v.

v = volume of test solution in ml.

It is suggested, however, to take the sensitivity data with reservations. The decimals are given mainly to permit calculation of the test volume. It must be kept in mind that limit of identification and limiting concentration are not exactly related via the test volume as the equations pretend. To obtain the limit of identification, slide tests are usually allowed to evaporate so that the test volume becomes an unknown quantity; in addition, a refined method of observation may be used. The equations silently assume that the techniques of working and the methods of observation retain their efficacy regardless of scale of work; this, of course, is not true.

The tests described represent merely the preferences of the author. Many others may be found in the literature.

P. 44   Group I A: Alkali Metals

No. 3: Lithium, 6.939

Flame Test (9.0).

Crimson red flame, see P. 38 and Table IX.

Hexamethylene and Potassium-Iron (2 or 3) Cyanide (5.7–7.2).

The test solution must not contain ammonium salts or metals other than the alkalies (854). Place a small droplet of test solution upon a slide and evaporate to dryness. If necessary, ignite for the removal of ammonium salts. Place a small drop of a 15% solution of hexamethylenetetramine upon the residue. When the latter has dissolved, take up the solution with a capillary pipet for transfer to another slide. There deposit it to give two drops of like size and next, but not too close, to one another. Place into one drop a small grain of $K_3Fe(CN)_6$ and into the other, one of $K_4Fe(CN)_6$. The ferricyanide gives yellow octahedra with lithium. These should appear upon addition of the reagent; on evaporation, also blank tests give octahedra. The ferrocyanide produces in presence of lithium short rods which may combine to give X-forms and radial clusters. The crystals seem to be isotropic.

Lithium-Zinc-Uranyl Acetate (3.4–6.4) (875).

The test should be used only after the lithium has been separated from sodium and the other alkali metals.

Reagent. Dissolve 10 g uranyl acetate dihydrate in 6 g 30% acetic acid and 49 ml water; dissolve 30 g zinc acetate trihydrate in 3 g 30% acetic acid and 32 ml water; mix the two solutions and filter after 24 hours (7).

Performance of Test. Place a droplet of the neutral test solution upon a microscope slide and next to it a droplet of like size of the reagent solution so that the two drops are about 1 mm apart. Connect the two droplets by drawing a narrow channel with a glass needle. Lithium-zinc-uranyl acetate separates immediately and mostly in granular shapes. Regularly developed octahedra are quite abundant and can hardly be confused with the rather characteristic, often large, elongated forms of the sodium-zinc-uranyl acetate. With the latter, sturdily developed prismatic crystals and elongated hexagons are frequent, whereas regularly developed octahedra are rarely found.

On evaporation of the test drop, the reagent itself gives crystals that might be mistaken for those of the triple acetate. They appear along the circumference of the drop. Between crossed polars (nicols), however, the crystals of the reagent appear bright and colored, whereas even large crystals of the triple acetate show only a light gray of the first order. Insertion of a first-order red selenite plate reveals the composite structure of many of the octahedral crystals of the triple acetate; some sectors of these crystals appear yellow while others, at the same time, display the blue of second order.

No. 11: Sodium, 22.9898

Flame Test (8.4).

Orange yellow flame absorbed by blue cobalt glass, *see* P. 38 and Table IX.

Sodium-Uranyl Acetate (1.9–5.2; Li: 2.7 (1.4–4.7; K: 2.7).

Obtain a neutral solution of the chloride, nitrate, or sulfate, which contains about 1 to 10 mg Na/ml. Deposit a small droplet of this solution upon a slide and add a small grain of ammonium-uranyl acetate. The isotropic tetrahedra (triangles) of the sodium-uranyl acetate appear immediately after adding the reagent. The corresponding lithium salt is rather soluble, but may crystallize along the circumference of the drop when the concentration rises because of evaporation. The lithium salt gives prismatic crystals, irregular hexagons, and forms which more or less imitate the appearance of the octahedron or the tetrahedron. The crystals of the lithium salt, however, are strongly anisotropic and easily distinguished from those of the sodium salt. The presence of 10 parts of lithium for 1 part of sodium does not prevent the instantaneous separation of the tetrahedra of the sodium-uranyl acetate (875).

No. 19: Potassium, 39.102

Flame Test (7?; Na: 2.3).

The bluish violet flame may be made visible in the presence of sodium and lithium by means of blue cobalt glass. *See* P. 38 and Table IX.

Potassium Chloroplatinate (3–7; Na: 1.2).

The test cannot be carried out in presence of NH_4, Rb, Cs, Tl(1), organic amines, and alkaloids, all of which are also precipitated. Iodide, sulfate, and nitrate interfere.

If Rb, Cs, Tl(1) may be assumed to be absent, the performance of the test is simple: Place upon a slide a solid sample or a volume of test solution corresponding to about 1 μg K. Evaporate the solution to dryness. If it seems advisable, ignite the evaporation residue or the solid sample for the removal of ammonium salts and organic substance. Treat the cold residue with 1 μl 5% H_2PtCl_6 solution. If potassium is present, yellow octahedra of K_2PtCl_6 separate immediately. The test is not reliable in a laboratory, the air of which is laden with NH_3.

The following procedure is recommended if the potassium has been separated from rubidium and cesium: the resulting fraction containing the potassium is evaporated and, if necessary, the residue is ignited to remove ammonium salts.

Prepare an aqueous solution which contains approximately 10 mg K/ml. Of this solution, deposit two separate drops, each of about 1-μl volume, side by side upon a slide. Place next to them in a systematic manner four more drops of like volume, two drops of pure KCl solution and two drops of pure RbCl solution, all of them containing 1 mg metal/ml. Gently heat the slide to evaporate all drops to dryness. When the slide has cooled to room temperature, treat three of the residues–one of each kind–with about 1 μl each of 5% H_2PtCl_6. Without delay treat the other three drops with about 1 μl each of a solution of chlorostannic acid obtained by dissolving 2 g $SnCl_4 \cdot 5 H_2O$ in 5 ml 13-F HCl and 15 ml water.

If pure potassium salt has been isolated, the residues of the test solution will behave exactly as the residues of the pure KCl solution: they will give a copious precipitation with chloroplatinic acid, but no precipitate with the stannic chloride solution. If the residue of the isolated fraction contains heavy alkali (Rb, Cs), comparison of the chlorostannate precipitates will show whether or not the heavy alkali may account for all of the chloroplatinate obtained with the unknown residue. Make the comparison of quantities with the unaided eye after placing the slide upon a dark paper (875).

No. 37: Rubidium, 85.47

Flame Test (6.7).

All alkalies and the strontium interfere so that their presence calls for the use of a spectroscope. Pure rubidium salts color the flame violet. See also P. 38 and Table IX.

Rubidium-Silver-Gold Chloride (3.7–7.0).

Ammonium must be absent; K, Li, Na, and the alkaline earths do not interfere. Rubidium may be recognized in presence of cesium which gives a more insoluble compound of the same type (118, 1008, 1012, 1013).

Reagent. Saturate with solid silver chloride a 5% solution of $AuCl_3 \cdot 2 H_2O$ in 13-F HCl.

Procedure. If a solid is to be tested, place a droplet of the reagent upon a slide and introduce a small particle of the solid into the edge of the drop. To test a solution, evaporate some of it upon a slide so that a few micrograms of solid residue remain behind in a very small area; then place upon the residue a droplet of the reagent.

If any cesium is present, it is immediately precipitated in finely granular form, but after some time black squares, hexagons, triangles, four- and six-pointed stars may be recognized with a magnification of 200 or more diameters. Well developed orthorhombic plates and prisms of the more soluble rubidium salt, $Rb_2AgAuCl_6$ (?), which vary in color from yellowish red through red to almost black depending upon their thickness, appear after most of the cesium salt has separated. Twinning may take place, and radial clusters of rods may form. In addition, the crystals of the rubidium salt may show a play of spectral colors on their brilliantly reflecting surfaces.

No. 55: Cesium, 132.905

Flame Test (6.0).

The violet coloration is hidden in presence of any one of the other alkalies or strontium, and a spectroscope must be used in such instances. *See* also P. 38 and Table IX.

Cesium Iodobismuthite (4.4–7.7; K, Rb: 2 or better).

Reagent. Dissolve 0.3 g Bi_2O_3 in 1 ml HI, sp. gr. 1.6, and dilute with 2 ml water (1159).

It is assumed that the test solution does not contain ammonium, organic amines, alkaloids, or heavy metals.

Place a droplet of the solution to be tested upon a slide and deposit an equal volume of the reagent close to it. By means of a glass thread, make a connecting channel. The separation of hexagonal plates or six-pointed stars which may be, depending upon thickness, yellow, orange, red, or nearly black, proves the presence of cesium. If it happens that the crystals do not separate immediately after adding the reagent, allow the test to stand until complete evaporation has taken place. Then search for the hexagonal stars or plates which are dark between crossed polars and remain dark during the full rotation of the stage. The reagent itself gives yellow to dark red crystals which are strongly birefringent and separate as needles, rods, prismatic plates, diamonds, dendrites, or in ornamental patterns (875).

Cesium-Silver-Gold Chloride. *See* under Rubidium.

Group II A: Alkaline Earths
No. 4: Beryllium, 9.0122

Test with Quinalizarin (5.5–6.8; alkalies, Ca, Sr, Ba, Al, Zn, Cd, Sn, Bi, Se, Te, Nb, Ta: 2.5; Fe, Mn, Co, Mo, W, As, Sb, Ag: 2.0).

In alkaline solution, the violet color of the dye is changed to a pure blue on adding beryllium (1144). Ammonium, magnesium, Y, Ti, Zr, Th, lanthanides, V, uranyl, and Cu must be absent. The sensitivity given above refers to the use of a white spot plate; it may be improved with the coloriscopic capillary, Expt. 55.

To a drop of test solution, which must not contain free acid, add a drop of a saturated solution of quinalizarin in ethanol (0.005%) and then a drop of 0.1-F NaOH. Perform a blank and compare the colors.

To test for beryllium in presence of magnesium which produces the same color change proceed as follows. To a drop of the "neutral" test solution add two drops of a 0.05% solution of the dye in 2-F NH_3 and then 1 ml saturated bromine water. The blue solution may become less intensely blue, but keeps the color if beryllium is present. Compare with a blank that contains some magnesium.

Test with p-Nitrobenzene-azoorcinol (5.3–6.7; alkalies, alkaline earths, Al, Y, lanthanides, Ti, Zr, Th, Nb, Ta, V, Mo, W, Fe, Bi, Ca, As, Sb, Sn, Se, Te: 2.5; Ag, Pb, Mn, Ni: 1.5; Mg, Cu, Zn, Co must be absent).

The yellow alkaline solution of the dye gives an orange red lake with beryllium; magnesium gives a brownish yellow lake that would interfere with the beryllium test.

Place a drop of 0.025% solution of p-nitrobenzene-azoorcinol in 1-F NaOH on filter paper. Take the "neutral" test solution into a capillary pipet and deliver it to the center of the yellow spot of dye. Finally add another drop of dye solution. Depending upon the amount of beryllium, either the whole spot or just its center becomes orange red.

No. 12: Magnesium, 24.312

Magnesium-Ammonium Phosphate (4.7–8.0; Na, K: 3; Ca: 1.7).

All cations that are precipitated by phosphate should be absent. If calcium is present, treat the test solution with some ammonium citrate. Mn, Zn, and Co give metal-ammonium phosphates of the same appearance, but the Zn and Co compounds are soluble in ammonia, and the Mn salt may be recognized by the formation of brown MnO_2 when treating the crystals with H_2O_2, W. BÖTTGER.

Transfer the test drop to a slide and evaporate to dryness. Dissolve the residue in such a volume of 2-F HNO_3 to give a 0.1 to 1% solution of magnesium salt. Invert the slide and place the hanging drop over the opening of a bottle with 12-F NH_3. After a few minutes, transfer the preparation to the stage of the microscope and push a small grain (0.2 µl volume, 0.6-mm diameter) of Na_2HPO_4 into the edge of the drop. Observe with a magnification of about 80 diameters. $MgNH_4PO_4 \cdot 6 H_2O$ forms dendrites, X shapes, and prismatic crystals that are insoluble in dilute ammonia. *See* also Expt. 40.

Test with Magneson (4.6–5.6; Ca, Sr, Ba: 2).

Ni, Co, and Cd give the same test and must be absent. Ammonium salts are best removed.

On the spot plate, in a microcone, or in a capillary treat some of the test solution (which should not contain free acid) with a like volume of a 0.1% solution of *p*-nitrobenzene-azo-resorcinol in 50% ethanol and a volume of 0.1-*F* KOH equal to the volume of test and dye solution. A blue coloration or precipitate indicates magnesium; a blank should remain red.

No. 20: Calcium, 40.08

Flame Test (4, ?).

Only the volatile salts, especially the chloride, color the flame which seems to become yellow in the interior and red on the outside. *See* also P. 38 and Table IX.

Calcium Sulfate Dihydrate (4.5–7.5; Mg: 1.3; Sr, Ba: 0) and (3.8–6.8; Mg: 2; Sr, Ba: 1).

Fe(3), Cr(3), Pb, Sc, lanthanides, and Th interfere with the test.

Place the test droplet upon a slide and evaporate to dryness. After cooling, dissolve the residue in a volume of 6-*F* HCl that promises a 0.1 to 1% solution of calcium. Place 1 µl 1-*F* H_2SO_4 close to the test drop, and connect the two drops by drawing a narrow channel. $CaSO_4 \cdot 2 H_2O$ separates as fine needles, sheaves of needles, and occasionally in rhomboids that combine to twins which imitate the outline of the tail of a swallow (or arrow). If very little calcium is present, the crystals form only on complete evaporation of the drop.

Other Tests for Calcium. The test with sulfuric acid is entirely satisfactory for the identification of calcium after separation of the alkaline earth group. In nature, barium and strontium occur only rarely in high concentration and will not interfere with the detection of calcium.

Quite recently, sensitive tests for calcium have been described, which permit the detection of calcium in presence of most other elements (342, 472). The spot test described by WEISZ (950), based upon involved reasoning, will certainly serve within the alkaline earth group.

No. 38: Strontium, 87.62

Flame Test (6). After lithium and potassium have been vaporized at low temperature, the purplish red coloration due to strontium may be perceived without difficulty. *See* also P. 38 and Table IX.

Potassium-Strontium-Cupric Nitrite (4.0–7.3; Mg: 2). V, Mo, W, Pd, Au, Pb, Se, Te must be absent; Ba inhibits the test if its quantity is ten times that of the Sr.

Reagent. Prepare the reagent before use by mixing equal volumes of aqueous solutions of 50 g KNO_2 per 100 ml and of 45 g sodium acetate and 10 ml glacial acetic acid per 100 ml. The reagent remains effective for at least two days; the KNO_2 solution should be renewed every year.

Place a droplet of the test solution upon a slide and evaporate to dryness. Treat the residue with a volume of 1% cupric nitrate or acetate solution that contains about five times as much copper as there is strontium in the residue. Again evaporate to dryness. After cooling, moisten the residue or part of it with a very small amount of reagent without touching the surface of the slide with the tool. Use transmitted light of high intensity to observe the green squares of the triple nitrite, which separate within a few minutes (875).

No. 56: Barium, 137.34

Flame Test. See P. 38 and Table IX.

Barium Chromate (5.0–8.0; Ca, Sr, Mg: 2). Ag, Hg, Pb, Bi, Tl must be absent.

Place the test droplet upon a slide and evaporate to dryness. After cooling, dissolve the residue in 1-F acetic acid to obtain an approximately 1% solution of barium ion. Push a small kernel of $K_2Cr_2O_7$ into the edge of the drop. $BaCrO_4$ separates as very small yellow squares, rectangles, and rhombs.

Barium Fluosilicate (3.8–6.8; Mg, Ca, Sr: 1) and (3.0–6.0; Mg, Ca, Sr: 2). Zr, Mo, Al, Mn must be absent.

Place the test droplet upon a slide and evaporate it to dryness. Dissolve the residue in 2-F HCl to obtain an approximately 1% solution of barium ion. Push into the edge of the drop a relatively large kernel of ammonium fluosilicate. The crystals of $BaSiF_6$ appear after a few minutes and have the shape of rather large lentils, spears, and bundles of spears.

Group III B: Scandium Group
No. 21: Scandium, 44.956

Cochineal Test (4.7–4.2; alkalies, alkaline earths, Be, Y, lanthanides, Nb, Ta, Cr, Mn, Co, Ni, Zn, Cd, Al, Tl, Pb: 1). V, Ti, Zr, uranyl, Cu, Ag, Au, Hg, Sn, and fluoride must be absent (137).

In a test tube, treat 5 ml of the solution to be investigated with a few drops of tincture of cochineal. Add drops of 2-F NaOH until a violet coloration is obtained, warm the solution a little, and then add a drop or two of glacial acetic acid. Scandium gives a blue coloration or precipitate (878, 911).

No. 39: Yttrium, and the Lanthanides, Nos. 57 to 71

The chemical behavior of yttrium and the lanthanides is sufficiently different from that of the other elements so that the "rare earths" may be isolated or recognized as a group. Inside this group, however, the chemical behavior is sufficiently uniform to make unprofitable the use of classical analytical procedures. The various forms of spectroscopy are suited for identification and estimation. Whereas these elements do not give colored flames, they give characteristic arc and spark spectra. In addition, absorption spectroscopy may be applied to solutions, and also the light reflected by minerals, oxide or oxalate mixtures, and colored salts gives useful spectra. Concerning separation on anion exchange resin see FARIS and WARTON (490). On the use of the fluorescence of borax beads, see P. 39. Useful contributions of the classical method are the following group tests.

Test for Lanthanides, Yttrium, and Thorium. These give a (colorless) crystalline precipitate when a drop of the test solution (which should be slightly acid with HCl or HNO_3) is treated with a saturated aqueous solution of oxalic acid (136).

Test for Elements 57 to 62, La to Sm, Inclusive. Transfer a drop of test solution to a slide and evaporate to dryness. Dissolve the residue in as little water as possible and add an excess of ammonium succinate in the form of a few crystals of it. Elements 57 to 62 give fine colorless needles (136).

Test for Elements 63 to 71, Eu to Lu, Inclusive. Place a drop of the test solution upon a slide and evaporate just to dryness. Dissolve the residue in a just sufficient amount of water and add an equal volume of 85% lactic acid of reagent quality. Sc, Y, and elements 63 to 71 give colorless prismatic crystals (136).

Salts and solutions of the following ions show color: reddish violet, Nd(3); pink, Eu(3), Er(3); reddish brown, Sm(2); orange yellow, Ce(4); yellow, Sm(3), Ho(3); greenish yellow, Dy(3); green, Pr(3); pale green, Tm(3) and Yb(2).

No. 58: Cerium, 140.12

Borax Bead. On heating in an oxidizing zone, the bead is orange yellow while hot and yellow when cold. The color may be seen with 1 part of cerium in 950 parts of borax when cold and in 4700 parts, when hot (1138).

Cerium Perhydroxide (5.2–6.4).

This test is specific inside the lanthanide group. Strongly colored ions interfere; the precipitation of ferric hydroxide may be prevented by adding Rochelle salt (tartrate).

Mix a drop each of test solution, of 3% hydrogen peroxide, and of 3-F NH_3 in a porcelain crucible and warm gently. A yellow to orange coloration or precipitate indicates $Ce(OH)_3OOH$ (121, 640, 722).

Test with Anthranilic Acid (4.0–7.3; rare earths and nearly all other metals: 2) and [3.5–6.8; Fe(3), Cr(3), Cu, Bi, As(5): 1].

Au, V, chromate, Sn, and chloride must be absent.

The test substance should be dissolved in about 3-F HNO_3. If iron is present, add strong H_3PO_4 to suppress the ferric color. On a spot plate, mix one drop of this solution with one drop of a 5% solution of anthranilic acid in ethanol and add some PbO_2. Ce causes a bluish black precipitate to appear, which again quickly dissolves to give a brown solution (136).

No. 63: Europium, 151.96

Cacotheline Test (5.5–5.5). Oxidants like NO_2^-, NO_3^-, ClO_3^-, and the metals Mo, W, U, Ti, V, Nb, Re and Sn(2) must be absent.

In a test tube, treat 1 ml of solution to be investigated with a few drops of 4-F HCl and a granule of metallic zinc. Then add a few drops of 0.25% aqueous solution of cacotheline. Presence of europium is indicated by a violet coloration (136).

No. 70: Ytterbium, 173.04

Naphthoresorcinol Test (5.3–5.0).

The test is specific inside the rare earth group. The interferences are about the same as in the cacotheline test *above*.

Reagents. (a) Add 0.5 cm³ (0.5 g) metallic sodium to 3 to 4 ml mercury and stir until solution is complete. (b) Aqueous saturated solution of oxalic acid. Boil it briefly before use to remove dissolved oxygen. (c) Solid 1,3-dihydroxynaphthalene.

To 1 or 2 ml of the test solution, which should be about 1-F with H_2SO_4, add a small piece of ice and 0.5 ml sodium amalgam. Then add 3 to 4 drops of boiling hot oxalic acid, about 3 mg naphthoresorcinol, and 3 to 4 drops 6-F HCl. Boil 2 minutes, cool, and then shake with 2 ml diethyl ether. The latter turns pink if ytterbium is present (137).

Group IV B: Titanium Group

No. 22: Titanium, 47.90

Bead Test, *see* P. 39 and Table X.

Test with Hydrogen Peroxide (5.0–5.0; Be, La, Ce, Zr, Th, Cr, U, Fe, Al, Ga, silicate: 2). Nitrate, chloride, bromide, and strongly colored ions decrease the sensitivity; acetate and formiate must be absent.

Use a solution of the material in 2-F sulfuric acid. Treat 1 ml of the solution with a drop of 3% H_2O_2. Titanium gives a yellow orange coloration,

$H_2 \cdot TiO_2(SO_4)_2$. Molybdenum and vanadium give similar colorations, but these persist when some solid NH_4F is added, whereas the dioxydisulfatotitanic acid is destroyed.

In presence of Mo and (or) V, the proof for titanium is merely the partial loss of color on adding fluoride. Since also the color of ferric iron fades upon adding fluoride, it becomes necessary to remove the ferric color before the fluoride treatment by adding some phosphoric acid.

Peroxide Test after Isolation of Titanium (6.0–6.3; in presence of large amounts of Be, La, Ce, Zr, Th, Cr, U, Fe, Al, etc.). Sulfate must be absent.

Titanium is separated from other ions by coprecipitation with zirconium arsenate and then tested with hydrogen peroxide.

Dissolve the test material in approximately 1-F HCl. In a centrifuge cone, treat 0.5 ml of this solution with 0.05 ml 20% arsenic acid and 0.01 ml 1% $ZrOCl_2$ in water. Heat in the steam bath, centrifuge, remove the solution, and wash the precipitate once with 1-F HCl containing some H_3AsO_4 if V, Mo, or much Fe are present. Mix 2 ml 2-F H_2SO_4 with 1 drop 3% H_2O_2 and add 0.2 ml of this mixture to the precipitate. If titanium has been present, a yellow solution is obtained, which turns colorless on adding some solid ammonium fluoride.

Test with Chromotropic Acid (4.7–6.0; in presence of Be, La, Ce, Zr, Th, Cr, U, Fe, Al, Ga, etc.).

Reagent. Solution of 20 mg 1,8-dioxy-3,6-disulfonic acid in 20 ml 18-F H_2SO_4.

On a spot plate, treat one drop of a solution of the test material in dilute sulfuric or hydrochloric acid with one drop of the reagent solution. A blood red coloration indicates Ti.

No. 40: Zirconium, 91.22

Rubidium Heptafluorozirconate (3.8–6.8; alkalies, alkaline earths, Be, V, Cr, W, U, Co, Ni, Pd, Pt, Rh, Au, Zn, Cd, Al, Tl, As, Sb, Bi: 1.3) and (2.8–5.8; rare earths, Ti, Th, Nb, Ta, Mo, Mn, Cu, Ag, Hg, Sn, Pb, Se, Te: 1.3).

The test is best performed upon a slide of plastic. Treat a droplet of the acid solution with a small grain of NH_4F and a like particle of RbCl. Rb_3ZrF_7 separates in colorless octahedra, and six-sided plates (118).

Test with Alizarin S (4.5–5.8; Y, lanthanides, V, Nb, Ta, Cr, Mo, W, Mn, Fe, Co, Ni, platinum metals, Ag, Au, Zn, Cd, Hg, Al, Sn, Pb, As, Sb, Bi, Se, Te: 2.0) and (3.7–5.0; Be, Sc, Ti, Th, U, Cu: 1.3). Fluorides, phosphate, and organic hydroxy acids interfere; the interference of sulfate may be eliminated by adding $BaCl_2$. The test material is best dissolved in HCl.

On a spot plate treat one drop of the weakly acidic test solution with one drop of 1% aqueous solution of sodium alizarin-3-sulfonate and one drop of 12-F HCl. Zirconium gives a red precipitate.

No. 90: Thorium, 232.038
radio-active

Thallium Salt of the Carbonate Complex (4.0–7.0; Na, K, Cr, W, Pd, Pt: 1.5) and (3.4–6.4; Li, alk. earths, Y, lanthanides, Zr, Nb, Ta, Mo, Mn, Co, Ni, Cu, Sn, Pb, Te: 1.0). Uranyl must be absent.

Place a drop of the neutral or slightly acid test solution upon a slide and mix with a small drop of 10% $(NH_4)_2CO_3$ solution. Add a small crystal of $TlNO_3$. The colorless crystals of $Tl_6Th(CO_3)_5$ are very small and have the shape of somewhat deformed rectangles, squares, and rhombs with slightly curved outlines. Some appear black because of their high refractivity. Some irregular shapes suggest twins.

Thorium Iodate (5.4–5.7; alkalies, alk. earths, Be, Y, lanthanides, V, Nb, Ta, Cr, Mo, W, U, Mn, Fe, Co, Ni, Pt, Ir, Cu, Zn, Hg-ic, Al, Tl, Pb, Se, Te: 1.3). Ti, Zr, Hg, and Sn give similar precipitates and must be absent.

In a centrifuge cone, treat 0.5 ml of the slightly acidic test solution with 25 μl 8-F HNO_3 and 10 to 20 μl of a saturated aqueous solution of KIO_3. Thorium gives a white precipitate of $KIO_3 \cdot 4 Th(IO_3)_4 \cdot 18 H_2O$.

Group V B: Vanadium Group
No. 23: Vanadium, 50.942

Bead Test. See P. 39 and Table X.

Quinolinol Fusion. This will detect 1 μg V in presence of 38000 μg MoO_3 and 7000 μg WO_3 (340). See also P. 39.

In a porcelain crucible treat some of the solid material (or residue of the evaporation of a test drop) with a few milligrams of solid 8-hydroxy-quinoline. Heat upon the steam bath. A brownish black melt indicates vanadium.

Test with Benzidine Acetate (5.3–7.3; Mo: 2.0) and (5.0–7.0; W: 3.3). Chromate and other oxidants must be absent.

Transfer a droplet of the test solution to filter paper and add a drop of a saturated solution of benzidine in 10% (1.7-F) acetic acid. Pentavalent vanadium oxidizes to "benzidine blue". With much vanadium, a green coloration results (136).

No. 41: Niobium, 92.906

Niobium and tantalum have in common that, because of the pronounced tendency to hydrolyze with the separation of the pentoxides, it is very difficult to keep their compounds in solution.

Sodium Metaniobate (3.4–6.1). Ta gives the same test; W and Fe do not interfere.

Transfer one droplet of the alkaline test solution to a slide; add 1 drop of 2-F NaOH and 1 drop of saturated aqueous solution of sodium acetate. Colorless monoclinic needles and prisms prove the presence of Nb or Ta (NaNbO$_3$ or NaTaO$_3$).

Zinc and Thiocyanate (4.3–4.3; Ta, Ti, and W do not interfere).

Place a few crystals of KCNS into a small test tube. Add 1 ml of the alkaline test solution, a few granules of metallic zinc, and 5 drops 12-F HCl. Depending upon the amount of niobium, the resulting coloration deepens from yellow to reddish brown (136). Perform a blank test.

No. 73: Tantalum, 180.948

See also Niobium, above.

Dipotassium Heptafluotantalate (4.0–6.7). Nb and Ti interfere and must be absent.

Transfer a drop of the alkaline test solution to a slide and add a drop of saturated aqueous solution of KF. The crystals of K$_2$TaF$_7$ or K$_2$NbF$_7$ appear after about ten minutes: colorless orthorhombic needles which are weakly birefringent.

Test with Tetraethylrhodamine (4.6–4.6; Nb, Ti: 2.5). Mo, W, Fe, Au, Hg, and Sb interfere and must be absent.

In a test tube, treat 1 ml of the test solution which contains the tantalum as fluoride (TaF$_5$ or K$_2$TaF$_7$) with 50 mg solid rhodamine B (tetraethylrhodamine). Tantalum gives a violet coloration or precipitate; a blank test must be carried out.

Group VI B: Chromium Group

No. 24: Chromium, 51.996

Bead Test. L. I. = 2 μg Cr. *See* P. 39 and Table X.

Silver Dichromate. L. I. = 0.06 μg CrO$_4$. All anions interfere, that give acid insoluble silver salts.

The test is an inversion of the silver test of Expt. 25. Add a small grain of solid silver nitrate to the test solution which should be about 3-F in HNO$_3$. A suitable test solution of chromate is obtained by dissolving an evaporation residue in HNO$_3$ of the given concentration.

Perchromic Acid (approx. 4.0–4.0; V: 0.7; As: 1.0; Ti, Fe-ous, Ir, Tl-ic, Bi, Te: 1.2; and in presence of nearly all other metallic ions: 2.0).

Place into a small test tube 1 drop 3% H$_2$O$_2$, 1 drop 2-F HCl, a few drops of diethyl ether, and finally 1 ml of the slightly acidic solution to be tested for chromate ion. Shake the mixture briefly and cool by running tap water over the outside of the test tube. The blue coloration given by CrO$_6^=$ is not stable and disappears soon. With much chromate, the whole ether turns blue; with small amounts of chromate, the blue coloration appears only at the interface of the liquids.

Test with Strychnine (6.0–7.3; most metal ions including Mo, W, Au, Rh, Pd, Ir: 2.0), (5.5–6.8; Th, U, Fe, Co, Pb, Se, Te: 1.5); stannous tin, ceric ion, and other strong oxidants must be absent.

Also this test, as all others, applies to chromate ion.

On a spot plate, treat one drop of the solution of the test substance in water or dilute H_2SO_4 with 1 drop of a 1% solution of strychnine in 18-F H_2SO_4. Chromate gives a pink coloration which may require 15 minutes to appear if only little chromate is present.

No. 42: Molybdenum, 95.94

Flame Test. See P. 38 and Table IX.

Borax Bead. See P. 39 and Table X. One part of MoO_3 may be recognized in 60 parts of borax.

Thallous Molybdate and Wolframate (5.5–8.5; As, Sb, Sn, selenite, tellurite, and tellurate do not interfere).

On a slide, treat one drop of definitely alkaline molybdate solution (2-F in NaOH) (or tungstate solution) with a small crystal of $TlNO_3$. The colorless or pale yellow crystals of Tl_2MoO_4 or Tl_2WO_4 form regular hexagons, six-pointed stars, skeletons, or rosettes. Dendrites may first grow out of the grain of reagent.

Test with Thiocyanate and Stannous Chloride (5.7–7.2; in presence of arsenite, phosphate, and antimony) and (4.3–5.8; wolframate: 2.0; also stannate and germanate may be present). Selenite, tellurite, and tellurate must be absent.

The test may be carried out in the centrifuge cone, on a spot plate, on spot test paper, etc. If necessary, the test solution is evaporated with HCl for the removal of nitrate. The residue is moistened with 12-F NH_3 which is then allowed to evaporate. The residue from the ammonia treatment is dissolved in 3-F HCl to give an about 0.1% Mo solution.

To perform the test, first add to the acid test solution a small volume of 10% (1-F) solution of KCNS. A red coloration is obtained in presence of ferric ion. Molybdate may give a yellow coloration. Now add a 5% solution of $SnCl_2$ in 3-F HCl. The red coloration given by ferric thiocyanate disappears, and any red color persisting or appearing is now due to the complex $Mo(CNS)_6^{--}$.

No. 74: Wolfram (Tungsten), 183.85

Bead Test. On the borax bead see Table X; concerning the differentiation between W and Ti by means of the phosphate bead, P. 39.

Thallous Wolframate, see Molybdenum.

Test with Stannous Chloride (4.0–5.7). Mo, Nb, Ta do not interfere. A wolframate solution is used for the test.

Place a large drop of 12-F HCl upon spot test paper and, without delay, add the test drop to the center of the wet spot. A yellow fleck of WO_3 may become visible. Now add 1 drop of a 25% solution of $SnCl_2$ in about 3-F HCl and 1 drop of 10% (1-F) KCNS. A blue spot in the center indicates W, and a red ring around it may suggest Mo.

[*No. 92: Uranium, 238.03*]

Bead Tests. See P. 39 and Table X. Uranium in a bead of NaF gives a violet fluorescence with ultraviolet light, which permits detection of 0.02 μg U; by using a spectrograph, this limit of identification may be improved to 1 ng U (119).

Sodium-Zinc-Uranyl Acetate. L. I. = 0.6 μg U. The test is specific for uranium.

Transfer a droplet of test solution to a slide and evaporate to dryness. Dissolve the residue in a volume of 3-F acetic acid, that promises a 1% uranium solution. Place into the edge of the drop first a grain of sodium acetate and then, at a point not too far away, a grain of zinc acetate. Search under the microscope for tetrahedra (triangles) of the sodium-uranyl acetate and for octahedra-like birefringent crystals of the triple salt. *Compare* under Lithium and Sodium in P. 44. Observe the fluorescence of the crystals in ultraviolet light.

Quinolinol Fusion (453). Probably quite specific.

Treat a residue of uranyl salt upon the slide with sufficient solid 8-hydroxyquinoline so that the space between the slide and a fragment of a cover slip will just fill with the melt. Place the piece of cover glass upon the reagent and heat upon the steam bath. When the quinolinol melts, at about 80° C, the uranyl salt separates in rods and willow leaf-shaped crystals having pleochroism (yellow–red). At about 120° C, these crystals dissolve in the melt, and upon cooling separate rectangular crystals having an oblique extinction of 8 degrees (both indices of refraction are higher than that of the melt at 35° C).

P. 50   Group VII B: Manganese Group

No. 25: Manganese, 54.9380

Flame Test. Only $MnCl_2$ gives a very fleeting green coloration to the flame, *see* Table IX.

Borax Bead. L. I. = 20 μg Mn. See P. 39 and Table X.

Chlorate Fusion. L. I. = 0.1 μg in presence of 3 μg Cu, Co, or Ni; 10 μg Fe, Cr, U; or 100 μg alk. earths, Be, Y, lanthanides, Ti, Mo, W, Re, Cd, Pb, As, Se.

On a platinum wire or foil–or in a platinum spoon or crucible–and in a non-reducing flame, fuse the solid particle to be tested (or the test drop) with a small amount of a solid mixture of 1 part (volume or weight) $NaClO_3$ with 5 parts Na_2CO_3. On cooling, the bead or melt will show the green color of manganate. Dissolve in a small volume of 2-F acetic acid to obtain the pink to purple permanganate coloration.

If the bead is dissolved upon the slide in 1-F $HClO_4$, adding solid KCl or RbCl into the edge of the drop will give mixed crystals of permanganate and perchlorate, see below.

Thermoluminescence Test. L. I. = 1 ng Mn; see Expt. 60.

Persulfate Oxydation (5.7–7.0; Cr: 2.5; Ce: 3.0). The test works in presence of most other metals without impairment of the sensitivity.

Place the test solution upon a slide and evaporate to dryness. Three times treat the residue with 3-F HNO_3 followed by evaporation to dryness (for the removal of chloride). Finally dissolve the residue in 1-F HNO_3 and transfer the solution to a centrifuge cone. Add a very small volume of 1% $AgNO_3$ and a small crystal of $K_2S_2O_8$. Mix and heat to about 60° C. The pink or purple color of the permanganate proves the presence of manganese. It may be necessary to centrifuge the mixture and to inspect the clear solution in the coloriscopic capillary, Expt. 55. The solution may finally be used for the rubidium perchlorate test. With the red crystals as well as with the coloriscopic capillary, the identification may be improved by the use of a spectroscopic attachment to the eyepiece of the microscope.

Rubidium Perchlorate-Permanganate (5–7.7; for most metals including Re: 3.0).

Transfer the solution of permanganate to a slide and push into the edge of the drop–at two places somewhat distant from one another–, a crystal of $NaClO_4$ and a crystal of $RbNO_3$ or RbCl. Prismatic mixed crystals of $Rb(ClO_4, MnO_4)$ are obtained, and the color varies from colorless to pink, red, and black depending upon the ratio of permanganate to perchlorate.

Manganese Cyanurate (4.5–7.8; alkaline earths, Cr, U, Fe, Co, Zn, chloride: 2.0; Be, Y, lanthanides, Ti, Mo, Re, Ni, Al: 2.5). **Magnesium** should be absent.

Place a droplet of the solution of manganous salt upon a slide and evaporate to dryness. After cooling, treat the residue with a droplet of a saturated solution of cyanuric acid, $(HNCO)_3 \cdot 2 H_2O$, in 1-F NH_3. Colorless needles and pointed rods indicate manganese. If only little manganese is present, hasten crystallization by covering the test drop with a small watch glass to prevent evaporation and warming slightly.

No. 75: Rhenium, 186.2

Flame Test. Greenish white flame.

Acridine Perrhenate (4.7–8.0; molybdate, wolframate: 2.0; alkaline earths, Be, Y, lanthanides, Cr-ic, Mn-ous, Fe, Co, Ni, Cu, Zn, Al, Sn: 2.7). Chromate, permanganate, and uranyl must be absent.

Transfer the neutral or slightly acidic perrhenate solution to a slide. Place upon the steam bath and evaporate the solution to dryness or nearly to dryness. Remove from the steam bath and add to the warm residue a small drop of a saturated aqueous solution of acridine hydrochloride. Perrhenate gives yellow pointed needles and loose sheaves of such (1202). Permanganate, chromate, and uranyl give similar crystallizations.

Reduction of Tellurate (6.3–7.6; alkaline earths, Be, Y, lanthanides, V, Nb, Ta, Ti, Zr, Th, Cr, U, Mn, Fe, Co, Ni, Ag, Zn, Cd, Al, Tl, Sn, Pb, As, Sb: 2.0), (4.0–5.3; tungstate, Se: 1.5; Pt, Ir, Cu: 2.2). Titanium, molybdate, Rh, Pd, Ag, Au, and Hg must be absent since they also give dark precipitates.

Perrhenate is reduced by $SnCl_2$ to Rh(3) which, in turn reduces tellurate to elemental tellurium.

On a spot plate, treat a drop of the slightly acidic test solution with 1 drop of 1% aqueous solution of Na_2TeO_4 and 1 drop of 50% $SnCl_2$ in 10-F HCl. Perrhenate causes precipitation of black Te.

Group VIII: Fe-Ni Triad

No. 26: Iron, 55.847

Bead Test. See P. 39 and Table X.

Quinolinol Fusion. L. I. = 40 ng Fe.

Perform the test as directed under Vanadium, P. 48. Iron gives a greenish black coloration.

Prussian Blue (4.5–5.8; alkaline earths, Be, Y, lanthanides, Nb, Ta, Pd, Au, Al, As, Sb, Te: 2.0), (4.0–5.3; Cr, Th, Mn, Ni, Pt, Ir, Rh, Ag, Zn, Cd, Bi, Pb, Se: 1.5; Ti, Zr, V, W, Co, Hg, Tl, Sb-5: 1.3), and (2.8–4.1; molybdate: 0.7; uranyl: 0.3). Fluoride, phosphate, and aliphatic hydroxy acids must be absent since they hold the iron in stable complexes.

If not certain that the iron is present in the ferric state, expose the test drop to bromine vapors and then evaporate just to dryness. Dissolve the residue in 0.2-F HCl and treat a drop of the solution upon the spot plate with 1 drop of 10% (2-F) potassium ferrocyanide solution. Ferric iron gives a dark blue precipitate or a light blue coloration with little iron. A blank is essential if only a coloration is obtained.

The sensitivity may be improved by using small drops, performing the test on paper or on a textile fiber impregnated with $K_4Fe(CN)_6$ or $K_2ZnFe(CN)_6$.

To discover about 2 µg of iron, place a small droplet of the test solution upon a slide and evaporate just to dryness. Allow to cool and then moisten the residue with a small volume of a mixture of equal parts of 0.5% (0.1-F) potassium ferrocyanide solution and 1-F HCl, which has been centrifuged just before use to remove any blue sediment. The "amorphous" blue precipitate should be observed with strong transmitted light and with reflected light before a white and before an orange background.

Insoluble iron compounds may be dissolved in a tiny bead of $K_2S_2O_7$ or of borax, which is then placed upon the slide and treated with the above acid reagent solution. Separation of flocks of Prussian blue may then be observed during the slow dissolution of the bead.

Ferric Thiocyanate (4.0–7.0). Fluoride, phosphate, arsenate, iodide, iodate, acetate, and complex forming organic oxy acids must be absent.

Place the test droplet upon a slide, treat with a small amount of a mixture of 1 volume 16-F HNO_3 with 8 volumes 12-F HCl, and evaporate just to dryness. Dissolve the residue in a small volume of a mixture of equal parts of 1-F KCNS and 6-F HCl. If necessary use the coloriscopic capillary for the observation of the red color, Expt. 55. The red color is discharged upon adding stannous chloride, see Titanium, P. 47.

o-Phenanthroline Test for Ferrous Iron (6.5–8.8; alkaline earths, Be, Sc, Y, lanthanides, Ti, Zr, Th, V, Nb, Cr, Mo, W, U, Mn, Ni, Ru, Rh, Pd, Os, Pt, Ag, Au, Zn, Cd, Hg, Al, Ga, In, Tl, Sn, Pb, As, Sb, Bi, Se, Te, fluoride, phosphate, aliphatic oxy acids: 3.7), (5.2–6.5; Co, Sb: 2.7), and (4.2–5.5; Cu: 1.7).

Ferrous iron gives a red cation, $Fe(C_{12}H_8N_2)_3^{++}$. Of course, the treatment with hydroxylamine must be omitted when testing for ferrous iron specifically. Additional precautions would be necessary.

Evaporate the test solution just to dryness, and dissolve the residue in a small volume of 0.1-F HCl. On a spot plate, treat with a kernel of $NH_2OH \cdot HCl$. Hasten the dissolution by stirring, and then treat with a like volume of 0.025-F aqueous solution of o-phenanthroline. If necessary, observe the red coloration in a coloriscopic capillary, Expt. 55.

No. 27: Cobalt, 58.9332

Borax Bead. The blue coloration may be recognized with 0.2 µg Co, see Expt. 59, P. 39, and Table X.

Cobalt-Mercuric Thiocyanate (6.0–10.0; Be, Y, lanthanides, Ti, Mn, Re, Zn: 2.3; alkaline earths, Zr, V, Cr, Mo, W, U, Fe, Ni, Rh, Os, Ir, Pt, Cu, Ag, Au, Zn, Cd, Hg, Sn, Pb, As, Sb, Bi, Se, Te: 2.0).

For the performance *compare* Expt. 37. To obtain highest sensitivity for cobalt, proceed as follows. Place a droplet of test solution upon a slide and evaporate to dryness. (At this time, zinc may be added to increase the sensitivity. If this seems desirable, add a droplet of 1% zinc solution to the residue and again evaporate.) Allow to cool to room temperature, then add as little reagent as possible with the use of a loop or capillary pipet. Be certain **not** to touch the surface of the slide with the tool since this may cause seeding and the separation of many tiny crystals tracing the scratches. Observe the blue needles and clusters of spear-shaped crystals with strong transmitted light. The crystals of $CoHg(CNS)_4$ are relatively thick and dark blue, but if zinc has been added, they will be lighter in color. Ferric ion gives a dark red solution which may be decolorized by adding NH_4F.

Cobalt Rubeanate (6.0–7.5; Cr, Re, Zn, Al: 2.0), (5.0–6.5; Mn, Fe-ic: 2.0). Ag, Ni, and Cu interfere.

Concerning the performance *see* Expt. 27.

No. 28: Nickel, 58.71

Bead Test. *See* P. 39 and Table X.

Molybdate Test (3.7–7.0; Co and perrhenate do not interfere). Al, Tl, Ce, Fe-ous, and Mn should be absent. The interference by ferric ion may be prevented by adding NaF.

Place the droplet of test solution upon a slide and evaporate to dryness. After cooling, dissolve the residue in a small volume of 0.1-F HCl and add a grain of NaF if iron is present. Next to the test drop, deposit a drop of a saturated aqueous solution of $(NH_4)_6Mo_7O_{24} \cdot 4 H_2O$ and draw a channel to connect the drops. Separation of colorless squares, rhombs, and wedges indicates nickel. Aluminum and ferrous iron give crystals of the same shape.

Nickel Dimethylglyoxime (5.0–8.5; Co: 1.7) and 0.3 ng Ni may be detected with darkfield illumination when the test is performed upon a slide. The test is quite specific.

Place the droplet of test solution upon a slide and evaporate to dryness. After cooling, dissolve the residue in 0.3 to 3 μl 6-F NH_3. Without delay, add a small crystal of dimethylglyoxime and cover with a small (1-cm diameter) watch glass to prevent evaporation of the ammonia. After five minutes inspect under the microscope. The nickel salt crystallizes as long, fine, red needles which show pleochroism. They appear red when they are parallel to the vibration direction of the polar (nicol). The far more soluble copper salt of similar color appears red when the vibration direction of the polar is vertical to the long edges of the prismatic crystals.

If the test is carried out on the spot plate or on spot paper, the sensitivities are specified as follows (5.5–6.8; Mn, perrhenate, chromate, Zn, Al: 3.0), (4.5–5.8; Fe: 3.3; Co, Pd-ous: 2.0), and (3.5–4.8; Cu: 1.0).

For reagents serve a saturated solution of dimethylglyoxime in 95% ethanol and 6-F NH_3.

Nickel Rubeanate (6.0–9.0; alkaline earths, V, Cr, Mo, W, Pt, Au, Zn, Cd, Al, Tl, Sn, Pb, As, Sb, Bi, Se, Te, perrhenate: 2.0), (5.0–8.0; ferric ion: 2.0). Cu, Ag, and mercurous ion must be absent.

For the performance *see* Expt. 27.

P. 52   Group VIII: Ru-Pd Triad

Catalytic Glow Test for Platinum Metals (697, 1003). The test detects all platinum metals with the exception of ruthenium and osmium. When using test drops of 1 μl volume, the limits of identification are: 20 ng Rh, 10 ng Pd, 180 ng Ir, and 40 ng Pt. The test is quite indifferent to the presence of other elements, and Mo, U, Fe, Co, and Cu may be present in thousandfold excess. For As_2O_3, which poisons the catalysts, the ratio may be 50 to 1 Pt. The procedure for the performance of the test has been recommended by HAHN (856). As a possible alternative, one might take up the test solution in the tip of a small asbestos triangle cut to give a fine point (881).

Cut asbestos paper of 0.5-mm thickness to a strip of 2 cm × 6 cm. Moisten one end of the paper and and make there a depression with the rounded end of a glass rod and the paper resting upon the flat surface of a cork having a pit that serves for matrix. Finally form at the lowest point of the cup a sharp conical depression by pressing with the other end of the glass rod, which is drawn out to a sharp point. The paper may be slightly pierced by this operation. Dry and then strongly ignite the shaped end of the strip by heating in the reducing zone of a Bunsen flame.

With a capillary pipet transfer the test solution to the point of the tiny cone. Ignite again and then, before the paper gets quite cold, expose it to a stream of pure and dry hydrogen gas which escapes from a capillary tube. The gas should be supplied with low pressure (20 cm water column) so that it gives at the opening of the capillary a flame of 5- to 10-mm height. Extinguish the flame, and hold the paper so that the gas blows into the cupshaped depression. The point of the conical tip will start glowing. Find the position which gives the strongest incandescence by varying the distance of the paper from the orifice of the capillary.

No. 44: Ruthenium, 101.07

Borax Bead. Black. *See* P. 39.

Test with Thiourea (136) (5.0–6.3). Osmium gives a red coloration and consequently interferes with the test. A solution of the test material in hydrochloric acid is preferred.

Place 1 drop 10-*F* HCl into a small test tube. Add a drop of the solution to be tested and a few small grains of thiourea. Warm gently. A blue coloration shows the presence of ruthenium.

Test with Rubeanic Acid (5.0–6.0). Co, Ni, and Cu supposedly interfere, but Os does not. Ruthenium shall be present in the trivalent state.

In a porcelain crucible, treat a drop of the test solution with 1 drop 2-F HCl and 1 drop of 0.2% rubeanic acid in glacial acetic acid. Heat with a microflame. A blue coloration is caused by ruthenium. Platinum and palladium give red precipitates; to render the blue coloration visible, transfer the mixture to a microcone, whirl in the centrifuge, and take the solution into a coloriscopic capillary, Expt. 55, if necessary.

No. 45: Rhodium, 102.905

Borax Bead: brown, slate gray in reflected light. See P. 39.

Potassium Pyrosulfate (162). A trace of metallic rhodium heated with $K_2S_2O_7$ on platinum foil or wire gives a melt which is red when hot and yellow when cold.

Tests with Aqueous Solutions of Rhodium. For the following tests, KÖNIG and CROWELL (883) treat the residue to be tested with oxidant and potassium salt to obtain the complex $RhCl_6^{--}$. Potassium rhodonitrite is repeatedly evaporated at 125° C with *aqua regia*, and this treatment is followed by evaporation with 1 drop HCl (12-F) and a crystal $KClO_3$. The residue is dissolved in water. An ethylxanthate precipitate is given just the chlorate treatment described above.

Test with Hexamethylenetetramine (3–5.3; Pd, Ir, Pt, Au: 1.0).

On a slide, treat a droplet of the test solution with a crystal of hexamethylenetetramine hydrochloride. After 5 to 10 minutes, bright red crystals will appear: rods crossed at right angles in presence of sodium, or thick hexagonal plates in presence of potassium (885). Controls are recommended.

If iridium is coprecipitated, the color of the crystals changes to brown, and this may already happen with quite small amounts of iridium (1 iridium : 50 rhodium).

Stannous Chloride Test (5.0–6.0). All platinum metals as well as Ag, Hg, and As are reduced to the metallic state, but only Au gives a red coloration, the purple of CASSIUS (883).

In a centrifuge cone, heat 2 drops of the test solution with 1 drop of a 40% solution of $SnCl_2$ in 30% (9.5-F) HCl for 5 to 10 minutes. A purplish red coloration which develops on cooling indicates rhodium.

No. 46: Palladium, 105.4

Borax Bead: black, *see* P. 39.

Test with Dimethylglyoxime (4.0–7.0). Only Au and Ni are also precipitated. The composition of the precipitate is analogous to that of the nickel dimethylglyoxime.

Place a drop of the weakly acidic test solution upon a slide and introduce a crystal of dimethylglyoxime. Yellow pointed needles, sheaves and stars of such, separate in a short time.

Test with Thioglycolic Acid (5.0–6.3; Ir, Pt, Au: 2.4; Rh: 1.4). Ru, Os, Cu, Ag, nitrite, and cyanide interfere. Simple as well as complex ions of bivalent and tetravalent palladium give the test.

Transfer a drop of the test solution to a white spot plate and add a drop of 10% aqueous solution of thioglycolic acid or of a correspondingly smaller volume of the commercially available solution of 80%. A yellow color develops immediately if palladium is present. High concentrations of palladium give a yellow precipitate that dissolves upon diluting to give the yellow coloration. Similarly, also Au, Pt, Ir (present in high concentration) give white amorphous precipitates; these also dissolve upon diluting with water but give colorless solutions (884).

No. 76: Osmium, 190.2

Borax Bead: brown and clear, *see* P. 39.

Potassium Nitrate Bead (119).

The bead is first completely black, but becomes brown on further heating and finally colorless since the metal is volatilized as OsO_4, the chlorine-like odor of which may be perceived with 20 ng of the compound per milliliter air. OsO_4 is known in two modifications: one yellow, m. pt. 41° C; the other white, m. pt. 39.5° C. The boiling point is 134° C. The solution in water, "osmic acid" or "perosmic acid", and the vapor are very poisonous and injurious to the eyes and mucuous membranes.

Potassium Osmate.

By using the volatility of OsO_4, the test like any other osmium test may be made specific.

Treat the solid test substance with some 8-F HNO_3 or treat a test solution with an equal volume of 16-F HNO_3. Heat the mixture gently in any suitable apparatus that permits absorbing the liberated OsO_4 in 2-F KOH. Apparatus shown in Figs. 75*b*, 15*a*, 18*c*, and 69 may be used. The KOH will become yellow to light red as it absorbs increasing amounts of OsO_4, and the resulting solution is used for confirmatory tests; RuO_4 is not liberated under the conditions.

Transfer a small volume of the distillate to a slide and introduce a small grain of KNO_2 to aid in the reduction to OsO_2. Garnet red, small but well developed, strongly birefringent (orthorhombic) "octahedra" of $K_2OsO_4 \cdot 2 H_2O$ separate (865).

Osmium Thioureacomplex (5.0–6.0) (136).

In a small test tube, treat 0.1 ml of the test solution, which may contain osmium of the valence $+8$ or $+4$ and which should be 1.5-F with HCl, with a few grains of thiourea. Warm the mixture. A red coloration indicates osmium.

Test with Thiocyanate (6.0–7.0). Au, Ir, Pt, Fe, Co, and various other elements interfere. The Os shall be in the valence state $+8$ (162).

In a small test tube, treat 0.1 ml of the solution, which shall be lightly acidified with HNO_3, with a few grains of KCNS or NH_4CNS. Osmium gives a blue coloration which may be extracted with ether or amyl alcohol.

No. 77: Iridium, 192.2

Borax Bead: brown and clear, see P. 39.

Potassium Chloriridate, see Platinum. The hexamethylenetetramine tests used by KÖNIG and CROWELL (883, 885) deserve consideration.

Test with Sulfuric Acid and Nitrate (3.9–5.2). Appreciable amounts of Ru, Rh, or Pd interfere with the test (162).

In a porcelain crucible, treat 1 drop of the test solution with 2 to 3 drops 18-F H_2SO_4 and heat to drive out HCl. Then add a few crystal grains of $AgNO_3$ and heat again. It is assumed that the blue coloration is due to some hydrate of IrO_2.

No. 78: Platinum, 195.09

Borax Bead: brown and turbid, L. I. $= 0.2\,\mu g$. See P. 39.

Alkali Chloroplatinate (For K_2PtCl_6: 4.0–7.5). Os and Ir interfere. A suitable test solution is obtained by evaporating with *aqua regia* and dissolving the residue in 2-F HCl. This procedure will also eliminate osmium.

Transfer a small droplet of the test solution to a slide and add a tiny crystal of KCl. Yellow isotropic octahedra of K_2PtCl_6 prove the presence of platinum. The corresponding iridium compound, K_2IrCl_6, is red. Solutions containing both, Pt and Ir, give mixed crystals, the color of which depends upon the ratio Pt : Ir.

The sensitivity of the test may be improved by substituting Rb, Cs, or Tl+ for the potassium. The solubility of the salts decreases in the given order.

Test with Rubeanic Acid (4.0–5.3). Ru, Pd, Au, and Bi interfere.

On a spot plate, treat 1 drop of the solution of the test substance in hydrochloric acid (see K_2PtCl_6 test above) with 1 drop of 0.02% rubeanic acid in glacial acetic acid. The precipitate of $Pt(C_2H_3N_2S_2)_2$ is reddish violet.

P. 54   Group I B: Copper Group

No. 29: Copper, 63.54

Flame Test. L. I. $= 0.3\,\mu g$ Cu. See P. 38 and Table IX and P. 60.

Bead Test. See P. 39 and Table X.

Copper-Mercuric Thiocyanate (5.0–8.0; V, Mo, W, Pt, Ag, Au, Cd, Tl, Pb, As, Sb, Bi, Se, Te: 2.0). Fe-3, Co, and Ni interfere and should not be present in appreciable amounts.

Proceed as directed in Expt. 37.

Copper-Ammonia Complex (4.0–6.0). Ni, Co, Cr, and other substances interfere, that give colored ammoniacal solutions.

Proceed as directed in Expt. 55.

Cupric Rubeanate (6.0–8.0; V, Mo, W, Pt, Au, Cd, Tl, Sn, Pb, As, Sb, Bi, Se, Te: 2.0). Ag, Hg, and in some measure Co and Ni interfere.

Proceed as directed in Expt. 27.

No. 47: Silver, 107.870

Borax Bead: yellow; L. I. $= 0.2\,\mu g$ Ag. See P. 39.

Silver Dichromate (3.2–6.7). Apparently no interferences that have practical consequences.

Proceed as directed in Expt. 31.

Silver Chloride (7.0–9.0). Mercurous ion interferes with the test but may be converted to mercuric ion. Aside from this, the test is specific when performed so that a small amount of hydrochloric acid is added to an already acid solution. Neither $PbCl_2$ (orthorhombic needles) nor TlCl (cubic) should separate under these conditions.

Test tube, centrifuge cone, microcone, capillary, capillary cone may be used, or the droplet may be spread between slide and fragment of cover slip or suspended in oil. For the detection of small amounts, it is essential to inspect the solution before and after adding the HCl with strong lateral illumination before a black background (Tyndall illumination or darkfield condenser).

Depending upon circumstances, dilute the test solution with water or acidify it with HNO_3 so that it will be about 2-F with free acid. Inspect the solution to make certain that it is clear. If necessary, centrifuge and use the clear solution after careful inspection for absence of turbidity. Treat the test solution with an amount of 1-F HCl, that will bring the chloride concentration of the test solution to about 0.001 to 0.01 formal. A white precipitate proves the presence of silver. Small amounts of silver will give only a turbidity which is made visible by suitable illumination: strong light from the side and black background or use of a darkfield condenser.

For recrystallization from NH_3, see Expt. 32, the precipitation of the AgCl is made complete by adding more HCl, if necessary, and heating on the steam bath for 5 to 15 minutes to hasten flocculation. The precipitate is collected and washed once with 0.1-F HNO_3. Without the aid of micromanipulation, about 0.1 µg AgCl may be recrystallized with success. The recrystallization from NH_3 establishes the identity of the AgCl.

No. 79: Gold, 196.967

Borax Bead: L. I.: 25 ng; ruby red by colloidal gold, see P. 39.

Metallic Gold (4.7–6; Cu: 1.3 if the bead of gold is taken for criterion) or (6.0–7.3; Cu: 2.0 if the pink line of colloidal gold serves this purpose). The test is based upon the experience that all compounds of gold are reduced to the metal when heated. Aside from Rh, Pd, Ir, and Pt, no other metal can interfere, and the anion to which the gold is bound is without consequence.

DUVAL and FAUCONNIER (912) give the following directions and suggest that the experimenter should not wear any objects of gold. From soft glass or Pyrex glass tubing, which has been properly cleaned, prepare a medicine dropper with a rather thick-walled capillary tip. Attach the rubber bulb and take about 0.05 ml of the sample solution into the tip. Try to do this so that the solution remains with the outer meniscus at the orifice of the tip. Holding the dropper horizontal and squeezing the bulb to hold the drop continuously at the opening, cause it to evaporate very slowly without spattering by warming cautiously over a microflame or under an infrared lamp. Finally, remove the bulb and heat the tip of the dropper with suitable rotating in the Bunsen flame to obtain a solid sphere of glass of a few millimeters in diameter. The glass sphere serves for magnifyer and will show a tiny globule of metallic gold if 1 µg of it or more was present. With less than this and down to 0.05 µg Au, a pink line of colloidal gold is visible.

Copper, if present in larger amounts than 18 µg, may give a metallic bead that could be mistaken for gold. If less copper is present, it will give green specks in lively contrast with the bead of gold or the red streak of colloidal gold.

Colloidal Gold Upon a Fiber (5.4–8.7). Platinum metals do not interfere. Alkalies should be absent, and significant amounts of Ag and Mo interfere.

Reagent. Use clean viscose-rayon fibers or fibers of raw true silk. Soak them in 10% (2-F) KOH for several hours and then wash them thoroughly. Dissolve 10 g $SnCl_2 \cdot 2 H_2O$ in 5 ml 12-F HCl; add 95 ml water and filter. In small portions and with continuous stirring, add to the filtrate 10 g white crystals of pyrogallol. Place the washed fibers into this solution and heat 10 minutes upon the steam bath. Wash the fibers with distilled water and then dry them by pressing between filter paper.

Keep them wrapped in clean filter paper until they are assuredly dry, and then store in a stoppered amber bottle (better in a dark closet).

Perform the test as described by DONAU (1002). Place the test drop upon a watch glass and evaporate to dryness on the steam bath. Cover the residue with 12-F HCl, and again evaporate. After cooling, extract the residue with 0.5 μl water, and transfer the extract to a slide by means of a capillary pipet. Using the technique of Expt. 57, insert the end of one of the prepared fibers into the droplet, and allow the latter to evaporate. If gold is present, the end of the fiber turns red, violet, or blue–depending upon the amount of gold.

Molybdates give a blue color without any reddish tinge. Silver imparts a brown color if present in large amount, but it does not conceal the color produced by gold unless very little of the latter is present.

Reduction with Stannous Chloride (5.4–5.7) (137).

Reagent. Stir 22.6 g $SnCl_2 \cdot 2 H_2O$ with 2.5 ml 12-F HCl and heat for two to three minutes. Cool to about 50° C, and then pour the solution into 400 ml water.

In a small test tube, treat 0.5 ml of the weakly acid test solution with 0.1 ml (3 drops) of the reagent and mix. The reduction to colloidal gold gives a yellow to yellowish brown coloration.

Pyridine Bromaurate (4.5–7.5; Mo, V, Pt, Cu, Cd, Hg, Sn, Bi, Se, Te: 2.0; W, As, Sb: 1.0). Mercurous mercury, silver, lead, and thallium should be absent (136).

Transfer to a narrow slide a drop of test solution that promises to contain between 0.1 and 1 μg Au. Evaporate upon the steam bath so that the residue is collected in a very small area. After cooling, add to the residue 0.3 to 1 μl of a solution of 1 volume of pyridine in 9 volumes 40% HBr. The thin, almost rectangular plates of the bromaurate appear yellow, brown, or dull red. They show oblique extinction (angle of 10 degrees) and strong pleochroism from colorless or yellow to brown when the long edges are at 80° to the vibration direction of the polar.

Use filter paper or a capillary pipet to remove the mother liquor. When the preparation has become dry, ignite it over a small Bunsen flame. After cooling observe the pseudomorphs of gold having the shape of the bromaurate crystals.

Group II B: Zinc Group

No. 30: Zinc, 65.37

Metallic Zinc. For the detection of zinc or cadmium in an alloy (150), place some of the latter into the sealed end of an 8-cm length of narrow glass tubing (4-mm i. d.) or of a capillary. Heat in the flame of a Bunsen burner. Cadmium (b. pt. 767° C) and zinc (b. pt. 907° C) will distil and give a condensate consisting partly of the metal and partly of the oxide (ZnO, white; and CdO, brown). The tube may be cut apart, and condensate and residue may be tested separately, which may save tedious separations. Interferences may be due to Hg (b. pt. 356.6° C) and As (sublimes copiously at 615° C), but will rarely happen.

Zinc and cadmium compounds may be heated with sodium oxalate to get reduction to the metals; if the heating is performed in a current of hydrogen or of inert gas, P. 27, 29, the condensate will consist of metal only.

Zinc-Mercuric Thiocyanate (5.5–8.8) and (4.0–7.3; alkalies, alkaline earths, Be, Y, lanthanides, Zr, Th, V, Nb, Ta, Cr, Mo, W, U, Fe, Ni, Rh, Pd, Ir, Pt, Cd, Hg, Al, Sn, Bi, Se, Te: 2.0; Mn, Co, Ag, Au, Tl, Pb, As, Sb: 1.0).

Proceed as directed in Expt. 37.

The test may be carried out upon a spot plate: Treat the test solution with some H_2SO_4 and evaporate to dryness. Dissolve the residue in such a volume of 0.2-F H_2SO_4 to obtain an approximately 0.1% solution. Transfer 1 drop of this solution to a spot plate and add 1 drop each of 0.1% cupric sulfate solution and a solution of 2.7 g $HgCl_2$ and 3 g NH_4CNS in 100 ml water. A chocolate brown or violet black precipitate of mixed crystals is obtained if zinc is present.

Rinnmann's Green (3.2–6.2). The absence of appreciable amounts of other metals is required. One part of zinc may be detected in the presence of 5 parts Cd, 2 parts Al, 1 part Ni or Ti, 0.5 part Co, or 0.1 part Mn. Aluminum gives a light blue ash; titanium, a dark blue one. Tin and antimony should be absent since they may give a green ash. The zinc may be quite simply isolated for the test by electrolytic precipitation from NaOH solution (1143).

Reagent Paper. Soak "ash-free" filter paper in a solution of 4 g $K_3Co(CN)_6$ and 1 g $KClO_3$ in 100 ml water. Dry at room temperature or at 100° C in the oven.

Procedure. Place a droplet of the test solution upon a watch glass and evaporate to dryness upon the steam bath. Moisten the residue with 6-F HNO_3 and again evaporate to dryness. Dissolve the residue in such a volume of 2-F HNO_3 as promises a 0.1 to 1% solution of zinc. Take up some of this solution into a capillary pipet and transfer it to the test paper by using the technique of Expt. 24. Dry the moist spot by holding the paper high above a small Bunsen flame. First, a yellow circle will appear and indicate the circumference of the drop. Continue drying until the center turns brown. Then grasp the far end of the paper with forceps, light the paper, and allow the ash to drop into a porcelain crucible or onto a watch glass held ready for the purpose. If zinc is present, some green ash (a solid solution of some CoO in ZnO: Rinnmann's green) will be visible at the center of the spot where the tip of the pipet touched the paper. This green spot (with little zinc, just a few green fibers) is usually surrounded by a circular zone containing very little ash, often only a delicate network of black fibers connecting to the surrounding sheet of black ash (403).

The test may be carried out in a more sensitive and convincing manner by first precipitating and inspecting the $KZnCo(CN)_6$ on a microscope slide (1143).

No. 48: Cadmium, 112.40

Cadmium Metal. *See* Zinc, above.

Test with Brucine Acetate and Sodium Bromide (4.0–7.0; Cu, Ag, Zn, Tl, Sn, Pb, As, Sb, Se, Te: 2.0; Mo, Pt, Au: 0.0). Mercury, Bi, V, and W interfere and should be absent.

Follow the directions given in Expt. 38.

Cadmium-Mercuric Thiocyanate (3.0–6.0; Co, Cu, Zn: 0.0).

Proceed as directed in Expt. 37.

Test with Cadion 3 B (6.0–9.0; Na, K, NH_4, Zn, Pb: 4.0; Mg, Be, Cr, Mn, Co, Ni, Cu, Al, As, Sb, Bi: 3.7; and Ca: 3.0).

Proceed as directed in Expt. 28.

No. 80: Mercury, 200.59

Flame Test: *see* Table IX.

Metallic Mercury and Conversion to Iodide (3.7–6.7; Ag: 2.0). The test is specific for mercury.

Proceed as directed in Expt. 49.

Cupric-Zinc-Mercuric Thiocyanate (4.3–7.3).

Proceed as directed in Expt. 39.

Mercuric Iodide (3.3–6.3); spot test specific in presence of all common metals.

Proceed as directed in Expt. 24.

P. 56 Group III A: Boron-Thallium Group

No. 5: Boron, 10.811

The following tests are given by boric acid and borates.

Flame Test: *see* P. 38 and Table IX.

FEIGL and SUTER (620) place a drop of the test solution into a porcelain crucible, add 1 drop 18-F H_2SO_4 and 5 drops methanol, warm the mixture, and light the escaping vapor. The volatile methyl borate gives a green flame. According to LENHER and WELLS (695), Ba, Cu, Tl, MoO_3, TeO_2, and H_3PO_4 may interfere by giving a similar coloration to the flame.

The interference by substances that do not form volatile ester may be eliminated as follows. Prepare a miniature gas washing bottle from a soft-glass test tube, about 16-mm i. d. and cut short to 8-cm length. Provide it with a 2-hole rubber stopper, an inlet tube that goes to the bottom of the test tube, and a short straight outlet tube ending in a capillary tip. Introduce into the test tube some of the solid sample, 0.5 ml 18-F H_2SO_4, and 3 ml methanol. Close the apparatus, mix by swirling, and blow a stream of air through the mixture and into a non-luminous Bunsen flame. Any methyl borate, $B(CH_3)_3$, carried along with the air will color the flame green. It should be possible to detect 20 μg boric acid or less. A plug of glass wool may be inserted into the outlet tube.

Isolation of Boric Acid via Methyl Borate. The volatility of methyl borate may be used to isolate boric acid for increasing the specificity of the tests performed with the distillate. Suitable apparatus are suggested by Figs. 75b, 15a, 18c, and 69.

Collect the methyl borate in a drop of water where it will undergo saponification. Treat the solid sample (solutions may be made alkaline, if necessary, and evaporated in the "pot" of the still) with like volumes of first 18-F H_2SO_4 and then methanol. Then heat the mixture by inserting the pot into water of 80° C.

Turmeric Test. L. I. = 0.5 ng B; 1 part of boron may be detected in presence of 100 parts of Mg, Ca, silicate, and phosphate; or 10 parts of ferric ion. Ti, Zr, Hf, Nb, Ta, and Mo give similar colorations in acid solution, but do not give the blue color with NH_3.

The limit of identification applies to the use of a textile fiber as outlined in Expt. 57. In most instances, use of turmeric paper will suffice; the test solution may be taken up with a hook or a loop, exposed to fumes of HCl to acidify it, and then transferred to the edge or a sharp point of the paper. For drying, the paper is then placed upon a watch glass (avoid borosilicate glass) and heated upon a steam bath.

Quinalizarine Test (7.0–8.4); 200 parts H_2GeO_3 to 1 of H_3BO_3 prevent the test; chromate, nitrate, and fluoride interfere to some extent.

Treat 2 volumes of test solution with 18 volumes of 18-F H_2SO_4. Mix and cool; then add 1 volume of a solution of 10 mg quinalizarine in 10 ml water and 90 ml 18-F H_2SO_4. Allow to stand for five minutes, and then compare the blue coloration with the color of a blank test.

No. 13: Aluminum, 26.9815

Potassium-Aluminum Alum (3.0–6.7; Fe-ic and Cr-ic: 1.3). Ferric iron, chromic ion, gallium, and thallic ion also give alums; Ba, Sr, W, Hg, and Te should be absent.

Seeded Cesium Chloride. In a mortar, grind 10 mg $KAl(SO_4)_2 \cdot 12\ H_2O$ with 0.1 g CsCl. Then grind 10 mg of the obtained mixture with another portion of 0.1 g CsCl, and repeat this four more times to obtain a cesium chloride containing 0.0001% alum.

Place the test droplet upon a slide and evaporate just to dryness. Dissolve the residue in a volume of 2-F HNO_3 which will give an approximately 1% solution of aluminum. Add a reasonably large crystal of $KHSO_4$ and cover with a small watch glass (10-mm diameter) to retard evaporation. Inspect after three minutes. $KAl(SO_4)_2 \cdot 12\,H_2O$ separates as isotropic octahedra.

Alum has a strong tendency to form supersaturated solutions. Thus if crystallization did not start by itself, try to get it going by scratching with a glass needle. If this does not help, first mix the drop and then add some seeded cesium chloride. The less soluble cesium alum separates immediately if some aluminum is present. The crystals are small and less regularly shaped than those of the potassium alum. If, however, sufficient time is allowed, they grow to satisfactory size and shape (isotropic octahedra).

To confirm the presence of aluminum, remove the mother liquor with a capillary pipet, and treat the alum crystals with a large drop of 12-F NH_3. Aluminum alum becomes white and opaque; gallium alum dissolves; ferric (and indium) alums turn brown and chromic alum, distinctly green. Duval.

Test with Alizarin S, Sodium Alizarinsulfonate (5.0–8.0; Be, Sc, lanthanides, Zr, Th, Cr, uranyl, Mn, Fe, Co, Ni, Cu, Zn: 2.0). Ti, Ga, and Bi interfere and should be absent.

Transfer 1 volume of test solution to spot test paper, add an equal volume of 0.2% aqueous solution of alizarin S, and expose the wet spot to fumes of NH_3. When the spot has turned violet, immerse the paper in 2-F acetic acid. A red or brownish red fleck indicates aluminum. Performance of a blank test is advisable.

No. 31: Gallium, 69.72

Flame Test. The chloride is quite volatile and gives a violet color to the flame. The spectral lines are weak. *See* P. 38 and Table IX.

Hydrous Oxide, Sulfide, Chloride. The commonly met most stable ions, Ga^{+++} and In^{+++}, behave very much like the aluminum ion, but there are differences that make separations possible. Confirmatory tests may be made conclusive by combining them with separations.

The hydrous oxides resemble $Al(OH)_3$, but that of gallium is soluble in ammonia at room temperature. The chlorides of gallium and indium collect in the ether layer if their solution in 6-F HCl is extracted with diethyl ether. Aluminum remains in the aqueous layer and precipitates as $AlCl_3 \cdot 6\,H_2O$ when the mixture of ether and aqueous layer is saturated with gaseous HCl. Sulfides of gallium and indium can be precipitated from aqueous solution: Ga_2S_3, white, and In_2S_3, yellow (under special conditions, a white sulfur compound of In seems to form).

Potassium-Gallium Alum (3.3–6.3). Fe-ic, Cr-ic, Al, In, Tl-ic give similar crystals. Only $K_2Ga(SO_4)_2 \cdot 12\,H_2O$ dissolves in NH_3.

Proceed as directed under Aluminum, *above* (873).

Gallium Ferrocyanide (Estimate: 4–5). Aluminum and chromic ion are not precipitated. If performed in a capillary cone, the test will detect probably less than 1 ng Ga.

Take 3 volumes of solution of the gallium salt (hydrous oxide) in 4-F HCl and add 1 volume 0.3-F $K_4Fe(CN)_6$ solution. The $Ga_4[Fe(CN)_6]_3$ separates as a white or bluish white precipitate.

Induction of the Manganese Ferricyanide Precipitation (3.9–5.6). The test may be applied without fear of interference in presence of alkalies, alkaline earths, Be, Y, lanthanides, Zr, Th, Nb, W, Ni, Rh, Pd, Ir, Pt, Zn, Cd, Hg-2, Al, In, Tl-3, Sn, As, Sb, Bi, and Te. The following, however, must be absent since they give the same test: Sc, Ce-4, Ti, V, Cr, Mo, U, Fe, Co, Cu, Tl-1, and Se.

Reagent. Treat 120 ml of a 0.5% solution of $MnCl_2 \cdot 4\,H_2O$ in 7-F HCl with 30 ml 0.25-F $K_4Fe(CN)_6$ and 1 ml 0.05-F $KBrO_3$.

Procedure. Place 1 drop of the reagent upon a spot plate and add a drop of the solution to be tested. Presence of Ga induces the oxidation to ferricyanide and subsequent precipitation of reddish brown manganese ferricyanide (121, 136).

No. 49: Indium, 114.82

Flame Test. Violet coloration of Bunsen flame, *see* P. 38 and Table IX.

Indium Sulfide (4.3–5.6). The test is characteristic when performed after separation from interfering elements, i. e., all elements that give a sulfide precipitate in 6-F acetic acid. When performed in a capillary cone, the test would have a respectable identification limit of 0.5 ng In.

Dissolve the solid test material (hydrous oxide) in 6-F acetic acid and saturate with hydrogen sulfide. Allow to stand for ten minutes. The In_2S_3 separates as distinctly yellow flocks.

Test with Hexamethylenetetramine and Ammonium Thiocyanate (3.5–6.5; alkalies, alkaline earths, Cr, W, Mn, Ni, Cu, Zn, Cd, Al, As: 2.0; Sc, Fe-3, Sn, Pb: 1.0). Cobalt gives a precipitate of like appearance and must be absent.

Upon a slide, treat a droplet of the weakly acid test solution with a crystal of hexamethylenetetramine and one of NH_4CNS. If indium is present, pink crystals (hexagons, crosses, and rosettes) separate. To start the crystallization with dilute solutions, it is suggested (136) to rub with a glass needle after warming the solution somewhat. Finally, it might be advisable to confirm the presence of indium further by conversion to the yellow sulfide.

No. 81: Thallium, 204.37

Flame Test: intensely green coloration, *see* P. 38 and Table IX.

Redox Reactions. The salts of the univalent thallous ion resemble partly those of potassium (or cesium rather) and partly those of lead. Thallous ion is not oxidized by boiling with nitric acid, but *aqua regia*, chlorine, and bromine readily convert it to trivalent thallic ion. Sulfur dioxide or HI reduce the Tl^{+++} to Tl^+.

Thallous Chloride (Estimate: 2.7–5.7). The test is specific if performed as follows.

Transfer a droplet of the test solution to a slide and evaporate just to dryness. Place upon the residue a droplet of a mixture of 1 volume 16-F HNO_3 and 4 volumes 12-F HCl; again evaporate to dryness. Place a droplet of 0.1-F HCl upon the residue. After a few minutes, transfer the clear solution to the center of another slide and expose it to an atmosphere of sulfur dioxide (which may be obtained by adding 0.5 ml conc. HCl to a bottle containing about 10 g metabisulfite). Crystals of TlCl will separate as cubes (from dilute solutions), crosses, star-like clusters, and hexagons. These are radiantly white in reflected light, but nearly black in transmitted light because of their high refractive index. The crystals dissolve again if the test drop is exposed to bromine fumes.

Thallous Bromide (Estimate: 4.0–7.0; Ag, Pb: 3.0 or better). The limit of identification is improved by the addition of KI and may become of the order of 10 ng. The use of KI, however, introduces the danger of the precipitation of Cu_2I_2, HgI_2, BiI_3 which may be present in solution if they are not excluded by a separation preceding the test.

In a centrifuge cone or in a capillary, treat the test solution with bromine and HBr (or bromide and HCl). Use the centrifuge to remove the precipitate. Treat the clear solution (on a slide) with SO_2 or (in a cone) by adding granules of $NaHSO_3$ or small portions of meta-bisulfite, $Na_2S_2O_5$. The thallous bromide will separate as a white precipitate. Finally, when an excess of SO_2 is assured by the odor, add some 1-F KI solution. The white TlBr is converted to yellow TlI. Furthermore, a precipitate may appear now in the clear solution with small amounts of thallium which would not be precipitated by bromide (870).

Group IV A: Carbon-Lead Group

No. 6: Carbon, 12.01115

Elemental Carbon

Complete lack of black particles or spots excludes the presence of elemental carbon. The black color and glowing upon heating in air suggest its presence but do not prove it. To obtain proof, the object is best heated in a current of oxygen which is finally passed through a solution of barium or calcium hydroxide for the collection of the CO_2 formed, *see* under carbonate, below.

Coal, charcoal, soot, and various types of "carbon blacks" burn readily to leave more or less of light colored ash (inorganic material of various kind). Graphite, which rarely occurs in black hexagonal plates but rather forms soft (brown, gray, or black) flakes or glistening and iridescent scales, burns slow and leaves much or little ash depending upon purity. Diamond is oxidized very slowly, even at high temperature, and leaves no or very little ash. *See* also CHAMOT and MASON (118).

Of course, also organic compounds burn to give CO_2, but they are not necessarily black, and they usually give also a condensate of water when burnt in oxygen.

Carbides

Carbides which are decomposed by water or acid will have been detected during the preliminary tests by the liberation of methane or acetylene, P. 36. Refractive carbides, nitrides, and borides are indicated by their hardness, P. 17. They too may burn when strongly heated in oxygen (cubic boron nitride, borazon, much slower than diamond) and leave an ash for identification. In general, it will be best to fuse them with either NaOH or a mixture of NaOH and Na_2O_2; the flux will contain carbonate, and for testing *see* below.

Carbon Monoxide

See P. 37. — Carbon monoxide may be confirmed by first passing the gas through Ascarite for the absorption of any CO_2, burning the gas in pure oxygen, and testing for CO_2 and water in the combustion gases. Only the former shall be formed.

Concerning an extremely sensitive test for the quick detection of CO in air *see* SHEPHERD (440).

Carbonate

Carbonate is generally recognized by the liberation of CO_2 when treated with a reasonably strong acid. The CO_2 is identified by its action upon barium or calcium hydroxide, *see* P. 36.

If lead, barium, strontium, or calcium are expected, one may prefer to use 4-F HNO_3 in place of the dilute sulfuric acid for the liberation of the carbon dioxide. Potassium permanganate may be added to this acid in order to prevent the liberation of SO_2 (from sulfites, thiosulfates), H_2S (from sulfides, thionates), HN_3 (from azides), and NO_2 (from nitrites). Oxydation by permanganate will, however, give CO_2 from cyanides, formates, and oxalates; it should not be added if these are present. FEIGL (121) suggests addition of $HgCl_2$ (binds H_2S and HCN), $AgNO_3$ (binds H_2S, HN_3 and HCN), zirconium salt (binds HF), and (or) hydrogen peroxide (oxidation of SO_2 and nitrite).

The testing of the carbonate precipitate for ready solubility in dilute acid (difference from fluoride) and for absence of sulfite should not be omitted, if possible, P. 36.

Testing the pH of a carbonate solution will suffice to distinguish between carbonate and bicarbonate.

Formic Acid

Reduction to Formaldehyde and Test with Chromotropic Acid (4.7–4.7). Hexamethylenetetramine interferes with the test, and glucose gives some formic acid (121, 137).

In a small test tube, treat 1 ml of the test solution with 1 ml 2-F HCl and a granule of magnesium metal. Wait until the liberation of gas stops and then pour 0.1 ml of the reaction mixture into another test tube. Treat this portion with 0.5 ml of a solution of 0.1 g chromotropic acid (naphthalene-1,8-disulfonic acid) in a mixture of 40 ml water and 60 ml 18-F H_2SO_4. Place for ten minutes into a water bath of 60° C. If formic acid was present, the solution assumes a violet color.

Acetic Acid and Acetate

Sodium-Uranyl Acetate (2.0–4.3). Specific test which is, however, prevented by presence of much ammonium ion, free alkali hydroxide, or free mineral acids. A satisfactory test solution may be obtained by evaporating the material to dryness, if necessary after making it alkaline, distilling the residue with strong H_3PO_4, and collecting the vapors in 0.1-F KOH.

Slightly acidify the test solution with formic acid and add a drop of a 10% solution of uranyl formate in saturated aqueous sodium formate. The characteristic yellow tetrahedra of the sodium-uranyl acetate are described under Sodium, P. 44.

Conversion to Indigo. L. I. = 60 μg CH_3COOH. Copper salts interfere, and the sensitivity suffers in presence of higher fatty acids. Chromate and MnO_2 have no adverse effect.

Mix some solid sample with calcium carbonate, or mix a drop of acid solution with $CaCO_3$ and evaporate to dryness. Then transfer the solid mix into a glass tube, 5-mm i. d., 8 cm long, and seal shut at one end. Place over the opening of the tube a square of filter paper which has been moistened with a saturated solution of *o*-nitrobenzaldehyde in 2-F NaOH. Cautiously heat the lower part of the tube to drive the vapors of acetone which form toward the moist paper. The acetone reacts to give indigo which colors the paper blue or green depending upon the amount formed. If there is a doubt concerning the color, remove the paper from the tube and moisten it with 3-F HCl. This discharges the original yellow color of the paper so that the blue of the indigo may show up clearly (121).

It should be possible to perform the test with much smaller quantities and a fiber or grain of porous tile inside a capillary in place of the paper.

Oxalic Acid and Oxalate

Test with Xanthocobaltic Chloride (2.7–5.7). Fluoride, thiosulfate, and dithionate give crystals of different appearance; other anions produce "amorphous" precipitates.

The test solution should be "neutral". On the slide, treat a droplet of the test solution with a droplet of 3% aqueous solution of nitropentamminocobaltic chloride. Oxalate gives dark yellow prisms, X forms, and H forms.

Formation of Aniline Blue. L. I. = 5 μg $(COOH)_2$. Formic, acetic, propionic, glycolic, glyoxylic, tartaric, citric, succinic, dihydroxymaleic, benzoic, and phthalic acid do not give the test.

Use a solid particle, the evaporation residue of the test solution, or a precipitate obtained with $CaCl_2$ from neutral or ammoniacal solution, which may also contain sulfate, sulfite, fluoride, tartrate, etc.

Place the dry solid substance into a small test tube. Add a like volume of diphenylamine and some 85% (15-F) H_3PO_4. Heat over a microflame to obtain conversion to the triphenylmethane dye known as aniline blue. The mixture may turn blue, but the color fades on cooling. Finally, take up the cold reaction mixture in ethanol. The dye dissolves to give a brilliant blue color. FEIGL and FREHDEN (867).

Tartaric Acid and Tartrate

Potassium-Hydrogen Tartrate (Estimate: 3.0–6.0). Boric acid should not be present in significant amounts.

Reagent. Dissolve 15 g KCl in 100 ml 2-F formic acid and then add 6-F KOH until the pH of the solution is 3. Test with indicator (Hydrion) paper.

Transfer a droplet of the test solution to a slide. (If it is not approximately neutral, add first very small amounts of KOH until just alkaline and then expose to fumes of HCl until acid.) Evaporate just to dryness by placing the slide upon the steam bath. After cooling, cover the residue with a droplet of the reagent solution and cover with a 1-cm watch glass to prevent evaporation. If crystallization does not start by itself within a few minutes, try to start it by rubbing with a glass needle. KOOC · · CHOH · CHOH · COOH separates in hemihedric forms of the orthorhombic class, long hexagons, pentagons, half hexagons?, prisms or plates with heavily shaded faces. All crystals should show parallel extinction.

Remove the mother liquor with filter paper or a capillary pipet and use the crystals for the test with gallic acid.

Heating with Sulfuric Acid and Gallic Acid. L. I. = 2 μg. Formaldehyde, glyoxylic acids, glycolic, glyceric, and tartronic acids give colored products and interfere. No colors are given by fatty acids, oxalic, malic, succinic, cinnamic, citric, and salicylic acids. Since oxidants might interfere with the test, it is best to first isolate the tartaric acid as potassium-hydrogen tartrate or as calcium tartrate.

Dissolve some of the solid sample, of a tartrate precipitate, or of the evaporation residue from a test solution in 1 droplet to 0.5 ml of a solution of 10 mg gallic acid, $C_6H_2(OH)_3COOH$, in 100 ml 18-F H_2SO_4. Heat upon the slide or in a small test tube to 120 to 150° C. Depending upon the amount of tartrate present, a yellowish green to blue color will develop. EEGRIWE (1156).

Citric Acid and Citrates

Calcium Citrate (Estimate: 4.0–4.0). All anions may interfere, that give insoluble calcium salts.

To 1 ml test solution add 3 drops of 1% (0.1-F) $CaCl_2$. The solution will stay clear if it is acid or neutral. Add 1 to 3 drops 2-F NaOH. A flocculent precipitate of tertiary calcium citrate may form. This should dissolve when solid NH_4Cl, in small portions, is dissolved in the reaction mixture. When the clear solution is boiled, a crystalline calcium citrate should precipitate (56). It is obvious that the test may be carried out on a smaller scale.

Conversion to Pentabromacetone (4.0–4.0). Acetone and all substances that furnish acetone may give the test.

Use a solution of the free acid or of any citrate dissolved in 1-F sulfuric or nitric (but not hydrochloric) acid. To 1 ml of the solution add 2 to 3 drops of 0.02-F $KMnO_4$ and place just the part of the tube filled with solution into a water bath of not more than 40° C. The citric acid, $HOOC \cdot C(OH) : (CH_2 \cdot COOH)_2$, is oxidized to acetonedicarboxylic acid, $CO : (CH_2 \cdot COOH)_2$, which gives pentabromacetone, $CHBr_2 \cdot CO \cdot CBr_3$, more readily than the acetone itself, that would be obtained by longer heating or higher temperature. Formation of the acetone would be undesirable also because of its volatility. Consequently remove the test tube from the bath as soon as the reaction mixture shows a turbidity ($MnO_2 \cdot x\, H_2O$) or turns brown. Add 1 or 2 drops of ammonium oxalate solution and 0.5 ml 1-F H_2SO_4, which will render the mixture clear and colorless. Now add a few drops of bromine water. If citric acid was present, white, crystalline pentabromacetone will precipitate. STAHRE (1136).

Hydrocyanic Acid and Cyanide

Silver Cyanide (4.2–7.2). The test, which has been described already in P. 36, is highly selective. Hydrazoic acid and acetylene interfere, but these will often be ruled out by the nature of the material under investigation. To avoid the slow liberation of HCN by the action of strong acids upon ferrocyanide, ferricyanide, or other stable cyanide complexes, one may treat the test material with $NaHCO_3$ or KH_2PO_4 solution instead of using a free acid. Much HCN is produced by the hydrolysis of alkali cyanides so that a strong test is obtained if the droplet of silver nitrate solution is exposed to the air from the bottle containing solid alkali cyanide.

According to BRUNSWIK (1086) proceed as follows. Liberate HCN on the floor of the gas reaction cell and use a drop of 0.5 μl 1% $AgNO_3$ solution on the cover glass. The separation of a white precipitate is usually seen with the unaided eye. When this happens, exchange the cover slip for another one with a fresh drop of $AgNO_3$ solution to collect more of the AgCN. Inspect the first precipitate. As a rule, well developed crystals will not be found. Allow the test drop to evaporate. Place on the residue a cover slip with a drop of 8-F HNO_3 hanging on the underside, which is just large enough to barely fill the whole space between slide and cover slip. Cautiously heat over the microflame until bubbles of steam become visible between slide and cover slip. Immediately stop the heating and place the slide upon a cork for slow cooling. Using a magnification of 80 to 100 diameters, and not too strong illumination, search the whole preparation systematically for the fine needles of AgCN. Be certain to try the test first with controls.

Mercuric cyanide is not decomposed by adding dilute acid. Thus, in presence of mercury, the test solution must be subjected to a preliminary treatment. Add dilute NaOH until just alkaline, then add Na_2S solution in small increments until the precipitation of black sulfides is complete. Add $CdSO_4$ until all sulfide ion is precipitated, centrifuge, and use the clear solution for the test.

Conversion to Thiocyanate (6.0–7.3). The test is also given by mercuric cyanide, and it may be performed with a precipitate of AgCN obtained in the preceding test. The test is specific for cyanide, but it is assumed that thiocyanate is not present in the test solution.

Treat the neutral or alkaline sample solution (solid particle or evaporation residue) with a little yellow ammonium sulfide and evaporate to dryness upon the steam bath. Extract the residue with a small volume of cold 2-F HCl, remove the clear solution from any insoluble residue, and treat it with a very small amount of 1-F $FeCl_3$. The thiocyanate formed by the reaction $CN^- + S_2^= \rightarrow CNS^- + S^=$ gives the familar blood red coloration of the ferric thiocyanate complex.

Cyanic Acid and Cyanate

HCNO is an unstable liquid of penetrating, disagreable odor, which immediately decomposes in aqueous solution:

$$HCNO + 2 H_2O \rightarrow HCO_2NH_2 + H_2O \rightarrow NH_4HCO_3$$

The salts are stable, but when their solution in water is acidified, the acid decomposes as indicated, and CO_2 is liberated by the decomposition of the ammonium bicarbonate.

There are no good "tests" for cyanate ion, and its presence is found, more or less by inference from the facts that $BaCl_2$ gives no precipitate with an aqueous solution of cyanate and that $AgNO_3$ gives a white, curdy precipitate which dissolves readily in dilute HNO_3 with the liberation of CO_2, whereafter a test for NH_4 may be obtained with the acidified solution.

Thiocyanic Acid and Thiocyanates

Thiocyanic acid, HCNS, is a colorless, unstable liquid of penetrating odor. In aqueous solution, HCNS is reasonably stable and ionized like a strong acid.

Ferric Thiocyanate (5.0–6.3). Azide must be absent since it gives the same coloration. Iodide interferes because of the liberation of I_2 by the ferric ion. Sulfide and thiosulfate give a precipitate of sulfur and require a larger amount of reagent than usually provided. The same holds for ferrocyanide, ferricyanide, and the anions that give ferric complexes: phosphate, fluoride, oxalate, and tartrate. The interference of the Prussian blue precipitate may be overcome by centrifuging.

In a microcone, treat the test solution with an equal volume of 6-F HCl and then with 1% (0.1-F) $FeCl_3$. The red color may be observed in the coloriscopic capillary.

Ferrocyanide

The anhydrous acid is a white crystalline solid, stable in dry air. The aqueous solution is slightly acidic and decomposes with the separation of Prussian blue.

Prussian Blue (5.0–6.3; thiocyanate: 3.5; iodide: 3.3).

Use filterpaper impregnated with 1% (0.1-F) $FeCl_3$ solution and dried. Place a drop of the acid (HCl) test solution upon the filter paper. A blue circle or ring indicates ferrocyanide; farther out, there may be a red concentric ring indicating thiocyanate.

If only a deep red fleck of thiocyanate or a deep brown fleck of I_2 is seen, add a drop of saturated thiosulfate solution. The Prussian blue becomes distinctly visible against the background of the white paper (121).

Ferricyanide

The anhydrous acid is a brown crystalline solid; the aqueous solution is strongly acid.

In the Prussian blue test (above), ferricyanide gives only a white or bluish white precipitate; a blue precipitate is obtained with $FeSO_4$ solution, so-called "Turnbull's blue" which may be identical with the Prussian blue.

A number of tests based upon the oxidizing action of ferricyanide are listed by FEIGL (121).

Test with LiCl and Hexamethylenetetramine (4.0–7.0). Specific for ferricyanide, but Ca and Mg should be absent since they give precipitates with hexamethylenetetramine in presence of ferricyanide and ferrocyanide.

Transfer a droplet of the neutral test solution to the slide and place next to it a droplet of a solution of 1.5 g hexamethylenetetramine in a 9% aqueous solution of lithium chloride (1 g LiCl + 10 ml water). Connect the drops by drawing a channel with a glass needle. If ferricyanide is present, yellow octahedra and 4- and 6-pointed stars form where the two solutions merge. Ferrocyanide gives colorless crystals of different shape.

No. 14: Silicon, 28.086

Bead Test. Many silicates leave an insoluble "skeleton" floating in the phosphate bead, but various other insoluble oxides behave in the same manner. *See* P. 39.

Sodium Fluosilicate (4.7–7.7). The test is specific and is given by elemental silicon and apparently also by all of its compounds. Borate prevents the test if present as more than a minor constituent of the sample and must be removed which may be done by adding 18-F H_2SO_4 and repeatedly evaporating with methanol. If one is afraid of a possible interference by germanium, this latter could be eliminated by first evaporating with 9-F HBr and some bromine, P. 68.

It should be considered that not all forms of SiO_2 are attacked by HF with equal ease. Quartz and vitreous silica react slowly, and it is necessary to grind them (in a steel mortar) to a fine powder as this is done with silicates. Materials that may contain sulfide or carbonate should be roasted first in air and then ignited for the removal of CO_2. If heavy metals are present, this should be done on a sheet of nickel (nickel crucible); if not, one will do it in the platinum crucible in which the test is to be performed.

Obviously, apparatus of glass, silica, or any kind of silicate must be avoided. The slide should consist of clear plastic; in an emergency, a square of clear cellophane, cemented upon a glass slide, will do. Plastic apparatus has been described by HAHN (624).

Use a platinum crucible of 1-ml capacity and provided with a platinum lid. If a solution is to be tested, evaporate it in the crucible. A solid sample, of not more than 1-mg weight, is simply placed upon the bottom of the crucible. Apply a small drop (1 μl) of water to the underside of the crucible cover. Then treat the sample in the crucible with 1 to 3 mg NH_4F (that should come from a plastic container) and 2 drops 18-F H_2SO_4. Place the cover upon the crucible without delay. For cooling, put a large drop of water on top of the crucible cover. Set the crucible on the center of a large watch glass and heat it in this manner on the steam bath. After two or three minutes, lift off the cover and touch its underside to the surface of the plastic slide. This will transfer the drop of condensate to the slide. Place a fresh drop of water on the underside of the cover and return the latter upon the crucible for the collection of more distillate. A series of distillates may be collected, and several of them should be tested if silicon is found to be absent in the first distillate.

Test the drop on the plastic slide by adding, with a platinum wire or a plastic tool, a kernel of sodium chloride. If silicon tetrafluoride has been absorbed by the drop of water, $3 SiF_4 + 3 H_2O \rightleftharpoons 2 H_2SiF_6 + H_2SiO_3$, the characteristic crystals of Na_2SiF_6 separate. They form hexagonal plates, short hexagonal prisms, or six-pointed stars that have a light pink color because of the Christiansen effect. Avoid strong illumination; otherwise the crystals are difficult to see since their refractive index is quite close to that of the solution.

If crystals of sodium fluosilicate are obtained, transfer the next drop of distillate to an ash-free filter paper. Add a drop of molybdate reagent (5 g ammonium molybdate dissolved in 100 ml cold water and poured into 35 ml 6-F HNO_3) and warm the moist spot gently (jet of steam or infrared lamp). Add a drop of benzidine solution (50 mg benzidine or its hydrochloride is dissolved in 10 ml glacial acetic acid and then diluted with water to 100 ml) and then expose to ammonia fumes. A blue spot (reduction of the silicomolybdate to molybdenum blue and oxidation of the benzidine to benzidine blue) indicates the presence of silica. The blue spot should be compared with a blank test.

No. 32: Germanium, 72.59

The silvery white, brittle metal is little affected by acids but dissolved by 3% H_2O_2. It is resistant against NaOH solution but reacts violently with the fused hydroxide to form germanate. The compounds of tetravalent germanium are more stable than those of the divalent. The white GeO_2 is amphoteric; dissolution in NaOH solution gives germanates; treatment with strong HF, HCl, HBr gives conversion to the volatile tetrahalides which, with the exception of GeF_4, are rapidly hydrolized by water. GeF_4 gives H_2GeF_6, the sodium salt of which is identical in appearance with Na_2SiF_6. To complete the similarity, germanate gives molybdate complexes in analogy to silicic acid, and the very weak germanic acid forms a strongly acidic complex with glycerol, glucose, and mannite similarly to boric acid. Obviously, the identification of germanium requires separation from boron (distillation of methyl borate) and from silicon (and also titanium and tin by distillation of $GeCl_4$ or $GeBr_4$). Germanium appears in the arsenic group of the hydrogen-sulfide scheme of separation and may also be isolated as the white GeS_2.

Rubidium Hexafluogermanate (123). The test may be used after separation from Si, Ti, Zr.

With a platinum loop transfer some solution of the test substance in HF to a plastic slide. With a platinum wire, put into the edge of the drop a grain of RbCl. The colorless hexagonal bipyramids are characteristic; also hexagonal plates form.

Mannite Complex (5.0–6.3; molybdate, arsenite, Sn, Sb-3, Te may be present). Borate gives the same test.

On a spot plate, mix a drop of the weakly acid test solution with a drop of 1% phenolphthalein in ethanol and add NaOH until the mixture turns light pink. Mix carefully. If the pink color persists, add a few small crystals of mannite (hexanhexol) and mix again. If germanate is present, the mixture should turn colorless. A blank is desirable, and a tiny square of pH test paper (Hydrion paper) might be briefly soaked in the test drop before adding the mannite and again after the addition.

Benzidine Blue (5.6–6.9). Sb-3, tellurate and tellurite reduce the sensitivity somewhat. Reducing substances, Si, P, and As must be absent.

Concerning the reagents *see above* the last paragraph of the section Silicon.

On a spot plate, treat a drop of the test solution with a drop of molybdate reagent, a drop of benzidine reagent, and a few drops of saturated aqueous solution of sodium acetate for the adjustment of the pH. A blue coloration (benzidine blue and molybdenum blue) indicates germanium.

No. 50: Tin, 118.69

Flame Test (6.2–7.5) (121). Niobium and arsenic are the only two elements that interfere; gold gives a green luminescence when present in high concentration (136).

In a porcelain crucible, treat 5 drops of sample solution with 5 ml 12-F HCl and a small piece of metallic zinc. Dip into this mixture the closed end of a test tube, filled with cold water, and then hold it into the reducing zone of a non-luminous Bunsen flame. Presence of tin is indicated by a mantle of blue flame around the test tube (1151, 1152).

FEIGL and KAPULITZAS (121) obtain the limit of identification quoted above by taking a drop of test solution on a magnesia stick and evaporating it at a low temperature (holding it near a Bunsen flame; an infrared lamp might be used or the radiation from some hot object). A droplet of 12-F HCl is applied to the residue which is then held into the reducing (inner?) portion of a (non-luminous?) microflame. A mantle of blue flame forms around the magnesia stick. If there is more than 0.25 μg of tin, the moistening with HCl and heating may be repeated several times and continues to give the test.

Rubidium Chlorostannate (4.5–8.0; Sb: 2.0).

Proceed as directed in Expt. 41.

Reduction of Phosphomolybdic Acid (3.0–6.0; Cu: –0.6).

Proceed as directed in Expt. 30.

No. 82: Lead, 207.19

Flame Test: *see* P. 38 and Table IX.

Potassium-Copper-Lead Nitrite (4.5–8.5; Cu: 2.5; Hg-ic: 2.0). "Large" quantities of Bi or Cd hinder the test.

Proceed as directed in Expt. 34. Addition of the cupric ion as cupric sulfate causes precipitation of $PbSO_4$ which then dissolves slowly in the reagent to give reasonably large crystals.

Lead Iodide Slide Test (3.0–6.7). All ions and combinations of such that give insoluble iodides will interfere. The test is useful after the isolation of the lead.

Proceed as directed in Expt. 33.

Lead Iodide Spot Test (3.7–6.7; Bi: –1.7). *See* also the preceding test. Perform the test as directed in Expt. 24.

P. 58 Group V A: Nitrogen-Bismuth Group
No. 7: Nitrogen, 14.0067

Ammonia and Ammonium Ion

Ammonia readily transfers from alkaline solutions via the gas phase to acid solutions, and this may be used to separate the ammonia of the ammonium ion from all cations except those that derive from volatile (organic) bases. Consequently, it is customary to combine tests for ammonium ion with the "distillation" of ammonia. As a rule, the reagent could be directly applied to the test solution if it were known that interfering substances are absent.

Ammonium Chloroplatinate (3.7–7.0 if performed in the gas reaction cell). Only volatile bases can interfere.

If much material is available for testing, treat some in a test tube with a small volume of 6-F NaOH (and some Na_2S if mercury is present) and, if it seems desirable, warm the mixture. It may be possible to perceive the characteristic odor of ammonia. Heavy white fumes are obtained on inserting into the gas space of the test tube a loop with 12-F HCl. A loop with a solution of chloroplatinic acid will contain a precipitate of chloroplatinate that may be transferred to a slide and inspected under the microscope. Other reagents may be brought into the gas space by means of the loop, and test papers may be introduced suspended from a hook.

For the detection of small amounts of ammonia follow the directions of Expt. 46. As an alternative, the NH_3 may be absorbed in a droplet of 3-F HCl (or other desirable acid), whereafter portions of this "distillate" may be tested by adding Na_2PtCl_6 or any other reagent or reagent paper desired.

Test with Nessler's Reagent (5.7–9.7). The alkali metals do not interfere. Sulfate, sulfide, and metal ions including Hg interfere; amine bases give the test.

Reagent. Dissolve 5 g HgI_2 and 3.65 g KI in 100 ml water. The solution is stable. Before use, mix some of it with an equal volume 3-F NaOH. This alkaline solution does not keep.

With a capillary pipet, transfer 0.1 μl of the alkaline reagent to a small square of filter paper and expose the latter to the vapors given off by the test solution (which has been treated with NaOH) in a gas reaction cell or equivalent closed apparatus. An orange fleck indicates the presence of NH_3.

The reagent may be added directly to the neutral or slightly acid solution of an ammonium salt; depending upon the amount of the latter, an orange precipitate or coloration will be obtained: $IHg-O-Hg-NH_2$.

Hydrazine

The free base, H_2N-NH_2, is a colorless liquid at room temperature (m. pt. = 1.4° C, and b. pt. 113.5° C). It is soluble in water (pK = 5.5) and gives salts with one and with two equivalents of acid. Hydrazine is a strong reducing agent which precipitates metallic gold, silver, and mercury from the aqueous solutions of their salts. Cupric ion is reduced to the cuprous state in acid solution, and to metallic copper in alkaline solution. The oxidation products of hydrazine are ammonia, hydrazoic acid, and nitrogen depending upon the oxidant and the conditions of the reaction; permanganate gives mostly ammonia; and hydrogen peroxide, mostly hydrazoic acid.

Salicylaldazine (5.7–7.0). Ammonia, ammonium salts, urea, thiourea, azide, nitrate, and nitrite do not interfere.

Reagent. Boil 1 g salicylaldehyde with a solution of 1 ml glacial acetic acid in 60 ml water until the oil disappears. Cool and filter for the elimination of undissolved aldehyde. The solution is stable, but must be filtered from time to time for the removal of separating aldehyde.

In a small test tube, treat a drop of the slightly acid test solution with a drop of the reagent. A white turbidity or a precipitate, depending upon the amount of hydrazine, appear after a short time.

Hydroxylamine

The free base, NH_2OH, forms colorless orthorhombic crystals that melt at 34° C. The freshly prepared aqueous solution (pK = 8.0) is odorless, but it gradually decomposes to give water, N_2, and NH_3. Salts are formed with one equivalent of acid. Hydroxylamine is a strong reducing agent and precipitates the metal from solutions of gold, silver, and mercury. Alkaline copper solutions, however, are reduced only to cuprous oxide. The hydroxylamine is oxidized to N_2, nitrogen oxides, or HNO_2, depending upon the conditions. Especially in alkaline solutions, hydroxylamine may also act as oxidant and become reduced to NH_3.

$$2 Fe(OH)_2 + NH_2OH + H_2O \rightarrow 2 Fe(OH)_3 + NH_3.$$

The reaction may be used to detect hydroxylamine in the presence of hydrazine.

Oxidation to Nitrous Acid and Griess Test (6.7–8.0). Hydrazine, azide, and urea do not interfere. Nitrite and oxime must be absent.

Reagents. (a) Dissolve 1.3 g I_2 in 100 ml glacial acetic acid; (b) dissolve 1 g sulfanilic acid in 75 ml water and 25 ml glacial acetic acid; (c) dissolve 0.3 g alpha-naphthylamine in 70 ml water and 30 ml glacial acetic acid.

On a spot plate, mix a drop of the test solution with a few milligrams sodium acetate, 1 to 2 drops of the solution of sulfanilic acid, and iodine (solution in glacial acetic acid) until the mixture remains brown. Allow the mixture to stand for three minutes, and then add 0.1-F $Na_2S_2O_3$ to the disappearance of the iodine color. Finally add 1 drop of the naphthylamine solution. The red color of the azo dye indicates the presence of NH_2OH. A blank test is desirable.

Hydrazoic Acid and Azides

The anhydrous acid, HN_3, is at room temperature a colorless, mobile liquid of penetrating odor (m. pt. = $-80°$; b. pt. = $37°$ C). The aqueous solution (pK = 4.6) is fairly stable, but slowly decomposes to nitrogen and an ammonium salt when boiled with a mineral acid. $KMnO_4$, HNO_2, or I_2 acting on HN_3 give H_2O and N_2. The salts of HN_3, with the exception of those of the alkalies and alkaline earths, are explosive.

Silver Azide (4.0–7.0). Interference by sulfide, sulfite, thiosulfate may be avoided by first oxidizing the neutral or alkaline test solution with H_2O_2 (121).

Proceed as outlined in P. 36. The use of a small gas reaction cell for the distillation will improve the limit of identification. As an alternative, the acid may be collected in a small droplet of 0.1-F NaOH which may be finally neutralized by adding dilute HNO_3, exposure to NH_3 fumes, and evaporation on the steam bath. Portions of the aqueous solution of the residue may then be used for tests. *See* next paragraph.

Ferric Azide and Cupric Azide. On a spot plate, treat a droplet of neutral azide solution with a droplet of 0.1-F $FeCl_3$; the mixture is red. The coloriscopic capillary may be used for observation.

Treat a droplet of neutral test solution with a droplet of 1% (1-F) $CuSO_4$ solution; a reddish brown precipitate of CuN_6 separates. The test may be performed in a capillary.

Be certain to **discard** all azide precipitates **promptly by rinsing** them **down the drain.** They become very dangerous if allowed to dry.

Nitrous Acid and Nitrites

Nitrous acid has not been isolated in the pure state. The aqueous solution (pK = 3.4) gradually decomposes to HNO_3, NO, and water. All normal salts are soluble in water.

Nitrous acid is an oxidant in acid solution toward HI, H_2S, SO_2, NH_4^+, and urea, but it reduces permanganate, chromate, chlorate, bromate, and iodate, etc. Metallic zinc and acid or aluminum and NaOH solution reduce to ammonia (ammonium salt), but this also happens with nitric acid. The reaction with ammonium ion (or urea) may be used to eliminate nitrite from solutions before testing for nitrate: $NO_2^- + NH_4^+ \rightarrow N_2 + 2 H_2O$.

Diazotation Test (6.7–8.7; nitrate: 2.0). The test is specific.

Reagents. See above under Hydroxylamine.

On a spot plate, treat a drop of the test solution with one drop each of the solutions of (first) sulfanilic acid and (then) alpha-naphthylamine. The red color of the azo dye is evidence for the action of nitrous acid.

Indole Test (6.0–8.0; nitrate: 2.0). The test is specific.

In a small test tube, treat 1 drop of test solution with 10 drops of a 0.015% solution of indole in 95% ethanol and 5 drops 7.5-F H_2SO_4. A reddish violet coloration is given by the nitroso indol formed.

Nitric Acid and Nitrate

Most tests for nitrate are also given by nitrite. Thus, reduction to ammonium ion by metallic zinc and dilute H_2SO_4 and subsequent test for ammonium are useful to prove the absence of both nitrate and nitrite (and ammonium), but does not show which of the two is present.

The test with diphenylamine is given by nitrite as well as by nitrate and by many other oxidants.

Nitric Oxide–Ferrous Complex. L. I. = 1 μg or less. Nitrite interferes, but gives a warning of its presence.

In a test tube, treat 1 ml of the test solution with 5 drops 6-F H_2SO_4 and an equal volume of freshly prepared saturated aqueous $FeSO_4$ solution. Mix and cool by running tap water over the outside of the tube. If the solution turns violet to brown, nitrite is present. If the solution remains pale green or colorless, add 1 or 2 ml 18-F H_2SO_4 so that it collects unmixed below the aqueous solution. This is easily accomplished by holding the tube at an angle of 60 degrees to the horizontal and allowing the acid to flow slowly down the side of the tube. A violet or brown zone at the interface of the two liquids is derived from the reaction, $3 Fe^{++} + NO_3^- + 4 H^+ \rightarrow 3 Fe^{+++} + 2 H_2O + NO$, and the dissolution of the NO in the excess of $FeSO_4$ solution with the formation of the labile compound $FeSO_4 \cdot x$ NO.

It should be possible to perform the test in a capillary with very small volumes of solution.

Janowski's Test (4.0–6.0). Nitration gives m-dinitrobenzene which, in turn, furnishes a violet compound with acetone and NaOH. The test is specific for nitrate, but the question of absence or presence of nitrate in a large amount of nitrite will remain academic until means are found to prevent the formation of some nitric acid during the performance of tests (56). Chloride does not interfere.

On the steam bath, evaporate to dryness 1 ml of the neutral or slightly alkaline test solution. Dissolve the residue in 0.3 ml of a solution of 1 ml nitrobenzene in 10 ml 18-F H_2SO_4. Transfer the solution to a test tube and then heat it for three minutes in the steam bath. Slowly add 5 ml purest acetone and mix thoroughly. Finally add 3 ml 40% (14-F) NaOH without mixing. If nitrate was present, a violet coloration will appear at the interface of acetone and aqueous solutions, become slowly stronger, and spread through the acetone layer.

No. 15: Phosphorus, 30.9738

Of the four allotropic modifications, the analyst may, as a rule drop from consideration the very poisonous white phosphorus (m. pt. = 44.1° C) since it is very unstable in contact with air. Sensitive tests are available for the use of the toxicologist (40, 55, 56, 186).

The non-poisonous red and black varieties melt above 300° C, but start burning before that temperature is reached. All forms are oxidized to phosphoric acid by nitric acid or *aqua regia*.

Phosphine and Phosphides

Phosphine forms when phosphides are treated with water or dilute acid or when any of the elemental forms of phosphorus reacts with hydroxide.

$$P_4 + 3\ NaOH + 3\ H_2O \rightarrow PH_3 + 3\ NaH_2PO_2.$$

It also forms when hypophosphorous acid or phosphorous acid is reduced with metallic zinc and dilute sulfuric acid.

On the identification *see* P. 36.

Hypophosphorous Acid and Hypophosphites

H_3PO_2 (m. pt. = 26.5° C) is usually met as a colorless syrupy liquid. It is a reasonably strong acid, but only one hydrogen ionizes, and only one series of salts is known.

On ignition, the acid as well as its salts give phosphate and phosphine.

$$2\ H_3PO_2 \rightarrow H_3PO_4 + PH_3,$$
$$2\ Ca(H_2PO_2)_2 \rightarrow Ca_2P_2O_7 + H_2O + PH_3.$$

There is no reaction with dilute H_2SO_4. Concentrated H_2SO_4 is reduced to SO_2 on warming.

H_3PO_2 and the hypophosphites are powerful reducing agents. $AgNO_3$ in approximately neutral solution precipitates white AgH_2PO_2 which is reduced to metallic silver even at room temperature; at the same time, H_2 may be liberated. In acid solution, $CuSO_4$ is reduced to metallic copper, which may be used for the separation of copper from cadmium; at 50° C a dark red precipitate of copper hydride may form first, and then decompose with the liberation of H_2 and $Cu°$ when heated to 100° C.

Permanganate Test (5.2–5.2); H_3PO_3 gives the same test.

In a test tube treat 1 ml of the neutral test solution with 2 drops 10% (2.5-F) NaOH and 1 drop 0.02-F $KMnO_4$. With hypophosphite or phosphite present, the solution assumes a greenish coloration within two to three minutes. Compare with a blank (3, 1203).

Phosphorous Acid and Phosphites

(Ortho) phosphorous acid, H_3PO_3, forms deliquescent needles that melt at 73.6° C and are readily soluble in water. The acid is quite strong, but only two hydrogens seem to ionize so that only two series of salts are known. The alkali salts, NaH_2PO_3 and Na_2HPO_3, are soluble in water; most other metals give water insoluble salts. H_3PO_3 is a strong reductant; in contact with air, it slowly changes to H_3PO_4. The ions of gold, silver, mercury, and copper are reduced to the metals. $HgCl_2$ is slowly reduced at room temperature to Hg_2Cl_2; if there is enough phosphite and the solution is heated, the reduction continues to the metal. Neutral test solutions give a white precipitate of Ag_2HPO_3 which, in strong solutions, changes at room temperature to $Ag°$ and H_3PO_4; in dilute solutions this reduction takes place only on heating.

On ignition, H_3PO_3 and phosphites undergo autoreduction-oxidation similar to hypophosphite.

$$4\ H_3PO_3 \rightarrow 3\ H_3PO_4 + PH_3,$$
$$8\ Na_2HPO_3 \rightarrow 4\ Na_3PO_4 + Na_4P_2O_7 + H_2O + PH_3.$$

There is no reaction with dilute H_2SO_4; concentrated H_2SO_4 is reduced to SO_2 on heating.

Alkaline Earth Phosphite. Barium, strontium, and calcium salt solutions precipitate white, water insoluble $BaHPO_3$, $SrHPO_3$, and $CaHPO_3$, respectively, all of which are soluble in acid. The corresponding hypophosphites are soluble.

Hypophosphoric Acid and Hypophosphates

The free acid, $H_4P_2O_6$, forms colorless needles that melt at 55° C. The melt decomposes at 70° to give HPO_2 and H_3PO_4. All four hydrogens may be replaced by metal, and as one would expect, the solution of $Na_4P_2O_6$ is distinctly alkaline due to hydrolysis.

$H_4P_2O_6$ is not reduced by zinc and dilute sulfuric acid and it is also decidedly more stable toward oxidants than H_3PO_2 and H_3PO_3. $AgNO_3$ gives a white precipitate that does not become dark. The aqueous solution is not oxidized by dilute chromic acid and only slowly by $KMnO_4$ at room temperature.

Thorium Hypophosphate. Acidify the test solution with HCl and then add an equal volume 6-F HCl and a few drops of a thorium salt solution. Hypophosphate gives a white precipitate of ThP_2O_6.

Phosphoric Acid and Phosphates

All phosphoric acids and acid solutions of phosphates revert on boiling of the aqueous solution to orthophosphoric acid, H_3PO_4, the colorless crystals of which melt at 42.4° C. At 213° C, the acid loses water and gives pyrophosphoric acid, $H_4P_2O_7$, m. pt. = 61° C, which is converted to metaphosphoric acid, HPO_3, at red heat. The latter vaporizes on further heating. In fact, there are three series of isopolyacids: the chain phosphates, $H_{n+2}P_nO_{3n+1}$; the ring or meta phosphates, $H_m(PO_3)_m$; and the extremely unstable, highly branched ultraphosphates. For n equal to 1, 2, 3, 4, and ∞, the chain polyphosphate formula gives the orthophosphoric acid, H_3PO_4; the pyrophosphoric acid, $H_4P_2O_7$; the tri(poly)phosphate, $H_5P_3O_{10}$; the tetraphosphate, $H_6P_4O_{13}$; and $(HPO_3)_n$ which up to this time was called metaphosphoric acid.

All of these isopolyacids are converted to H_3PO_4 by evaporation to just dryness (if aqua regia is present or an equivalent oxidant, all compounds of phosphorus will be converted to H_3PO_4), dissolving the residue in 1-F HNO_3, and boiling gently for ten to twenty minutes.

Just a few hints shall be given concerning the identification of particular acids. Only orthophosphates give a yellow precipitate when treated with acid molybdate reagent at 4° C; neither poly (chain) phosphates nor meta (ring) phosphates do this. At pH 3 to 4, $AgNO_3$ gives a yellow precipitate with orthophosphate, a white precipitate with polyphosphates (chain phosphates), and no precipitate with metaphosphates (ring phosphates) if the solution is not too concentrated. Controls are indicated, and this holds also for the albumin coagulation test, for which the reagent is a saturated solution of fresh egg white in distilled water. A very small amount of the phosphate to be tested is added to 3 ml 0.2-F acetic acid, and this solution is mixed with 2 ml of the albumin solution. A turbidity or flocculation is obtained only with chain phosphates of high molecular weight [including, of course, $(HPO_3)_n$], but not with H_3PO_4, $H_4P_2O_7$, $H_5P_3O_{10}$, or ring metaphosphates (3).

The following tests are given by H_3PO_4 and orthophosphates (and also by all isopolyacids that have been converted to H_3PO_4 by boiling of the aqueous solution).

Heating of the Solid Salts. Assuming that the cation is stable at high temperature, tertiary phosphates remain unchanged on heating, secondary phosphates form pyrophosphates, and primary phosphates become glassy long-chain polyphosphates.

Ammonium Phosphomolybdate (4.0–7.3). Arsenate and possibly silicate and vanadate give the same test. Reducing agents must be absent, but the test may be carried out in presence of all common cations and anions.

Molybdate Reagent. Dissolve 5 g ammonium molybdate in 100 ml cold water and pour the solution into 35 ml 6-F HNO_3.

Transfer one drop (50 μl) of the test solution to a watchglass and evaporate just to dryness on the steam bath. Moisten with 16-F HNO_3 and again evaporate to dryness for the removal of halide. Repeat this once. Dissolve the residue in one drop of 6-F HNO_3 and transfer the solution to a centrifuge cone. Add 2 drops of molybdate reagent and place the tube for ten minutes into water of about 65° C. If there is a yellow precipitate, transfer some of it by means of a capillary to a slide (a cover slip may be applied) and inspect under the microscope. $(NH_4)_3PO_4 \cdot 12\ MoO_3 \cdot 2\ H_2O$ forms highly refractive isotropic disks, octahedra, and occasionally cubes and pentagon dodecahedra.

Wash the precipitate in the centrifuge cone with some 2-F HNO_3. Dissolve the washed precipitate in a volume of 6-F NH_3 to obtain an about 1% solution and use this solution for the following test with magnesium acetate.

Magnesium-Ammonium Phosphate Hexahydrate (4.0–7.7). Arsenate gives the same test. Cations that may be precipitated by ammonia should be absent. Polyphosphates give "amorphous" precipitates. Phosphite gives tiny six-pointed stars and rosettes, dendrites reminiscent of butterflies, and elongated plates with parallel sides and irregularly acute ends.

Take a volume of test solution representing a few micrograms of phosphate and transfer it to a slide. Evaporate just to dryness, and dissolve the residue in 1 μl 2-F HNO_3. Expose the droplet to the fumes of concentrated NH_3 to render it strongly ammoniacal. Finally add a crystal of magnesium acetate (about 0.1-μl volume or 0.6-mm diameter) to the edge of the droplet. The crystals of $MgNH_4PO_4 \cdot 6\ H_2O$ form clusters of feathery dendrites if the precipitation occurs instantaneously. Slow growth gives prismatic crystals. Characteristic X-shapes may nearly always be found.

The conversion to silver phosphate as described in Expt. 40 permits differentiation between phosphate and arsenate.

Benzidine Blue (6.0–7.3; arsenate: 3.0). Silicic and germanic acids behave similarly. The interference of arsenate is suppressed by the procedure of the test. H_2O_2, fluoride, and oxalic acid interfere because of reactions with the molybdate reagent (the first gives permolybdates, the latter complexes).

Reagents. a) Ammonium molybdate *see* above under Ammonium Phosphomolybdate. b) Dissolve 50 mg benzidine in 10 ml glacial acetic acid and dilute with water to 100 ml.

Place a drop of the molybdate reagent upon filter paper (best No. 589 of Schleicher and Schuell) and place into a drying oven. When the spot has dried, put into its center a drop of the test solution and add 1 drop of benzidine solution and 1 drop of saturated aqueous sodium acetate. A blue fleck or ring will form depending upon the amount of phosphate.

Arsenic acid does not interfere since it gives the heteropolycomplex only very slowly in the cold and the reaction with benzidine leading to the mixture of molybdenum blue and benzidine blue is given by the complex only.

No. 33: Arsenic, 74.9216

Flame Test. *See* P. 38 and Table IX. If the substance is heated in the upper reducing zone (d, Fig. 10), all arsenic compounds are reduced to elemental arsenic which vaporizes (sublimes above 600° C). The vapor burns in the oxidizing zone with a bluish white flame to As_2O_3, whereby a garlic-like odor is given off. A cold glazed porcelain surface inserted close above the heated sample collects a brownish black mirror of elemental arsenic; if it is inserted some distance above the heated object, it collects white As_2O_3. The As mirror dissolves readily in a drop of sodium hypochlorite solution, $As_4 + 10\,NaClO + 6\,H_2O \rightarrow 4\,H_3AsO_4 + 10\,NaCl$. The As_2O_3 deposit is changed to yellow Ag_3AsO_3 if moistened with $AgNO_3$ solution and exposed to fumes of NH_3 (excess of which dissolves the yellow precipitate).

The same phenomena of reduction and oxidation take place when arsenic compounds are heated on charcoal, Expt. 17, and also the garlic-like odor will be perceived.

Liberation and Detection of Arsine (4.4–2.7 to 7.4–5.7 depending upon the reducing agent). The action of acid or alkali on metal, liberating hydrogen, converts arsenic compounds to AsH_3 which accompanies the hydrogen. Interferences depend upon the choice of metal and reagent and upon the method used for the detection of arsine. Interfere may all substances able of giving a volatile hydride: Sb, (Bi), Ge, P, reducible compounds of phosphorus, sulfur, and selenium. Mercuric salts and fluorides may interfere with the formation of AsH_3. The tests are known under the names Berzelius-Marsh and Gutzeit and are widely used in toxicology and in the testing of food and all kinds of articles of daily use. For details see the literature (3, 40, 56, 162, 186).

Using pipets so as not to wet the inside wall, place a few pieces of metallic aluminum, 1 ml test solution, and 1 ml 2-F KOH on the bottom of a test tube. Push into the upper third of the tube a wad of cotton that has been treated with some saturated aqueous solution of lead acetate (for the absorption of H_2S), and place upon the opening of the tube a small disk of filter paper which has been dipped into 3% aqueous $HgCl_2$ solution. Warm the reaction mixture at the bottom of the tube. A yellow to brown spot on the filter paper indicates the presence of trivalent arsenic. If arsenate is present, it should be reduced by adding a drop of aqueous SO_2 and warming before the test is started. Antimony is not reduced to SbH_3 and does not interfere (4.0–4.0; V, Mo, Cd, Hg, Tl, Pb, Sn, Sb, Se: 2.0; W, Pt, Cu, Ag, Au, Tl, Bi: 0.0) and (3.0–3.0; W, Pt, Cu, Ag, Au, Bi, Te: 2.0).

Bettendorff Test (4.7–7.7; Sb, Sn: 4.0). Platinum metals, Au, Ag, Hg, Se, Te are also reduced to the elemental state and interfere with the test.

Proceed as directed in Expt. 50.

Magnesium-(Calcium-)Ammonium Arsenate Hexahydrate (4.0–7.3). The crystals are isomorphous with those of the corresponding phosphate precipitate. All substances interfere that give with NH_3 a precipitate.

Proceed as directed in Expt. 40. If $CaCl_2$ is added in place of the magnesium salt, the crystals of the calcium salt separate, $CaNH_4AsO_4 \cdot 6\ H_2O$.

Silver Arsenate and Arsenite (Estimate: 5.0–5.0). The test is useful for the identification of the valence state, but requires absence of all other ions that would give a precipitate with $AgNO_3$ in neutral solution.

In a test tube, treat 1 ml of the test solution with a few drops of $AgNO_3$ solution and then add strong NH_3 dropwise until the solution is clear again. Holding the test tube at an angle of 45 degrees, run 16-F HNO_3 down the inside wall of the tube so that it collects in a layer on the bottom of the test tube. The precipitate will appear in the neutral zone between the acid and the ammoniacal solution: Ag_3AsO_4 is chocolate brown; Ag_3AsO_3 and Ag_3PO_4 are yellow. For performance as a spot test, see Expt. 29.

No. 51: Antimony, 121.75

Flame Test. See P. 38 and Table IX. The mechanism is the same as explained under Arsenic. The flame is a pale greenish white; the garlic-like odor is missing. The antimony mirror is black; the deposit of Sb_2O_3 turns black when moistened with $AgNO_3$ and exposed to NH_3 fumes: $Sb_2O_3 + 4\ AgNO_3 + 4\ NH_3 + 2\ H_2O \rightarrow Sb_2O_5 + 4\ Ag + 4\ NH_4NO_3$. — The antimony mirror is not affected by NaClO.

Thermoluminescence Test. L. I. = 1 ng Sb. See Expt. 60.

Stibine and Its Detection. Stibine forms under essentially the same conditions as arsine, see under Arsenic, above. In absence of the latter, the test described there may be used to detect antimony. Some changes are necessary, however. Granulated zinc is used in place of aluminum turnings, and 1 ml of the test solution is treated with 1 ml 8- to 10-F H_2SO_4.

Group VI A: Oxygen-Polonium Group

The filter paper is treated with $AgNO_3$ solution and a black precipitate indicates antimony, see also P. 36.

Cesium Iodoantimonite (4.0–8.0). All substances interfere that react with iodide, i. e., Hg, Pb, Tl, nitrite, etc. Bismuth gives the same test, but may be readily recognized by adding stannite reagent.

Proceed as directed in Expt. 35.

Quinine and Cinchonine Iodoantimonite (4.8–6.3). The interferences are the same as with Cesium Iodoantimonite.

Proceed as directed in Expt. 26.

No. 83: Bismuth, 208.980

Flame Test. Heated in the reducing zone, bismuth compounds give a pale greenish or bluish white flame. The mechanism is the same as explained under Arsenic, *above*. A barely visible deposit of Bi_2O_3 may be caught upon the glazed underside of a porcelain dish (filled with cold water) held above the oxidizing flame. Fumes of HI may be obtained by warming a drop of the acid in a porcelain crucible. On exposing the condensate to the fumes, it forms red $HBiI_4$. On breathing upon the deposit, the red color fades, but reappears when the moisture evaporates. Exposure to fumes of NH_3 converts to orange NH_4BiI_4.

Thermoluminescence Test. L. I. = 0.1 ng Bi. See Expt. 60.

Precipitation of Sulfide and Conversion to Chromate and Metallic Bismuth. L. I. = 8 ng Bi. See Expt. 58.

Cesium Iodobismuthate (4.5–8.0). Sb gives the same test, and all substances that react with iodide will interfere. Even in presence of antimony, bismuth will be recognized by the treatment of the precipitate with stannite reagent.

Proceed as outlined in Expt. 35.

Bismuth Cobalticyanide Pentahydrate (4.5–8.0; Tl: 1.3; Pb: 1.1; Sn-ous: 0.7; Cu, Ag, Cd, Zn, Hg-ic, As, Sb: 0.0). 1% chloride ion in solution prevents the test. Antimony does not give this test.

Proceed as outlined in Expt. 36.

Quinine and Cinchonine Iodobismuthate (5.5–7.0). Substances that react with iodide interfere with the test.

Proceed as directed in Expt. 26.

P. 59   Group VI A: Oxygen-Polonium Group

No. 8: Oxygen, 15.9994

More or less pure oxygen is recognized by the fact that it rekindles a glowing splint. In mixtures with other gases it may be recognized by the fact that it gives brown NO_2 when colorless NO is added to the gas mixture. It also gives a deep red coloration when the gas mixture is allowed to act upon an alkaline solution of pyrocatechol and $FeSO_4$ (603).

Ozone

The presence of relatively small percentages of ozone in an atmosphere is already indicated by the characteristic odor. Its presence may be confirmed by allowing the gas to act upon a neutral solution of KI in water; O_3 will liberate I_2 and the solution becomes alkaline, which may be proven with a pH paper. H_2O_2 does not react with neutral iodide solution.

For additional proof, a slightly acid solution of titanium sulfate may be exposed to the gas; it will remain colorless if O_3 is present.

Oxidation of Metallic Silver. Heat some silver foil to about 250° C and then expose it to the gas to be tested: steel blue spots with violet edges develop immediately. Silver which has been polished with emery paper (and thus contaminated) does give the test at room temperature.

Water

Water is most convincingly isolated as such by vaporization followed by condensation. The condensate may then be identified by testing with pH paper and determination of boiling point (Expt. 21) and then freezing point, p. 173.

Decomposition of Potassium Tetraiodoplumbate. Place upon spot test paper a drop of 20% K_2PbI_4 in anhydrous acetone and dry in an oven of 105° C. Then add a drop of the organic liquid that shall be tested. A yellow spot of PbI_2 is obtained with 0.05% H_2O in methanol and with 0.1% H_2O in acetone. Peroxides interfere since they produce a brown coloration.

Solid substances might be treated with a suitable solvent (anhydrous acetone) which would extract water held as "moisture" and adsorbed water.

Hydrogen Peroxide

Pure H_2O_2 melts at $-2°$ C to a colorless, volatile, strongly acid, and explosive liquid that strongly irritates the skin. As a rule, only more or less dilute solutions in water are met. These act as oxidants more or less comparable to the action of various per acids. Characteristic for solutions of H_2O_2 is the liberation of oxygen on adding catalysts like MnO_2. In addition, H_2O_2 reacts at times as oxidant and at other times as reductant. It liberates I_2 from acidified (H_2SO_4) solution of KI; and it reduces $KMnO_4$ in dilute H_2SO_4 to Mn-2 ion.

Dioxy-Disulfato-Titanic Acid. Fuse a small amount of TiO_2 in a porcelain crucible with 15 times as much $K_2S_2O_7$ and dissolve the melt in cold 2-F H_2SO_4. On a spot plate, treat a drop of this solution with a drop of the test solution. A yellow to orange coloration [$H_2TiO_2(SO_4)_2$ or $TiO_2 \cdot H_2O_2$ or H_2TiO_4] indicates H_2O_2.

Reduction of Ferricyanide (6.0–6.0). Persulfate does not give the test. Perborate and percarbonate give a positive test. Iodide and sulfite must be absent.

In a centrifuge tube combine 3 drops 3% $AgNO_3$, 2 ml 8-F NH_3, and 3 drops freshly prepared aqueous 6% $K_3Fe(CN)_6$. Mix and centrifuge. Transfer 1 ml of the clear supernate to 1 ml of the test solution. H_2O_2 produces a white precipitate.

No. 16: Sulfur, 32.064

Depending upon the allotropic modification present, elemental sulfur melts at 120° ("amorphous"), 119.0° (monoclinic), or 112.8° C (orthorhombic). The boiling point of the dark red liquid is 444.6° C. In contact with air, the liquid sulfur ignites at 248° C and burns with a blue flame to give SO_2 of characteristic odor.

Crystalline sulfur (both modifications) is soluble in CS_2; octahedral crystals are obtained upon evaporation of the solution. All forms of sulfur are insoluble in water, but dissolve in hot caustic alkali,

$$4\,S + 6\,NaOH \rightarrow 2\,Na_2S + Na_2S_2O_3 + 3\,H_2O$$

in solutions of sulfide, especially on warming,

$$Na_2S + n\,S \rightarrow Na_2S_{n+1} \text{ where } n = 1, 2, 3, 4$$

and in solutions of sulfite,

$$Na_2SO_3 + S \rightarrow Na_2S_2O_3.$$

Bromine and HCl, $KClO_3$ and HCl, aqua regia, and hot concentrated HNO_3 slowly oxidize sulfur to H_2SO_4.

In qualitative testing, sulfur is frequently met as a condensate of liquid or solidified droplets. The following two tests may be carried out with parts of such a condensate.

Test for Elemental Sulfur (5.7–5.0). Selen and phosphorus do not interfere; halogens, carbon disulfide, hydrocarbons, chloroform, etc. must be absent.

Treat the dry test substance in a test tube with 5 ml pyridine, reagent-grade and freshly distilled. All forms of sulfur are soluble in this solvent. Filter into another test tube and treat the colorless filtrate with one tenth of its volume (0.5 ml) of 4-F NaOH. Depending upon the amount of sulfur present, the mixture turns blue, olive green, to reddish brown. Also polysulfides give the test since they liberate sulfur.

Oxidation to Sulfate (Estimate, L. I. $= 0.1\,\mu g$ S). Elemental sulfur as well as nearly all compounds of sulfur, organic and inorganic, give the test.

Transfer a tiny particle of the material to be tested to a slide and add 0.5 μl 1% (0.1-F) $CaCl_2$ solution. Invert the slide and place it upon the opening of a bottle with bromine water so that the test drop is exposed to the vapor. After five minutes, inspect the test drop under the microscope. The characteristic sheaves of needles of $CaSO_4 \cdot 2\,H_2O$ will appear along the edge of the drop as evaporation takes place if only little sulfur is present (150).

Hepar Test. This test, which may be about as sensitive as the preceding one, is given by elemental sulfur and all compounds of sulfur, selenium, and tellurium.

When performing a charcoal test by heating the test material with Na_2CO_3, it is customary to finally place some of the melt upon a silver coin and to moisten it with a drop of water. Sulfur compounds are reduced to sulfide during the heating upon charcoal. In contact with the silver, the sulfide gives a brown to black spot,

$$2\,Na_2S + 4\,Ag + 2\,H_2O + O_2 \rightarrow 2\,Ag_2S + 4\,NaOH.$$

A blank test is recommended. On the performance of the test with acid soluble sulfides such as CdS, ZnS, lazurite *(lapis lazuli)*, and synthetic ultramarines *see* BONTINCK (877).

Hydrogen Sulfide and Sulfides

Hydrogen sulfide is a colorless, poisonous gas, the rotten-egg odor of which may be recognized when 1 volume of H_2S is present in 700 000 volumes of air. Water saturated at 20° C and 1 atm. pressure is about 0.13-F with respect to H_2S.

For the detection of H_2S gas, one uses filter paper moistened with a solution of lead acetate or lead plumbite [solution of $Pb(OH)_2$ in NaOH]. The latter will give a black spot with 1 ng H_2S. If the paper is treated with an alkaline solution of sodium nitrosopentacyanoferrate, $Na_2Fe(NO)(CN)_5$, a transient purple color is obtained. *See* also P. 36.

Water soluble sulfides give the Hepar test, above, without preceding reduction on charcoal. Acidifying with HCl or dilute H_2SO_4 liberates H_2S that may be recognized by the odor and the tests indicated above. Polysulfides, when treated with acid, give a white precipitate of sulfur in addition to the H_2S. Sulfides that are not attacked by acid may be briefly fused with Na_2CO_3 in a covered porcelain crucible; in spite of some oxidation, the melt will contain enough soluble alkali sulfide to give a test or to liberate H_2S upon adding dilute HCl.

When heated in absence of air, some sulfides sublime and some decompose with the liberation of sulfur, e. g., FeS_2, pyrite $\rightarrow$ FeS + S. In contact with air, the sulfides of the heavy metals are oxidized to give SO_2 and oxide or the metal.

Antimony Trisulfide (4.0–5.3). The test is specific, especially when H_2S is first liberated and allowed to act upon the test paper.

Place upon spot test paper a drop of a 5% solution of potassium-antimonyl tartrate [tartar emetic, $K(SbO)C_6H_4O_6$] and expose to the gas liberated from the acidified sample (or add a drop of the test solution and expose to fumes of HCl). An orange fleck of Sb_2S_3 indicates sulfide.

Methylene Blue (7.7–7.7). The test which has been recommended by EMIL FISCHER (601) should be specific. The test solution may contain free H_2S or a sulfide that liberates H_2S on adding acid.

To 1 ml of the test solution add 0.1 ml 12-F HCl, about 1 mg dimethylparaphenylenediamine sulfate, $NH_2 \cdot C_6H_4 \cdot N(CH_3)_2 \cdot H_2SO_4$, and–when it has dissolved–a drop of 1% (0.1-F) aqueous $FeCl_3$. With very little sulfide, the color of the methylene blue may require one hour to develop. If a red coloration is obtained, it is due to the action of the reagent upon the $FeCl_3$ and may be discharged by adding some more HCl.

Sulfurous Acid and Sulfites

Sulfur dioxide is a colorless gas, very destructive to the vegetation, the suffocating odor of which may be perceived with 1 volume of SO_2 in 200000 volumes of air. Its saturated solution in water of 20° C, 1 atm., is about 1.6 formal and contains mostly SO_2 and only a very small amount of H_2SO_3 and its ions. H_2SO_3 is a moderately strong acid and a good reducing agent.

The alkali sulfites and the acid sulfites of the alkaline earths are soluble in water. All other sulfites are dissolved by dilute solutions of strong acids with the liberation of SO_2. The sulfite ion has the ability to give soluble complexes with many of the heavy metals.

SO_2 is recognized by its odor, the precipitation of $BaSO_3$, $SrSO_3$, and $CaSO_3$ from solutions of the corresponding hydroxides, and the reduction of permanganate, chromate, and iodine, etc.; see also P. 36. One should be aware of the fact that especially $CaSO_3$ (not $BaSO_3$) dissolves readily in an excess of SO_2 by forming bisulfite.

Heated in absence of air, alkali sulfites are converted by auto oxidation-reduction to sulfide and sulfate; the other sulfites give SO_2 and the oxide or metal.

Strontium Sulfite (Estimate: 4.5–5.5.; thiosulfate: 3.0). All substances interfere that give a precipitate with $SrCl_2$ in neutral aqueous solution.

In a centrifuge cone treat the neutral or slightly alkaline test solution with 2% (0.1-F) aqueous $SrCl_2$ solution. If a white precipitate is obtained, remove the supernate, wash the precipitate once with a little water, and then treat it with little 6-F HCl. SO_2 is liberated if the precipitate was $SrSO_3$, and the solution should stay clear when heated upon the steam bath (absence of thiosulfate).

Test with Nitroprusside and Zinc Sulfate (4.2–5.5). The test is not given by thiosulfate (121).

On a spot plate, treat a drop of cold saturated zinc sulfate solution with one drop 0.25-F $K_4Fe(CN)_6$ and one drop 1% (0.04-F) $Na_2Fe(CO)(CN)_5$.

Add to the mixture containing a white precipitate of zinc ferrocyanide a drop of the neutral test solution. The precipitate turns red if sulfite is present.

Test for Sulfite in Presence of Sulfide and Thiosulfate (3.5–4.7; sulfide: 1.9; thiosulfate: 1.7). SO_2 is liberated after destruction of sulfide and thiosulfate and detected by its action on zinc nitroprusside paste (121). The test may be carried out with the use of test tube and glass loop, Fig. 75, or in apparatus shown in Fig. 18.

Reagent. Add an excess of $ZnCl_2$ to a boiling solution of $Na_2Fe(NO)(CN)_5 \cdot H_2O$. Filter and wash the precipitate, and store it moist in an amber bottle.

Mix a drop of the test solution with 2 drops of saturated aqueous $HgCl_2$:

$$S^= + Hg^{++} \rightarrow HgS,$$

$$S_2O_3^= + Hg^{++} + H_2O \rightarrow HgS + H_2SO_4.$$

After one minute add a drop of 3-F H_2SO_4 and test for escaping SO_2 with a small amount of zinc nitroprusside paste. When it may be assumed that the SO_2 has acted, expose the paste briefly to NH_3 fumes. It will assume a red color. Concerning a more sensitive reagent for this test see SENISE (936).

Thiosulfuric Acid and Thiosulfates

The strong thiosulfuric acid probably forms when a solution of thiosulfate is acidified, but it decomposes quickly into $SO_2 + H_2O + S$.

In absence of air, the ignition of alkali thiosulfate gives sulfate, polysulfide, sulfide, and sulfur which distills off. Most thiosulfates are soluble in water; only those of Ba, Ag, and Pb are just slightly soluble. All thiosulfates are dissolved and decomposed by dilute acids with the liberation of SO_2 and precipitation of sulfur. Thiosulfate ion has, furthermore, a distinct tendency to give soluble complexes with heavy metals.

Like sulfite, thiosulfate reduces permanganate, chromate, and iodine in acid solution.

Reduction of Ferric Chloride. Adding a few drops of $FeCl_3$ solution to the neutral or slightly acid test solution produces a violet coloration that disappears after a short time to give a solution containing ferrous ion and tetrathionate.

$$2\ S_2O_3^= + 2\ Fe^{+++} \rightarrow S_4O_6^= + 2\ Fe^{++}.$$

Test with Nitroprusside. Add test solution to a solution of $K_2Fe(NO)(CN)_5$ which has been treated with a few drops of $K_3Fe(CN)_6$ solution and a few drops of 3-F NaOH. Thiosulfate gives a blue coloration. It darkens on standing, heating, or adding of some more ferricyanide.

Tri-Ethylenediamminonickel Thiosulfate (3.0–4.3). Dithionates give a similar precipitate; sulfide, sulfite, and tetrathionate do not interfere; the plates of persulfate cannot be mistaken for a thiosulfate precipitate.

Reagent. Dissolve 4 g $Ni(NO_3)_2 \cdot 6 H_2O$ in 10 ml water and dropwise add ethylenediamine until the solution is violet.

Place a droplet of the neutral or slightly alkaline test solution upon a slide and add one or two droplets of the reagent solution. $Ni(H_2N \cdot C_2H_4 \cdot NH_2)_3S_2O_3$ forms strongly anisotropic prisms and needles.

Thionic Acids and Thionates

None of the acids $H_2S_nO_6$, where $n = 2, 3, 4, 5$, has been isolated in the pure state. Even the aqueous solutions and the solutions of the salts are unstable. Heating of a solution of dithionic acid gives $H_2S_2O_6 \rightarrow$ $\rightarrow H_2SO_4 + SO_2$. All other thionic compounds decompose with the separation of some sulfur.

Sulfuric Acid and Sulfates

For the detection of SO_4 ion in acid insoluble sulfates, Luis (957) heats a particle of the solid substance with 85% H_3PO_4 in a glass tube that is drawn out to a capillary. The H_2SO_4 is distilled into the capillary and identified by the tests given below. As little as 20 ng SO_4, given in the form of insoluble sulfate, may be detected.

Calcium Sulfate Dihydrate, Gypsum (2.7–6.2). The test seems to be specific.

Transfer a droplet of test solution to a slide and acidify it, if necessary, by exposing it to the fumes of 12-*F* HCl. Next to the test drop, deposit a like volume of 1% (0.1-*F*) $CaCl_2$ solution and draw a channel to connect the drops. Fine needles, sheaves of needles, rhombic plates, and arrow-head (swallow-tail) twins of monoclinic $CaSO_4 \cdot 2 H_2O$ usually exhibit oblique extinction.

Barium Sulfate (Estimate: 6.0–7.3). The test is specific.

In a small centrifuge tube treat 50 μl of the nearly "neutral" test solution with 15 μl 12-*F* HCl and 10 μl 1-*F* $BaCl_2$. Mix, centrifuge, and inspect the tip of the cone under magnification. A finely crystalline, white precipitate indicates the presence of sulfate.

If other sulfur compounds and reducing substances are absent, the test may be performed with the addition of $KMnO_4$ to make the solution dark red before adding the $BaCl_2$. Because of coprecipitation, the sulfate precipitate will appear violet and clearly visible after destruction of the permanganate excess in solution by adding H_2O_2.

Persulfuric Acid and Persulfates

Solutions of persulfates in water decompose slowly when cold and more rapidly on warming.

$$2 S_2O_8^= + 2 H_2O \rightarrow 4 HSO_4^- + O_2.$$

Consequently the freshly prepared solution does not give a precipitate with $BaCl_2$, but on standing for some time and more quickly on boiling, $BaSO_4$ will precipitate.

Persulfates are strong oxidizing agents and precipitate from neutral or slightly acid solutions of Mn, Co, Ni, and Pb the corresponding black peroxides. Black silver peroxide, Ag_2O_2, is precipitated from $AgNO_3$ solution. If a concentrated solution of ammonium persulfate is treated with NH_3 and a little $AgNO_3$, nitrogen is liberated and the solution becomes boiling hot.

Oxidation of Iodide. Persulfates react with neutral solutions of KI to liberate iodine. Percarbonates and perborates do not.

Oxidation of Benzidine (4.7–6.0; bromate: 3.0; iodate: 3.8). Alkali peroxides, perborates, and H_2O_2 do not give the test. Chromates, permanganates, ferricyanides, and hypohalogenites give the same test and must be absent.

On a spot plate, treat a drop of the neutral (or acidified with acetic acid) test solution with a drop of a 2% solution of benzidine in 2-F acetic acid. A blue coloration indicates presence of persulfate.

No. 34: Selenium, 78.96

Like sulfur, selenium occurs in several allotropic modifications: red amorphous, melting at 50° C; crystalline, melting at 220° C; and metallic, melting at 217.4° C. The boiling point is 688° C. The element and all of its compounds are severely toxic.

Flame Test. When heated in the upper reducing zone of the Bunsen flame, selenium and its compounds give a blue coloration, and an odor of rotting radishes will be perceived. Red selenium will condense on a cold porcelain surface brought into the blue flame. A drop of 18-F H_2SO_4 placed upon the deposit will dissolve the selenium to form a green solution from which red selenium will again separate upon adding water.

All selenium compounds will give the odor of rotten radishes and a blue flame when heated with Na_2CO_3 upon the charcoal. The melt gives the Hepar test, see under Sulfur, above.

Hydrogen Selenide and Selenides

H_2Se is a colorless gas with properties and odor similar to H_2S. It is more soluble in water and a stronger acid than the sulfur compound, and it forms acid selenides, neutral selenides, and polyselenides. The selenides of the heavy metals are strongly colored, insoluble in water, and some even insoluble in acid.

In general, selenides may be decomposed by dilute HCl or H_2SO_4 with the liberation of H_2Se which may be (a) lighted to give a blue flame and a Se deposit on cold porcelain; (b) passed through a glass tube which is heated in one spot to give a red selenium mirror; or (c) absorbed in water from which red selenium will separate on standing or, faster, on bubbling air through it.

Selenous Acid and Selenites

SeO_2 forms white needles that sublime at 317° C and dissolve in water to give H_2SeO_3 which is very soluble in water and forms hexagonal prisms that decompose on heating. H_2SeO_3 is a weak, dibasic acid; the salts are mostly colorless and soluble in acid if insoluble in water. Characteristic of all selenium compounds is the tendency to decompose, more or less slowly, with the separation of the element. The identification and separation is mostly based on some method for reduction to elemental Se.

Barium chloride precipitates from neutral solutions white $BaSeO_3$ which is soluble in dilute HCl; cupric sulfate gives a greenish blue, crystalline precipitate with neutral selenite solution, whereas selenate solution is not precipitated. Hydrogen sulfide precipitates a yellow to orange mixture of selenium and sulfur, soluble in ammonium sulfide and in alkali sulfide and hydroxide solutions.

Reduction with Hydriodic Acid (4.4–5.9). Sn, Sb-3, and tellurate do not interfere. Molybdate must be absent; Ge and arsenite prevent the test when present in hundredfold excess. Selenates do not give the test.

Place upon filter paper one drop of concentrated HI (or 12-F HCl in which some KI has been dissolved) and put a drop of the acid test solution into the center of the moist spot. Add a large drop of 5% $Na_2S_2O_3$ solution. A reddish brown fleck that does not dissolve indicates selenite.

Reduction with Sulfur Dioxide (5.2–5.2). As-3, Sb-3, Ge, Sn-4, Te do not interfere; Sn-2 must be absent.

In a test tube treat 1 ml of the test solution with 1 ml 18-F H_2SO_4 and saturate with SO_2 (which may be obtained by dropwise adding of concentrated H_2SO_4 to solid sodium sulfite). Boil 1 minute and then allow to cool. A red coloration or precipitate indicates selenium (also selenate gives this test).

Selenic Acid and Selenates

H_2SeO_4 forms hexagonal prisms that melt at 58° C and are very soluble in water.

$BaCl_2$ gives a white precipitate which is insoluble in dilute acids; it dissolves, however, on boiling with HCl.

$$BaSeO_4 + 4\ HCl \rightarrow BaCl_2 + H_2SeO_3 + Cl_2.$$

H_2S gives no precipitate with selenic acid, unless the solution is boiled with HCl which reduces to selenous acid.

No. 52: Tellurium, 127.60

Tellurium is silvery white with a metallic luster and very brittle, m. pt. = 452° C, b. pt. = 1390° C. From solution, it is precipitated as a voluminous brown powder. When heated in air, it will burn with a bluish green flame to TeO_2.

Alkali tellurides are colorless, but their aqueous solutions quickly become red in contact with air due to the formation of polytelluride. The tellurides of the heavy metals are dark in color. Treatment with acid gives H_2Te, a colorless gas with the odor of H_2S, that is very soluble in water; in contact with air, the solution rapidly decomposes with deposition of elemental tellurium. The gas burns with a bluish flame to TeO_2, and deposits (mirrors) may be obtained as with H_2Se.

The white, crystalline dioxide sublimes at 450° C without melting. It is slightly soluble in water which then contains the slightly soluble H_2TeO_3 that cannot be prepared in pure form since it starts losing water at 40° C to finally become TeO_2. The corresponding tellurites resemble the normal sulfites and selenites.

The white solid H_2TeO_4 starts losing water at 160° C to end up as yellow TeO_3. The very weak acid is quite soluble in water which seems to contain orthotelluric acid, H_6TeO_6.

Halides, such as $TeCl_2$ and $TeCl_4$, hydrolyze in aqueous solution.

Flame Test. Metallic tellurium is formed by heating tellurides in the upper reducing zone of the Bunsen flame. A pale green coloration is obtained, and a black deposit may be collected on the glazed surface of porcelain or upon the outside surface of a test tube filled with cold water. The deposit dissolves in a drop of concentrated H_2SO_4 to give a carmine red solution. Adding water again precipitates black tellurium. *See* also P. 38 and Table IX.

Tellurium compounds fused with Na_2CO_3 upon charcoal give the Hepar test, *see* under Sulfur, *above*.

Treatment with Hydrogen Sulfide precipitates from acid solutions of Te-4 the reddish black TeS_2 which is readily soluble in ammonium sulfide, etc. The same precipitate is obtained with hot solutions of Te-6, but cold solutions of tellurate are not precipitated by H_2S, which provides a simple means of separating heavy metals from tellurium.

Acidifying with Hydrochloric Acid precipitates quite difficultly soluble H_2TeO_3 from solutions of tellurites. Solutions of tellurate give no precipitate in the cold; on boiling, the soluble H_6TeO_6 is converted to the slightly soluble H_2TeO_3 with liberation of chlorine.

Reduction with Sulfur Dioxide. From solutions of Te-4 in dilute HCl, Te is precipitated completely as a black powder. From 12-F HCl, tellurium is not precipitated by SO_2, not even on boiling; this provides a means of separating Se and Te.

Cesium Chlorotellurite and Iodotellurite (3.0–6.5). Selenium does not interfere, but interferences must be expected from Ag, Pt, Sn, Pb, Sb, Bi, and possibly others.

Tellurate may be converted to tellurite by boiling with strong HCl. Acidify selenite solution with HCl and evaporate upon the steam bath just to dryness. Dissolve the residue in 3-F HCl to obtain a solution containing not less than 1% Te. Transfer 0.5 to 1 µl of this solution to a slide and introduce a large grain of CsCl into the center of the drop. There will be a dense, granular precipitate close around the reagent, but beyond this zone, moderately large, well developed, light yellow octahedra of Cs_2TeCl_6 may separate. Still further out, thin triangles and hexagons may form. On adding KI to the drop, the crystals are converted to black Cs_2TeI_6.

Reduction with Stannite Reagent (4.6–5.9). Tellurite and tellurate give the test. Selenite and selenate do not interfere; molybdate, Cu, Ag, Hg, Sn-4, arsenite, Bi, and Sb-3 interfere and should be absent.

Reagent. Dissolve 1 g $SnCl_2 \cdot 2 H_2O$ in 1 ml 12-F HCl and dilute with water to 10 ml.

Transfer 1 drop 6-F NaOH to the spot plate, and add first 1 drop of the stannous chloride solution and then 1 drop of the alkaline test solution, preferably a soda extract which would not contain the interfering metals. Presence of Te is indicated by the separation of black flocks. With low concentrations, only a gray coloration is obtained and performance of a blank becomes necessary (121).

Group VII A: Halogen Group
No. 9: Fluorine, 18.9984

Pure $(HF)_n$ is a colorless liquid that boils at 19.4° C to give an intensely corrosive gas. The aqueous solution, hydrofluoric acid, is a weak acid which is able to penetrate tissue deeply, without any immediately visible discoloration of the skin, to inflict extremely painful burns that heal slowly. The vapor is as dangerous as the liquid.

The following tests apply to fluorides and hydrofluoric acid.

Calcium Fluoride (Estimate: 6.0–7.3); all anions interfere, that give calcium salts insoluble in dilute acetic acid: borate, oxalate, silicate, phosphate, and arsenate.

Treat a drop of the neutral test solution (neutralized soda extract) with a drop of 4-F acetic acid and a drop of 1% (0.1-F) $CaCl_2$ solution. CaF_2 separates as a white, slimy precipitate that is difficult to filter, but may be collected and washed with the use of a centrifuge tube and used for the alizarin test below.

Etch Test According to Mannheimer. Heat a rod of lead glass in the reducing flame until it turns black by the separation of lead on the surface. Then draw out a supply of fine threads.

Using the technique of the fiber tests, Expt. 57, insert the end of such a thread into a test drop obtained by treating a sample (solid or liquid) on a platinum sheet (foil) or on a plastic sheet with an equal volume of

9-F H$_2$SO$_4$. Move the supporting sheet from time to time to obtain a stirring action. After five minutes remove the thread from the test drop, rinse it in water, and mount it in cedarwood oil between slide and cover glass for microscopic inspection. HF removes the black surface layer. Observe the boundary line where the thread ermerged from the solution. Silicate and borate in the sample will interfere with the test.

Sodium Fluosilicate (4.0–7.0; phosphate, sulfate, chloride, hypochlorite, chlorate, bromide, hypobromite, bromate, iodide, iodate, periodate: 2.0). Borate hinders the test and should be absent.

The test may be carried out in a 1-ml platinum crucible with lid, in a small, 1-ml, porcelain crucible covered with a slide, in a test tube, Fig. 75, or in apparatus shown in Fig. 18.

If a solution is to be tested, transfer a sample of it to the "pot of the still" and evaporate it there to dryness. If the sample is a solid, simply transfer it to the same location. Add a small amount of finely powdered glass. Before adding a drop of 18-F H$_2$SO$_4$, get ready the loop with a drop of water in it or the cover of the crucible with a droplet of water hanging on its underside. Add the acid, get the water drop into position (cover the crucible), and warm the "pot" (steam bath or jet of steam). After two to three minutes transfer the drop of absorbing water to a slide and continue the distillation with a fresh drop of water in place, *compare* P. 57 under Silicon.

The drop of water may contain a white precipitate resulting from:

$$\text{"pot"}, \ SiO_2 + 4\ F^- + 4\ H^+ \rightarrow SiF_4 + 2\ H_2O,$$
$$\text{drop}, \ 3\ SiF_4 + 3\ H_2O \rightarrow H_2SiO_3 + 2\ H_2SiF_6.$$

A heavy precipitate indicates also a high concentration of fluosilicic acid; in such instances dilute the drop of "distillate" with four volumes of water and use only a fraction for the following test.

If the drop seems clear, do not dilute it, just place a grain of NaCl (of appropriate size) into the edge of the drop. Light pink hexagonal plates, short prisms, and six-pointed stars of Na$_2$SiF$_6$ indicate the presence of fluoride. If the test remains negative and the sodium chloride does not dissolve in the test drop, the latter may be too concentrated and adding some water may start the crystallization. *See* also P. 57 under Silicon. Use the distillate also for the test with molybdate and benzidine described under Silicon. The presence of silica is indirect proof for the presence of fluoride in the sample.

The test will not discover small amounts of fluoride in silicates. To this end, it will be necessary to perform a Na$_2$CO$_3$ fusion and to test for the fluoride in the aqueous extract, P. 72.

Action Upon the Alizarin Lake of Zirconium (4.0–5.3; borate, chloride, chlorate, perchlorate, bromide, bromate, iodide, periodate: 2.0). Ferricyanide, oxalate, nitrate, phosphate, thiosulfate, sulfate, hypochlorite, and hypobromite interfere with the test and must be absent.

Reagent. To 0.1% $ZrOCl_2$ solution in 5-F HCl add a 0.2% solution of alizarin S (alizarin-3-sulfonic acid) in ethanol until an excess of the dye is indicated by the fact that ether, shaken with a sample of the mixture, becomes yellow. Heat the solution for ten minutes on the steam bath. Dip filter paper into the solution and then dry it.

Place upon the reagent paper first a drop of 8-F acetic acid and then add a drop of the test solution. If fluoride is present, a yellow spot appears on the violet paper. The reaction may be hastened by exposing the spot to a jet of steam.

$$\text{Red Zr-lake} + 6\ F^- \rightarrow ZrF_6^= + \text{yellow dye.}$$

The test appears useful in the presence of large amounts of borate. The interferences by nitrate and the halogen anions may be removed by first precipitating CaF_2 (and possibly $BaSO_4$) and performing the test by mixing the precipitate with a small droplet of the violet reagent solution (121).

$$\text{Red Zr-lake} + 3\ CaF_2 \rightarrow ZrF_6^= + 3\ Ca^{++} + \text{yellow dye.}$$

Additional Tests of interest have been recently described by BALLCZO and WEISZ (941) and by FINE and WYNNE (776). Plastic apparatus has been recently described by HAHN (624).

No. 17: Chlorine, 35.453

Flame Test. Organic and probably most inorganic compounds of chlorine, bromine, and iodine (but not fluorine) give a blue or green coloration to the Bunsen flame if heated on a copper spatula, *see* Expt. 18. The test is not specific since also cyanides, thiocyanates, and various organic nitrogen compounds produce the phenomenon. If inorganic substances are present, one has to add all the interferences by elements giving green and blue flame colorations. If the sample does not give a green or blue flame when heated upon the oxidized copper spatula, one may assume that compounds of chlorine, bromine, and iodine are probably absent. The Beilstein test may be performed with very small amounts of material (919).

Elemental Chlorine

The gas is recognized by its yellowish green color, its characteristic odor, its bleaching action upon litmus and indigo, and its ability to displace bromine from bromides and iodine from iodides. These tests are also given by water that contains chlorine dissolved.

Eosine. L. I. = 1 μg Cl_2 (estimate); bromine gives the same test.

Reagents. (a) 0.1% solution of fluorescein in 50% ethanol which has been made slightly alkaline by adding KOH; (b) mix part of solution (a) with an equal volume 0.05-F KBr in 0.1-F aqueous KOH.

Cut two strips of filter paper, about 5 mm × 5 cm. Dip the end of one strip into reagent solution (a), and the end of the other strip into reagent solution (b). Expose the yellow ends of both strips to the gas. The presence of chlorine is proven if only one of them turns pink. If both become pink, bromine is present and chlorine may be present.

Hydrochloric Acid and Chlorides

Liberation of Hydrogen Chloride (3.3–6.3).

Most solid chlorides give HCl gas if warmed with concentrated H_2SO_4 or H_3PO_4. *See* P. 37. The liberated HCl is permitted to act upon $AgNO_3$ solution (test tube and glass loop or gas reaction cell, etc.) and the precipitated AgCl is recrystallized from NH_3.

Liberation of Chlorine (5.3–5.3).

Using suitable apparatus (test tube and glass hook, gas reaction cell, etc.), liberate chlorine by treating the solid material or the test solution with 6-F H_2SO_4 and a grain of MnO_2. Warm the mixture and test the gas with both fluorescein reagent papers described above under Elemental Chlorine.

Chromyl Chloride and Silver Dichromate (Estimate: 4.0–5.3; chlorate, perchlorate, bromide, bromate, iodide, iodate, periodate: 2.0). Fluoride, nitrite, and nitrate must be absent. Fluorine gives a volatile chromium compound, and nitrite and nitrate may interfere by converting the chloride to chlorine and nitrosyl chloride.

Place the material to be tested on the bottom of a 1-ml porcelain crucible (glass cup of similar dimensions). If a solution is to be tested, evaporate a sample of it in the crucible. Add about 1 mg finely powdered $K_2Cr_2O_7$ to the test material. Place upon the crucible a microscope slide with a droplet of 1 μl 4-F HNO_3 hanging on the underside. Lift the slide for a moment, and from a pipet, add one large drop 18-F H_2SO_4 to the bottom of the crucible. Heavy, brownish red fumes of CrO_2Cl_2 (b. pt. = 117.6° C) may be seen in the crucible. Warm the bottom of the crucible by placing it on a steam bath or cautiously playing a microflame on it. After one or two minutes remove the slide and replace it by another one. The droplet of acid may have become orange yellow, and there may be a dark deposit with a metallic luster around the droplet.

Place a kernel of $AgNO_3$ into the drop. If crystals of $Ag_2Cr_2O_7$ do not form (*see* Expt. 25), use a glass needle to mop up with the drop the whole circular area that was exposed to the atmosphere in the crucible. The crystals may appear then or on evaporation of the drop. The presence of chromate (from $CrO_2Cl_2 + 2\ H_2O \rightarrow H_2CrO_4 + 2\ HCl$) on the slide proves the presence of chloride in the test material. The curdy precipitate of silver halide does not contribute to the argument since it may be AgBr.

Silver Chloride (Estimate: 8.0–9.3). All anions interfere, that give a silver salt insoluble in acid.

Onto a black spot plate place one drop of the neutral or slightly acid test solution and two (separate) drops of 1% (0.05-F) $AgNO_3$ in 3-F HNO_3. Use strong lateral light and make certain that all drops are clear. Then combine the drop of test solution with one of the drops of reagent. Keep the second reagent drop for a blank. A white turbidity or precipitate in the test drop indicates halide. The turbidity should clear up when the test drop is exposed to fumes of ammonia.

To improve the sensitivity, the test could be performed under the microscope with the use of darkfield illumination.

To eliminate the interference of iodide, bromide, and thiocyanate, WEISZ (923) suggests the following procedure. Treat the test drop with 1 drop of a 2% solution of 8-hydroxyquinoline in 4-F acetic acid, 1 drop of a mixture of 2 volumes 6% H_2O_2 with 1 volume 4-F acetic acid, and a small drop of 8-F HNO_3. Heat four minutes to get complete reaction of the liberated halogen with the oxine, but do not evaporate to dryness. Transfer the drop of the clear, honey yellow liquid to the spot plate and add $AgNO_3$. The turbidity of AgCl is still visible with dilutions 1 : 20000, or with the adopted form of notation: (4.3–5.6; iodide, bromide, thiocyanate: 1.7).

Hypochlorous Acid and Hypochlorites

The very weak hypochlorous acid is assumed to be present in aqueous solutions of chlorine monoxide, an orange yellow gas (b. pt. = 3.8° C) which decomposes explosively into Cl_2 and O_2 when heated.

$$2\ HClO \rightleftharpoons Cl_2O + H_2O.$$

All hypochlorites are soluble in water and decomposed by acids, including carbonic acid,

$$4\ HClO \rightarrow 2\ Cl_2 + O_2 + 2\ H_2O.$$

The chlorine may be detected by the reagents given under Chlorine, *above*. In general, the detection must be based upon the observation of the oxidizing action and of the products of decomposition. Alkaline solutions of hypochlorites bleach blue KI-starch paper, litmus paper and indigo solution, but they do not reduce permanganate. Auto oxidation-reduction takes place when boiling solutions of hypochlorite,

$$3\ ClO^- \rightarrow ClO_3^- + 2\ Cl^-.$$

The removal of a product of the reaction will obviously aid it; thus adding silver nitrate makes the reaction complete at room temperature,

$$3\ ClO^- + 2\ Ag^+ \rightarrow ClO_3^- + 2\ AgCl.$$

After testing the supernatant solution for chlorate, it remains to prove that chlorate was not present previous to adding $AgNO_3$.

There is a possibility to test for hypochlorite in the presence of chlorine (56), but the problem seems rather academic in nature, for the two are inseparable: alkaline solutions of chlorine always form hypochlorite, and acid solutions of hypochlorite usually contain some chlorine.

Chloric Acid and Chlorates

A 40% chloric acid, $HClO_3 \cdot 7 H_2O$, is a colorless and odorless liquid which starts decomposing at 40° C. The acid and its salt are strong oxidants and give explosive mixtures with combustible substances (S, P, C, organic substances, etc.). All chlorates are soluble in water.

Dilute Sulfuric Acid. Addition of H_2SO_4 to the solution of a chlorate causes slow decomposition:

$$3\ HClO_3 \rightarrow HClO_4 + H_2O + Cl_2 + 2\ O_2.$$

KI-starch paper will turn blue in the acid solution.

Concentrated Sulfuric Acid. In a test tube, treat a tiny particle of the test substance with 1 drop of 18-F H_2SO_4. Chlorates react with the liberation of orange yellow chlorine dioxide gas which colors the sulfuric acid yellow and has a chlorine-like odor.

$$3\ KClO_3 + 3\ H_2SO_4 \rightarrow 3\ KHSO_4 + HClO_4 + 2\ ClO_2 + H_2O.$$

The mixture is dangerous and may violently explode on warming. From a capillary, add a few microliter of a saturated aqueous solution of aniline sulfate. A blue coloration is given by chlorate but not by nitrate.

Reduction to Chloride. In a centrifuge tube, acidify 0.5 ml of the test solution with 1 drop 4-F HNO_3 and then add dropwise 1% (0.1-F) $AgNO_3$ until it does no longer give a precipitate. Centrifuge and treat the clear supernate with a kernel of reagent-grade bisulfite or metabisulfite. If chlorate was present, it is reduced to chloride which gives a white precipitate or turbidity of AgCl with the excess of $AgNO_3$ in solution.

Ignition of Chlorates. Oxygen is given off, and the residue contains chloride.

Perchloric Acid and Perchlorates

Anhydrous perchloric acid, $HClO_4$, is a colorless liquid that boils at about 130° C and decomposes first slowly and finally explodes. An acid of about 70%, $HClO_4 \cdot 2 H_2O$, is sufficiently stable for practical use as dehydrating agent and oxidant. When hot, the action of the 70% acid is comparable to chromic-sulfuric acid; when cold, the acid as well as its salts are remarkably stable. Acid solutions are not reduced by SO_2 and do not react with KI-starch paper or bleach indigo solution. Dilute and concentrated H_2SO_4 have no visible action.

With the exception of the salts of K, Rb, Cs, NH_4, and organic bases, all perchlorates are soluble in water.

Rubidium Perchlorate–Permanganate (3.0–6.0). The test seems to be specific for perchlorate.

Evaporate, if necessary, and then dissolve enough material to get a 0.1 to 1% solution of perchlorate, which may be acid or neutral. Spread 1 μl of it to an elliptical drop on a microscope slide. Introduce into one end of the drop a mere speck of $KMnO_4$ and into the other, a grain of RbCl. Mixed crystals of rubidium perchlorate and permanganate separate as orthorhombic prisms, bipyramids, and combinations, and the color deepens from pink to deep red depending upon the permanganate content.

Test with Methylene Blue and Zinc Sulfate (3.3–4.6). Only persulfate gives the same test.

Treat 20 volumes of about neutral test solution with 5 volumes of saturated zinc sulfate solution, 5 volumes of 20% (2-F) $NaNO_2$, and 1 to 3 volumes of 0.03% aqueous methylene blue. Mix. A reddish violet coloration indicates perchlorate.

Ignition of Salts. At high temperature, chlorides are formed and oxygen is given off.

No. 35: Bromine, 79.909

Bromine is a brown red, heavy (D = 3.12 g/ml) liquid that boils at 58.8° C and gives off heavy brown fumes at room temperature. The characteristic odor resembles that of chlorine, and 1 volume of bromine in 100 000 volumes of air has still an irritating effect upon the mucuous membrane of the nose. From the aqueous solution, bromine may be extracted with CS_2 or CCl_4; the extract is yellow to dark brown depending upon the amount of bromine.

Flame Test. *See* Chlorine, above.

Eosine (5.0–5.0 if the bromine is liberated from aqueous solution). Use procedure and reagent (a) described *above* under Elemental Chlorine.

2,4,6-Dibromo-1,3-diaminobenzene (4.0–4.0 if the bromine is liberated from aqueous solution). The other halogens do not interfere.

Expose a droplet of 5% *m*-phenylenediamine in 0.05-F H_2SO_4 to the gas or vapor to be tested. If the droplet is suspended from a slide, microscopic examination will not require a transfer. Bromine gives fine, colorless needles which often combine to form sheaves or stars.

Hydrobromic Acid and Bromides

Hydrogen bromide is a heavy, colorless gas of suffocating odor which, like HCl, gives heavy white clouds when getting in contact with NH_3 vapors. It is very soluble in water. The solution, hydrobromic acid, turns soon brown because of the action of atmospheric oxygen.

$$4 HBr + O_2 \rightarrow 2 H_2O + Br_2.$$

The bromides behave similarly to the chlorides. Adding chlorine water, liberates bromine which may be extracted with CS_2 or CCl_4 or identified with fluorescein or m-phenylenediamine as described above.

AgBr. Silver nitrate precipitates from neutral and acid solutions of soluble bromides pale yellow AgBr which is less soluble than AgCl. It dissolves only slightly in NH_3 and gives crystals resembling those of AgCl on evaporation of the NH_3.

Liberation of Hydrogen Bromide. Similar to chlorides, most solid bromides give HBr when warmed with concentrated H_3PO_4. If concentrated H_2SO_4 is used, some bromine forms and the gas appears brown.

$$2\ HBr + H_2SO_4 \rightarrow Br_2 + SO_2 + 2\ H_2O.$$

The liberated gas may be tested with $AgNO_3$ solution, *see* P. 37.

Oxidation to Bromine. L. I. $= 0.5\ \mu g$ Br. Cyanide interferes with the test.

Place the test substance or solution into a 1-ml porcelain crucible. Treat acid solutions with small portions of 6-F NaOH until they are alkaline. Evaporate to dryness. If much iodide is present, add $NaNO_2$ and 3-F acetic acid, and again evaporate to dryness. Dissolve the residue in 3-F HNO_3 and add a small crystal of $KMnO_4$. Stir until the solution assumes the color of permanganate. Cover the crucible with a slide having on the underside a droplet of m-phenylenediamine solution, *see above*, or a little square of filterpaper moistened with fluorescein solution (a), *see* under Elemental Chlorine. Heat the bottom of the crucible to about 50° C with a microflame. The presence of bromide is proven if both bromine tests are positive. Needles with the phenylenediamine but no red coloration with fluorescein would indicate cyanide. Fluoride, chloride, and small amounts of iodide do not interfere. Sulfide, sulfite, thiosulfate, and thiocyanate are destroyed by the permanganate.

Bromic Acid and Bromates

$HBrO_3$ is only known in the form of solutions which are colorless, odorless, and strongly oxidizing. All normal bromates are more or less soluble in water.

On adding $AgNO_3$ to not too dilute bromate solution (not less than 1%), $AgBrO_3$ crystallizes as colorless crosses and short, pointed, tetragonal prisms. The crystals are moderately birefringent and appear almost black in transmitted light because of their high refractive index.

All bromates decompose upon ignition. The alkali and alkaline earth bromates give oxygen and bromide. The salts of heavy metals give oxygen and bromine.

When a bromate solution is acidified with dilute H_2SO_4, the mixture remains colorless for some time. As soon as the gradual decomposition has furnished enough HBr to start the reaction $5\ HBr + HBrO_3 \rightarrow 3\ Br_2 +$

\+ 3 H_2O, the solution turns suddenly brown and added CS_2 extracts bromine. If bromide is present in the test solution, the separation of bromine takes place immediately upon acidifying. Hence, if acidifying the cold test solution does not give a yellow color to a drop of CS_2 present, but the color appears immediately on adding bromide, this proves the presence of bromate, especially if no other oxidant can be found. Furthermore, if a bromide solution is treated with a droplet of CS_2 and dilute H_2SO_4, bromate must be absent if bromine does not separate immediately (40).

No. 53: Iodine, 126.9044

Iodine forms bluish black scales that melt at 114° C and begin to vaporize already at room temperature, b. pt. = 184° C. The vapor is violet and has a characteristic odor.

Iodine is slightly soluble in water and more soluble in KI solutions because of the formation of triiodide ion I_3^-. As a general rule, to which there are exceptions, solvents containing oxygen dissolve iodine with a yellow to brown color; solvents containing no oxygen, with a violet color. To increase the sensitivity of the test, one extracts with a small drop of solvent, $CHCl_3$; absence of iodide improves the sensitivity.

A very delicate test for free iodine may be based upon the blue color of the absorption compound of starch and triiodide ion since only one starch grain need be exposed to show the color under the microscope. The presence of iodide is necessary and improves the sensitivity of this test.

Hydriodic Acid and Iodides

Hydrogen iodide is a heavy, colorless gas of suffocating odor, that is readily soluble in water. The aqueous solution is usually brown because of the dissolved iodine.

$$4\ HI + O_2 \rightarrow 2\ I_2 + 2\ H_2O.$$

The iodides of the alkalies and the alkaline earths are colorless and soluble in water. The iodides of the heavy metals are usually colored (yellow to red); some of them are insoluble in water but readily dissolved by excess iodide with the formation of soluble complexes.

Iodides are readily oxidized to iodine.

Flame Test: *see above* under Chlorine.

Silver Iodide. $AgNO_3$ precipitates from neutral and acid solutions the iodide AgI which is practically insoluble in ammonia. The pure silver iodide is pale yellow; a distinctly deep yellow color indicates that either bromide or chloride are present and that traces of them have been co-precipitated.

Dilute Sulfuric Acid. Iodine may be liberated at room temperature or upon warming.

Concentrated Acid. Concentrated H_2SO_4 oxidizes iodide to iodine, whereby the sulfuric acid is reduced to SO_2, S, or H_2S depending upon the amount of iodide present. Use of concentrated H_3PO_4 gives HI which may be recognized by its action on $AgNO_3$ solution. *See* also P. 37.

Oxidation with Chlorine. Cyanide interferes.

In a test tube, treat 0.5 ml of the neutral or slightly acid test solution with a drop of CS_2 and then add 1 drop of chlorine water and mix. In presence of iodide, the carbon disulfide becomes violet. Continue the dropwise addition of chlorine water with shaking after each addition. The iodide is oxidized and the carbon disulfide becomes colorless (or yellow if bromide is present also).

$$I_2 + 6\ H_2O + 5\ Cl_2 \rightarrow 2\ HIO_3 + 10\ HCl.$$

Oxidation with Nitrite (5.8–6.1). Bromide and reducing agents interfere.

In a test tube, treat 0.5 ml test solution with 1 drop CCl_4, 1 drop 0.5-F H_2SO_4, and 1 drop 0.1-F KNO_2. Agitate thoroughly for 10 seconds. Iodine is recognized by the violet to brown color of the solvent.

Palladous Iodide (4.7–6.0; fluoride, chloride, chlorate, perchlorate, bromide, bromate, iodate, periodate: 2.0). Hypochlorite, hypobromite, sulfide, cyanide, and thiocyanate interfere and must be absent; azide, ferro- and ferricyanide interfere when present in excess.

On spot test paper treat 1 drop of test solution with 1 drop of 1% aqueous $PdCl_2$ solution. PdI_2 precipitates and gives a dark brown spot.

Also the precipitation of the very characteristic PbI_2 may be tried, *see* Expt. 33.

Iodic Acid and Iodates

HIO_3 is a white crystalline solid that can be dehydrated by slow heating to 110° C. The white I_2O_5 formed decomposes at about 300° C to I_2 and O_2. The acid is readily soluble in water, and the saturated solution (at 20° C) is about 14-F HIO_3.

The alkali iodates are soluble in water; the other iodates are difficultly soluble to insoluble, which distinguishes iodate from chlorate and bromate.

Action of Sulfuric Acid. Neither dilute nor concentrated sulfuric acid decompose iodate if reducing substances are absent.

Silver Iodate. $AgNO_3$ precipitates white, curdy $AgIO_3$ which is difficultly soluble in HNO_3, but dissolves readily in NH_3. From solutions that are made slightly acid with HNO_3, $AgIO_3$ may be obtained as pointed needles, single and in radial clusters, which show strong birefringence and parallel extinction. When performed upon the slide, L. I. about $= 1\ \mu g$.

Barium Iodate. The white precipitate of $Ba(IO_3)_2$ dissolves only slowly in dilute acids. When precipitated upon the slide, separation of curved needles seems characteristic for the iodate, but also prisms, rhombs, crosses are obtained, which are similar to the crystals of the more soluble $Ba(BrO_3)_2$.

Reduction to Iodine (3.9–5.4).

Place upon starch paper 1 drop of 5% (0.5-F) KCNS solution and add a drop of acid test solution. A blue stain indicates iodate (121).

$$6\ IO_3^- + 6\ H^+ + 5\ CNS^- + 2\ H_2O \rightarrow 5\ HSO_4^- + 5\ HCN + 3\ I_2.$$

Ignition of the Salts. All iodates are decomposed on ignition. Deflagration occurs on charcoal or in the presence of other oxidizable matter. This is similar to the behavior of chlorates and bromates.

Periodic Acid and Periodates

The periodic acids, HIO_4 and $HIO_4 \cdot 2\ H_2O$ or H_5IO_6, are described as colorless, crystalline solids that are soluble in water and decompose at about 120° C.

Most periodates are only slightly soluble in water, but they are readily soluble in dilute HNO_3. The composition of the precipitate obtained with $AgNO_3$ depends upon the acidity of the solution; the color varies from yellow to orange, red, and gray. Boiling the silver precipitate with water causes it to become dark red.

Periodates are powerful oxidants. A reaction that is not given by iodate is the oxidation of manganous salt to permanganate which may be performed as follows.

Mix equal volumes of 2-F acetic acid and 10% (0.5-F) solution of $MnCl_2$. On a spot plate, mix a drop of the mixture with a drop of the neutral test solution. Appearance of the color of permanganate indicates the presence of periodate. Chlorates, bromates, and iodates do not interfere with this test. Persulfate should be absent. To intensify the color effect, the acetic acid used for the reagent mixture may be saturated with "tetrabase" (121) which is converted to a blue dye by the permanganate. In this instance, the sensitivity is improved to (5.0–6.3).

Separations

If the orientation tests have shown that the material is of a complex nature, i. e., probably contains several elements (cations or anions), the use of a scheme of separation is indicated. If only little material is available, the question arises whether or not a qualitative analysis with approximate estimation of the quantities will be able to solve the problem; a quantitative analysis or a series of quantitative determinations may better serve the purpose.

Even with relatively simple substances, use of a scheme of separation may become necessary if the orientation tests fail to give sufficient evidence for definite identification. The performance will not be tedious if only a few elements or radicals are present; only few analytical groups will be represented, and these will contain only one or two members so that further separation may not be necessary.

P. 62 Systematic Schemes for the Detection of Cations

Adherence, in principle, to the classical hydrogen sulfide scheme is recommended. For this scheme, the behavior of all elements is best known, an advantage that should not be dismissed if the presence of uncommon elements is possible. In addition, the hydrogen sulfide scheme permits the use of gaseous reagents for the separation of the five principal groups. These reagents may be obtained free from metallic impurities and they may be again removed from the reaction mixture by the simple expedient of evaporation and ignition. Finally, the very carefully tested scheme of Noyes and Bray (49), which includes the less common elements, is essentially an extension of the hydrogen sulfide scheme.

There may be good and sufficient reason to eliminate rare elements from consideration or to consider only a certain group of elements. In such instances some other scheme of separation may be preferable. Swift's version of the classical procedure (53) permits semiquantitative estimation; the scheme of West and Mukherji (473), the outline of which is readily accessible, uses liquid-liquid extraction for the separation of groups which are rather small and quite different from those of the classical scheme.

Whatever procedure may be used, the experimenter should keep in mind that there is no scheme in existence–and probably never will be–that has been tested for all possible combinations and all possible ratios. Even the most carefully tested schemes may fail to reveal the "whole truth" with certain materials even when the operator is on the lookout for signs of abnormal behavior. Ammonia and ammonium ion are not considered by the schemes and must be found in a separate test, Expt. 46.

P. 63 *The Classical Scheme*

The classical scheme is applied to the neutral or acid solution of the material. Such a solution will fail to contain niobium and tantalum so that these elements need not be considered. Wolfram may be present if the solution contains arsenic or phosphorus.

The classical scheme is recommended if it is simple to obtain a neutral or acid solution of the material under investigation or when only the common metals need be considered and complete dissolution may be obtained by means of a simple fusion procedure. The scheme of Noyes and Bray (49) is recommended if the material refuses to dissolve without leaving a residue; it is also recommended for the safe detection of arsenic, selenium, germanium, osmium, and ruthenium. For the analysis of complex silicates, expert advice will be found in Applied Inorganic Analysis (7).

Characteristic for the classical scheme is the use of hydrogen sulfide, ammonium sulfide, and ammonium carbonate for the separation of the analytical groups. There are many modifications of it, but none of them describes the reliable separation of all metallic elements, common and rare. Consequently it becomes the task of the analyst to recognize the presence of odd elements from unusual phenomena observed in the course of separation. This implies that he is familiar with the behavior of the constituents considered by the scheme. The ability of correctly interpreting descriptions of the phenomena is only a poor substitute for this familiarity unless the precaution is taken to compare the questionable phenomenon with that obtained with a material that has been prepared to have the composition assumed for the fraction under investigation.

The following outline in the form of a flow sheet gives the key numbers under which the directions and the observations (notes) are given.

P. 64   *Outline for the Separation of the Analytical Groups*

Acidic Solution of the Sample
   + HCl ⟶ Ppt. 1: Silver Group. $AgCl, Hg_2Cl_2, PbCl_2, TlCl$
 ↓
Supernate 1:
  + H_2S ⟶ Ppt. 2: H_2S Group.
         Extracted with $(NH_4)_2S_x$

     Residue 3: Cu Group.   Extract 3: As Group.
     Black: $CuS, HgS, PbS,$ Black: $(CuS), MoS_3,$
      $Bi_2S_3, Re_2S_7(?), RuS_x,$ $Re_2S_7, RuS_x, Ir_2S_3,$
      $Rh_2S_3,$ PdS, $OsS_4,$ $PtS_2, Au_2S_2;$
      $PtS_2;$ brown: SnS, Te;
     golden: Au;   orange: $Sb_2S_{3.5}$, Se;
     yellow or orange: CdS, yellow: $As_2S_{3.5}, SnS_2,$
      (Se)   Se;
                white: S, GeS_2

Supernate 2:
 + NH_3 + H_2S or $(NH_4)_2S$ ⟶ Ppt. 4: $(NH_4)_2S$ Group.
             Black: $FeS, CoS, NiS, UO_2S, Tl_2S;$
             blue: $Cr(OH)_3, Nd(OH)_3;$
             green: $Cr(OH)_3, Pr(OH)_3, MnS;$
             tan: MnS;
             yellow: $Sa(OH)_3, In_2S_3;$
             white: $ZnS, Ga_2S_3, In_2S_3, Al(OH)_3,$
              $Be(OH)_2$, (Sc, Y, La, Ce, Gd, Tb) ·
              · $(OH)_3$, (Ti, Zr, Th) · (OH) ,
              $MgNH_4PO_4 · 6 H_2O$, $(Ca, Sr, Ba)_3 · (PO_4)_2$

Supernate 4:
 + $(NH_4)_2CO_3$ ⟶ Ppt. 5: Alkaline Earths Group.
          (Ca, Sr, Ba) · CO_3

Supernate 5:
Mg and the alkalies.

P. 65

Procedure

The amounts of reagents suggested in the procedure are based upon the assumption that one gram of solid sample has been used for the preparation of the solution to be used for the separation. If n gram of solid sample have been taken, multiply all volumes and weights mentioned in the procedure by n.

Sample Solution and Treatment with HCl. The sample solution is prepared according to the experiences made in testing the solubility of the material under investigation. The solution must be neutral or acidic, and its volume should be between 10 ml and 50 ml. Prolonged boiling with HCl or HBr may cause loss of As, Sb, Hg, Se, and Ge, which may be prevented by using an efficient reflux condenser.

One might assume that a solution containing HCl will not contain silver; since, however, AgCl is somewhat soluble in strong HCl (or HBr), a side test should be made with such solutions to see whether or not diluting with water gives a slight turbidity of silver halide (a heavy precipitate upon diluting suggests the presence of Sb or Bi).

If HCl (HBr) has not been used in preparing the sample solution, it is suggested to first add just one drop (0.05 ml) of 1-F HCl. A white precipitate or turbidity suggests the presence of silver. If the solution remains clear, heat for five minutes in a steam bath and then again cool to room temperature; if no turbidity appears, silver is obviously absent.

Whatever the effects of the first drop may be, finally add 3 ml 12-F HCl and mix.

Precipitation of the Hydrogen Sulfide Group. If the unknown has been dissolved in nitric acid, it may be best to remove the excess of HNO_3 by evaporation just to dryness and then adding 10 ml 6-F HCl before proceeding to treatment with H_2S. Of course, HNO_3 and other oxidants interfering with the H_2S treatment might be destroyed by addition of a suitable reductant: SO_2, N_2H_4HCl, or $NH_2OH \cdot HCl$.

As a rule, the supernate 1 from the silver group is strongly acid corresponding to 3-F HCl or stronger. Saturate the strongly acid solution at room temperature with H_2S; heat to about 95° C (steam bath); saturate with H_2S while the mixture is cooling; dilute with 50 ml water; saturate with H_2S at room temperature; heat to about 95° C; dilute with another 50 ml water and again saturate with H_2S at room temperature; finally repeat diluting with 50 ml water and saturating with gas until the hydrogen ion concentration in the solution has dropped to 0.3-F (pH = 0.5). This state is most conveniently established by adjusting the volume of the mixture; if the latter contains a total of m mval (milliequivalents) of acid (added in dissolution and precipitation of the silver group), the volume should be approximately $= m/0.3$ ml, usually about 200 ml.

Flocculation indicates that precipitation is complete at the prevailing acidity; more sulfide may come down when the mixture is further diluted, but dilution should not go beyond pH 0.5 since this would lead to precipitation of ZnS.

Test for complete precipitation by saturating the supernate or a sample of it with H_2S and heating to about 95° C. There should be no more precipitate.

Wash the precipitate with three 10-ml portions of 0.1-F HCl that has been saturated with H_2S at room temperature. Combine the washings with supernate 2.

Separation of the Copper and Arsenic Groups. Treat Ppt. 2 with 10 ml 6-F $(NH_4)_2S_{1.2}$*; with occasional stirring digest for five minutes at 40° C; separate the extract from the residue; wash residue 3 with three 10-ml portions of 2-F $(NH_4)_2S_{1.2}$ and reject the washings.

To obtain the sulfides of the arsenic group from extract 3, first dilute with 30 ml water and then add with stirring (about 30 ml) 2-F H_2SO_4, about 5 ml at a time, until the mixture is distinctly acid (pH 3 to 1; acid to Congo red). Remove and reject the liquid, and wash the precipitate once with 0.1-F H_2SO_4 that has been saturated with H_2S at room temperature.

Precipitation of the Ammonium Sulfide Group. Treat supernate 2 with 2-ml portions of 3-F NH_3 (about 5 portions should suffice) until ammoniacal as indicated by pH paper and odor. Heat to about 60° C and, with stirring, slowly add 6 ml 2-F $(NH_4)_2S$. (Alternative: treat the ammoniacal mixture with 8 ml 3-F NH_3; saturate with H_2S; warm to 60° C and add another 8 ml 3-F NH_3.) Separate solution and precipitate, and wash the latter with two 10-ml portions of 0.1-F $(NH_4)_2S$. Combine the washings with supernate 4.

Analysis of Supernate 4. Treat the supernate with 5-ml portions of 2-F HCl with mixing until acid. Separate from the precipitate which, in addition to S, may contain NiS, V_2S_5, and (or) WS_3. On the steambath, evaporate the clear solution to about 10 ml. Then add 5 ml 16-F HNO_3 and evaporate to dryness. Finally heat gently (direct flame or heating block) to vaporize any ammonium salt that might be left behind. After cooling to room temperature, dissolve the residue in 10 ml 1-F HCl. Separate from any insoluble matter and treat the clear solution with 2 ml 6-F NH_3 and 20 ml 1-F $(NH_4)_2CO_3$. Heat just to boiling, then allow to cool to room temperature. Separate the solution from the precipitate, and wash the latter once with 5 ml water. The washing may be combined with supernate 5.

---

* Saturate 50 ml 12-F NH_3 with H_2S, add 50 ml more 12-F NH_3, and dissolve 0.6 g powdered sulfur.

Analysis of Supernate 5. A side test for magnesium may be carried out with a small part of supernate 5. Before testing for alkalies, however, it is preferable to remove the ammonium salts. To this end treat supernate 5 with 6 ml 12-F HCl and 3 ml 16-F HNO_3. Evaporate to dryness on the steam bath, and gently ignite the residue for the removal of any ammonium salt left. After cooling, estimate the amount of residue left and then dissolve it in the suitable volume of the desired solvent.

P. 66 *Observations and Notes*

Precipitate 1: Silver Group. Silver chloride is somewhat soluble in strong HCl, and very small amounts of silver may be missed when adding a large amount of HCl at the start. On the other hand, silver and mercurous mercury may be completely separated from lead by adding small portions of dilute HCl until these ions are precipitated completely, whereas the solubility product of lead has not yet been reached.

AgCl (curdy) and $PbCl_2$ (distinctly crystalline) are usually recognized by their appearance.

Supernate 1. If barium and strontium are of interest, a tenth of supernate 1 may be tested with the dropwise addition of 2-F H_2SO_4. A white precipitate may be $BaSO_4$, $SrSO_4$, $PbSO_4$, or $CaSO_4 \cdot 2\ H_2O$. After recording the volume (comparison with a known amount of $BaSO_4$), the precipitate is isolated, washed, and tested accordingly. $PbSO_4$ may be extracted with a solution of ammonium acetate.

If thallium may be present, it is recommended to saturate supernate 1 with SO_2. Reduction to thallous ion is followed by precipitation of TlCl which separates as a fine, crystalline powder. Also reduced are permanganate, chromate, and ferric ion as may be observed from the color changes. $BaSO_4$, etc. may precipitate since SO_4 is formed. Before adding H_2S, the excess of SO_2 is removed from the solution by boiling and bubbling CO_2 or air through it.

Precipitate 2: Hydrogen Sulfide Group. The presence of oxidants (SO_2) in supernate 1 will delay precipitation of sulfides since the H_2S is destroyed. Large amounts of white sulfur may separate, and the color of the solution may change because of the reduction of colored ions (preceding paragraph). Oxidation to sulfate may also cause the separation of sulfates of the alkaline earths. It is undesirable to have so much strong oxidant present that a violent reaction with liberation of heat and separation of molten or plastic sulfur ensues. Normally, sulfur will appear, depending upon its amount, as a turbidity or as a white milk which passes through filter paper and which may have to be boiled for some time to obtain flocculation.

Some authors adjust the acidity by adding ammonia. This must not be done, however, when the solution once contains H_2S. Regardless of how efficient the mixing is, local excess of ammonia and alkaline reaction will last long enough so that CoS and NiS precipitate locally. When the solution is again uniformly mixed and acid throughout, these sulfides may not dissolve again and remain in Ppt. 2.

The sulfides separate in the order of their solubilities. First separate the sulfides of mercury and arsenic (yellow); mercury may first give a dense white precipitate of Hg_2SCl_2 (?) which with more H_2S gives first distinctly yellow $Hg_3S_2Cl_2$ (?) that then changes through brown to black HgS. The sulfides of mercury and arsenic separate even from 12-F HCl; the orange sulfide of antimony (which may turn grayish black) is not precipitated until the concentration of the acid is below 6-F. The black sulfides of the common heavy metals separate from still less acid solutions, and the yellow (from acid solutions more often orange) CdS will not appear before the acid concentration has dropped to 0.5-F.

Lead "sulfide" may separate with a dull red color from solutions containing much chloride.

The black MoS_3 is very insoluble but does not precipitate readily. It is characteristic that saturating with H_2S followed by heating gives some black precipitate and a bluish supernate (probably a colloidal suspension of MoS_3). Repeated treatments with H_2S produce additional portions of precipitate. To obtain complete precipitation, the solution must be saturated with H_2S and heated (steam bath) in a sealed vessel to prevent escape of the gas. Similar trouble give platinum metals, of which Ir is most difficult to precipitate. In the instance of ruthenium, the solution becomes sky blue (sensitive and characteristic test) before the separation of the brown sulfide starts.

Brown metallic gold is precipitated from hot solutions.

Residue 3: Copper Group. The residue will contain some white sulfur and it may hold sulfates of the alkaline earths.

If it is intended to separate HgS from the other members of the group by extraction with nitric acid, residue 3 must be washed free of chloride ion which otherwise would react with the acid to give chlorine that dissolves the HgS.

For dissolution in HNO_3, residue 3 should be thoroughly mixed with cold dilute acid and then the mixture warmed with continuous stirring until dissolution occurs. This procedure favors the liberation of H_2S without much oxidation of it and avoids the separation of plastic sulfur that could envelop sulfides and later prevent their dissolution. The residual HgS may become converted to white $HgS(NO_3)_2$ giving the impression that only sulfur is left behind (mixed possibly with the sulfates of the alkaline earths). It is suggested to dissolve the HgS (or the white mixed sulfide) in a small volume of sulfide-hydroxide reagent (solution 2-F in

Na_2S and 1-F in NaOH); this may leave a residue of CdS and (or) ZnS (which give solid solutions in HgS) and also Au, PtS_2, RuS_x, and Rh_2S_3.

The sulfides of Os and Pd dissolve in HNO_3.

Precipitate 4: Ammonium Sulfide Group. If the ammonia or the ammonium sulfide contains carbonate or the precipitation is exposed to the laboratory air for a significant time, alkaline earths will be precipitated as carbonates and accompany this group even in the absence of phosphate.

To obtain subgroups, Ppt. 4 may be treated with 20 to 60 ml 0.5-F ammonium citrate solution. The hydroxides and phosphates dissolve, and the sulfides remain insoluble.

Supernate 4. If supernate 4 is brown, it probably contains some nickel sulfide in colloidal dispersion. A brilliant violet red color indicates vanadium. When the supernate is acidified, the following may separate: some white sulfur, a small amount of black NiS, black V_2S_4 or V_2S_5, and light brown WS_3. The precipitation of V and W is expected at this point since they are precipitated by neither H_2S nor $(NH_4)_2S$, but separate when the solution is treated with the latter and then acidified.

Heating with HNO_3 and HCl destroys the ammonium salts so that very little solid is left when the solution goes to dryness. If HNO_3 is applied in excess (3 HCl + HNO_3 → 2 H_2O + NOCl + Cl_2), the residue consists mostly of nitrates which are not inclined to decrepitation when heated.

Precipitate 5: Alkaline Earths. For various reasons indicated above, the alkaline earths may be carried down with ppts. 2 and 4 so that only little or nothing of them may appear in this place.

Supernate 5: Magnesium and the Alkalies. Magnesium need not interfere when testing for Na and Li, but ammonium salts must be removed before testing for K, Rb, and Cs.

Satisfactory proof for the presence of small amounts of potassium or sodium is difficult since they may be derived from the reagents and apparatus used in the preceding operations. They also may be lost by coprecipitation, especially with precipitate 4. A simplified procedure is desirable for the safe detection of alkalies.

If the material is only partly soluble in water, it is recommended to separately analyze the aqueous extract. Otherwise the separation may be simplified by treating supernate 2 with 1 ml 9-F H_2SO_4, evaporating, and finally heating (300 to 400° C) until fumes of SO_3 are no longer given off. The residue is extracted with 20 ml hot water, and the clear extract is treated with a saturated solution of $Ba(OH)_2$ until precipitation is complete (one drop may suffice, but not more than 70 ml should be required). After separation, the clear solution is precipitated with $(NH_4)_2CO_3$ like supernate 4 to remove barium, whereafter the filtrate (centrifugate) is evaporated and ignited for the removal of ammonium salts. The clear aqueous extract of the residue may contain the alkalies as carbonates and will have to be acidified or neutralized depending upon the requirements of the tests.

Insoluble materials (silicates) are best treated according to J. LAWRENCE SMITH as described on p. 925 of Applied Inorganic Analysis (49). The procedure of BERZELIUS-KRISHNAYYA, p. 930 (49), is preferable if significant amounts of boric acid are present.

Whatever procedure is used, the purity of all reagents must be watched. Apparatus is best of platinum, plastic, or clear vitreous silica.

P. 67 Analysis of Metals and Alloys Attacked by Nitric Acid

Treatment with nitric acid is by no means obligatory. Some metals and alloys do not dissolve in HNO_3. Aluminum may be advantageously dissolved in NaOH solution, whereby copper, iron, etc. remain as metals in the residue.

In general, the treatment with HNO_3 has the advantage that the "anion" of sulfides, phosphides, arsenides, antimonides, and silicides is oxidized to a corresponding oxygen acid instead of being lost as H_2S, AsH_3, etc. as may happen when the sample is dissolved in HCl.

The following procedure is essentially that of A. A. NOYES (48).

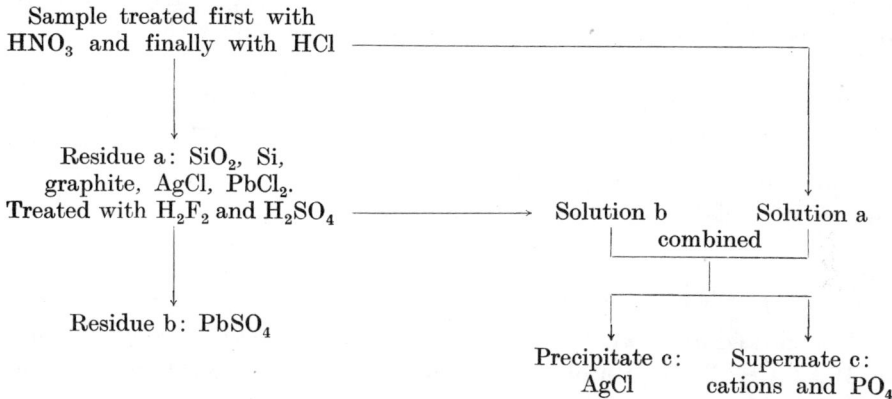

Procedure

Dissolution of Sample. Treat 0.5 g of filings, borings, or turnings with 5 ml 6-F HNO_3 and heat on the steam bath until the reaction ceases. If some metal remains undissolved, add 1 ml 16-F HNO_3 and continue heating. Finally evaporate to 2 ml. Add 5 ml 12-F HCl and evaporate slowly just to dryness. To render insoluble most of the silicic acid, heat the residue to 120° C in a drying oven or heating block. Using a stirring rod and a pestle, loosen the residue from the walls of the apparatus and rub it to a fine powder. Add 5 ml 6-F HCl, cover, and allow to digest for five minutes at 80° C. (If dissolution does not seem complete, add 2 ml 12-F HCl and evaporate slowly almost to dryness. Again treat with 5 ml 6-F HCl, as before.) Add 10 ml water, heat to just boiling, and separate the solution from any residue while still boiling hot. Wash any residue with 5 to 10 ml 2-F HCl and then with five 5-ml portions of hot water. Reject all washings. Reserve solution a.

Residue a. Graphite will be recognized by its color; *see* also P. 57 under carbon and silicon.

Estimate the amount of precipitate and then use the whole residue for a test for silica, P. 57. When working on the gram scale, use a platinum crucible of at least 10-ml capacity. Use 2 ml water to transfer the bulk of residue a to the platinum crucible. Evaporate the slurry to dryness and then add 1 ml 18-F H_2SO_4 and 0.3 ml (6 drops) 48% H_2F_2. Test for silica as outlined in P. 57. If the test is positive, add 2 ml 48% H_2F_2, cover the crucible, and digest 15 minutes on the steam bath. Regardless whether or not the test is positive, evaporate the mixture until the heavy white fumes of SO_3 appear. Allow to cool and then pour the contents of the crucible into 30 ml water (adding water to the acid is permissible when working upon the microscale). Mix and boil gently for five minutes. After cooling to room temperature, remove solution b and wash any residue b with 0.5-F H_2SO_4; reject the washings (Ag_2SO_4 may crystallize out when the mixture is finally cooled; inspection under the microscope will identify it).

Residue b may be $PbSO_4$ and (or) carbon (graphite).

Solution b. Add to solution a. A precipitate of AgCl may separate.

Supernate c contains phosphate and the cations. Proceed as with the supernate 1 of the classical scheme, P. 64.

P. 68 *Separation of the Analytical Groups of Noyes and Bray*

The procedure of A. A. NOYES and W. C. BRAY (49) is recommended if (1) As, Ge, Se, Os, Ru may be present or are of special interest; (2) the material resists dissolution; (3) the material is very complex in composition; or (4) presence of the less common elements is probable and these are of interest.

A complete outline of the scheme of NOYES and BRAY with emphasis on later modifications suited for work on the milligram scale may be found in *Anorganische qualitative Mikroanalyse* (162), and the numbering used there for residues, extracts, solutions, and precipitates has been adopted also for the following outline to facilitate finding the directions for intergroup separations whenever they should be needed.

The following description assumes that 1 g of solid material (0.5 g of metal or alloy) is taken for analysis. If n gram (0.5 n gram) is taken instead, all weights and volumes of the directions[1] are to be multiplied by n.

The scheme of NOYES and BRAY has been developed with the assumption that the sample for analysis will not contain more than 0.5 g of any common element and not more than 0.2 g of any rare element. Under these conditions, it should be possible to detect 0.001 g of any element; frequently even 0.0001 g may be discovered, but in a few instances the limit is 0.003 g. This implies limiting proportions of from 1 : 170 to 1 : 5000 (usually 1 : 500) if a common element is present in the maximum amount of 0.5 g and of from 1 : 70 to 1 : 2000 (usually 1 : 200) if a rare element predominates to the limit of 0.2 g.

The reasons for the selection of the methods of separation may be found in the original papers of NOYES and BRAY (49), CROWELL et al. (420, 883), and BENEDETTI-PICHLER et al. (413, 415, 421, 866, 869, 870).

The material for analysis should be reduced to a fine powder to facilitate attack of the solvents. In the instance of metals and alloys, borings and turnings may be preferable to the finer filings since the last may be more seriously contaminated by material from the tool. Of course, any lubricant should be removed by washing with solvent and drying; steel particles from tools may be removed from the dry material with a magnet, provided that the sample itself is not attracted.

NOYES and BRAY separate into 16 analytical groups: I – selenium group, II – wolfram group, III – tantalum group, IV – gold group, V – thallium group, VI – tellurium group, VII – copper group, VIII – ammonium hydroxide group, IX – ammonium sulfide group, X – nickel group, XI – zirconium group, XII – aluminum group, XIII – chromium group, XIV – lanthanide group, XV – alkaline earths group, and XVI – alkali group. Fractions of groups VIII and IX are combined to give groups X to XIV.

Aside from one exception (tantalum group), the intergroup separations are not included in this outline since one may hope that it may be possible to obtain confirmatory tests without preceding separations if only one or two members of the group are present.

A glance at the flow sheets may be discouraging, but it should be realized that even with complex materials not more than half of the analytical groups will be represented; with relatively simple compounds, only a few precipitates will be obtained, and most of the solutions will be empty.

---

[1] *Anorganische qualitative Mikroanalyse* gives ge (*Gewichtseinheit* = unit of weight) and ve (*Volumeinheit* = unit of volume) which may be translated as g (gram) and ml.

Outline of Separation into Analytical Groups

Solid Sample + HBr + Br$_2$

↓ ↓

Residue A 1: insolubles
+ H$_2$F$_2$ and heat, → SiF$_4$ ↑ BF$_3$ ↑
+ HClO$_4$ and evaporate, → HF ↑

Extract A 1: acid soluble material

Residue A 3: add to extract A 1

↓

Residue A 3 + Extract A 1 distilled

↓ ↓

Residue A 4: solution and undissolved matter + HNO$_3$ and distil

Distillate A 4: Selenium Group, Se, Ge, As, (Sb, Sn)

→ **Distillate A 5:** OsO$_4$

Residue A 5 + HClO$_4$ and distil → **Distillate A 6:** RuO$_4$

Residue A 6. Reflux with formic acid → **Solution A 7.** Evaporate to a small volume and cool

↓ ↓

Crystallizate A 9: perchlorates of K, Rb, Cs

Supernate A 9: groups V to XVI and PO$_4$

Residue A 7. Heat with H$_2$F$_2$ → **Extract A 8.** Heat with H$_2$SO$_4$ to fumes of SO$_3$, add H$_2$O, NH$_3$, and (NH$_4$)$_2$S and heat in pressure bottle

↓

Precipitate II 1: Tantalum Group, Ta$_2$O$_5$, Nb$_2$O$_5$, TiO$_2$; Bi$_2$S$_3$; phosphates and vanadates of Ti and Zr

Solution II 1. Add dilute H$_2$SO$_4$

→ **Precipitate II 2:** Wolfram Group, WS$_3$, MoS$_3$, V$_2$S$_5$, TeS$_2$, Sb$_2$S$_3$, SnS$_2$

Supernate II 2: PO$_4$, (WO$_4$)

Residue A 8

Outline Continued

Residue A 8: noble metals, AgBr, S, fluorides of Ca, Th, Pb, Bi. Boil with Na$_2$CO$_3$ ⟶ Extract A 10: F, SO$_4$

↓

Residue A 10: Acidify with HClO$_4$ ⟶ Extract A 11: perchlorates of Ba, Sr, Ca, Pb, (Cr)

↓

Residue A 11 + HNO$_3$ + HCl ⟶ Extract A 12: Gold Group, Au, Hg, Pt, Pd, (Ir, Rh)

↓

Residue A 12 extracted with NH$_3$ ⟶ Extract A 13: Ag(NH$_3$)$_2^-$

↓

Residue A 13: C, SiC, Al$_2$O$_3$, Cr$_2$O$_3$, FeO·Cr$_2$O$_3$, TiO$_2$, SnO$_2$, ZrO$_2$·SiO$_2$, phosphates of Th and the lanthanides, ThF$_4$, some silicates, Pt metals and alloys, alloys of Fe–Ni–Cr–Mo–W–Si. Fused with K$_2$S$_2$O$_7$ and melt extracted ⟶ Extract A 14: metals of all groups

↓

Residue A 14: sulfates of Ba, Sr, Ca, Pb, Cr; oxides of Si, W, Ta, Nb, Sn; refractive silicates, metals, and alloys. + H$_2$F$_2$ and digested ⟶ Extract A 15: fluoride complexes of W, Ta, Nb, Si. It is treated like extract A 8

↓

Residue A 15 is boiled with Na$_2$CO$_3$ solution ⟶ Extract A 16: F, SO$_4$ ions. Is rejected

↓

Residue A 16 is extracted with HClO$_4$ ⟶ Extract A 17: perchlorates of Pb, Ba, Sr, Ca

↓

Residue A 17: SnO$_2$, Cr$_2$(SO$_4$)$_3$, some silicates, Ir. It is brought into solution by fusion with Na$_2$O$_2$

Treatment of Supernate A 9

Supernate A 9 + HBr ⟶ Precipitate V 1: Thallium Group, AgBr, TlBr, PbBr$_2$

Supernate V 1 + H$_2$S ⟶ Precipitates VI 1 + VI 2. The combined sulfide ppts. are dissolved in HNO$_3$ + HCl, and the solution is saturated with SO$_2$

Precipitate VI 3: Te

Supernate VI 3 is extracted with ether

Ether extract VI 4: Mo

Aqueous Layer VI 4: Copper Group (Cu, Cd, Bi, Pb) and Ir and Rh

Supernate VI 2 is heated after adding ferric nitrate and ammonium acetate ⟶ Precipitate VIII 1: Ammonium Hydroxide Group

Supernate VIII 1 + NH$_3$ + H$_2$S ⟶ Precipitate IX 1: Ammonium Sulfide Group

Supernate IX 1 is evaporated with HCl and the solution of the residue treated with (NH$_4$)$_2$CO$_3$ ⟶ Precipitate XV 1: Alkaline Earths Group, Mg, Ca, Sr, Ba

Supernate XV 1: Alkali Group, Li, Na, K, Rb, Cs

Systematic Schemes for the Detection of Cations

Resolution of the Ammonium Hydroxide and Sulfide Groups

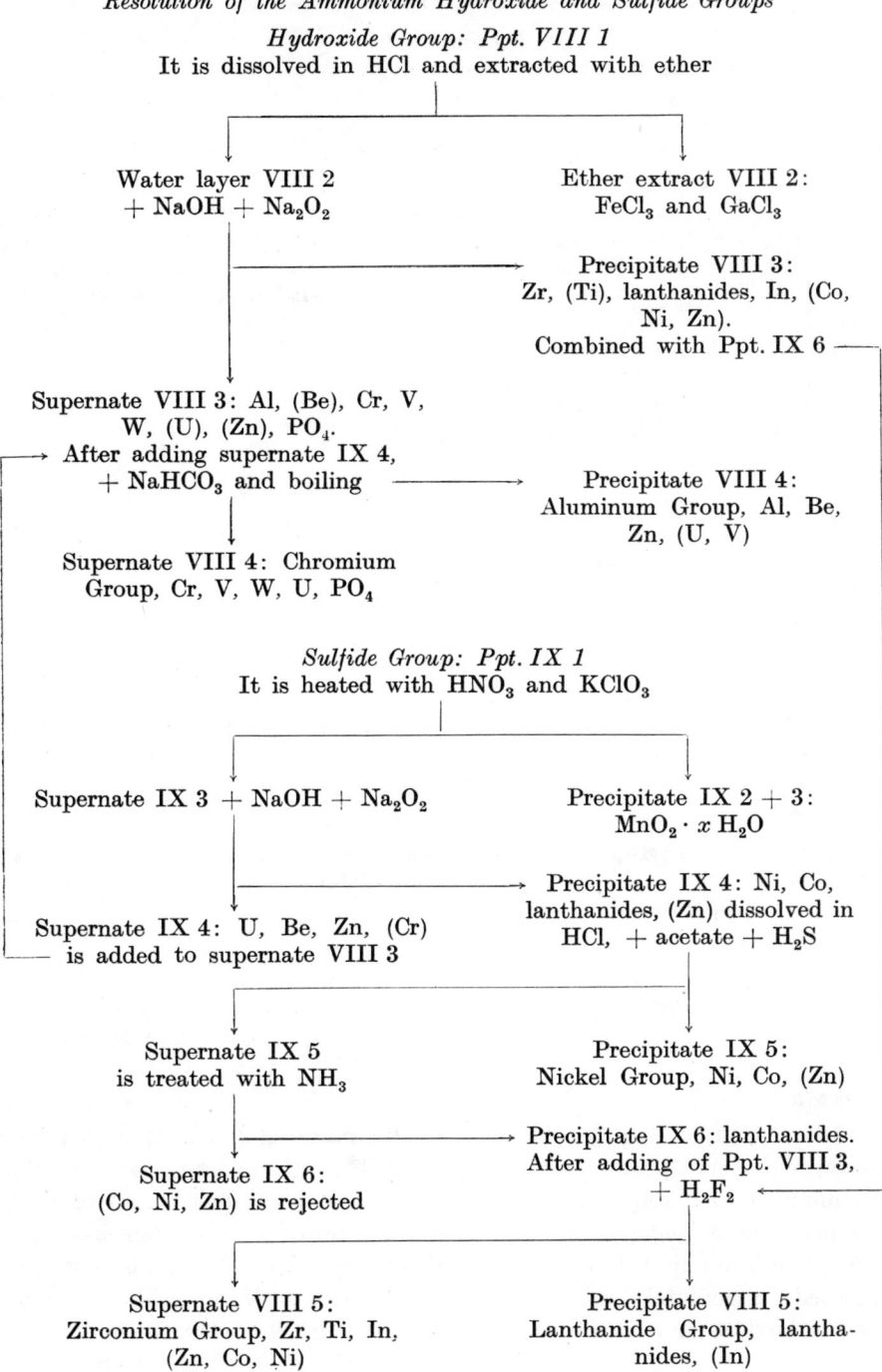

Analysis of the Tantalum Group

Precipitate II 1 is boiled with a solution of Na salicylate

↓

Residue III 1: Ta_2O_5, Nb_2O_5, ZrO_2, $Zr(HPO_4)_2$, $(Bi_2S_3$, other sulfides, and SiO_2 from the apparatus). For removal of SiO_2 and the oxidation of sulfides and organic substance, H_2F_2, HNO_3, and H_2SO are added and evaporated. The residue is fused with K_2CO_3, and the melt extracted with cold water

Extract III 1: Ti salicylate, VO_3, PO_4. After removal of salicylic acid by extraction with ether, the aqueous solution is treated with NaOH

Precipitate III 3: $TiO_2 \cdot x\,H_2O$, (Zr, V)

Supernate III 3: VO_3, PO_4, (Mo, W)

Residue III 5: ZrO_2, Bi_2O_3, (PbO, Fe_2O_3) is fused with $K_2S_2O_7$, and the melt extracted with water. The extract is treated with H_2S

Extract III 5: tantalate, niobate, PO_4, (Mo, W, V) is diluted with water, saturated with SO_2, and boiled

Precipitate III 6: Bi_2S_3, (PbS, ZrO_2, Ta_2O_5, Nb_2O_5, Pt from the apparatus)

Supernate III 6: $ZrOSO_4$

Supernate III 7: PO_4, (Mo, W, V)

Precipitate III 7: Ta_2O_5 and Nb_2O_5

Procedure

Dissolution of Sample. Treat 1 g solid material (0.5 g of a metal or alloy) with small portions of 9-F HBr until 10 ml have been added. Boil 10 minutes with refluxing. If dissolution is not complete, cool to room temperature and add 0.5 ml Br_2. Heat 10 minutes on the steam bath and add more bromine if its color should disappear. If dissolution is complete, proceed immediately to distillation, *see* below. If a residue remains undissolved, separate and wash it first with 2 ml 9-F HBr (combine with extract A 1) and then thoroughly with hot water which is rejected.

Residue A 1. Dry it in a drying oven or heating block. Add 2 ml 16-F HNO_3 and evaporate just to dryness. Add little water, stirr to obtain a slurry, and transfer to a platinum crucible or dish. Evaporate to dryness. Add 3 to 10 ml 27-F HF, cover, and digest on the steam bath for 15 minutes. Rinse the cover into the crucible with 3 ml 9-F $HClO_4$ and 2 ml 16-F HNO_3. Cautiously (infrared lamp) heat the contents of the crucible until white fumes of $HClO_4$ appear. Cover again, if necessary, and continue heating to light fumes for 5 to 10 minutes; if necessary, add acid to keep the volume at 2 ml. Finally evaporate to dryness and transfer the residue with 4 to 6 portions of 0.5 ml water to the distilling apparatus already containing extract A 1.

Distillation of Extract A 1 Combined with Residue A 3. Place 5 ml bromine water into the receiver and distill until about 3 ml of the charge are left in the distilling apparatus.

Distillate A 4. Treat with 1-ml portions of 3-F $NH_2OH \cdot HCl$ until the solution is colorless. Then add 5 ml more of the reductant, close the apparatus, and heat 5 minutes in the steam bath. Selenium precipitates; the supernate is treated with a like volume of 12-F HCl and then with H_2S for the precipitation of the sulfides of As and Ge. The filtrate from the sulfide precipitate should be added to the solution of the tungsten (wolfram) group.

Distillation Residue A 4. If Os and Ru cannot be present, add 4 ml 16-F HNO_3 and boil slowly until the volume is reduced to 3 ml. Add 5 ml 9-F $HClO_4$ and again boil slowly until the volume is 3 ml; the boiling of the concentrated $HClO_4$ must last at least 2 minutes for the dehydration of the acids of the wolfram and tantalum groups. Then proceed as with residue A 6, *below*.

If Os or (and) Ru may be present, proceed as follows. Treat residue A 4 in the distilling apparatus with 5 ml 16-F HNO_3. Place into the receiver 10 ml 6-F NaOH and cool it with ice. Distill slowly until the bromine vapor is no longer noticeable in the delivery tube. Gradually add to the cold distillate 1 ml Na_2O_2 powder. Continue the distillation until only 4 ml of the charge are left in the still pot. On the milligram scale follow the directions of RACHELE (413).

Distillate A 5. Os is indicated by a yellow color of the distillate. Add 2 ml ethanol, allow to stand for 5 minutes, separate from any precipitate, and saturate the clear solution with H_2S: black osmium sulfide.

Residue A 5. Treat the residue in the still with 5 ml 9-F $HClO_4$ and place 12 ml 6-F NaOH into the receiver which is cooled with ice water. Distil until only 3 ml remain in the still pot. An orange to red color of the distillate indicates Ru (a red precipitate, indicating mercury, should be separated out). If the distillate is colored, add 2 ml more of 9-F $HClO_4$

to the residue in the pot and distil into 5 ml 6-F NaOH until again only 3 ml remain in the pot. If it is colored, add the second distillate to the first.

Distillate A 6. Proceed as with distillate A 5; the precipitate is ruthenium sulfide.

Residue A 6. If this seems advisable, transfer the residue from the distillation to apparatus which facilitates the separation of solid from liquid. Treat the residue with 10 ml 12-F HCOOH (which also may be used for rinsing the still into the apparatus to which the residue has been transferred). Heat and keep gently boiling for 15 minutes. (If dissolution is complete and no precipitate forms, then absence of Ta, Nb, Au, Hg, Pt, and Pd is proven; proceed as with solution A 7.) If a solid is present after the indicated boiling period, remove the supernatant solution A 7 while it still hot. Wash the solid with three 4-ml portions of hot water, and combine the washings with solution A 7.

Solution A 7. Evaporate to a volume of 10 ml, cool with tap water, separate any crystalline solid from the supernate A 9, and wash it thoroughly with a small amount of ice cold 3-F HClO$_4$. Reject the washings. Save supernate A 9 which contains the analytical groups V to XVI.

Crystallizate A 9 proves the presence of large amounts of K, Rb, or Cs. Reserve it, and add it to supernate XV 1—or better XVI 2—so that it will be ignited with ammonium salt (acidify with HCl before evaporating) for the destruction of perchlorate: $3 \text{KClO}_4 + 8 \text{NH}_4\text{Cl} \rightarrow 3 \text{KCl} + 8 \text{HCl} + 4 \text{N}_2 + 12 \text{H}_2\text{O}$.

Residue A 7. Use a small amount of water to transfer the residue into suitable platinum apparatus (or substitute). Evaporate the water used in transfer. Add 5 to 10 ml 27-F HF and heat 3 to 10 minutes on the steam bath. Allow to cool and then add a volume of water equal to 2.5 times the volume of the mixture. Mix thoroughly with a platinum tool to bring into solution any tin fluoride that may have separated. Separate extract A 8 and transfer it to suitable platinum apparatus. Thoroughly wash residue A 8 with water, but combine only the first (intentionally small) portion of wash liquid with extract A 8. Reject the rest of the washings.

Extract A 8. Add 4 ml 9-F H$_2$SO$_4$ and evaporate to fumes of SO$_3$. Use 10 ml water (and possible the reagent solutions to be added) to transfer the concentrate to apparatus suitable for heating under pressure and separation of the liquid and solid phases. There add 15-F NH$_3$ until the mixture is ammoniacal (about 6 ml will be needed) and then 10 ml 6-F (NH$_4$)$_2$S which has been freshly prepared. Mix, close the apparatus, and immerse it in boiling water. If a clear solution results, treat it as solution II 1. If a small precipitate forms, cool and remove the supernate. If, however, a large precipitate separates, add 10 ml more of 6-F (NH$_4$)$_2$S, close the apparatus, and heat for 30 minutes in boiling water. Then allow to cool and separate solution II 1 from precipitate II 1. Wash the precipitate with 5- to 10-ml portions of hot water; combine the first washing with solution II 1.

Precipitate II 1. *See* p. 420.

Solution II 1. Under a fume hood, add the solution in small portions and with continuous mixing to 40 ml 3-F H_2SO_4. The mixture must remain acid when all of solution II 1 has been added. Finally stir for 2 minutes and then separate precipitate II 2 from the supernate. Wash the precipitate with hot water and finally dry it by placing the apparatus containing it into an oven or a heating block at 120° C.

Supernate II 2. It may contain significant amounts of wolfram only when phosphate is present. Consequently test it first for phosphate, and if this is present, test it for wolfram.

Residue A 8. Use 50 ml 1.5-F Na_2CO_3 for transferring the residue to some apparatus (large test tube) which returns the condensate of the steam to the solution. Heat to boiling, add 10 g solid Na_2CO_3, and boil gently for 15 minutes (so that the steam condenses in the upper part of the tube). Separate extract A 10 from residue A 10 while hot, and wash the residue thoroughly with hot water to remove all sulfate and fluoride. Reject extract A 10 and the washings.

Residue A 10. Add 10 ml 3-F $HClO_4$, heat a few minutes on the steam bath, and then separate the extract A 11 from the residue. Wash the latter thoroughly with hot water, but add only the first washing to the extract.

Extract A 11 may still contain some sodium derived from the Na_2CO_3 and, consequently, should not be combined with the main solution. Treat the extract as directed for supernate V 1 and analyze it separately, but combine the group precipitates with those of the of the main solution (VI 1, VIII 1, IX 1, and XV 1).

Residue A 11. Add 4 to 12 ml of a mixture of 3 volumes 16-F HNO_3 with 1 volume 12-F HCl and heat on the steam bath for 10 minutes. Evaporate to a volume of 0.5 to 1 ml, add 12 ml 0.025-F HCl, heat to boiling, and separate extract A 12 containing the gold group. Wash the residue A 12 with three 3-ml portions of 0.025-F HCl. Combine the first washing with the extract, and reject the rest.

Residue A 12. Add 10 ml 15-F NH_3, stir for 2 minutes, and then remove extract A 13. Test one droplet of the extract for the presence of silver as in Expt. 32. If crystals of AgCl form, wash residue A 13 with 6-F NH_3 until free of silver, and combine the washings with extract A 13.

Extract A 13. Heat on the steam bath until nearly all NH_3 has been expelled. Then add 1 ml 6-F HCl and, if not yet acid, some HNO_3. Heat 5 minutes on the steam bath. Collect the AgCl and estimate the amount of silver.

Residue A 13. If it is purely metallic, treat it as directed for residue A 17. If it is partly or wholly non-metallic, fuse with 10 g $K_2S_2O_7$ in a crucible or test tube of vitreous silica (when working on the gram scale, introduce small portions of the dried residue into the fused salt; when working on a small scale transfer the residue as a slurry with water and evaporate the latter before adding the solid salt). Heat for 20 minutes at incipient red heat so that SO_3 escapes only slowly (on the small scale, it may be necessary to repeatedly add H_2SO_4, *see* p. 112). Allow to cool and extract the melt with 25 ml cold water (on the gram scale, transfer the solidified melt to a mortar and crush it to a fine powder; on the small scale, try to spread the solidifying melt to a thin layer and observe the subsequent dissolution with a magnifier). Wait until the melt is completely broken down by the action of the water, mix, and remove the clear extract A 14 as completely as possible. Wash the residue with 2 ml cold water and combine this washing with the extract. Treat the residue with 5 to 10 ml 12-F HCl and heat on the steam bath until the attack seems to be complete. Evaporate to a volume of 1 ml, add 5 ml water, mix, and transfer the clear solution to extract A 14. Wash any residue A 14 with cold water and add the first two washings to extract A 14.

Extract A 14. Do not combine with the main solution (A 9) since this would make impossible the testing for potassium. Precipitate with H_2S as directed in P. 65 of the classical scheme. Treat the sulfide precipitate as directed for the solid sample of the scheme of NOYES and BRAY, and proceed with the filtrate from the precipitated sulfides as with supernate VI 2, *below*. This implies going through the whole scheme of separations, but simplifications will be possible. If the appearance of any group precipitate is similar to the corresponding group precipitate of the main solution, the two may be combined; if they are quite dissimilar, it will be advantageous to analyze them separately.

Residue A 14. Treat with H_2F_2 as described for residue A 7.

Extract A 15. Treat as directed for extract A 8.

Residue A 15. Treat as directed for residue A 8.

Residue A 16. Treat as directed for residue A 10.

Extract A 17. Proceed as directed for supernate V 1, *below*. Treat the sulfide precipitate like precipitate VI 1 + 2 and the supernate, like supernate IX 1.

Residue A 17. Perform a Na_2O_2 fusion with apparatus of nickel, iron, or platinum. Since all of these materials are attacked, use the one which is least objectionable as an impurity. Platinum is strongly attacked, but this may not matter (aside from the damage to the apparatus) if platinum metals need not be considered in the analysis. For manipulation adopt one of the alternatives mentioned for the treatment of residue A 13.

Fuse residue A 17 with 3 to 5 g Na_2O_2 at a temperature just sufficiently high to keep the melt in the liquid state (incipient red heat). Maintain the temperature until the residue is completely dissolved, which may take from 3 to 20 minutes. Allow to cool and then transfer with the use of 30 to 35 ml water into glass apparatus suitable for the separation of solid and liquid phase. Caution is required on the gram scale since adding water liberates heat and may be accompanied by the violent expulsion of steam and gas.

Continuously stirring, heat the resulting mixture for 10 minutes on the steam bath. Then cool with tap water and neutralize with continuous cooling to prevent the escape of OsO_4. Add 12-F HCl, first in portions of 1 ml and finally in portions of 0.1 ml until the mixture is just acid. Add 1 ml ethanol (to reduce the OsO_4) and 2.5 ml 12-F HCl. Heat on the steam bath and stir until dissolution is complete. (If there should be an insoluble residue, separate it, add 5 ml 12-F HCl and 1 ml 16-F HNO_3, heat on the steam bath for 5 to 10 minutes, evaporate to dryness, and transfer the residue—solid and liquid—with the use of 3 ml 1-F HCl to the bulk of the solution. If platinum apparatus has been used, yellow Pt_2O_3 may separate when the melt is dissolved and may be discarded.)

Saturate the solution of the melt (together with some insoluble residue) with H_2S, close the apparatus, and heat for one hour in the steam bath. When the apparatus is opened after cooling, the odor of H_2S must be discernible (otherwise saturate and heat again). Finally add 50 ml water and saturate the cold mixture with H_2S. Separate and wash the solid with hot water.

Treat the supernate as directed for supernate VI 2, and analyze the sulfide precipitate like any solid sample, starting with the dissolution in HBr and Br_2.

Supernate A 9 (Main Solution). Add 2 ml 2-F HBr, free from bromine (mix 1 ml 9-F HBr, 1 ml 12-F HCOOH, 2.5 ml water and heat a few minutes in the steam bath), cool with tap water, stir a few minutes, and allow the precipitate to settle. Add another 2-ml portion of the HBr and continue this until precipitation is complete. Finally separate supernate V 1 from the precipitate V 1, wash the latter with 5 ml cold 1-F HBr (free from bromine), and add the washing to supernate V 1.

Supernate V 1. Add 3 ml 3-F NH_4Cl and evaporate to dryness on the steam bath. Dissolve the residue by heating with 30 ml 1-F HCl. Transfer the solution to suitable apparatus and saturate it with H_2S while still warm. Add 70 ml water, saturate with H_2S, close gas-tight, and heat for 15 minutes in the steam bath (immersion in boiling water). Allow the precipitate to settle, cool, and again saturate the supernatant solution with H_2S. Close again and heat 10 more minutes in the steam bath. Remove supernate VI 1, wash precipitate VI 1 once with 10 ml hot water, and add the washing to supernate VI 1.

Supernate VI 1 (not indicated on flow sheet). If Ir and Mo cannot be present, treat it as directed for supernate VI 2. Otherwise evaporate to a volume of 10 ml. Add 20 ml 6-F HCl, heat 10 minutes on the steam bath, cool to room temperature, saturate with H_2S, close the apparatus gas-tight, and immerse for 30 minutes into a bath of boiling water. After cooling, saturate again with H_2S, close, and heat for 30 more minutes. Separate supernate VI 2 from precipitate VI 2, and wash the latter with 5 to 10 ml hot water. Reject the washings.

Precipitates VI 1 and VI 2. Heat precipitate VI 1 with 20 to 40 ml 12-F HCl for five minutes on the steam bath. Add 4 to 8 ml 16-F HNO_3 and keep at 70° C until dissolution is complete.

If a precipitate VI 2 has been obtained, heat it in like manner first with 5 to 10 ml 12-F HCl and then after addition of 1 to 2 ml 16-F HNO_3. Add the solution to that of precipitate VI 1, and add also two rinse portions of 10 ml 12-F HCl.

Evaporate to dryness (steam bath) the solution of precipitate VI 1 or the combined solutions of precipitates VI 1 and VI 2. Dissolve the residue in 10 to 20 ml 12-F HCl and saturate the solution with SO_2. (If Se precipitates, separate it out and continue with the clear supernate.) Add water (35 to 70 ml) to lower the HCl concentration to 2.7 formal. Saturate with SO_2, close gas-tight, and heat for 15 minutes in the steam bath. Open and, while hot, separate supernate VI 3 from precipitate VI 3. Wash the precipitate with two 10-ml portions of hot water and combine the washings with the supernate.

Supernate VI 3. Concerning the detection of rhenium in this solution see CHRISTINA C. MILLER (724). For the separation of the other elements, evaporate supernate VI 3 to dryness on the steam bath. Add 1 ml 16-F HNO_3, evaporate to dryness, and repeat this with another 1-ml portion of the acid to be certain of complete oxidation of molybdenum. Cool to room temperature, dissolve the residue in 5 ml 6-F HCl, and extract with three 50-ml portions of diethyl ether. Wash the combined ether extracts once with 5 ml 6-F HCl, and add this washing to the aqueous solution VI 4.

Supernate VI 2. Use one twentieth of the supernate for a side test for Fe and PO_4. Boil to expel H_2S. Then add 0.5 ml 3-F HNO_3 and evaporate nearly to dryness. After cooling, dissolve the residue in 1 ml 12-F HCl. Add 5 ml 6-F HCl and 6 ml diethyl ether for the extraction of $FeCl_3$. — Allow the ether extract to evaporate. Treat the residue with 0.5 ml 6-F HCl, 10 ml water, and 2 ml 1-F KCNS. Use a light red coloration for the estimation of the amount of iron. If the solution turns dark red, add 2 ml 6-F NaOH and estimate the quantity from the volume of $Fe(OH)_3$. — Evaporate the aqueous layer nearly to dryness, then add 3 ml water and 10 ml of a mixture of equal volumes of 6-F HNO_3 and molybdate reagent [dissolve 90 g $(NH_4)_2Mo_7O_{24} \cdot 4 H_2O$ in 100 ml 6-F NH_3, add 240 g NH_4NO_3, and dilute with water to 1 liter). Immerse for 15 minutes into a water bath at 70° C. Estimate the amount of PO_4 from the volume of the yellow precipitate.

Treat the main part of supernate VI 2 as follows. Boil to remove the dissolved H_2S (passing a stream of air through the solution will hasten the process). Add bromine until color or odor of the solution indicate presence of a slight excess (this oxidizes pale green Fe^{++} and blue $V_2O_4^{--}$). Heat and pass air through the solution to expel the excess of bromine. Add 4 ml 6-F NH_3 and 15 ml 3-F ammonium acetate (if phosphate is absent, heat to near boiling—if no precipitate forms, the amounts of Fe, Al, Ga, Ti, and Zr must be less than 0.002 g—and then cool again to room temperature). Treat the cold solution (mixture) with 1-ml portions of 1-F $Fe(NO_3)_3$ until its color becomes deep brown due to the hydrolysis of an excess of ferric ion. Add 3 ml more of the ferric nitrate solution, heat to incipient boiling, maintain this for 2 minutes, and then place into a water bath at 90° C. If a precipitate does not settle out, add 2 ml 6-F NH_3, and heat again to near boiling for 2 minutes. If this does not cause settling, add 5 ml 0.5-F $(NH_4)_2HPO_4$ and again heat for 2 minutes. Keep above 90° C all the time. If the solution should assume a brown color, add 1 ml 6-F NH_3 (test with litmus paper and add 6-F acetic acid, if necessary, to restore the acid reaction) and heat again for 2 minutes near boiling. Repeat the treatment with NH_3 if necessary.

Separate precipitate VIII 1 from the supernate when the condition is satisfied that the precipitate settles out and the solution does not show a coloration indicating ferric ion. Separate solution and precipitate while hot, and wash the precipitate with two 50-ml portions of boiling water. Add the washings to supernate VIII 1.

Precipitate VIII 1. Without delay, treat with 5 to 10 ml 6-F HCl and evaporate on the steam bath to a volume of 1 ml. Continue as directed on p. 418.

Supernate VIII 1. Treat with 1-ml portions of 6-F NH_3 until the mixture has the odor of NH_3, and then add 2 more milliliter 6-F NH_3. Heat just to boiling. Pass H_2S into the mixture until it has the odor of H_2S. Test with litmus and, if necessary, add NH_3 to restore alkaline reaction. Heat to incipient boiling and remove the supernate. Boil the supernate for half a minute (or longer if it has a dark color) and separate the clear supernate IX 1 from any small amount of precipitate which may have formed. Pass some H_2S into 50 ml 0.06-F NH_3 (test with litmus to be certain of the alkaline reaction), and use this solution for washing; with the first portion transfer the small amount of precipitate obtained on boiling the supernate to the main portion of precipitate IX 1. Combine the first washing with supernate IX 1.

Precipitate IX 1. Without delay dissolve precipitate IX 1 in 10 ml 6-F HNO_3, evaporate just to dryness on the steam bath, and reserve the residue for further treatment as outlined below.

Supernate IX 1. Evaporate a small portion of the supernate on the steam bath to dryness, add to the residue another small portion, evaporate again, and continue until the whole centrifugate has been evaporated. Treat the residue with 25 ml 12-F HCl, evaporate to dryness, and repeat this treatment once. Extract the residue with three 50-ml portions of water. Evaporate this extract in suitable platinum apparatus (dish, crucible, sheet) to dryness on the steam bath (again it is advised to evaporate small portions). Moisten the residue with 10 ml 6-F NH_4Cl and evaporate to dryness. Heat for 10 minutes at 140° C (drying oven or heating block) and then heat with a small flame until fumes are no longer coming off. Finally heat to incipient dark red heat and quickly remove the flame. Extract the residue with several 10-ml portions of water, and evaporate the clear extracts to dryness. Dissolve the residue in 10 ml water and separate out any turbidity. Treat the clear solution with 15 ml of carbonate reagent (25 g freshly pulverized ammonium carbonate dissolved in 6-F NH_3 to obtain a volume of 100 ml) and 15 ml 95% ethanol. If the precipitate is large, add a second installment of 15 ml carbonate reagent and 15 ml ethanol. Finally stir, allow to stand for 10 minutes and aid crystallization by occasional shaking or exposure to vibrations of high (but not supersonic) frequency (413). Separate supernate XV 1 from precipitate XV 1, and wash the latter with a very small volume of a mixture of equal parts of carbonate reagent and ethanol.

Precipitate VIII 1. Treat the 1 ml of the solution of the precipitate with 10 ml 6-F HCl and (regardless of any precipitate) 10 ml diethyl ether (shake 40 ml ether, immediately before use, with 80 ml 6-F HCl to make it suitable for this extraction). Repeat the extraction with two more 10-ml portions of the ether. The combined ether extracts are extract VIII 2. A white precipitate forming on addition of ether indicates presence of a large amount of Zr or W; a yellow coloration of the aqueous layer suggests Ti.

Water Layer VIII 2. Evaporate to a volume of 1 ml. Add 2 ml 16-F HNO_3 and evaporate nearly to dryness. Dissolve in 10 ml water and, if necessary, transfer to suitable apparatus of either nickel, platinum, or vitreous silica (the last is most convenient; Vycor or Pyrex glass may be used if contamination with aluminum is not objectionable). Add small portions of pure 6-F NaOH (tested for absence of Al if small amounts of Al shall be detected in the sample) until the mixture is alkaline. If the mixture is thick with precipitate, add 10 to 20 ml more water. Cool the apparatus with ice water, and treat its contents by slowly adding, with continuous stirring, 1 to 3 ml solid Na_2O_2 (free from Al and assuredly undecomposed). Continuous liberation of oxygen indicates that enough of the reagent has been added. Boil gently for 5 minutes, dilute with water to 40 ml, heat to a brief boil, cool, and separate supernate VIII 3 from precipitate VIII 3 (hardened filter paper may be used, but centrifuging and decantation are preferable). Wash the precipitate thoroughly with hot water.

Supernate VIII 3. Cool the supernate and add 0.5- to 1-ml portions of 6-F HNO$_3$ until the solution is acid. Set aside for combining with supernate IX 4.

Precipitate VIII 3. Without delay add 5 to 10 ml 6-F HCl and set aside for combining with precipitate IX 6.

Precipitate IX 1. Treat the residue from the evaporated solution of precipitate IX 1 with 2 ml 16-F HNO$_3$ and again evaporate to dryness. Treat the residue with 15 ml 16-F HNO$_3$ and 1 ml solid KClO$_3$, and stir for 5 minutes while heating in the steam bath. Add another 1-ml portion of solid KClO$_3$, heat and stir for 2 minutes, and repeat this with a third 1-ml portion of the solid reagent. Remove supernate IX 2 from the precipitate IX 2. Treat the latter with 10 ml 16-F HNO$_3$ and 1 ml solid KClO$_3$, heat two minutes with stirring, separate, and transfer the supernate to supernate IX 2.

Supernate IX 2 (not mentioned on the flow sheet). Evaporate just to dryness, and dissolve the residue in 20 ml water. Separate from any solid (Ppt. IX 3) to obtain the clear supernate IX 3.

Precipitates IX 2 and IX 3 are combined for the estimation of the amount of Mn.

Supernate IX 3. Proceed exactly as directed for water layer VIII 1. Obtained are supernate IX 4 and precipitate IX 4. Wash the latter with three 10-ml portions of hot water and add the first washing to supernate IX 4. Without delay treat precipitate IX 4 with 5 to 20 ml 6-F HCl and set it aside for later use.

Supernate IX 4. Cool and add 0.5- to 1-ml portions of 6-F HNO$_3$ until acid, and then add to supernate VIII 3.

Combined Supernates VIII 3 and IX 4. With mixing add 0.5-ml portions of solid NaHCO$_3$ until Congo red paper remains red. Then add 2 g solid NaHCO$_3$, 40 ml water, and 2 ml 3% H$_2$O$_2$. Close the apparatus air-tight, and place it in boiling water for 20 minutes. Cool, open the apparatus, and separate supernate VIII 4 from precipitate VIII 4. Wash the latter with one 20-ml portion of cold water and combine the washing with supernate VIII 4. Without delay treat precipitate VIII 4 with 5 to 15 ml 6-F HCl.

Precipitate IX 4. Evaporate it with the acid (which has been added) nearly to dryness (steam bath). Treat the residue with 30 ml cold water, and add 1-ml portions of 6-F NH$_3$ until the mixture has the odor of ammonia. Add 2 ml more of 6-F NH$_3$ and mix. (If no precipitate appears at this stage, lanthanides are apparently not present in this solution which may be treated immediately with H$_2$S for the precipitation of Co, Ni, and Zn.) Add 2-ml portions of 6-F acetic acid until the mixture is acid to litmus (separate a precipitate "LA" which refuses to dissolve, wash it once with hot water, treat it with 5 ml 6-F HCl, and reserve the mixture for combining with precipitate IX 6). Mix into the clear solution 10 ml 3-F ammonium acetate, heat to about 75° C, and pass into it for 5 minutes a slow stream of H_2S. Mix and heat 5 minutes on the steam bath. Cool to room temperature, saturate with H_2S, heat again to about 75° C, saturate again with H_2S, and then separate supernate IX 5 from precipitate IX 5. Wash the latter with cold water that has been saturated with H_2S. Attend immediately to the precipitate.

Precipitate IX 5. If the precipitate is pure white, it consists of ZnS and is used for the estimation of the amount of Zn. Otherwise add 10 to 30 ml cold 1-F HCl, stir for 5 minutes, and without delay separate supernate X 1 (containing Zn with little Co and Ni) from residue X 1 consisting of CoS and NiS.

Supernate IX 5. Heat on the steam bath and stir to remove the dissolved H_2S. Remove any precipitate of sulfur, and treat the clear solution with small portions of 6-F NH_3 until the mixture has a slight odor of NH_3. Finally add 2 ml more of 6-F NH_3, and separate supernate IX 6 from a precipitate IX 6. Washing is not necessary.

Precipitate IX 6. Add the acid solutions of precipitate VIII 3 and of a precipitate "LA". Transfer the mixture to suitable apparatus of platinum or plastic (also vitreous silica and Pyrex glass may serve with the obvious limitations) and evaporate to dryness. Treat the residue with 2 ml water, 0.5 ml 27-F HF, 1 ml 6-F HCl, and finally again 9 ml water. If a large residue is left, add 0.1-ml portions of the HF (but not more than a total of 0.5 ml) until there is no more dissolution (the volume of the residue remains constant). Heat 5 minutes on the steam bath, and then remove supernate VIII 5 from the precipitate VIII 5. Wash the latter with two very small portions of a mixture of 10 ml water, 0.5 ml 27-F HF, and 1 ml 6-F HCl.

Without delay treat supernate VIII 5 as described in the literature. [Add small portions of 6-F NH_3 until alkaline, then add 0.3 ml portions of 3-F HF until acid, add 0.3 ml more of the H_2F_2, saturate with H_2S, allow to stand for 5 to 10 minutes, and separate In_2S_3 (CoS, NiS, ZnS) from the supernate containing Ti and Zr.]

Analysis of the Tantalum Group

Precipitate II 1. Treat with 400 ml hot water which contains dissolved 15 g salicylic acid and 4.7 g Na_2CO_3. Boil the mixture for two hours (but stop the boiling immediately if complete dissolution should occur), and replace water which has vaporized to keep the volume constant. Separate extract III 1, while boiling hot, from residue III 1. Wash the latter thoroughly with boiling hot water.

Extract III 1. A slowly developing violet coloration is caused by the presence of iron; 0.0003 g Ti give a yellow and 0.0025 gTi an orange coloration.

Evaporate to a volume of 40 to 60 ml. Add 6 ml 9-F H_2SO_4, cool with tap water, add 40 ml diethyl ether, mix, and separate the layers. Repeat the extraction with two 20-ml portions of ether. Reject the ether extracts, but treat the remaining aqueous layer as follows. Evaporate to a volume of 15 to 20 ml. Add small portions of 6-F NaOH until the mixture is alkaline, and then add 10 ml more 6-F NaOH. Mix and heat for 5 minutes on the steam bath. Remove supernate III 3 while still hot, and wash precipitate III 3 thoroughly with hot water.

Precipitate III 3 is suited for the estimation of the amount of Ti; it will not contain more than 0.001 g Zr and possibly a trace of V.

Supernate III 3. Use one fifth of it for estimating and confirming phosphate by means of molybdate reagent, and use the rest of it for confirming and estimating vanadium.

Residue III 1. Add water, make a slurry, and transfer it to suitable platinum apparatus (dish, crucible, sheet). Evaporate the water and treat with a mixture of 2 ml 9-F H_2SO_4, 0.5 ml 16-F HNO_3, and 1 ml 27-F HF. Evaporate to the appearance of fumes of SO_3. Add 0.5 ml 16-F HNO_3, evaporate again to fumes of SO_3, and repeat the treatment until all organic matter is destroyed. Finally evaporate to dryness. Use the amount of residue to get an estimate of the total amount of Ta, Nb, Zr, and Bi. Then add 5 g of finely powdered anhydrous K_2CO_3 which contains 2% KNO_3. Fuse and maintain the liquid state for 5 minutes. After cooling, extract the melt with 10 ml cold water. Stir until the melt is completely disintegrated, and then remove extract III 5 and wash residue III 5 with several 5-ml portions of cold 50% K_2CO_3 and finally with 5 ml water. Combine the first washing with extract III 5.

Extract III 5. Dilute with water to a volume of 40 ml and saturate with SO_2. Heat 30 minutes on the steam bath while passing SO_2 through the solution. Remove supernate III 7 while hot, and wash precipitate III 7 with hot water.

Supernate III 7. Boil it for the removal of SO_2 before testing for phosphate with molybdate reagent.

Precipitate III 7. The formation of the precipitate indicates the presence of either Ta or (and) Nb. If necessary, transfer as a slurry with water to suitable apparatus, evaporate the water, and dry by placing into a drying oven. Add 5 g $K_2S_2O_7$, heat, and keep the melt in the liquid state for 5 to 10 minutes. After cooling, add 100 ml 0.2-F $(NH_4)_2C_2O_4$, heat on the steam bath and stir until the melt is completely disintegrated.

To test the clear solution of the melt for Ta and Nb, add 5-ml portions of 2% tannin solution. Heat after each addition on the steam bath, and then centrifuge (avoid to stir up any precipitate which has been already collected). If no precipitate is obtained, add some solid NH_4Cl and 2 ml $2\text{-}F$ NH_3. Continue adding portions of tannin solution as long as a precipitate forms.

The yellow Ta compound precipitates first (without requiring NH_4Cl and NH_3). Later (after adding the ammonia buffer) an orange mixture of Ta and Nb compound separates and finally, the red Nb compound. A small amount of Ti which, however, should not get into ppt. III 7, gives a red ppt. on adding the first portion of tannin. MILLER (723) gives advice concerning this difficulty.

Residue III 5. Perform a pyrosulfate fusion as directed for precipitate III 7, *above*. After cooling, extract the melt with 20 ml water by heating on the steam bath and stirring until the melt is completely disintegrated. After cooling to room temperature, saturate with H_2S, remove supernate III 6, and wash precipitate III 6 with some water saturated with H_2S.

Supernate III 6. Boil to remove H_2S, and then test the clear solution for Zr. [Add 2 ml $1\text{-}F$ H_2O_2 and 10 ml $0.5\text{-}F$ Na_2HPO_4. Allow to stand for 1 hour: 0.001 g Zr still gives a white ppt. of $Zr(HPO_4)_2$.]

Precipitate III 6. Of interest is only the presence of Bi. Not more than 0.002 to 0.005 g of it may be expected in this place, but it may happen that all of a small amount of Bi collects here. A small amount of black precipitate may also be PtS_2 derived from the use of platinum apparatus.

To test a black precipitate III 6 for Bi, first estimate its amount and then dissolve it by heating with 10 ml $2\text{-}F$ HNO_3. Treat the clear solution as directed for the solution of ppt. VI 5, *see* the literature.

P. 69   Systematic Search for Anions

The search for cations is essentially ultimate or elemental analysis, i. e., an investigation of the absence or presence of elements without regard to their binding. In the instance of anions, however, one customarily tests for "molecular" species: Cl^-, ClO^-, ClO_3^-, etc. This task would be very difficult if a systematic separation were to include all known inorganic anions, among others such as $SnO_3^=$, $SnCl_6^=$, $Mn(CN)_6^{==}$, $Co(NO_2)_6^{--}$, $Co(NH_3)(NO_2)_5^=$. There is no need for such a scheme, however, since knowledge of elemental composition, of radicals present, and of the appearance and physical properties usually permits drawing the necessary conclusions.

If the tests with dilute and concentrated sulfuric acid (P. 36 and P. 37) have not yet been performed, they should be carried out now. Table XI gives a list of the "acids" that should be considered. The inclusion of some organic compounds and of the iron complexes may be justified by their frequent occurrence and the possibility that their organic nature may not be recognized when the substance is ignited. Italics indicate the acids that may be recognized by testing with sulfuric acid. The presence of Se,

Table XI. *Acids Commonly Considered in Testing for Anions*

HF					
HCl	*HClO*	*HClO$_2$*	*HClO$_3$*	**HClO$_4$**	
HBr			**HBrO$_3$**		
HI			HIO$_3$	HIO$_4$	
H$_2$S	*H$_2$SO$_3$*	*H$_2$S$_2$O$_3$*	*H$_2$S$_x$O$_6$*	H$_2$SO$_4$	**H$_2$S$_2$O$_8$**
H$_2$Se	H$_2$SeO$_3$			H$_2$SeO$_4$	
H$_2$Te	H$_2$TeO$_3$			H$_2$TeO$_4$	
HN$_3$	*HNO$_2$*			**HNO$_3$**	
H$_3$P	*H(H$_2$PO$_2$)*	*H$_2$(HPO$_3$)*	*H$_2$(H$_2$P$_2$O$_6$)*	H$_3$PO$_4$	Poly
H$_3$As		H$_3$AsO$_3$		H$_3$AsO$_4$	
H$_4$C	*H(HCOO)*	*H$_2$(COO)$_2$*	*H$_2$CO$_3$*	**H(CH$_3$COO)**	
	HCN	*HCNO*	*HCNS*	H$_4$Fe(CN)$_6$	H$_3$Fe(CN)$_6$
(H$_4$Si)	**H$_2$SiO$_3$**	*H$_2$SiF$_6$*			
(H$_3$B)	**HBO$_2$**	*HBF$_4$*			

Italics indicate the acids that may be discovered when testing a sample with dilute or concentrated sulfuric acid.

Boldface indicates those which are not precipitated by barium or silver nitrate (or may escape detection by these tests) so that special tests are required.

Te, and As is discovered in the search for metals. In addition, the following direct ultimate tests may be used to determine presence or absence of nonmetals:

F: volatilization as SiF$_4$ and precipitation of Na$_2$SiF$_6$, P. 60.

Cl, Br, I: strong ignition of a solid sample with sufficient Na$_2$CO$_3$ to maintain alkalinity and then testing of the solution of the melt with HNO$_3$ and AgNO$_3$. To exclude the interference of cyanide, recrystallize from NH$_3$.

S: oxidation to sulfate or Hepar test, P. 59.

P: oxidation by heating with conc. HNO$_3$, boiling the aqueous solution, and molybdate test, P. 58.

Si: volatilization of SiF$_4$ and precipitation of Na$_2$SiF$_6$, P. 57.

B: tests given in P. 56.

When finally the solubility of the material and its metallic constituents are considered, a large number of possibilities may be eliminated from the list of anions so that the remaining choice may be decided by a few confirmatory tests (P. 43); these have to be selected so that the substances known to be present and also those that might be present will not interfere. To this end, it may be necessary to remove the metals (P. 71, 72) or to perform some simple separations. Obviously, if the material under investigation is quite uniform in appearance and seems to represent only one substance, the number of anions will probably not exceed three.

With complex materials, it may be desirable or necessary to do some more preliminary testing (P. 70) before proceeding to the confirmatory tests.

In general, it is advisable to follow a slight modification of the reasoning of A. A. NOYES (48) and to adjust the procedure depending upon the type of material: (I) materials that contain only alkali metals or NH_4, P. 70; (II) such that are readily dissolved or decomposed by water or acids, P. 71; and (III) refractive materials which are insoluble or only with difficulty attacked by acids, P. 72.

P. 70   *I. Only Alkali Metals or (and) Ammonium are Present*

If the material is difficultly soluble in water, the choice becomes quite limited; inspection under the microscope and a few tests should solve the problem.

Otherwise dissolve 1 g of the material in 2 ml water. With the use of 4-F HNO_3 and 6-F NaOH (best a freshly prepared solution of pure reagent) adjust the pH to about 5 to 9 and then dilute with water to 12 ml and mix to obtain the sample solution for orientation and confirmatory tests. Portions of this solution will serve for orientation and confirmatory tests (if n grams are taken for work on a small scale, multiply by n all weights and volumes given in the following directions, P. 70–72).

For orientation, it is suggested to use the classification with barium-calcium nitrate and silver nitrate—and some of the tests for oxidizing and reducing agents given *below*. The observations should reduce the number of possibilities to an extent permitting conclusion of the search with a few confirmatory tests, P. 43. A preliminary report by Ly (1263) indicates that classification by barium-calcium and silver nitrate will detect 0.0004 g (1% of about 0.04 g total anion) or less of most of the anions listed. It is suggested to separate the supernates from the precipitates for additional testing; this will aid in saving sample.

Classification by Barium-Calcium Nitrate

Dilute 1 ml of the test solution with 1 ml water. Add 0.1-ml portions of 6-F acetic acid until litmus turns red and then as much more as has already been added. Treat with 1 ml 0.5-F $Ba(NO_3)_2$ and 3 ml 0.5-F $Ca(NO_3)_2$, heat nearly to boiling, and let stand for 10 minutes: precipitate (1) and supernate (1).

Precipitate (1) may contain:
pale yellow $BaCrO_4$ soluble in 6-F HCl,
white S from polysulfide, thiosulfate, thionate; insoluble in acid,
$BaSO_4$ insoluble in acid,
$BaSeO_4$ and $BaTeO_4$ dissolved by boiling 12-F HCl with liberation of chlorine,
CaF_2 and $BaSiF_6$ (the latter distinctly crystalline) and both dissolving slowly in 6-F HCl,
$CaSO_3$ and CaC_2O_4 dissolving readily in 6-F HCl,
silicate which may or may not dissolve in HCl.

Supernate (1). Add 0.05-ml portions of 12-F NH_3 until the odor of NH_3 indicates a slight excess. Warm and then allow to stand: precipitate (2) and supernate (2).

Precipitate (2) may contain: iodate, sulfite, possibly thiosulfate, selenite, phosphite, hypophosphate, phosphate (ortho or poly), arsenite, arsenate, borate, and possibly some carbonate. It may be dissolved in 8-F HNO_3 and tested with $AgNO_3$, *below*.

Supernate (2) may contain: any of the halogen ions excepting fluoride (and only little iodate), persulfate, little sulfide or sulfite, azide, nitrite, hypophosphite, and possibly hypophosphate. There is much acetate present, that could give a crystalline precipitate with $AgNO_3$, soluble in hot water. The supernate may be tested with $AgNO_3$, *below*.

Classification by Silver Nitrate

To 1 ml of the test solution add 5 ml water, 1 ml 1-F $AgNO_3$, and 2 ml 6-F HNO_3: precipitate (3) and supernate (3).

Precipitate (3) may contain:
black Ag_2S from sulfide, thiosulfate, thionate,
  Ag_3P (possibly a small amount of it),
  $Ag°$ from H_3PO_2 or H_3PO_3,
  Ag_2O_2 from $H_2S_2O_8$,
slate colored $Ag_3H_2IO_6$, from periodate,
dark red $Ag_2H_3IO_6$ from periodate,
orange $AgIO_4$ and $Ag_3Fe(CN)_6$,
yellow AgI,
pale yellow $AgBr$,
white $AgCl$, $AgCN$, $AgCNS$, $Ag_4Fe(CN)_6$, and possibly $AgIO_3$.

If the precipitate is black, boil it briefly with 12-F HNO_3 and dilute with four volumes of water to see if a light colored or white precipitate persists.

Supernate (3): test 1 ml of it by adding 0.05 ml (drop) of 1-F $NaNO_2$.[1] A white precipitate may be $AgCl$ derived from hypochlorite or chlorate, or $AgBr$ from bromate.

Treat the rest of the supernate (3) by adding 0.05-ml portions of 6-F NaOH to obtain a pH of 7 to 9. A precipitate may contain:
black $Ag°$ from phosphite or hypophosphite,
  Ag_2O_2 from persulfate,
reddish brown Ag_2CrO_4 or Ag_3AsO_4,
yellow Ag_3PO_4 or Ag_3AsO_3,
white, turning brown to black: quickly $Ag_2S_2O_3$, AgH_2PO_2,
              slowly Ag_2HPO_3, $AgBO_2$,

---

[1] Must be free from chloride; it suffices to treat 1 ml of the reagent with 1 drop $AgNO_3$, to centrifuge, and to use the clear supernate.

white $AgIO_3$,
   Ag_2SO_3, little and possibly derived from $H_2S_2O_3$,
   $Ag_2S_2O_3$,
   Ag_2SeO_3,
   $Ag_4P_2O_6$, $Ag_2C_2O_4$, and $AgBO_2$.

Tests for Oxidants and Reductants

The absence of oxidants (reductants) cannot be considered conclusively proved unless also the test for reductants (oxidants) has a negative outcome. This still assumes that oxidant and reductant are not present in amounts to produce complete cancellation of both their actions.

Test for Oxidants

a) NOYES (48): To 0.5 ml of test solution add 2 ml saturated solution of $MnCl_2$ in 12-F HCl, and heat nearly to boiling. A dark brown to black coloration may be caused by:

ClO^-, ClO_3^-, NO_2^-, NO_3^-, $Fe(CN)_6^{--}$, $CrO_4^=$, MnO_4^-.

b) SWIFT (53): Treat 0.5 ml test solution with 0.5 ml CCl_4, 0.5 ml 3-F H_2SO_4, and five 0.1-ml portions of a freshly prepared solution of 1 g KI in 2.5 ml water, mixing thoroughly after each addition. A red or violet color of the CCl_4 may be caused by:

BrO_3^-, IO_3^-, IO_4^-, $S_2O_8^=$, NO_2^-, $Fe(CN)_6^{--}$, $CrO_4^=$, MnO_4^-.

Chlorate and nitrate give the coloration after standing for 30 minutes.

Tests for Reductants

c) MIDDLETON and WILLARD (44): Add 0.02-F $KMnO_4$ to 1.5-F H_2SO_4 until a stable pink coloration is obtained. Mix 1 ml of this acid with 0.5 ml test solution. If the mixture becomes colorless, the following may be present:

Br^-, I^-, CNS^-, $Fe(CN)_6^{--}$, $S^=$, $SO_3^=$, $S_2O_3^=$, NO_2^-, AsO_3^{--}, $C_2O_4^=$.

d) MIDDLETON and WILLARD (44): Mix 2 ml water, 2 ml 0.3-F $Fe(NO_3)_3$, 1 ml freshly prepared saturated solution of $K_3Fe(CN)_6$, and 4 ml 3-F HCl. Mix 2 ml of the freshly prepared reagent with 0.5 ml test solution and compare the color of the mixture with that of the reagent. A blue or green coloration may be brought about by:

I^-, $Fe(NCN)_6^{--}$, $S^=$, $SO_3^=$, $S_2O_3^=$, $C_2O_4^=$.

(A red color indicates thiocyanate.)

It should be understood that a test for nitrate is meaningless if nitric acid has been used in the neutralization of the test solution. The search is concluded by the performance of confirmatory tests, and it should not be forgotten to test for the acids indicated by boldface in Table XI. If much borate is present, it is recommended to test the Ba-Ca precipitate (1) for fluoride.

P. 71 *II. Nonmetallic Materials Readily Dissolved or Decomposed by Water or Acids*

Into this class belong industrial products of all kinds as well as soluble minerals and rocks.

It is assumed that the original material has been already tested for sulfide and carbonate. The search for the other anions is carried out with the soda extract which is prepared to eliminate most of the possibly interfering cations. The procedure for the preparation of the soda extract will have to be somewhat modified depending upon the nature of the sample, *see below*. If part of the material is insoluble in acid, prepare the soda extract from the acid solution and treat the insoluble residue according to P. 72. Proceed with the soda extract as outlined in P. 70.

The Material is Solid or a Paste. Treat 1 g of the finely powdered material with 10 ml 1.5-F Na_2CO_3 and heat on the steam bath for 10 minutes with continuous stirring. Remove the clear supernate, dilute it to 12 ml, and mix. It is the soda extract for the anion search. Wash the residue once with water and save it for possible later use.

The Material is an Aqueous Solution of pH 5 to 10. Take a volume of it, which corresponds to approximately 1 g solid residue, and evaporate to about 2 ml. Add 10 ml 1.5-F Na_2CO_3 and heat for 5 minutes on the steam bath with continuous stirring. Remove the clear supernate, dilute with water to 12 ml, and mix; this is the soda extract.

The Material is an Acid Solution. Take a volume corresponding to 1 g solid residue and add small portions of freshly prepared 6-F NaOH until alkaline. Evaporate to 2 ml and then add 10 ml 1.5-F Na_2CO_3 and continue as directed in the preceding paragraph.

The Material is an Alkaline Solution. Take a volume corresponding to 1 g solid residue and saturate with CO_2 (which may have to be washed with bicarbonate and then water). Evaporate to 2 ml, add 10 ml 1.5-F Na_2CO_3, and continue as directed for an aqueous solution, *above*.

P. 72 *III. Nonmetallic Refractive Materials*

This class comprises insoluble minerals and ores, ceramic products, glass, slags, glazes, and abrasives. Their origin at high temperature excludes many anions, and they have to be tested only for fluoride, chloride, nitride, phosphate, sulfide, sulfate, carbide, carbonate, cyanide, boride, and borate. Also the elements C, Si, S, Se, and Te should be considered, and obviously oxides may be present.

Perform a fusion with 0.5 g of the material or with the acid insoluble residue of 1 g of it. In platinum or nickel apparatus, fuse the finely pulverized material with 10 to 20 ml of anhydrous Na_2CO_3 (or a 1 : 1 mixture of Na_2CO_3 and K_2CO_3). An electric oven or an alcohol blast flame will not introduce sulfur. Keep in the molten state for 10 to 20 minutes. If dark particles remain undissolved after this time, add 0.3 ml solid $NaNO_3$ and again heat until the whole mass is liquid. After cooling, add 40 to 60 ml water and boil until the whole melt has crumbled. The clear supernate is the soda extract which is tested as outlined in P. 70. Test precipitate (1) for fluoride and also consider the following tests.

Test for Silicate and Borate. Use one third of the soda extract obtained. Acidify with HCl, evaporate to dryness on the steam bath, moisten the residue with 12-F HCl, again evaporate to dryness, and heat the residue at 130° C (oven or heating block) until completely dry. After cooling, add 6 ml 6-F HCl and warm the mixture to about 60° C. A white residue consists of partly dehydrated H_2SiO_3 and may be used for estimation of the amount and confirmatory tests. Remove the supernate and test it for boric acid.

P. 73 Final Review of Observations and Report

For all practical purposes, the important part of the report is the final conclusion. In the simplest instance, this will be a statement on presence or absence of some substance. The conclusion may state the identity of the material under investigation in more or less definite terms, or it may only indicate the general type of matter.

The observations from which the final conclusion is drawn may be made a part of the report. Their meaning increases in the same degree as the final conclusion becomes indefinite or vague. If the facts indicate a "new" substance which apparently has not yet been described, so that identification or recognition is not possible, the report must by necessity become a tabulation of the physical and chemical properties observed.

If the material has been recognized, the identification should explain all observations such as state, shape, color, odor, hardness, density, behavior on heating and toward reagents, etc. It may happen, however, that it is impossible to explain an unusual color or odor by the observed facts since the responsible trace of impurity or incidental contamination cannot be detected by the common means of qualitative analysis. If explanation of such deviations from the norm is essential, it will be necessary to turn to the literature on the material involved, its investigation, and to methods of trace determination. By all means, any features and facts not explained by the identification should be emphasized in the report.

Checking positive findings by blanks and negative findings by suitably performed controls becomes the more essential, the smaller the percentages are, which have to be considered; in a search for traces, the performance of blanks and controls may constitute the largest and most important part of the task.

It is occasionally stated that synthesis is the final proof of analysis. There are, however, limitations to synthesis on the basis of a qualitative analysis even when the latter provides good estimates of the amounts. To mention just one example, the preparation of an alloy may not succeed even when the correct elemental composition is duplicated. It may be necessary to perform an extensive investigation to discover the treatment of the alloy (tempering, quenching, cold and hot working, and surface treatment) which produces the microscopic or submicroscopic structure that gives the observed properties.

Appendix

Test Solutions

To obtain test solutions containing 10 mg of the substance of interest (metal, cation, anion) per milliliter solution dissolve the following weights to obtain 100 ml solution (if work is done on a semimicro or smaller scale, it is recommended to prepare only 10 ml of the solutions).

Dissolve in water:

(test solutions for cations, metals): 6.1 g $LiCl$; 2.5 g $NaCl$; 2.6 g KNO_3; 1.4 g $RbCl$; 1.3 g $CsCl$; – – – 19 g $BeSO_4 \cdot 4 H_4O$; 11 g $Mg(NO_3)_2 \cdot 6 H_2O$; 5.9 g $Ca(NO_3)_2 \cdot 4 H_2O$; 2.4 g $Sr(NO_3)_2$; 1.9 g $Ba(NO_3)_2$; – – – 4.3 g $Y(NO_3)_3 \cdot 6 H_2O$; 3.1 g $La(NO_3)_3 \cdot 6 H_2O$; 3.1 g $Ce(NO_3)_3 \cdot 6 H_2O$; 3.0 g $Nd(NO_3)_3 \cdot 6 H_2O$; – – – 3.0 g $Th(NO_3)_4 \cdot 12 H_2O$; – – – 7.7 g $Cr(NO_3)_3 \cdot 9 H_2O$; 1.8 g $(NH_4)_6Mo_7O_{24} \cdot 4 H_2O$; 1.8 g $Na_2WO_4 \cdot 2 H_2O$; 2.1 g $UO_2(NO_3)_2 \cdot 6 H_2O$; – – – 5.2 g $Mn(NO_3)_2 \cdot 6 H_2O$; – – – 7.3 g $Fe(NO_3)_3 \cdot 9 H_2O$; 5.0 g $Co(NO_3)_2 \cdot 6 H_2O$; 5.0 g $Ni(NO_3)_2 \cdot 6 H_2O$; – – – 2.0 g $RhCl_3$; 1.3 g OsO_4, fume hood; 2.7 g $H_2PtCl_6 \cdot 6 H_2O$; – – – 3.8 g $Cu(NO_3)_2 \cdot 3 H_2O$; 1.7 g $AgNO_3$; 2.0 g $HAuCl_4 \cdot 3 H_2O$; – – – 4.6 g $Zn(NO_3)_2 \cdot 6 H_2O$; 2.7 g $Cd(NO_3)_2 \cdot 4 H_2O$; 1.4 g $HgCl_2$; – – – 14 g $Al(NO_3)_3 \cdot 9 H_2O$; 1.3 g $TlNO_3$; – – – 1.1 g $Pb(NO_3)_2$; – – – 1.5 g As_2O_5; – – – 1.4 g SeO_2;

(test solutions for anions): 2.5 g $Na_2B_4O_7 \cdot 10 H_2O$; – – – 2.0 g $Na_2CO_3 \cdot H_2O$; 1.9 g $NaCN$; 2 g $KCNO$; 1.7 g $KCNS$; 2.0 g $K_4Fe(CN)_6 \cdot 3 H_2O$; 1.5 g $K_3Fe(CN)_6$; 1.5 g $NaHCOO$; 1.5 g $Na_2C_2O_4$; 2.3 g $NaCH_3 \cdot COO \cdot 3 H_2O$; 3.7 g $Na_2SiO_3 \cdot 9 H_2O$; – – – 1.5 g NaN_3; 1.8 g KNO_2; 1.7 g KNO_3; 1.6 g $NaH_2PO_2 \cdot H_2O$; 2.7 g $Na_2HPO_3 \cdot 5 H_2O$; 2.0 g $Na_2H_2P_2O_6 \cdot 6 H_2O$; 1.4 g KH_2PO_4; 1.5 g $NaAsO_2$; 2.3 g $Na_2HAsO_4 \cdot 7 H_2O$; – – – 8.0 g $Na_2S \cdot 9 H_2O$; 1.3 g $Na_2S_2O_5$; 2.2 g $Na_2S_2O_3 \cdot 5 H_2O$; 1.8 g K_2SO_4; 1.4 g $K_2S_2O_8$; 2.5 g $Na_2SeO_4 \cdot 10 H_2O$; – – – 2.0 g NH_4F; 1.7 g $NaCl$; 1.5 g $CaCl \cdot ClO$ + 1.5 g Na_2CO_3;[1] 1.3 g $NaClO_3$; 1.2 g $NaClO_4$; 1.3 g $NaBr$; 1.3 g $KBrO_3$; 1.3 g KI; 1.2 g KIO_3; 1.1 g $NaIO_4$.

Dissolve in 100 ml 3-F HCl:

16 ml 20% $TiCl_3$;[2] – – – 2.1 g $RuCl_3$; 1.7 g $PdCl_2$; 1.6 g $IrCl_3$; – – – 1.0 g Ge;[3] 1.9 g $SnCl_2 \cdot 2 H_2O$; – – – 1.9 g $SbCl_3$; 1.3 g $BiOCl$; – – – 1.3 g TeO_2.[4]

---

[1] Separate from the precipitated $CaCO_3$.

[2] Pass air through the solution until it is colorless.

[3] First warm with 5 ml 6-F HCl and 1 ml 6-F HNO_3, then dilute with 3-F HCl to 100 ml.

[4] First warm with 5 ml 6-F HCl until dissolved, and then dilute with 3-F HCl to 100 ml.

Treat with 5 ml 6-F HNO_3; when dissolved, dilute with water to 100 ml:
  1.5 g **Sc_2O_3**; – – – 4.7 g **$Zr(NO_3)_4 \cdot 5 H_2O$**; – – – 1 g **Ga**; 1 g **In**.
Dissolve in 100 ml 0.3-F NaOH:
  2.3 g NH_4VO_3.

Preparation of Unknowns

As a rule, all information needed by the student is supplied in the text of the experiment, and additional information restricting the field of the search should not be given.

Experiment 42. The experiment is described for the milligram scale, but it should also be carried out when working on either the gram or the centigram scale. The directions are readily converted by multiplying all volumes by 1000 or 10, respectively. The preparation of the unknown is described with the experiment, but the volumes taken should not be disclosed and should be varied from sample to sample. A relative agreement of ± 0.3 between estimated and given amount should be considered satisfactory.

Experiment 63. Directions are given with the experiment. The ratio should be varied and it should not be disclosed before all reports have been handed in.

Experiment 65. Test solution is diluted up to 10 times with the solvent used in its preparation, *see* preceding section.

Experiment 66. The quantity to be given is discussed under Expt. 66. Suitable unknowns are the various oxides, sulfides, carbonates, insoluble halides and sulfates of Ag, Hg(1, 2), Pb, Bi, Cu, Cd, As(3, 5), Sb(3, 5), and Sn(2, 4). A mixture of two modifications of a compound may be given, but each sample should contain only one reasonably pure compound.

Experiment 67. Two different solid samples in the amount specified in Expt. 66. Recommended are water-insoluble compounds, oxides, hydroxides, sulfides, sulfites, carbonates, phosphates of Fe(2, 3), Al, Cr(3), Ni, Co, Mn, Zn, and the metals listed for Expt. 66. Minerals and pigments fitting into this list are also recommended.

Experiment 68. Two solid samples in the amount specified in Expt. 66. Recommended are simple inorganic substances (hydrated salts), melting below 300° C and having the m. pt. listed in the handbook used, and reasonably pure organic solids, having a good melting point, which are included in the identification tables of KOFLER (159, 160) or McCRONE (163). These tables must be available for students' use in addition to some handbook of chemistry.

Experiment 69. Two solid samples as in Expt. 68. This time, it is suggested to include abrasives (various forms of carbon, carbides, quartz, corund) and to select otherwise compounds which will suitably round out the experience of the student.

Experiment 70. The study may be extended in any desired direction by including inorganic and organic liquids, metals, alloys, complex compounds, mixtures, substances which defy identification beyond recognition of the group of materials to which they belong, substances which have not been described in the literature, and organized material derived from living matter. It is understood that such work will require access to the literature.

Reagents

The necessary reagents are listed in connection with the experiments and directions. The concentrations are usually specified in terms of formality according to the definition that a one-formal, 1-F, solution contains one gram formula weight (corresponding to the formula given) per liter of solution.

Murrayite and Xyrax, obtainable from laboratory supply houses, may be substituted for Canada balsam. Dekadhese (Arthur H. Thomas Co., Philadelphia, Pa.) is recommended for cementing glass, varnishing labels, etc.

Table 1
Color of Some Inorganic Substances

Black (gray): metal powders, B, C, Si, As, Se, Te, I_2 – – V_2O_3, Mo_2O_3, MoO_2, UO_2, Mn_2O_3, Mn_3O_4, MnO_2, Fe_3O_4, $FeO \cdot Cr_2O_3$, Co_2O_3, Co_3O_4, Ni_2O_3, Ni_3O_4, CuO, Tl_2O_3, Pb_2O, and many oxidic minerals such as TiO_2 (anatase, brookite, rutile), (Th, U)O_2 (thorianite), pitchblende, Fe_2O_3 (martite), SiO_2 (quartz), SnO_2 (cassiterite) – – V_2S_2, V_2S_3, MoS_2, MoS_3, US_2, U_2S_3, UO_2S, MnS_2, FeS, CoS, NiS, Cu_2S, Ag_2S, Au_2S, Au_2S_3, Hg_2S, HgS, Tl_2S, Tl_2S_3, PbS, Sb_2S_3, various sulfidic minerals – – $FeCl_3$, $CuBr_2$, various triple salts, etc. – – permanganates – – boron carbide and the refractory nitrides, carbides, and borides of Si, Ti, Zr, Hf, Mo, and W – – various silicates, obsidian, lava, slags – – opaque materials (of any surface color) in transmitted light.

Dark brown: WO_2, Mn_2O_3, Mn_3O_4, $Fe_2O_3 \cdot x\,H_2O$, CoO, CdO, Tl_2O_3, PbO_2, Bi_2O_3 – – MoS_3, CoS.

Red: Se, P_4, Cu – – CrO_3, Fe_2O_3, HgO, Pb_3O_4 – – Cd(S, Se), HgS, Pb_2SCl_2 (?), As_2S_2, As_2S_3 – – chromates, compounds of Cr(3) and Co(2), iodides, HgI_2.

Pink: As_2S_2, MnS – – compounds of Cr(3), Mn(2), Co(2), Nd, Eu, Er.

Orange (light brown): Cu, bronze – – Cu_2O, HgO, PbO – – Fe(OH)$_3$, Tl(OH)$_3$, TlO(OH) – – CdS, As_2S_2, As_2S_3 – – compounds of Fe(3), Ce(4), and of platinum metals.

Yellow: Au, brass, bronze, S, P_4, As – – Na_2O_2, WO_3, HgO, In_2O_3, PbO, Pb_2O_3, Sb_2O_5, Bi_2O_3, $Bi_2O_4 \cdot 2\ H_2O$ – – CdS, Hg_2SCl_2 (?), SnS_2, As_2S_3, As_2S_5, FeS_2 (pyrite) – – Ag_3PO_4, $AgAsO_2$, – – – Compounds of V, uranyl, Fe(3), platinum metals, Au, Tl(3), Sm, Dy, Ho – – chromates, ferrocyanides, ferricyanides, bromides, iodides.

Green: U_3O_8, MnO, NiO, (Zn, Co)O known as Rinman's green – – MnS – – compounds of Cr(3), Fe(2), Ni, Cu, Pr – – manganates.

Blue: $Al_2O_3 \cdot x$ CoO known as Thénard's blue, Nd_2O_3 – – ultramarine, Prussian blue – – compounds of Cr(3), Co(2), and Cu.

Violet and *purple:* I_2 – – ultramarine – – compounds of Ti(3), Cr(3), Co, Nd – – permanganates.

Metallic luster: Metals, Te – – $FeTiO_3$ (ilmenite), titanates, tantalates, $MnO_2 \cdot x\ H_2O$ (pyrolusite), Fe_3O_4 (magnetite), (Fe, Mn, Zn)O $\cdot Fe_2O_3$ (franklinite), FeO $\cdot Cr_2O_3$ (chromite), $FeWO_4$ (ferberite), (Fe, Mn)WO_4 (wolframite), CuO (tenorite) – – most sulfide, selenide, telluride, arsenide, and stibnide minerals (ores) – $Ca(F, Cl)_2 \cdot Ca_4(PO_4)_3$ (apatite) – – mica – – tungsten bronzes.

Table 2
Substances Crystallizing in the Cubic System
Elements

Li, Na, K – – – Ca – – – Ce – – – Ti, Zr, Th – – – V, Nb, tantalum: Ta – – – Cr, Mo, W – – – ferrite: Fe; Co, Ni – – – Rh, rhodium gold: (Rh—Au), palladium: (Pd—Pt—Ir), platinum: Pt, iridium: (Ir—Fe—Pt) – – – copper: Cu, silver: Ag, gold: Au, perpezite: (Au—Pd) – – – amalgam: (Ag—Hg) – – – Al – – – diamond: C; Si, Ge, Sn, Pb – – – P_4 (red), As_4.

Inorganic Compounds

Hydrides: LiH, NaH, KH, $N_2H_4 \cdot 2\ HCl$, PH_4Cl.

Halides, Cyanides, Thiocyanates: LiF, LiCl, LiBr, LiI; villiaumite: NaF, **halite:** NaCl; NaBr, NaI, NaCN; KF, KHF_2, **sylvite:** KCl; KBr, KI, KCN; RbF, RbCl, RbBr, RbI; CsF, CsCl, CsBr, CsI; sal ammoniac: NH_4Cl; NH_4Br, NH_4I, NH_4CN – – – **fluorite:** CaF_2; $CaCl_2$, $Ca(CN)_2$; SrF_2, $SrCl_2$; BaF_2, $BaCl_2$ – – – yttrofluorite: $CaF_2 \cdot n\ YF_3$, yttrocerite: $CaF_2 \cdot n$ (Y, Ce)F_3 – – – TiI_4 – – – VOBr – – – WBr_2, WCl_6, UCl_4 – – – $MnCl_2$ – – – $Fe(CNS)_3 \cdot 3\ H_2O$ – – – $OsCl_3$, $PdCl_2$ – – – Cu_2Cl_2, Cu_2I_2; AgF, cerargyrite: AgCl, bromyrite: AgBr, embolite: Ag(Cl, Br) – – – ZnI_2; CdF_2, $CdCl_2$; HgF, HgF_2, eglestonite: Hg_4ClO – – – ralstonite: 3 Al(F, OH)$_3 \cdot$ (Na$_2$Mg)$F_2 \cdot 2\ H_2O$; TlF, TlCl, TlBr, TlI – – – SiI_4; $GeBr_4$, GeI_4; SnI_4 – – – $PSBr_3$.

Oxides: K_2O, Rb_2O, Rb_2O_2 – – – periclase: MgO; CaO, SrO, BaO – – – CeO_2, Y_2O_3, Dy_2O_3, Er_2O_3, Yb_2O_3 – – – ThO_2 – – – NbO – – – WO_2;

thorianite: $(Th, U)O_2$ - - - manganosite: MnO; Mn_2O_3 - - - martite: Fe_2O_3, **magnetite:** Fe_3O_4; CoO, Co_3O_4, bunsenite: NiO - - - cuprite: Cu_2O; CuO; Ag_2O - - - CdO - - - In_2O_3 - - - cristobalite: SiO_2; SnO - - - arsenolite: As_2O_3; Sb_2O_3; Bi_2O_3.

Hydroxides and Acids: $Pb_3O_2(OH)_2$, $H_7P(Mo_2O_7)_6 \cdot 28\,H_2O$, $H_2TeO_4 \cdot 2\,H_2O$.

Sulfides, Selenides, Tellurides: Li_2S, Na_2S_4 - - - BeS, MgS, CaS, SrS, BaS - - - alabandite: MnS, hauerite: MnS_2, $MnSe$ - - - **pyrite:** FeS_2; CoS_2, cobaltpyrite: $(Co, Fe)S_2$, linnaeite: Co_3S_4, cobaltite: $Co_2As_2S_2$; polydymite: Ni_3S_4, gerstorffite: $Ni_2As_2S_2$; $NiSe$ - - - laurite: $(Ru, Os)S_2$ - - - Cu_2S, berzelianite: Cu_2Se; argentite: Ag_2S, aquilarite: Ag_2S-Ag_2Se; Ag_2Se, eucairite: $Cu_2Ag_2Se_2$, naumannite: $(Ag_2Pb)Se$; hessite: Ag_2Te, petzite: $(Ag, Au)_2Te$ - - - **sphalerite:** ZnS; $ZnTe$, $CdTe$, tiemannite: $HgSe$ - - - **galena:** PbS, clausthalite: $PbSe$, altaite: $PbTe$, stannite: $FeS \cdot Cu_2S \cdot SnS_2$, ullmannite: $NiSbS$, bornite: Cu_5FeS_4, tennantite: $5\,Cu_2S \cdot 2\,ZnS \cdot 2\,As_2S_3$, tetrahedrite: $3\,(Cu_2, Fe, Zn)S \cdot Sb_2S_3$, argyrodite: Ag_8GeS_6, germanite: $5\,Cu_2S \cdot 12\,(CuFe)S \cdot As_2S_3 \cdot 2.GeS_2$.

Nitrides, Amides, Phosphides, Arsenides: $LiNH_2$ - - - arsenoferrite: $FeAs_2$, smaltite: $CoAs_2$ - - - sperrylite: $PtAs_2$ - - - Cu_5As_2 - - - Zn_3P_2, Zn_3As_2, Cd_3As_2 - - - BN (borazon).

Carbides, Silicides: Be_2C, Mg_2Si - - - TiC - - - WC - - - Mn_2Si - - - $FeSi$ (octahedral?).

Borides: CaB_6, BaB_6.

Chlorates, Bromates, Iodates: $NaClO_3$, $NaBrO_3$, $RbIO_3$ - - - $Co(ClO_3)_2 \cdot 6\,H_2O$, $Co(BrO_3)_2 \cdot 6\,H_2O$, $Co(IO_3)_2 \cdot 6\,H_2O$, $Ni(BrO_3)_2 \cdot 6\,H_2O$ - - - $Cu(ClO_3)_2 \cdot 6\,H_2O$, $Cu(BrO_3)_2 \cdot 6\,H_2O$ - - - $Zn(BrO_3)_2 \cdot 6\,H_2O$, $Hg(BrO_3)_2 \cdot 6\,H_2O$.

Thiosulfates, Sulfates: $K_2S_2O_3$ - - - $Cr_2(SO_4)_3 \cdot 18\,H_2O$ - - - $NiSO_4$ - - - (the alums:) $(K, Rb, Cs, NH_4, Tl)\,(Al, In, Ga, Tl, Fe, Cr, V, Rh)\,(SO_4)_2 \cdot 12\,H_2O$; tschermigite: $NH_4Al(SO_4)_2 \cdot 12\,H_2O$.

Nitrates, Hypophosphites, Phosphates, Pyrophosphates, Arsenates, Antimonates: (phosphamic acid:) $(OH)_2PNH_2$ - - - $2\,NaSbO_3 \cdot 7\,H_2O$, $Na_3SbS_4 \cdot 9\,H_2O$, $RbNO_3$ - - - $Ca(NO_3)_2$, $Sr(NO_3)_2$, nitrobarite: $Ba(NO_3)_2$; $Ba_3(PO_4)_2$ - - - TiP_2O_7 - - - $Ni(H_2PO_2)_2 \cdot 6\,H_2O$ - - - Ag_3PO_4, Ag_3AsO_4 - - - $TlNO_3$.

Silicates: hackmanite, sodalite: $6\,NaAlSiO_4 \cdot 2\,NaCl$, noselite: $6\,NaAlSiO_4 \cdot NaSO_4$, lazurite: $3\,NaAlSiO_4 \cdot Na_2S$, hauynite: $3\,NaAlSiO_4 \cdot CaSO_4$, analcime: $NaAlH_2Si_2O_7$, chalcolamprite, endeiolite: $Na_4Ca_2Nb_2-(F, OH)_2SiO_9$ (?), pollucite: $Cs_4Al_4H_2(SiO_3)_9$ - - - helvite: $3\,(Be, Mn, Fe)SiO_4 \cdot MnS$, danalite: $3\,(Be, Mn, Fe, Zn)SiO_4 \cdot ZnS$, **garnet:** $(Mg, Ca, Mn, Fe)_3(Cr, Fe, Al)_2(SiO_4)_3$, beckelite: $Ca_3(Ce, La, Dy)_4Si_3O_{15}$, schorlomite: $Ca_3(Fe, Ti)_2 \cdot 3\,(Si, Ti)O_4$ - - - rowlandite: $Fe(YF)_2(Y, Ce, La)_2(Si_2O_7)_2$, bodenbenderite: $Y-Al-Mn-$etc. titanosilicate; greenalite: $Fe_3H_2(SiO_3)_4 \cdot H_2O$ - - - zunyite $Al_2[Al(OH, F, Cl)_2]_6(SiO_4)_3$ - - - eulytite: $Bi_4(SiO_4)_3$.

Borates, Aluminates: boracite: $6 MgO \cdot MgCl_2 \cdot 8 B_2O_3$, spinel: $Mg(AlO_2)_2$, picotite: $(Mg, Fe) \cdot (Al, CrO_2)_2$; $Ca(BO_2)_2 \cdot 2 H_2O$, rhodizite: $(Li, Na, K)_8 \cdot Be_7Al_6B_{14}O_{39}$ (?) – – – ceylonite: $(Mg, Fe)O \cdot (Al, Fe)_2O_3$, hercynite: $FeO \cdot Al_2O_3$ – – – gahnite: $Zn(AlO_2)_2$.

Salts of other Oxygen Acids: pyrochlore: $(Na_2, Ca)_2(Nb, Ti)_2(OF)_7$, knopite, perovskite: $(Ca, Ce, Fe)TiO_3$, dysanalyte: $(Ca, Fe)_7Nb_2(TiO_4)_6$ (?), uhligite: $Ca(Zr, Ti)_2O_5 \cdot (Al_2O)TiO_4$, *davidite:* rare earths—V—Cr—U—Fe titanate; *samiresite:* U—Pb—etc. niobate-titanate; *ellworthite:* Ca—U—Ti—etc. niobate; *betafite:* hydrous Ca—U—etc. titanate-niobate; *mendelyeevite:* Ca—U titano-niobate; *zirkelite:* Ca—Fe titano-thorio-zirconate, *thorianite:* $(Th, U)O_2$ – – – *microlite:* hydrous Na—Ca—rare earths niobotantalate-fluoride; koppite: $Ca_2Ta_2O_7$ with Na, Ce, Fe, F, etc., neotantalite: hydrous Mn—Fe niobotantalite, *djalmaite:* U niobotantalate – – – chromite: $Fe(CrO_2)_2$; K_3CrO_8, $K_2W_4O_{13} \cdot 8 H_2O$, $(NH_4)_2W_4O_{13} \cdot 8 H_2O$; (tungsten bronzes:) Me_xWO_3 – – – *cleveite, nivenite, uraninite:* complex uranates with Y, lanthanides, Zr, Th, Pb, N, He, A; *broggerite:* U—Th—Pb uranate – – – $Sr(MnO_4)_2 \cdot 3 H_2O$ – – – franklinite: $(Mn, Fe, Zn)(FeO_2)_2$ – – – $K_2OsO_4 \cdot 2 H_2O$ – – – lewisite: $Ca_5Ti_2(SbO_4)_6$; mauzeliite: Ca—Pb titano-antimonate.

Fluoborates, Fluoraluminates, Fluosilicates: cryolithionite: $Li_3Na_3(AlF_6)_2$; KBF_4, hieratite: K_2SiF_6; Rb_2SiF_6, Cs_2SiF_6, $(NH_4)_2SiF_6$ – – – $Ag_2SiF_6 \cdot 4 H_2O$ – – – $Tl_2SiF_6 \cdot 2 H_2O$.

Chloro-, Bromo-, Iodo-, and Cyanocomplexes: K_2PdCl_4, $(K, NH_4)_2PdCl_6$; $(K, NH_4)_2OsCl_6$, K_2OsBr_6; $(K, NH_4)_2IrCl_6$; $(K, Rb, Cs, NH_4)_2PtCl_6$, $(K, NH_4)_2Pt(Br, I)_6$ – – – $KAg(CN)_2$ – – – $CsHgCl_3$ – – – $(K, NH_4)_2SnCl_6$, $(K, NH_4)_2PbCl_6$ – – – $(NH_4)_2SeBr_6$.

Neutral Complex: $Pd(NH_3)_2(OH)_2$.

Organic Compounds

Calcium oxalate, anh., CaC_2O_4 – – – carbon tetraiodide, CI_4, red, dec. – – – cinchonine bisulfate, octahedral – – – mercuric fulminate, $Hg(CNO)_2$ – – – morphine sulfate pentahydrate, $250°$ dec. – – – tri-iodoethane, $CH_3 \cdot CI_3$, yellow, octahedral, $95°$ dec.

Table 3
Substances Crystallizing in the Hexagonal System
Elements

Cs – – – Be, Mg – – – Y, Ce – – – Hf – – – Re – – – Ru, Os, ruthenosmiridium: Ru—Os—Ir alloy, iridosmine: Ir—Os with Pt, Rh, Ru, etc. – – – zinc: Zn; Cd – – – graphite: C – – – P_4, arsenic: As, antimony: Sb, bismuth: Bi – – – Se, tellurium: Te.

Inorganic Compounds

Halides: NH_4F − − − nocerite: $Mg_3Ca_3O_2F_8$, chlormagnesite: $MgCl_2$; $MgBr_2 \cdot 6\,H_2O$; tachyhydrite: $2\,MgCl_2 \cdot CaCl_2 \cdot 12\,H_2O$; yttrocalcite: $Ca_5Y_2F_{16}$; $CaCl_2 \cdot 6\,H_2O$, $CaBr_2 \cdot 6\,H_2O$, $BaI_2 \cdot 6\,H_2O$ − − − ZrF_4 − − − LaF_3, $CeF_4 \cdot H_2O$, NdF_3, fluocerite: $(Ce, La, Dy)F_3$ − − − VCl_2 − − − UF_3, UCl_3 − − − $CrCl_3$, $CrBr_3$, $CrCl_3 \cdot 6\,H_2O$, $CrBr_3 \cdot 6\,H_2O$ − − − $FeCl_2$, $FeBr_2$, FeI_2, $FeCl_3$; $CoF_2 \cdot 5\,HF \cdot 6\,H_2O$, CoF_3, $CoI_2 \cdot 6\,H_2O$; $NiF_2 \cdot 5\,HF \cdot 6\,H_2O$ − − − AgI − − − CdI_2, kleinite: $Hg_4O_3Cl_2$ − − − AlF_3, $AlCl_3$, $AlCl_3 \cdot 6\,H_2O$; Tl_2Cl_3, $TlCl_3$ − − − Si_2I_6; PbI_2 − − − PI_3, PI_2Cl_3; AsI_3, BiI_3.

Oxides: $Na_2O_2 \cdot 8\,H_2O$ − − − BeO − − − Nd_2O_3 − − − Ti_2O_3 − − − Cr_2O_3 − − − **hematite:** Fe_2O_3, hogbomite: chiefly $(Al, Fe)_2O_3$, MgO, TiO_2 − − − zincite: ZnO − − − corundum: Al_2O_3; In_2O_3, Tl_2O_3 − − − amethyst, **quartz,** tridymite: SiO_2.

Hydroxides, Acids: brucite: $Mg(OH)_2$; $Ca(OH)_2$ − − − H_2MoO_4 − − − $Mn(OH)_2$ − − − $Cd(OH)_2$ − − − $Tl(OH)_3$ − − − H_2SeO_3, H_2SeO_4.

Sulfides, Selenides, Tellurides: molybdenite: MoS_2 − − − FeS, Fe_3S_4, pyrrhotite: Fe_5S_6 to $Fe_{16}S_{17}$; millerite: NiS, melonite: $NiTe_2$ − − − covellite: CuS − − − wurtzite: ZnS; $ZnSe$; greenockite: CdS; $CdSe$; cinnabar: HgS − − − Al_2S_3 − − − SnS_2 − − − Ag_3AsS_3, pyrargyrite: Ag_3SbS_3; tetradymite: $Bi_2(S, Te)_3$.

Nitrides, Arsenides, Antimonides: NaN_3 (azide) − − − $CrAs$ − − − niccolite: $NiAs$; $NiSb$ − − − Cu_3As − − − BN (white graphite); Al_2N_3.

Carbides: Be_2C − − − CeC_2, NdC_2, SmC_2 − − − moissanite: SiC.

Chlorates, Bromates, Iodates, Perchlorates, Periodates: $LiClO_4 \cdot 3\,H_2O$; $NaClO_3$, $NaClO_4 \cdot H_2O$, $NaIO_4 \cdot 3\,H_2O$; $KBrO_3$ − − − $Ba(ClO_4)_2$, $Ba(ClO_4)_2 \cdot 3\,H_2O$ − − − $Y(BrO_3)_3 \cdot 9\,H_2O$, $La(BrO_3)_3 \cdot 9\,H_2O$, $Ce(BrO_3)_3 \cdot 9\,H_2O$, $Pr(BrO_3)_3 \cdot 9\,H_2O$, $Nd(BrO_3)_3 \cdot 9\,H_2O$, $Sm(BrO_3)_3 \cdot 9\,H_2O$, $Dy(BrO_3)_3 \cdot 9\,H_2O$ − − − $Co(ClO_4)_2 \cdot 6\,H_2O$; $Ni(ClO_4)_2 \cdot 5\,H_2O$, $Ni(ClO_4)_2 \cdot 5\,H_2O$, $Ni(IO_3)_2 \cdot 4\,H_2O$.

Sulfites, Dithionates, Sulfates, Selenates: Na_2SO_3, Na_2SO_4, hanksite: $9\,Na_2SO_4 \cdot 2\,Na_2CO_3 \cdot KCl$; $K_2S_2O_6$ − − − $MgSO_4 \cdot 6\,H_2O$; $CaS_2O_6 \cdot 4\,H_2O$, $SrS_2O_6 \cdot 4\,H_2O$ − − − $La_2(SO_4)_3 \cdot 9\,H_2O$, $Yb_2(SeO_4)_3 \cdot 8\,H_2O$ − − − $Fe_2(SO_4)_3 \cdot 9\,H_2O$ − − − alunite: $K_2Al_6(OH)_{12}(SO_4)_4$ − − − $Sn(SO_4)_2 \cdot 2\,H_2O$; $PbS_2O_6 \cdot 4\,H_2O$.

Nitrites, Nitrates, Hypophosphites, Phosphates, Arsenites and Arsenates: $LiNO_3$, $Li_3PO_4 \cdot 12\,H_2O$; nitratite: $NaNO_3$; $Na_3PO_4 \cdot 12\,H_2O$, $Na_3AsO_4 \cdot 12\,H_2O$; KNO_3, KH_2PO_2, $K_2HAsO_4 \cdot H_2O$; $RbNO_3$; $CsNO_3$ − − − $Ca(NO_2)_2 \cdot H_2O$, **apatite:** $Ca_5(F, Cl)(PO_4)_3$; $Ba(NO_2)_2 \cdot H_2O$ − − − buttgenbachite: hydrous copper chloronitrate; Ag_2HPO_4 − − − $AlPO_4$; stiepelmannite: $(Y, La, Ce, Pr, Yb)Al_3(OH)_6(PO_4)_3$; florencite: $Ce_2Al_6(OH)_{12}(PO_4)_4$; $TlNO_3$ − − − $Pb(NO_3)_2$, $Pb_3(PO_4)_2$, pyromorphite: $3\,Pb_3(PO_4)_2 \cdot PbCl_2$; plumbogummite: $PbAl_3(OH)_5(PO_4)_2 \cdot H_2O$; $Pb(AsO_3)_2$.

Carbonates: $Na_2CO_3 \cdot 7\,H_2O$ - - - magnesite: $MgCO_3$; **calcite:** $CaCO_3$; **dolomite:** $MgCa(CO_3)_2$; codazzite: $(Mg, Ca, Ce, Fe)CO_3$; parisite: $Ca(La, Ce, Dy)_2F_2(CO_3)_3$; $BaCO_3$, cordylite: $Ba(La, Ce, Dy)_2F_2(CO_3)_3$ - - - basnäsite: $(La, Ce, Dy)FCO_3$ - - - $Na_4UO_2(CO_3)_3$, andersonite: $Na_2CaUO_2(CO_3)_3 \cdot 6\,H_2O$ - - - rhodochrosite: $MnCO_3$ - - - stichtite: $Mg_6Cr_2(OH)_{16}CO_3 \cdot 4\,H_2O$ - - - **siderite:** $FeCO_3$; ankerite: $MgCa_2Fe(CO_3)_4$; $CoCO_3$ - - - smithsonite: $ZnCO_3$; $CdCO_3$ - - - $Pb_3(OH)_2(CO_3)_2$.

Silicates, Stannates: eucryptite: $LiAlSiO_4$; **tourmaline:** complex Li—Na—Mg—Fe—Al borosilicate; Na_2SiO_4, nepheline: $(Na, K)AlSiO_4$; gmelinite: $(Na_2, Ca)Al_2H_4(SiO_4)_3 \cdot 4\,H_2O$; chabazite: $(Na_2, Ca)Al_2(SiO_3)_4 \cdot 6\,H_2O$; cancrinite: $Na_8CaAl_8(SiO_4)_9(CO_3)_2$; bazzite: Na—Sc—rare earths—Fe silicate; eucolite, eudialyte: Na—Ca—Ce—Zr—Fe basic chloride-silicate; *steenstrupine:* hydrous Na—rare earths—Ti—Zr—Th—Mn—Fe—Al silicate; Na_2SnO_3; milarite: hydrous K—Be—Ca—Al silicate; $Na_2SnO_3 \cdot 3\,H_2O$; $K_2SnO_3 \cdot 3\,H_2O$ - - - phenakite: Be_2SiO_4; **beryl:** $Be_3Al_2(SiO_3)_6$; helidor: $Be_3(Fe, Al)_2(SiO_4)_3\,(?)$; levynite: $CaAl_2H_4(SiO_4)_3 \cdot 3\,H_2O$; abukumalite: Ca—Y phosphate-silicate; melanocerite: Ca—Y—La—Ce—Dy—Ta borate-silicate; *caryocerite:* similar to melanocerite with Th added; $SrSiO_3$; benitoite: $BaTi(SiO_3)_3$; cappelenite: Ba—Y borate-silicate - - - buszite: Pr—Eu—Er—Nb silicate - - - dioptase: $CuSiO_3 \cdot H_2O$ - - - willemite: $ZnSiO_4$.

Borates: $NaBO_2$; $K_4B_2O_7 \cdot 5\,H_2O$.

Salts of Other Oxygen Acids: geikielite: $(Mg, Fe)TiO_3$; pyrophanite: $MnTiO_3$; senaite: $(Mn, Fe, Pb)TiO_3$; **ilmenite:** $FeTiO_3$ - - - $Na_4V_2O_7$; vanadinite: $3\,Pb_3(VO_4)_2 \cdot PbCl_2$ - - - *hatchettolite:* $U(Nb, Ta)O_3 \cdot H_2O$ - - - ferritungstite: hydrous ferric tungstate - - - $BaMnO_4$ - - - mimetite: $3\,Pb_3(AsO_4)_2 \cdot PbCl_2$.

Fluosilicates, Fluogermanates: Na_2SiF_6; K_2SiF_6, K_2GeF_6 - - - tritomite: Ca—Ba—Y—La—Ce—Dy—etc. fluosilicate - - - $MnSiF_6 \cdot 6\,H_2O$ - - - $CoSiF_6 \cdot 6\,H_2O$; $NiSiF_6 \cdot 6\,H_2O$ - - - $ZnSiF_6 \cdot 6\,H_2O$.

Chloro-, Bromo-, and Iodocomplexes: $MgPdCl_6 \cdot 6\,H_2O$, $NiPdCl_6 \cdot 6\,H_2O$; $Li_2PtCl_6 \cdot 6\,H_2O$, $MgPtCl_6 \cdot 6\,H_2O$, $MgPtBr_6 \cdot 12\,H_2O$, $MnPtCl_6 \cdot 6\,H_2O$, $MnPtBr_6 \cdot 12\,H_2O$, $MnPtI_6 \cdot 9\,H_2O$, $FePtCl_6 \cdot 6\,H_2O$, $CoPt(Cl, Br)_6 \cdot 12\,H_2O$, $CoPtI_6 \cdot 9\,H_2O$ - - - $K_3Cu(CN)_4$, $KAg(CN)_2$, $K_3Ag(CN)_4$, $KAu(CN)_2$ - - - $MgSnCl_6 \cdot 6\,H_2O$, $CoSnCl_6 \cdot 6\,H_2O$.

Organic Compounds

Aconitine hydrobromide, sint. 160° - - - barium acetate monohydrate - - - benzil, $(C_6H_5=CO)_2$, 95° - - - benzoylcarbinol, $C_6H_5 \cdot CO \cdot CH_3OH$, 85–86° - - - camphor (d), 93–94° - - - cyclotetramethylenetetranitramine IV, I at 279° - - - glycerol diphenylether, 80–81° - - - iodoform, yellow, 119° - - - methionine (l), 283° dec. - - - methylnaphthoquinone monoxime, 166–168° - - - mudarol, $C_{30}H_{47}O(OH)$, 176° - - - threonine (d), $C_4H_9O_3N$,

225–227° dec. – – – triethylamine hydrobromide, 248° – – – triethylphosphine sulfide, $(C_2H_5)_3PS$, 94° – – – triphenylacetic acid, dec. – – – tryptophane (l), 289° – – – valdivin (in), glucoside, 230° dec. – – – ytterbium acetate tetrahydrate, $Yb(CH_3 \cdot COO)_3 \cdot 4\,H_2O$.

Table 4
List of Common Inorganic Compounds in the Order of Their Melting Points

Temperatures are given in degrees Celsius (centigrade); in general, only those compounds are listed, which melt below 900. In addition, the boiling points (b. pt.), transition points (tr.) and temperatures of sublimation (subl.) or volatilization (vol.) are given. The color refers to the solid substance.

Melting Point,

15.5	$NaOH \cdot 3.5\,H_2O$, monoclinic
16.83	SO_3, prisms, b. pt. 44.6
17	$Na_2MnO_4 \cdot 10\,H_2O$, green, monoclinic
17	MoF_6, crystals, b. pt. 35
19.2	$H_2SnCl_6 \cdot 6\,H_2O$, deliquescent
19.9	$Na_2CrO_4 \cdot 10\,H_2O$, yellow, monoclinic, deliquescent
22	$NaHS \cdot 3\,H_2O$, orthorhombic, decomposes
22	$Pb(CH_3COO)_2 \cdot 10\,H_2O$, orthorhombic
22.5	P_2O_3, monoclinic, deliquescent, b. pt. 173
25	PBr_8Cl_3, brown needles
25	$SrS_4 \cdot 6\,H_2O$, pink
25	$TlCl_3$, hexagonal plates, decomposes
25.5	RuO_4, yellow, orthorhombic
25.7	$BaI_2 \cdot 6\,H_2O$, hexagonal
25.8	$Mn(NO_3)_2 \cdot 6\,H_2O$, rose, monoclinic, b. pt. 129.5
26	$H_2SeO_4 \cdot H_2O$, needles, b. pt. 205
26.1	$GeBr_4$, gray, octahedra
26.5	H_3PO_2, decomposes
27	$HAuBr_4 \cdot 5\,H_2O$, brown
27	$FeBr_2 \cdot 6\,H_2O$, red, orthorhombic
27	$FeBr_3 \cdot 6\,H_2O$, red
27.2	ICl, red, cubic, b. pt. 97
28	$Na_2HAsO_4 \cdot 12\,H_2O$, monoclinic, $-12\,H_2O$ at 100
28.5	Cs, silvery, hexagonal, b. pt. 670
29.75	Ga, gray, b. pt. 1983
29.88	$LiNO_3 \cdot 3\,H_2O$
29.92	$CaCl_2 \cdot 6\,H_2O$, trigonal, $-6\,H_2O$ at 200
30	N_2O_5, orthorhombic, b. pt. 47

30	$Bi(NO_3)_3 \cdot 5 H_2O$, triclinic, decomposes at 30, $-5 H_2O$ at 80
30	$MnSO_4 \cdot 4 H_2O$, pink, orthorhombic, monoclinic, $-1 H_2O$ at 30, $-4 H_2O$ at 450
31.0	$SnBr_4$, orthorhombic, deliquescent, b. pt. 202
31	$AsBr_3$, prisms, b. pt. 221
32.4	$NaSO_4 \cdot 10 H_2O$, monoclinic, $-10 H_2O$ at 100
33	ICl_3, yellow to red-brown, orthorhombic, b. pt. 77, decomposes
33	$KFe(SO_4)_2 \cdot 12 H_2O$, colorless or violet, cubic
34	NH_2OH, orthorhombic, deliquescent
34	$Na_2CO_3 \cdot 10 H_2O$, monoclinic, $-H_2O$ at 106
34	$CdCl_2 \cdot 2.5 H_2O$, monoclinic, tr. at 34
34.6	$Na_2HPO_4 \cdot 12 H_2O$, monoclinic, $-12 H_2O$ at 180
35	$Fe(NO_3)_3 \cdot 6 H_2O$, orthorhombic, deliquescent, decomposes
35	$H_2S_2O_7$, crystals, decomposes
35	$Mg(ClO_3)_2 \cdot 6 H_2O$, deliquescent, boils and decomposes at 120
35.5	$H_3AsO_4 \cdot 0.5 H_2O$, hygroscopic, $-H_2O$ at 160
36	NH_4CN, cubic
36.4	$Zn(NO_3)_2 \cdot 6 H_2O$, tetragonal, $-6 H_2O$ at 105
36.5	$Cr(NO_3)_3 \cdot 9 H_2O$, purple prisms, decomposes at 100
37	$FeCl_3 \cdot 6 H_2O$, brown, deliquescent, b. pt. 280
37	$TlCl_3 \cdot 4 H_2O$, needles, $-4 H_2O$ at 100
37.5	$La_2(BrO_3)_6 \cdot 18 H_2O$, hexagonal, $-14 H_2O$ at 100
37.7	$SnCl_2 \cdot 2 H_2O$, triclinic, decomposes
38	$PSBr_3$, yellow, cubic, decomposes at 175
38	$H_4P_2O_5$, needles, decomposes at 130
38.2	$CaBr_2 \cdot 6 H_2O$, monoclinic, boils 149 to 150
38.5	Rb, silvery, b. pt. 700
39	$TiBr_4$, amber, deliquescent, b. pt. 230
39	$ZnSO_4 \cdot 7 H_2O$, orthorhombic, transition at 39, $-7 H_2O$ at 280
40	$TlBr_3 \cdot 4 H_2O$, yellow, deliquescent
40	$SeSO_3$, green prisms, decomposes with lib. of SO_2 at 40
40	$NH_4Fe(SO_4)_2 \cdot 12 H_2O$, violet, cubic
40	H_2TeO_3, orthorhombic, monoclinic, gives off H_2O at 40
40	OsO_4, monoclinic, b. pt. 135
40	$La(NO_3)_3 \cdot 6 H_2O$, triclinic, deliquescent, boils at 126
41	$KF \cdot 2 H_2O$, monoclinic prisms
41	$Ba(CH_3COO)_2 \cdot H_2O$, triclinic prisms, $-H_2O$ at 41
41.5	$3 CdSO_4 \cdot 8 H_2O$, monoclinic, transition at 41.5
41.7	$SeOBr_2$, yellow crystals, b. pt. about 220
42	$2 NaH_2PO_3 \cdot 5 H_2O$, monoclinic prisms, $-5 H_2O$ at 100
42	$CaI_2 \cdot 6 H_2O$, deliquescent, $-H_2O$ at 160
42 ±	BrI, dark gray crystals, b. pt. about 116
42.35	H_3PO_4, orthorhombic, $-0.5 H_2O$ at 213

42.7	$Ca(NO_3)_2 \cdot 4 H_2O$, monoclinic, $-H_2O$ at 42.7
43	BI_3, plates, b. pt. 210
44	$LiBr \cdot 2 H_2O$, prisms
44.1	P_4, yellow, hexagonal, b. pt. 280
45	KI_3, dark blue, monoclinic, deliquescent, decomposes at 225
45 to 46	$Be(stearate)_2$, waxy
47	$Fe(NO_3)_3 \cdot 9 H_2O$, pale violet, monoclinic, deliquescent, decomposes at 100
47	$Na_2SiO_3 \cdot 9 H_2O$, orthorhombic, $-6 H_2O$ at 100
48	$NH_2OH \cdot HNO_3$, crystals, decomposes below 100
49	$Ce(BrO_3)_3 \cdot 9 H_2O$, hexagonal, melts with decomposition
50	Se_8, red powder, amorphous, b. pt. 688
50	$HClO_4 \cdot H_2O$, needles, melts with decomposition
50	$(SO_3)_2$, silky needles
50	$Na_2S \cdot 9 H_2O$, deliquescent crystals, melt decomposes
50.7	$NaBr \cdot 2 H_2O$, monoclinic
51	$Co(CO)_4$, orange crystals, decomposes at 52
53	$Na_2HPO_3 \cdot 5 H_2O$, orthorhombic, deliquescent, $-5 H_2O$ at 120
53.3	$NiSO_4 \cdot 6 H_2O$, green monoclinic or blue tetragonal, transition at 53.3, $-6 H_2O$ at 280
55	$H_4P_2O_6$, vitreous, deliquescent, decomposes at 70
56	$POBr_3$, plates, b. pt. 190
56.5	$Pr_2(BrO_3)_6 \cdot 18 H_2O$, green, hexagonal, $-14 H_2O$ at 100
56.7	$Ni(NO_3)_2 \cdot 6 H_2O$, green, monoclinic, boiling at 136.7
57	$Co(NO_3)_2 \cdot 6 H_2O$, red, monoclinic, melt decomposes
57	$S_2O_3Cl_4$, orthorhombic, melt decomposes, sublimate
57	$NaBO_2 \cdot 4 H_2O$, monoclinic
58.0	$MnCl_2 \cdot 4 H_2O$, rose, monoclinic, deliquescent, $-H_2O$ at 106, $-4 H_2O$ at 200
58	H_2SeO_4, hexagonal prisms, b. pt. 260
58	$NaCH_3COO \cdot 3 H_2O$, monoclinic, $-3 H_2O$ at 120
58.5	$NaHSO_4 \cdot H_2O$, monoclinic, deliquescent, melt decomposes
59.4	$Cd(NO_3)_2 \cdot 4 H_2O$, needles, $-H_2O$ at 132
below 60	$H_2S_2O_8$, hygroscopic crystals
60	$H_2PtCl_6 \cdot 6 H_2O$, red to brown, deliquescent
60	$Be(NO_3)_2 \cdot 4 H_2O$, deliquescent, decomposes at 100
60	$Zn(ClO_3)_2 \cdot 6 H_2O$, monoclinic, deliquescent, melt decomposes
60	$NaH_2PO_4 \cdot 2 H_2O$, orthorhombic
60	$KBr \cdot IBr$, orthorhombic, decomposes at 180
60	$K_2S \cdot 5 H_2O$, orthorhombic, deliquescent, $-3 H_2O$ at 150
60	$KCl \cdot ICl$, monoclinic, decomposes at 215
60.2	$UO_2(NO_3)_2 \cdot 6 H_2O$, yellow, orthorhombic, boils at 118
60.5	$Fe(NO_3)_2 \cdot 6 H_2O$, crystals

List of Common Inorganic Compounds in the Order of Their Melting Points 441

61	$Co(ClO_3)_2 \cdot 6\ H_2O$, red, cubic, decomposes at 100
61	$Co(IO_3)_2 \cdot 6\ H_2O$, red, cubic, melt decomposes
61	PI_3, red, hexagonal, deliquescent, melt decomposes
61	$NaAl(SO_4)_2 \cdot 12\ H_2O$, cubic
61	$H_4P_2O_7$, needles
61	$SrCl_2 \cdot 6\ H_2O$, orthorhombic, $-4\ H_2O$ at 61, $-6\ H_2O$ at 100
62.3	K, b. pt. 760
62.3	$Al(BrO_3)_3 \cdot 9\ H_2O$, hygroscopic, melt decomposes at 100
64	$FeSO_4 \cdot 7\ H_2O$, green, monoclinic, $-7\ H_2O$ at 300
64.3	$MnBr_2 \cdot 4\ H_2O$, red, monoclinic, deliquescent, melt decomposes
65	$Cu(ClO_3)_2 \cdot 6\ H_2O$, green, cubic, deliquescent, decomposes at 100
65	$N_2H_4 \cdot HN_3$ deliquescent
65	$LiClO_3 \cdot 0.5\ H_2O$, tetragonal, deliquescent, $-0.5\ H_2O$ at 90
66.7	$Nd_2(BrO_3)_6 \cdot 18\ H_2O$, red, hexagonal, $-18\ H_2O$ at 150
70	Se_2I_2, steel gray crystals, melt decomposes at 100
70	AsI_5, brown, monoclinic
70	$K_4Fe(CN)_6 \cdot 3\ H_2O$, yellow, monoclinic, $-3\ H_2O$ at 70
70	$H_2MoO_4 \cdot H_2O$, yellow, monoclinic, $-H_2O$ at 70, $-2\ H_2O$ at 200
70	$LiCH_3COO \cdot 2\ H_2O$, orthorhombic, melt decomposes
70	$MgSO_4 \cdot 7\ H_2O$, orthorhombic, melts with decomposition
70	$ZnSO_4 \cdot 6\ H_2O$, monoclinic, $-5\ H_2O$ at 70
70.7	$N_2H_4 \cdot HNO_3$, crystals, sublimate at 140
70 to 80	$NaK \cdot$ tartrate $\cdot 4\ H_2O$, orthorhombic, $-4\ H_2O$ at 215
71.9	$Sr(HCOO)_2$, orthorhombic
72	$Au(NO_3)_3 \cdot HNO_3 \cdot 3\ H_2O$, yellow, triclinic, cubic, melt decomposes
73	NO_2HSO_3, orthorhombic, melt decomposes
73	$Al(NO_3)_3 \cdot 9\ H_2O$, orthorhombic, deliquescent, decomposes at 134
73	$LiI \cdot 3\ H_2O$, monoclinic, $-3\ H_2O$ at 300
73.4	$Na_3PO_4 \cdot 12\ H_2O$, trigonal, $-11\ H_2O$ at 100
73.4	$SbCl_3$, orthorhombic, deliquescent, b. pt. 220.2
74	H_3PO_3, decomposes at 200
74	$Y(BrO_3)_3 \cdot 9\ H_2O$, hexagonal prisms, $-6\ H_2O$ at 100
75	$Sm_2(BrO_3)_6 \cdot 18\ H_2O$, yellow, hexagonal, $-14\ H_2O$ at 100, $-18\ H_2O$ at 150
75	$Na_2B_4O_7 \cdot 10\ H_2O$, monoclinic, $-10\ H_2O$ at 200
75	$SiSCl_2$, prisms, b. pt. 185
75	$Pb(CH_3COO)_2 \cdot 3\ H_2O$, monoclinic, $-3\ H_2O$ at 75
75.5	$GaCl_3$, deliquescent needles, b. pt. 215
75.5	NbF_5, monoclinic, b. pt. 229
77.9	$Ba(OH)_2 \cdot 8\ H_2O$, monoclinic, $-8\ H_2O$ at 550

78	$H_7P(Mo_2O_7)_6 \cdot 28\,H_2O$, yellow, cubic, $-25\,H_2O$ at 140
78	$Dy(BrO_3)_3 \cdot 9\,H_2O$, yellow, hexagonal, $-6\,H_2O$ at 110
78	$CaS_2O_6 \cdot 4\,H_2O$, trigonal, $-4\,H_2O$ at 78, decomposes at 110
79	SbI_5, dark brown, b. pt. 400.6
79	$Hg(NO_3)_2 \cdot 0.5\,H_2O$, melt decomposes
79	$NaNH_4HPO_4 \cdot 4\,H_2O$, monoclinic, melt decomposes
80	SeI_4, dark gray crystals, $-4\,I$ at 100
80	$Ni(ClO_3)_2 \cdot 6\,H_2O$, green crystals, melt decomposes
80	$Mg(CH_3COO)_2 \cdot 4\,H_2O$, monoclinic prisms
80	$Na_4Ca(SO_4)_3 \cdot 2\,H_2O$, $-H_2O$ at 80
80	$Cd(ClO_3)_2 \cdot 2\,H_2O$, deliquescent
80	$In(ClO_4)_3 \cdot 8\,H_2O$, deliquescent, melt decomposes at 200
84.6	$Na_2Cr_2O_7 \cdot 2\,H_2O$, orange, monoclinic, $-2\,H_2O$ at 84.6, anhydrous salt melts at 356, decomposes at 400
85	$N_2H_4 \cdot 0.5\,H_2SO_4$, deliquescent plates
86	$CoCl_3 \cdot 6\,H_2O$, red, monoclinic, $-6\,H_2O$ at 110
86.3	$Na_3AsO_4 \cdot 12\,H_2O$, hexagonal
86.5	$Al_2(SO_4)_3 \cdot 18\,H_2O$, monoclinic, melt decomposes
88	$SrBr_2 \cdot 6\,H_2O$, $-6\,H_2O$ above 180
88.5	$Mg(stearate)_2$
89	$KCr(SO_4)_2 \cdot 12\,H_2O$, violet or green, cubic
89.3	$Tb(NO_3)_3 \cdot 6\,H_2O$, monoclinic
91	$TlAl(SO_4)_2 \cdot 12\,H_2O$, cubic
91	$Gd(NO_3)_3 \cdot 6.5\,H_2O$, triclinic
92	$KAl(SO_4)_2 \cdot 12\,H_2O$, cubic, $9\,H_2O$ may be lost at 64.5
92.6	$N_2H_4 \cdot HCl$, crystals
93	$SiSBr_2$, plates
93.5	$NH_4Al(SO_4)_2 \cdot 12\,H_2O$, cubic, $-10\,H_2O$ at 120, $-12\,H_2O$ at 200
94	$LiHCOO \cdot H_2O$, orthorhombic, $-H_2O$ at 94
95	$LiClO_3 \cdot 3\,H_2O$, hexagonal, $-2\,H_2O$ at 100, $-3\,H_2O$ at 150
95	Si_2Br_6, orthorhombic, b. pt. 240
95	$Mg(NO_3)_2 \cdot 6\,H_2O$, monoclinic, deliquescent, $-5\,H_2O$ at 330
96.6	$SbBr_3$, orthorhombic, b. pt. 280
96.8	$CoSO_4 \cdot 7\,H_2O$, red, monoclinic, $-7\,H_2O$ at 420
96.8	TaF_5, tetragonal, b. pt. 230
97.5	Na, silvery, b. pt. 880
97.5	$AlBr_3$, trigonal, b. pt. 268
98	$MoOF_4$, pale blue, b. pt. 180
98 to 100	$NiSO_4 \cdot 7\,H_2O$, green, orthorhombic, $-6\,H_2O$ at 103
99	$RbAl(SO_4)_2 \cdot 12\,H_2O$, cubic
below 100	$H_2PtBr_6 \cdot 9\,H_2O$, red, monoclinic, deliquescent, melt decomposes

below 100	PBr_5, yellow, orthorhombic, decomposes at 106
below 100	$LiNO_2 \cdot H_2O$, needles, melt decomposes
100	$PtO_2 \cdot H_2O$, black, $-H_2O$ at 100
100	$Co(OH)_3$, black, $-1.5\,H_2O$ at 100
100	$CuCr_2O_7 \cdot 2\,H_2O$, black, triclinic, $-2\,H_2O$ at 100
100	$Zn(MnO_4)_2 \cdot 6\,H_2O$, dark brown needles, $-5\,H_2O$ at 100, decomposes at higher temperature
100	$PtO_2 \cdot 2\,H_2O$, brown, $-2\,H_2O$ at 100
100	$Na_2PtCl_4 \cdot 4\,H_2O$, red, melt decomposes
100	$PtCl_4 \cdot 8\,H_2O$, red, monoclinic, $-4\,H_2O$ at 100
100	$Na_2PtCl_6 \cdot 6\,H_2O$, red, triclinic, $-6\,H_2O$ at 100
100	$CoBr_2 \cdot 6\,H_2O$, red, deliquescent, $-4\,H_2O$ at 100, $-6\,H_2O$ at 130
100	$Cr(NO_3)_3 \cdot 7.5\,H_2O$, purple, monoclinic, melt decomposes
100	$K_2Ni(CN)_4 \cdot H_2O$, red to yellow, monoclinic, $-H_2O$ at 100
100	$Co_3H_6(AsO_3)_4 \cdot H_2O$, rose, $-H_2O$ at 100
100	$FeF_3 \cdot 4.5\,H_2O$, yellow crystals, $-3\,H_2O$ at 100
100	$Pt(OH)_2 \cdot 2\,H_2O$, yellow, $-2\,H_2O$ at 100
100	$BaPt(CN)_4 \cdot 4\,H_2O$, yellow, monoclinic, $-2\,H_2O$ at 100, $-4\,H_2O$ at 150
100	H_2WO_4, yellow, orthorhombic, $-0.5\,H_2O$ at 100, melts at 1473
100	$IBr_3 \cdot 4\,H_2O$, olive green crystals, $-3\,H_2O$ at 100
100	$NH_4Cr(SO_4)_2 \cdot 12\,H_2O$, green or violet, cubic, melt decomposes
100	$Cr(OH)_3 \cdot 2\,H_2O$, green, $-2\,H_2O$ at 100
100	$Pr_2(CO_3)_3 \cdot 8\,H_2O$, green plates, $-6\,H_2O$ at 100
100	$FeF_2 \cdot 8\,H_2O$, green, $-8\,H_2O$ at 100
100	$(NH_4)_3Fe(C_2O_4)_3$, light green crystals, $-3\,H_2O$ at 100; decomposes at 165
100	$Na_3Fe(C_2O_4)_3 \cdot 5.5\,H_2O$, green, monoclinic, $-4\,H_2O$ at 100, $-5.5\,H_2O$ at 200
100	$K_3Fe(C_2O_4)_3 \cdot 3\,H_2O$, monoclinic, $-3H_2O$ at 100, decomposes at 230
100	$Cr_2(SO_4)_3 \cdot 15\,H_2O$, violet, $-10\,H_2O$ at 100
100	$CrPO_4 \cdot 6\,H_2O$, violet, triclinic, $-3.5\,H_2O$ at 100
100	Na_2SO_4, orthorhombic, tr. to monoclinic at 100, to hexagonal at 500, melting at 884
100	$(NH_4)_2W_4O_{13} \cdot 8\,H_2O$, cubic or $(NH_4)_6W_7O_{24}$, orthorhombic, $-H_2O$ at 100
100	$Na_2CO_3 \cdot H_2O$, orthorhombic, $-H_2O$ at 100
100	$NaH_2PO_4 \cdot H_2O$, orthorhombic, $-H_2O$ at 100, decomposes at 200
100	$KSbO \cdot$ tartrate $\cdot 0.5\,H_2O$, orthorhombic, $-H_2O$ at 100
100	$K_2TiF_6 \cdot H_2O$, monoclinic, $-H_2O$ at 100

100	$CaO_2 \cdot 8\,H_2O$, tetragonal, pearly, $-H_2O$ at 100
100	$Ca(H_2PO_4)_2 \cdot H_2O$, triclinic, $-H_2O$ at 100, decomposes at 200
100	$BaS_2O_3 \cdot H_2O$, orthorhombic, decomposes at 100
100	$Y_2(CO_3)_3 \cdot 3\,H_2O$, $-H_2O$ at 100, $-3\,H_2O$ at 130
100	$ThF_4 \cdot 4\,H_2O$, powder, $-H_2O$ at 100, $-2\,H_2O$ at 150
100	$Li_2SiF_6 \cdot 2\,H_2O$, monoclinic, $-2\,H_2O$ at 100
100	$Na_2WO_4 \cdot 2\,H_2O$, orthorhombic, $-2\,H_2O$ at 100
100	$CaSO_3 \cdot 2\,H_2O$, $-2\,H_2O$ at 100, decomposes at 650
100	$BaCl_2 \cdot 2\,H_2O$, monoclinic, $-H_2O$ at 100
100	$BaBr_2 \cdot 2\,H_2O$, monoclinic, $-2\,H_2O$ at 100 and decomposition
100	$ZnSO_3 \cdot 2.5\,H_2O$, monoclinic, $-2.5\,H_2O$ at 100, decomposes at 200
100	$Zn(H_2PO_4)_2 \cdot 2\,H_2O$, triclinic, melt decomposes
100	$Zn_3(AsO_4)_2 \cdot 8\,H_2O$, monoclinic, $-2\,H_2O$ at 100
100	$Zn(salicylate)_2 \cdot 3\,H_2O$, needles, $-2\,H_2O$ at 100, $-3\,H_2O$ at 150
100	$CdSeO_4 \cdot 2\,H_2O$, orthorhombic, decomposes at 100
100	$Cd(H_2PO_4)_2 \cdot 2\,H_2O$, triclinic, decomposes at 100
100	$MgHPO_4 \cdot 7\,H_2O$, monoclinic, $-3\,H_2O$ at 100
100	$Mg_2P_2O_7 \cdot 3\,H_2O$, amorphous, $-3\,H_2O$ at 100
100	Mg-tartrate $\cdot 5\,H_2O$, monoclinic, $-3\,H_2O$ at 100 and decomposition
100	$MgCO_3 \cdot 3\,H_2O$, orthorhombic, $-H_2O$ at 100
100	$Ca(lactate)_2 \cdot 5\,H_2O$, $-3\,H_2O$ at 100
100	$Sr(CNS)_2 \cdot 3\,H_2O$, deliquescent, $-3\,H_2O$ at 100, decomposes at 160
100	$Y(NO_3)_3 \cdot 6\,H_2O$, triclinic, $-3\,H_2O$ at 100
100	$Ce(NO_3)_3 \cdot 6\,H_2O$, deliquescent, $-3\,H_2O$ at 100, decomposes at 200
100	$In_2(SO_4)_3 \cdot 8\,H_2O$, crystals, $-3\,H_2O$ at 100, $-8\,H_2O$ at 260
100	$SrS_2O_3 \cdot 5\,H_2O$, monoclinic, $-4\,H_2O$ at 100
100	$Yb(CH_3COO)_3 \cdot 4\,H_2O$, hexagonal plates, $-4\,H_2O$ at 100
100	$ZnF_2 \cdot 4\,H_2O$, orthorhombic, $-4\,H_2O$ at 100
100	$In(NO_3)_3 \cdot 4.5\,H_2O$, deliquescent needles, $-4.5\,H_2O$ at 100
100	$MgHAsO_4 \cdot 7\,H_2O$, monoclinic, $-5H_2O$ at 100
100	$Na_2H_2P_2O_6 \cdot 6\,H_2O$, monoclinic, $-6\,H_2O$ at 100, the anhydrous salt melts at 250
100	$KNaCO_3 \cdot 6\,H_2O$, monoclinic, $-6\,H_2O$ at 100
100	$Zn(BrO_3)_2 \cdot 6\,H_2O$, $-6\,H_2O$ at 200
100	$Tl_2(SO_4)_3 \cdot 7\,H_2O$, $-6\,H_2O$ at 100
100	$SrO_2 \cdot 8\,H_2O$, $-8H_2O$ at 100
100	$BaO_2 \cdot 8\,H_2O$, pearly scales, $-H_2O$ at 100
100	$Li_3PO_4 \cdot 12\,H_2O$, trigonal

100	$Cr_2(SO_4)_3 \cdot 18\ H_2O$, violet, cubic, $-12\ H_2O$ at 100
above 100	$K_2OsO_4 \cdot 2\ H_2O$, violet, cubic, $-H_2O$ above 100
above 100	P_2O_4, orthorhombic, deliquescent, b. pt. 180
above 100	LiH_2PO_4
above 100	$Ca(ClO_3)_2 \cdot 2\ H_2O$, monoclinic, deliquescent, $-H_2O$ above 100
above 100	$Sr(NO_3)_2 \cdot H_2O$, $-H_2O$ above 100, decomposes at 240
102	$P_4O_6S_4$, tetragonal, deliquescent, b. pt. 295
103	CuS, blue, hexagonal, monoclinic, tr. at 103, decomposes at 220
104	$N_2H_4 \cdot 2\ HNO_3$, needles, melt decomposes
105	$Zn_3(PO_4)_2 \cdot 4\ H_2O$, orthorhombic, triclinic, tr. at 105, 140, and 163
105	$Sm_2(SO_4)_3 \cdot 8\ H_2O$, monoclinic, $-3\ H_2O$ at 105, $-8\ H_2O$ at 450
108	$CdSO_4 \cdot H_2O$, monoclinic, tr. at 108
110	P_2I_4, orange, triclinic
110	WOF_4, plates, b. pt. 187
110	$Tl \cdot CH_3COO$, silky needles
110	$Li_2Cr_2O_7 \cdot 2\ H_2O$, deliquescent, orange, $-2\ H_2O$ at 110
110	$UO_2(CH_3COO)_2 \cdot 2\ H_2O$, yellow, orthorhombic, $-2\ H_2O$ at 110
110	$UO_2(ClO_4)_2 \cdot 4\ H_2O$, yellow, melt decomposes
110	$CuCl_2 \cdot 2\ H_2O$, green, orthorhombic, $-H_2O$ at 110 and decomposition
110	$CuSO_4 \cdot 5\ H_2O$, blue, triclinic, $-4\ H_2O$ at 110, $-5\ H_2O$ at 250
110	$K_2HAsO_4 \cdot H_2O$, trigonal, $-H_2O$ at 110
110	$PbHIO_5 \cdot H_2O$, amorphous, $-H_2O$ at 110
110	$K \cdot$ benzoate $\cdot 3\ H_2O$, $-3\ H_2O$ at 110
110	$Mg(benzoate)_2 \cdot 3\ H_2O$, powder, $-3\ H_2O$ at 110
110	$YCl_3 \cdot 6\ H_2O$, orthorhombic, deliquescent, $-5\ H_2O$ at 110
110	$Gd_2(C_2O_4)_3 \cdot 10\ H_2O$, monoclinic, $-6\ H_2O$ at 110
110	$Ce_2(C_2O_4)_3 \cdot 9\ H_2O$, powder, $-9\ H_2O$ at 110
112.8	S_8, pale yellow, orthorhombic, b. pt. 444.6
113.5	I_2, blue black, orthorhombic, b. pt. 184.35
114	$NH_4 \cdot CH_3COO$, hygroscopic, melt decomposes
114	$(NPCl_2)_3$, orthorhombic
114.5	$Cu(NO_3)_2 \cdot 3\ H_2O$, blue, deliquescent, $-HNO_3$ at 170
114 to 116	$NH_4 \cdot HCOO$, monoclinic, deliquescent, decomposes at 180
115	$TlO \cdot OH$, red brown crystals, $-H_2O$ at 115
115	$Cu(CH_3COO)_2 \cdot H_2O$, dark green, monoclinic, boils at 240 with decomposition
115	$PrCl_3 \cdot 7\ H_2O$, green crystals
115	$TlHSO_4$, trimorphous, melt decomposes
115	$Ba(NO_2)_2 \cdot H_2O$, hexagonal, decomposes at 115
115	$Ce_2(CH_3COO)_6 \cdot 3\ H_2O$, needles, $-3\ H_2O$ at 115 and decomposition

117	$CsAl(SO_4)_2 \cdot 12\ H_2O$, cubic, melt decomposes
118	$MgCl_2 \cdot 6\ H_2O$, monoclinic, deliquescent, melt decomposes
119.0	S_8, pale yellow, monoclinic, b. pt. 444.6
120	$Bi_2O_5 \cdot H_2O$; red, $-H_2O$ at 120, $-O_2$ at 357
120	$BaCr_2O_7 \cdot 2\ H_2O$, yellow, $-H_2O$ at 120
120	$MgCrO_4 \cdot 7\ H_2O$, yellow, orthorhombic, $-3\ H_2O$ at 120
120	S, pale yellow, amorphous, b. pt. 444.6
120	$Sr(ClO_3)_2$, orthorhombic, melt decomposes
120	$Sr(BrO_3)_2 \cdot H_2O$, monoclinic, $-H_2O$ at 120, decomposes at 240
120	$Ba(ClO_3)_2 \cdot H_2O$, monoclinic, $-H_2O$ at 120
120	$Ca(salicylate)_2 \cdot 2\ H_2O$, cubic, $-H_2O$ at 120
120	$Sr(lactate)_2 \cdot 3\ H_2O$, $-3\ H_2O$ at 120
120	$Al_2F_6 \cdot 7\ H_2O$, $-4\ H_2O$ at 120, with boiling $-6\ H_2O$ at 250
120	$Mg_3(PO_4)_2 \cdot 8\ H_2O$, monoclinic prisms, $-5\ H_2O$ at 120, $-8\ H_2O$ at 400
120	$Na_2Mo_3O_{10} \cdot 7\ H_2O$, needles, $-6\ H_2O$ at 120
120.5	SiI_4, cubic, b. pt. 290
above 120	$(NH_4)_2Mg(SO_4)_2$, monoclinic
above 120	$CdHAsO_4 \cdot H_2O$
above 120	$Tl_4P_2O_7$, monoclinic
123	$NH_4H_2PO_3$, deliquescent, melt decomposes at 150
124	$NdCl_3 \cdot 6\ H_2O$, red, orthorhombic, $-6\ H_2O$ at 160
125	$Na_2HAsO_4 \cdot 7\ H_2O$, monoclinic
127	HgI_2, red tetragonal, tr. to yellow at 127, melts at 259, b. pt. 354
128	$N_2H_4 \cdot 2\ HCOOH$, cubic
128	$CaSO_4 \cdot 2\ H_2O$, $-1.5\ H_2O$ at 128, $-2\ H_2O$ at 163
129	$LiClO_3$, melt decomposes at 270
129	KNO_3, orthorhombic, tr. at 129, melts at 333, decomposes at 400
130	AsI_2, red, b. pt. 380
130	$Li_2CrO_4 \cdot 2\ H_2O$, yellow, orthorhombic, $-2\ H_2O$ at 130
130	$VF_3 \cdot 3\ H_2O$, dark green, orthorhombic, $-3\ H_2O$ at 130
130	$Li_2SO_4 \cdot H_2O$, monoclinic, $-H_2O$ at 130
130	$SrHAsO_4 \cdot H_2O$, orthorhombic needles, $-H_2O$ at 130, $-1.5\ H_2O$ at 360
130	$Ba(IO_3)_2 \cdot H_2O$, monoclinic, $-H_2O$ at 130
130	$3\ PbO \cdot H_2O$, cubic, $-H_2O$ at 130
130	$Ca_3(citrate)_2 \cdot 4\ H_2O$, needles, $-2\ H_2O$ at 130, $-4\ H_2O$ at 185
130	$Cd(CH_3COO)_2 \cdot 2\ H_2O$, monoclinic, $-H_2O$ at 130
130	$H_2TeO_4 \cdot 2\ H_2O$, cubic, monoclinic, $-2\ H_2O$ at 130
130	$Gd_2(SeO_4)_3 \cdot 8\ H_2O$, pearly, monoclinic, $-8\ H_2O$ at 130
132	$NH_4SO_3NH_2$, melt decomposes at 160
135	$CoI_2 \cdot 6\ H_2O$, red, hexagonal, $-6\ H_2O$ at 135
140	$P_3O_8I_6$, red crystals, melt decomposes

140	$Co(CH_3COO)_2 \cdot 4 H_2O$, red-violet, monoclinic, $-4 H_2O$ at 140
140	$CuCl_2 \cdot 2 CuO \cdot 4 H_2O$, blue-green, $-3 H_2O$ at 140
140	$MnHAsO_4 \cdot H_2O$, $-H_2O$ at 140
140	$K_2SnO_3 \cdot 3 H_2O$, trigonal, $-3 H_2O$ at 140
143	$Co(ClO_4)_2 \cdot 6 H_2O$, red, hexagonal
143.5	SnI_4, red, orthorhombic, cubic, b. pt. 340
144	GeI_4, orange, cubic, deliquescent, b. pt. 375
above 145	K_2S_4, red-brown, melt decomposes at 850
146	AsI_3, red, hexagonal, b. pt. 403
146.9	NH_4HSO_4, orthorhombic, boils at 490
147	$Mg(ClO_4)_2 \cdot 6 H_2O$, hygroscopic, $-6 H_2O$ at 250
148	$KH(CH_3COO)_2$, deliquescent needles or plates, melt decomposes at 200
149	$Ni(ClO_4)_2 \cdot 5 H_2O$ or $6 H_2O$, blue-green, hexagonal
149.6	NH_4CNS, monoclinic, melt decomposes at 170
150	TiI_4, red, cubic, boils above 360
150	$NbBr_5$, purple-red, b. pt. 270
150	$K_3OsCl_6 \cdot 3 H_2O$, red crystals, $- 3 H_2O$ at 150
150	$Mn(H_2PO_2)_2 \cdot H_2O$, rose, monoclinic, $-H_2O$ at 150 with decomposition
150	$Na_2PtO_3 \cdot 3 H_2O$, yellow, $-3 H_2O$ at 150
150	$Dy_2(CrO_4)_3 \cdot 10 H_2O$, yellow crystals, $-3.5 H_2O$ at 150
150	Ag_3AsO_3, yellow, melt decomposes
150	$Sc(NO_3)_3$
150	ReO_4, globules, melt decomposes
150	$SrC_2O_4 \cdot H_2O$, $-H_2O$ at 150
150	$BaHAsO_4 \cdot H_2O$, orthorhombic, monoclinic, $-H_2O$ at 150, $-1.5 H_2O$ at 225
150	$Na_3 \cdot$ citrate $\cdot 2 H_2O$, $-2 H_2O$ at 150
150	$Na_2SO_3 \cdot 7 H_2O$, monoclinic, $-7 H_2O$ at 150, decomposition
150	$Ba_3(citrate)_2 \cdot 7 H_2O$, $-H_2O$ at 150
150	$2 Na_3 \cdot$ citrate $\cdot 11 H_2O$, orthorhombic, $-11 H_2O$ at 150
above 150	$In(OH)_3$, gelatinous, $-H_2O$ above 150
150 to 155	$YbCl_3 \cdot 6 H_2O$, green, orthorhombic, $-6 H_2O$ at 180
151	$NH_2OH \cdot HCl$, monoclinic, melt decomposes
153	$HgICl$, red, orthorhombic, b. pt. 315
155	In, soft metal, b. pt. 1450
160	$AuBr_3$, dark brown, melt decomposes
160	$YCl_3 \cdot H_2O$, decomposes at 170 to 180
160	$Cd(IO_3)_2 \cdot H_2O$, tr. at 160
160	$Mn(CNS)_3 \cdot 3 H_2O$, deliquescent, $-3 H_2O$ at 160
161.4	$RbNO_3$, hexagonal, cubic, triclinic, and orthorhombic; tr. at 161.4, 219, and 310, melts above 700

163	$BiCl_2$ (?), black needles, decomposes at 300
163	$CaSO_4 \cdot 0.5\ H_2O$, $-0.5\ H_2O$ at 163
164	$GaCl_2$, deliquescent crystals, b. pt. 535
165	$MgBr_2 \cdot 6\ H_2O$, hexagonal, deliquescent, melt decomposes
167	SbI_3, orange, orthorhombic, monoclinic, b. pt. 401
167.5	KHCOO, orthorhombic, melt decomposes
169.6	NH_4NO_3, tetragonal, orthorhombic, monoclinic, cubic; decomposes at 210
170	SbOCl, monoclinic, melt decomposes
170	$NH_2OH \cdot 0.5\ H_2SO_4$, monoclinic, melt decomposes
170	$Ba(BrO_3)_2 \cdot H_2O$, monoclinic, $-H_2O$ at 170, decomposes at 260
170	$(NH_4)_2Pr_2(SO_4)_4 \cdot 8\ H_2O$, $-8\ H_2O$ at 170
170.5	$LiHSO_4$, prisms
172.3	KCNS, monoclinic, deliquescent, decomposes at 500
172.5	P_4S_3, yellow, orthorhombic, b. pt. 408
174	BeOH(stearate), powder
175	$TeCl_2$, black crystals, b. pt. 324
175	Ag_2S, black, cubic, tr. at 175
175	$Hg(OH)_2$, $-H_2O$ at 175
177	FeI_2, gray, hexagonal
180	$Li_2PtCl_6 \cdot 6\ H_2O$, orange-red, hexagonal, $-6\ H_2O$ at 180
180	$CsBr_3$, orthorhombic
180	$3\ K_2S_2O_3 \cdot H_2O$, monoclinic, deliquescent, $-H_2O$ at 180 and decomposition
180	$K_4P_2O_7 \cdot 3\ H_2O$, deliquescent, $-2\ H_2O$ at 180, $-3\ H_2O$ at 300
185	H_3BO_3, triclinic, melt decomposes
185	$NaAlCl_4$
186	Li, silvery metal, b. pt. 1336
190	$SeBrCl_3$, yellow-brown crystals
190	RbI_3, orthorhombic
190	PBr_2N, orthorhombic
190	$BeSO_4 \cdot 4\ H_2O$, tetragonal, $-4\ H_2O$ at 190, decomposes at 550
191	AlI_3, brown crystals, b. pt. 382
194	$MoCl_5$, black crystals, b. pt. 268
194	$NbCl_5$, yellow needles, deliquescent, b. pt. 240.5
195	$NH_4Cl \cdot MgCl_2 \cdot 6\ H_2O$, orthorhombic, deliquescent, $-4\ H_2O$ at 195
195.5	$CsBrI_2$
197	CrO_3, red, orthorhombic, melt decomposes
198	$N_2H_4 \cdot 2\ HCl$, cubic
198	$NH_4 \cdot$ benzoate, orthorhombic
199	InI_3, yellow deliquescent crystals
200	$K_2RuO_4 \cdot H_2O$, black, orthorhombic, $-H_2O$ at 200

List of Common Inorganic Compounds in the Order of Their Melting Points   449

200	$MnHPO_3 \cdot 3\ H_2O$, pink, $-H_2O$ at 200
200	$MnHPO_4 \cdot 3\ H_2O$, pink, orthorhombic, $-3\ H_2O$ at 200 and decomposition
200	$CaCrO_4 \cdot 2\ H_2O$, yellow, monoclinic, $-H_2O$ at 200
200	$DyPO_4 \cdot 5\ H_2O$, yellow, $-5\ H_2O$ at 200
200	$Dy_2(SeO_4)_3 \cdot 8\ H_2O$, yellow needles, $-8\ H_2O$ at 200
200	$NiBr_2 \cdot 3\ H_2O$, green, deliquescent, $-3\ H_2O$ at 200
200	$Ni(CN)_2 \cdot 4\ H_2O$, green plates, $-4\ H_2O$ at 200 and decomposition
200	$AuOH$, violet, $-H_2O$ at 200
200	$NH_4H_2PO_2$, orthorhombic, decomposes at 240
200	$NaH_2PO_2 \cdot H_2O$, monoclinic, $-H_2O$ at 200
200	$Li_2B_4O_7 \cdot 5\ H_2O$, $-2\ H_2O$ at 200
200	$Zn(IO_3)_2 \cdot 2\ H_2O$, $-2\ H_2O$ at 200
200	$Rb_2S \cdot 4\ H_2O$, deliquescent crystals, $-4\ H_2O$ at 200
200	$K_2TeO_4 \cdot 5\ H_2O$, orthorhombic, deliquescent, $-H_2O$ and $-O$ at 200
200	$Mg(BrO_3)_2 \cdot 6\ H_2O$, cubic, $-6\ H_2O$ at 200
200	$MgSO_3 \cdot 6\ H_2O$, trigonal, $-6\ H_2O$ at 200
above 200	$AgIO_3$, orthorhombic, melt decomposes
205	NH_2SO_3H, orthorhombic, melt decomposes
206	K_2S_5, orange
206	$TlNO_3$, cubic, orthorhombic, trigonal, transitions at 75 and 145; boils at 430
207.5	CsI_3, orthorhombic
210	$NaNH_2$, olive green, b. pt. 400
210	$KHSO_4$, orthorhombic, monoclinic, melt decomposes
210	Cs_2S_5
210	$Mg(IO_3)_2 \cdot 4\ H_2O$, monoclinic, $-4\ H_2O$ at 210
210 to 215	SnF_2, monoclinic
211	$WOCl_4$, red needles, b. pt. 227.5
212	$AgNO_3$, orthorhombic, decomposes at 444
215.5	$SnBr_2$, yellow, orthorhombic, b. pt. 620
217	Cs_2S_3, orange, boils above 800
217	$Ba(NO_2)_2$
217.4	Se_8, steel gray, b. pt. 688
218	Ag_2CO_3, yellow powder, melt decomposes
218	$BiBr_3$, yellow, b. pt. 453
220	Se_8, gray, trigonal, b. pt. 684.8
220	Re_2O_7, yellow crystals, sublimation at 450
221	$TaCl_5$, yellow prisms, b. pt. 242
224	$TeCl_4$, yellow, deliquescent, b. pt. 414
225	Rb_2S_5, red, orthorhombic, deliquescent

229	HgIBr, yellow, orthorhombic, b. pt. 360
230	$AgClO_3$, tetragonal, decomposes at 270
230	$BiCl_3$, b. pt. 447
231.85	Sn, transition from gray cubic at 18 to silvery tetragonal, b. pt. 2260
236	$LiClO_4$, deliquescent, decomposes at 410
237	$HgBr_2$, orthorhombic, b. pt. 322
237	$Zn(CH_3COO)_2 \cdot 2\,H_2O$, monoclinic
240	$TaBr_5$, yellow crystals, b. pt. 320
240	$Cu(IO_3)_2 \cdot H_2O$, blue, triclinic, $-H_2O$ at 240, decomposes at 290
242	$Zn(CH_3COO)_2$, monoclinic, sublimes in vacuum
244.7	NH_4SO_3F
246.8	$SnCl_2$, orthorhombic, b. pt. 623
248	WCl_5, dark green, deliquescent, b. pt. 275.6
248	$NaClO_3$, cubic, trigonal, melt decomposes
250	$Au(OH)_3$, brown-black, $-H_2O$ at 250
250	$Co_3(PO_4)_2 \cdot 8\,H_2O$, red powder, $-8\,H_2O$ at 250
250	$Na_2H_2P_2O_6$
250	AgN_3, prisms, explodes at 297
250	Si_2I_6, hexagonal plates, melt decomposes
250 to 300	H_2UO_4, yellow powder, $-H_2O$ at 250 to 300
251.8	Na_2S_5, yellow
252	K_2S_3, yellow-brown
253	NaHCOO, monoclinic
254	$N_2H_4 \cdot H_2SO_4$, orthorhombic
254	$AuCl_3$, red, deliquescent, melt decomposes, sublimation at 265
256	KH_2PO_4, tetragonal, deliquescent, melt decomposes
256	$Cd(CH_3COO)_2$, melt decomposes
259	TeI_4, gray crystals
259	HgI_2, yellow, orthorhombic, b. pt. 354
260	$CuCrO_4 \cdot 2\,CuO \cdot 2\,H_2O$, yellow-brown, $-2\,H_2O$ at 260
261	$LiNO_3$, trigonal
265	$KMgCl_3 \cdot 6\,H_2O$, orthorhombic, deliquescent
266	WO_2Cl_2, yellow tablets
267	As_2S_2, red, transition at 267, melting at 307, b. pt. 565
271	Bi, b. pt. 1450
271	$NaNO_2$, pale yellow, orthorhombic, decomposes at 320
272	CsOH, deliquescent
272	Tl_2CO_3, monoclinic
275	Na_2S_4, yellow, cubic
275	WCl_6, dark blue, cubic, b. pt. 346.7
276	WBr_5, violet brown needles, b. pt. 333
276	P_2S_5, yellow, deliquescent, b. pt. 530

List of Common Inorganic Compounds in the Order of Their Melting Points 451

277	$WOBr_4$, black, deliquescent, b. pt. 327
277	$HgCl_2$, orthorhombic, b. pt. 304
280	$Co(CN)_2 \cdot 2 H_2O$, buff, $-2 H_2O$ at 280, decomposes at 300
280	Rb_2O_4, yellow, $-O$ at 500
280	$Ba(NH_2)_2$, gray-white
280	$Pb(CH_3COO)_2$
above 280	K_2O_4, orange-yellow
282	$FeCl_3$, brown-black, hexagonal, deliquescent, b. pt. 315
283	$ZnCl_2$, deliquescent, b. pt. 732
287	NaCNS, orthorhombic, deliquescent
288	KH_2AsO_4, tetragonal
290	P_4S_6, grayish yellow, b. pt. 490
292	KCH_3COO
292	SbF_3, orthorhombic, sublimes
297	KNO_2, prisms, decomposes at 350
298	P_3S_6, yellow needles, boils at 400?
300	$MnPO_4 \cdot H_2O$, gray crystals, $-H_2O$ at 300 and decomposition
300	As_2S_3, red or yellow, monoclinic, b. pt. 707
300	Tl_2O, yellow, deliquescent, $-O$ at 1865
300	$U(SO_4)_2 \cdot 4 H_2O$, green, orthorhombic, $-4 H_2O$ at 300
300	$Na_2(NO)_2$, melt decomposes
300	$NaReO_4$, plates
300	$K_2S_2O_7$
300	RbOH, deliquescent
300	$Al(OH)_3$, monoclinic, $-H_2O$ at 300
300	$Hg_2NCl \cdot 3 NH_4Cl$, reddish crystals
300	$Na_6W_7O_{24} \cdot 16 H_2O$, triclinic, $-16 H_2O$ at 300
above 300	$NaPO_3$, glass, crystallizes above 300
303.5	Tl, bluish white metal, b. pt. 1650
307	As_2S_3, red, monoclinic, b. pt. 565
308	$NaNO_3$, trigonal, decomposes at 380
308	$Ce(CH_3COO)_3$, melt decomposes
310	B_2S_3, white crystals
311.1	KSO_3F
315	As_2O_3, vitreous
above 315	$NaHSO_4$, triclinic, melt decomposes
318.4	NaOH, deliquescent, b. pt. 1390
320	SnI_2, red, orthorhombic, monoclinic, b. pt. 720
320	AgCN, $-(CN)_2$ at 320
320.9	Cd, b. pt. 767
323	$Mg(CH_3COO)_2$
324	$NaCH_3COO$, monoclinic
327.3	Pb, b. pt. 1620

333	KNO_3, trigonal, decomposes at 400
334	TlN_3, pale yellow, tetragonal, explodes at 430
338	KNH_2, yellowish green, sublimes at 400
340	Tl_2Se, dark gray leaflets
340	SeO_2, tetragonal, sublimes at 317
above 340	$Tl(OH)_3$, brown, hexagonal
350	$IrO_2 \cdot 2\,H_2O$, dark blue, $-2\,H_2O$ at 350
350	$Cd(NO_3)_2$
350	$KReO_4$, tetragonal
356	$Na_2Cr_2O_7$, orange, decomposes at 400
357	$KAuCl_4$, yellow, monoclinic
360	Ru_2O_5, black crystals, $-0.5\,O$ at 360
360	As_2Se_3, dark brown
360	$CuOH$, yellow, $-0.5\,H_2O$ at 360
360	$Dy_2(SO_4)_3 \cdot 8\,H_2O$, yellow crystals, $-8\,H_2O$ at 360
360	$LiSO_3F$
368	$KClO_3$, monoclinic, decomposes at 400
370	$KBrO_3$, trigonal, melt decomposes
373	$PbBr_2$, orthorhombic, b. pt. 918
375	$Sr(OH)_2$, deliquescent
375	$Eu_2(SO_4)_3 \cdot 8\,H_2O$, pale red, $-8\,H_2O$ at 375
380	KOH, orthorhombic, deliquescent, b. pt. 1320
380	$TeBr_4$, orange prisms, b. pt. 421
380	$Zr(SO_4)_2 \cdot 4\,H_2O$, orthorhombic, $-4\,H_2O$ at 380
381	$NaBrO_3$, cubic
383	$Ag_4V_2O_7$
384	$NaBF_4$, orthorhombic, melt decomposes slowly
385	CdI_2, brown, hexagonal, b. pt. 713
388	$K_2WO_4 \cdot 2\,H_2O$, monoclinic, transitions at 388 and 555, melting at 921
390	$LiNH_2$, cubic, b. pt. 430
390	B_2S_5, white crystals
394	$ZnBr_2$, orthorhombic, b. pt. 650
398	$K_2Cr_2O_7$, orange, triclinic, decomposes at about 500
400	Cs_2O_3, chocolate brown
400	$Ba(ClO_4)_2 \cdot 3\,H_2O$, hexagonal, decomposes at 400
400	$Th(SO_4)_2 \cdot 9\,H_2O$, monoclinic prisms, $-9\,H_2O$ at 400
400 to 500	Tl_2Cl_3, yellow, hexagonal, melt decomposes
above 400	K_2PtBr_6, red, cubic, decomposes above 400
above 400	Tl_2SeO_4, orthorhombic
402	PbI_2, yellow, hexagonal, b. pt. 954
402	SrI_2, plates, melt decomposes
408	$Ba(OH)_2$, monoclinic

412	Tl_2Te
414	$CsNO_3$, hexagonal, melt decomposes
414	$Ba(ClO_3)_2$
419.5	Zn, silvery, b. pt. 907
420	$CsBrO_3$
above 420	Zn_3P_2, steel gray, cubic, b. pt. 1100
422	Cu_2Cl_2, cubic, b. pt. 1366
424	$TlVO_3$, dark gray crystals
430	PtO_2, black
430	$RbBrO_3$
430	$TlCl$, cubic, b. pt. 806
434	$AgBr$, pale yellow, cubic, decomposes at 700
435	AgF, yellow, cubic, deliquescent
439	BiI_3, red to black, hexagonal, boils at 500 with decomposition
440	TlI, yellow orthorhombic or red cubic, b. pt. 824
440	Ru_4O_9, black crystals, $-O$ at 440
440	$BeCl_2$, deliquescent needles, b. pt. 547
445	$LiOH$, crystals, boils at about 925
446	LiI, cubic, deliquescent, b. pt. 1190
446	ZnI_2, cubic, b. pt. 624
448	Tl_2S, bluish black, tetragonal, melt decomposes
450	FeS_2, yellow, transition from orthorhombic to cubic at 450
450	$Y_2(SO_4)_3 \cdot 8H_2O$, monoclinic, $-8H_2O$ at 450, decomposes at 700
452	Te, metallic, hexagonal, b. pt. 1390
454	$Tl_4V_2O_7$
455	$AgCl$, cubic, b. pt. 1550
455	$LiSO_3$, powder, melt decomposes
455	KHS, orthorhombic, deliquescent
460	Ce_2S_3, dark red, amorphous, deliquescent, decomposes above 800
460	$TlBr$, pale yellow, cubic, b. pt. 815
470	Rb_2O_3, black
471	K_2S, brown, deliquescent
471	K_2S_2, orange
474.5	$Cu_2(CN)_2$, monoclinic, melt decomposes
482	$NaClO_4$, orthorhombic, melt decomposes
482	$AgPO_3$
490	$BeBr_2$, deliquescent needles, sublimes at 450
498	$CuBr_2$, black, monoclinic, deliquescent
498	$CuCl_2$, brownish yellow powder, becomes Cu_2Cl_2 at 993
500	$PdCl_2$, brown, cubic, melt decomposes
500	$Fe(OH)_3$, brown, $-1.5H_2O$ at 500
500	Na_2SO_4, monoclinic, transition to hexagonal at 500, melting at 884

501	$PbCl_2$, orthorhombic, b. pt. 954	
501	$TlClO_4$, orthorhombic, melt decomposes	
504	Cu_2Br_2, cubic, b. pt. 1345	
505	$Ba(ClO_4)_2$, hexagonal	
510	BeI_2, needles, b. pt. 590	
513	$(NH_4)_2SO_4$, orthorhombic, melt decomposes	
520	CdF_2, cubic, boils above 1200	
524	$PbCl_2 \cdot PbO$, tetragonal, melt decomposes	
529.5	KBF_4, cubic, orthorhombic, melt decomposes	
547	$LiBr$, cubic, deliquescent, b. pt. 1265	
550	Sb_2S_3, black, orthorhombic	
550	$Zr(OH)_4$, gelatinous, $-2\ H_2O$ at 550	
555	PtO, violet-black	
560	KIO_3, monoclinic	
561	$Ca(NO_3)_2$, cubic	
563.7	$NaCN$, cubic, b. pt. 1496	
566	ThI_4, prisms, b. pt. 837	
568	$CdCl_2$, cubic, b. pt. 960	
570	HgF, yellow, cubic?	
570	$Sr(NO_3)_2$, cubic	
573	Bi_2Te_3	
575	CaI_2, deliquescent plates, b. pt. 718	
577	B_2O_3, vitreous, boiling above 1500	
580	$CdBr_2$, plates, b. pt. 963	
580	$Cd(OH)_2$, hexagonal, $-H_2O$ at 580	
582	KIO_4, tetragonal, $-O$ at 300	
585	$Ag_2P_2O_7$	
588	$TbCl_3$, red	
592	$Ba(NO_3)_2$, cubic, melt decomposes	
593	P_4, violet, monoclinic	
600	Rb_2O_2, yellow, cubic	
600	Cs_2O_4, yellow crystals, melt decomposes	
605	Cu_2I_2, reddish brown, cubic, b. pt. 1290	
611	Sb_2S_3, gray	
612	$Na_2Mo_2O_7$, needles	
614	$LiCl$, cubic, deliquescent, b. pt. 1360	
618	Li_2CO_3, monoclinic, melt decomposes	
621	CsI, cubic, b. pt. 1280	
627.6	$NaPO_3$, transition at 500 from the insoluble to the soluble modification	
628	$GdCl_3$, monoclinic	
629	Sb_2Te_3, gray	
630	$Ce_2(SO_4)_3 \cdot 8\ H_2O$, triclinic, $-8\ H_2O$ at 630	

630.5	Sb, b. pt. 1380
632	Tl_2SO_4, orthorhombic, melt decomposes
634.5	KCN, cubic, deliquescent
636	CsBr, cubic, b. pt. 1300
642	RbI, cubic, b. pt. 1300
643	$SrBr_2$, needles, melt decomposes
645	Ce, steel gray, b. pt. 1400
645	Ni_2S, yellow crystals
645	HgF_2, cubic, melt decomposes
646	CsCl, cubic, deliquescent, b. pt. 1290
650	$MnCl_2$, rose, cubic, deliquescent, b. pt. 1190
650	$SnSe_2$, white or brown
651	Mg, b. pt. 1110
651	NaI, cubic, b. pt. 1300
652	Sb_2O_3, cubic
652	Ag_2SO_4, orthorhombic, melt decomposes
654	$Na_4V_2O_7$, hexagonal
655	$DyCl_3$, yellow plates
656	Sb_2O_3, orthorhombic, b. pt. 1570
660	Al, b. pt. 2057
680	LiH, cubic
682	RbBr, cubic, b. pt. 1340
683	CsF, cubic, b. pt. 1250
685	Bi_2S_3, brown, orthorhombic, melt decomposes
below 686	YCl_3, plates
686	$SmCl_3$, greenish yellow crystals
687	Cu_3Sb, light gray
687	Na_2MoO_4
692	Na_2WO_4, orthorhombic
693	$PbCl_2 \cdot 2\, PbO$, colorless or yellow, orthorhombic
700	$MnSO_4$, pink, decomposes at 850
700	$MgBr_2$, deliquescent
704	Bi_2O_3, yellow, cubic, orthorhombic or tetragonal, transition at 704
710	$BiSe_3$, black, orthorhombic, melt decomposes
712	$MgCl_2$, hexagonal, b. pt. 1412
715	RbCl, cubic, b. pt. 1390
715	Mg_3Bi_2, metallic
721	Cd_3As_2, dark gray, cubic
723	KI, cubic, b. pt. 1330
728	$RaBr_2$, monoclinic, sublimes at 900
730	KBr, cubic, b. pt. 1380
730	K_2GeF_6, hexagonal, b. pt. 835

740	$BaI_2 \cdot 2\,H_2O$, monoclinic, water lost at 539, decomposition at 740
741	$Na_2B_4O_7$
755	NaBr, cubic, b. pt. 1390
759	Tl_2O_3, brown-black, hexagonal, $-2\,O$ at 875
760	RbF, cubic, b. pt. 1410
760	$CaBr_2$, deliquescent needles, b. pt. 810
766	$PbSiO_3$, monoclinic
772	$CaCl_2$, cubic, deliquescent, boils above 1600
775	Na_3Bi, bluish violet
780	SnTe, gray crystals
784	$NdCl_3$, violet prisms
790	KCl, cubic, b. pt. 1500
792	Na_2CrO_4, yellow, orthorhombic
795	MoO_3, orthorhombic, sublimes
797	NiS, black trigonal, yellow hexagonal
800	Sr, silvery, b. pt. 1156
800	PdS_2, gray, melt decomposes
800	BeF_2, vitreous, sublimes above 800
800	V_2O_5, orange, orthorhombic, decomposes at 1750
above 800	VF_3, green, orthorhombic, sublimes
800.4	NaCl, cubic, b. pt. 1413
802	$Pb_2As_2O_7$, orthorhombic
807	KPO_3
810	Ca, silvery, boils at about 1200
812	K_2Sb, yellowish green
815	$Pb_2P_2O_7$, orthorhombic, melt decomposes
820	$ThCl_4$, orthorhombic, deliquescent, b. pt. 921
822	$CeCl_3$, deliquescent
823	$PrCl_3$, green needles
825	Ag_2S, black, orthorhombic, melt decomposes above 830
826	La, gray, b. pt. 1800
830	Cu_3As, hexagonal
837	Li_3PO_4, orthorhombic
837	Rb_2CO_3, deliquescent, decomposes at 740
840	Nd, yellowish metal
844	$PbCrO_4$, yellow, monoclinic, melt decomposes
845	Li_3N, black
849	Ag_3PO_4, yellow, cubic
851	Na_2CO_3, melt decomposes
855	PbF_2, orthorhombic, b. pt. 1290
856	MnF_2, red prisms
856	Na_3Sb, deep blue

861	SnSe, steel gray prisms
866	Na_3VO_4, needles
872	ZnF_2, monoclinic, triclinic (?)
873	$SrCl_2$, cubic
above 875	Na_2Se, deliquescent crystals
880	SnS, brownish black, orthorhombic, b. pt. 1150
880	Ag_2Se, gray, cubic
880	KF, cubic, b. pt. 1500
884	Na_2SO_4, hexagonal
950	Na_2S, amorphous
966	$NaBO_2$, hexagonal prisms, b. pt. 1400
992	NaF, tetragonal, b. pt. 1704

Table 5

Inorganic Substances that Sublime, Arranged According to Color

Approximate
Temperature
°C

56.2	UF_6, black needles, volatilizes
—	NiI_2, black, deliquescent, sublimes
446	HgS, black amorphous, sublimes
—	HgSe, gray cubic plates, sublimes
—	$Mo_2O_3Cl_5$, dark brown, deliquescent, sublimes
—	$MoCl_4$, brown, deliquescent, volatilizes
—	AsP, red, sublimes and decomposes
—	HgI_2, red tetragonal, transition at 127 to yellow orthorhombic, m. pt. 259, b. pt. 354, yellow sublimate turning red on cooling (and scratching)
580	HgS, red, hexagonal, sublimes
—	$FeBr_3$, dark red, deliquescent, sublimes and decomposes
250	Ni-dimethylglyoxime, red needles, sublimes
—	$CoCl_3$, red crystals, sublimes
135	N_4S_4, orange-red, monoclinic, sublimes, melts at 178
135	Sb_2S_5, golden, decomposition and "sublimate" of sulfur
—	$(NH_4)_2CS_3$, yellow, sublimes
615	As_4, yellow cubic, transition at 358 to black amorphous or metallic hexagonal, sublimes
—	HgI_2, yellow orthorhombic, see above
140	HgI, yellow tetragonal, melts at 290, decomposes at 310
—	$NbOBr_3$, yellow crystals, sublimes
about 900	$NiCl_2$, yellow, deliquescent, sublimes
—	MoO_2Br_2, yellow, deliquescent, sublimes

—	MoO_2Cl_2, pale yellow, sublimes
—	$(NH_4)_3UO_2F_5$, yellow, tetragonal, volatilizes
250	$HgIO_3$, yellowish white, volatilizes
—	$SeCl_4$, yellow or white deliquescent crystals, sublimes and decomposes at 288
below 100	$MoOCl_4$, green, deliquescent, sublimes and melts below 100
below 100	$MoOCl_3$, green, sublimes
83	$CrCl_3 \cdot 6\,H_2O$, violet or green hexagonal plates, sublimes
high temp.	$CoCl_2$, blue crystals, sublimes
110	HIO_4, colorless crystals, sublimes
450	TeO_2, tetragonal, orthorhombic, sublimes
—	NH_4HCO_3, $NH_2 \cdot CO_2 \cdot NH_4$, sublimes
520	NH_4Cl, cubic, sublimes
542	NH_4Br, cubic, sublimes
551	NH_4I, cubic, sublimes
120	NH_4HS, orthorhombic, sublimes
230	NH_4BF_4, orthorhombic, sublimes
160	PCl_5, tetragonal, deliquescent, sublimes
—	PH_4Cl, cubic, sublimes
250	P_2O_5, amorphous, deliquescent, sublimes
—	HPO_3, sublimes
—	As_2O_3, cubic, monoclinic, sublimes
—	Na_2O, sublimes
178	$AlCl_3$, hexagonal, deliquescent, sublimes
400	$NbOCl_3$, needles, sublimes
—	ZrF_4, hexagonal, sublimes
610	$ThBr_4$, sublimes
400	Hg_2Cl_2, tetragonal, sublimes
345	$HgBr$, tetragonal, sublimes
440	$InCl_3$, plates, sublimes

Table 6

Inorganic Solids which Burst into Flame when Heated in Air
Ignition Temperatures in Centigrades

Metals of the alkali and alkaline earths groups, aluminum foil

The elements sulfur, carbon, and boron

Phosphorus, P_4, yellow hexagonal ignites at 34; black rhombohedral at 200; red cubic at 725

$(P_4H_2)_3$, yellow at 200

P_2Se_5, red needles

BP, maroon powder, at 200

KH_2PO_2, hexagonal, deliquescent

ThS_2, yellow, brown or black
ThC_2, yellow crystals
V_2O_2, light gray crystals
WO_2, brown, cubic
WP, gray prisms
U_2S_3, gray to black
$K_2Pt_4S_6$, blue to gray

Table 7
List of Solids which Explode on Heating[1]
Detonation temperatures in centigrades

Inorganic Compounds

Peroxides:
$CaO_2 \cdot 8 H_2O$, pearly, tetragonal, at 275
ZnO_2, yellow, at 212

Ammonium Salts:
NH_4ClO_3, monoclinic, at 100
NH_4IO_4, tetragonal
NH_4NO_2, needles
NH_4MnO_4, orthorhombic, at 60

Nitrides:
NI_3, black
NH_3NI_3, red, orthorhombic
Se_2N_2, orange or yellow, at 200
Hg_3N_2, brown powder
Ge_4N_4, orange or yellow, at 200

Azides:
BaN_6, orthorhombic, above 200
$BaN_6 \cdot H_2O$
AgN_3, prisms, at 297
TlN_3, pale yellow, tetragonal, at 430
HgN_3, at 245
PbN_6, needles, at 350

Chlorites and Chlorates:
$Pb(ClO_2)_2$, yellow, monoclinic, at 126
$HgClO_3$, orthorhombic, at 250

Nitrates:
$HgNO_3 \cdot H_2O$, monoclinic, below 70

---

[1] Like all others, this table is not complete. Thus failure to find a substance in this table does not prove that it is harmless.

Various Compounds of Ammonia with Silver, Gold, and the Platinum Metals:
$AgNH_2$
$Au \cdot NH \cdot Cl + Au \cdot NH \cdot NH_2$, yellow

Oxalates:
$Ag_2C_2O_4$, at 140
HgC_2O_4, below 160

Cyanides and Fulminates:
$Hg(CN)_2 \cdot HgO$
$Ag_2(NCO)_2$, needles
$Hg(NCO)_2$, cubic

Acetylides:
$Cu_2C_2 \cdot H_2O$, reddish brown
Ag_2C_2

Various Instable Compounds:
$(SiOOH)_2$, white, amorphous
SiI_2, explodes in air

Organic Compounds

Iodocompounds:
Iodosobenzene, yellow, amorphous, at 210
Iodoxybenzene, needles, at 236 to 237

Peroxides:
Benzoylperoxide, orthorhombic, m. pt. 108, melt explodes

Chlorates, Perchlorates, Nitrates of Organic Bases.

Nitric Esters:
Glycerol dinitrate, oil
Glyceryl trinitrate, oil, at 270
Erythritol tetranitrate, m. pt. 61, melt explodes
Mannitol hexanitrate, needles, m. pt. 112 to 113

Azides and Oximes:
Benzazide, plates, m. pt. 32, melt explodes
Chloramine T, yellowish white powder, at 175 to 180
Benzohydroxamic acid, orthorhombic, m. pt. 132, melt explodes
Phloroglucinol trioxime, powder, at 155

Nitrocompounds:
Nitro uracil, needles
o-Nitrophenylpropiolic acid, needles, at 155
Phenylnitroamine, leafy, m. pt. 46, explodes at 98
Dinitroresorcinol, yellow leaves, m. pt. 148, melt explodes

Nitranilic acid, yellow plates, after loss of 1 H_2O melts at 100, explodes at 170
3,4-Dinitro-o-xylene, needles, m. pt. 82, explodes at 413
3,5-Dinitro-o-xylene, yellow needles, m. pt. 76, explodes at 438
Trinitromethane, crystals melt at 23 and explode
2,4,6-Trinitrotoluene, TNT, crystals melt at 80.1, explode at 280
2,3,4-Trinitrotoluene, crystals melt at 112, melt explodes at 290 to 310
2,4,5-Trinitrotoluene, yellow plates melt at 104, melt explodes at 290
2,4,6;1,3,5-Trinitrotrimethylbenzene, triclinic, m. pt. 232, explodes at 415
Trinitro-p-xylene, monoclinic, m. pt. 140, explodes at 410
2,4,6-Trinitrophenol, yellow, orthorhombic, m. pt. 121.8, explodes above 300
Trinitro-m-cresol, yellow needles, m. pt. 109.5, explodes at 150
Trinitro orcinol, yellow needles, m. pt. 163, melt explodes
Trinitro-naphthol, yellow needles, m. pt. 190, melt explodes
Trinitro aniline, yellow, monoclinic, m. pt. 188 to 190, melt explodes
Trinitro acetonitrile, waxy, m. pt. 41.5, explodes at 220
Tetranitrophenol, light yellow, m. pt. 140, melt explodes
α-Tetranitronaphthalene, light yellow, m. pt. 259, melt explodes
β-Tetranitronaphthalene, needles, m. pt. 203, melt explodes
Tetranitrodiphenyldisulfide, yellow needles explode above 280

Azocompounds:
Dichloro-azodicarbonamide, yellow needles, at 155
Diazo uracil, red or yellow plates
Diazobenzene chloride, hygroscopic crystals
Diazobenzene nitrate, needles
Diazobenzenesulfonic acid, needles
Diazosalicylic acid, yellow crystals, at 155
Diazo-aminobenzene, yellow leaves, m. pt. 96 to 98, melt explodes

Table 8

Inorganic Solids Moderately Soluble in Water at Room Temperature

LiF, cubic; Na_2SiF_6, pink, hexagonal; (Na, Li) · (Mg, Zn)-uranyl acetates, octahedral, yellow, green fluorescence.

Perchlorates (orthorhombic), permanganates (orthorhombic), fluorosilicates (cubic, hexagonal), chloroplatinates and chloriridates (cubic), acid tartrates (orthorhombic), and picrates of K, Rb, Cs, NH_4; $K_2TiF_6 \cdot H_2O$, monoclinic; KBF_4, cubic, orthorhombic; KIO_4, tetragonal; $K_2TeO_4 \cdot 5 H_2O$, orthorhombic; Cs alum, cubic; $CsAuCl_4$, yellow, monoclinic; NH_4VO_3.

BeCO$_3$ · 4 H$_2$O; Ca(OH)$_2$, hexagonal; Sr(OH)$_2$ · 8 H$_2$O, tetragonal; BaO$_2$ · · 8 H$_2$O, pearly scales; BaF$_2$, cubic; Ba(BrO$_3$)$_2$ · H$_2$O, monoclinic; Ca(IO$_3$)$_2$, triclinic; CaSO$_4$ · 2 H$_2$O, monoclinic; BaS$_2$O$_3$ · H$_2$O, orthorhombic; CaWO$_4$, tetragonal; SrWO$_4$, tetragonal; Ca(BO$_3$)$_2$ · 2 H$_2$O, cubic; BaSiF$_6$, prismatic; Sr tartrate tetrahydrate, monoclinic.

MoO$_3$, orthorhombic; H$_2$MoO$_4$ · H$_2$O, yellow, monoclinic.

PtBr$_4$, dark brown.

Cu(IO$_3$)$_2$, green, monoclinic; Cu(IO$_3$)$_2$ · H$_2$O, blue triclinic; AgNO$_2$, orthorhombic; AgBrO$_3$, tetragonal; Ag$_2$SO$_4$, orthorhombic; AgMnO$_4$, purple to black, monoclinic; Ag acetate, needles; Ag tartrate, scales.

HgBr$_2$, orthorhombic; Hg(BrO$_3$)$_2$ · 2 H$_2$O, crystals; mercurous formate and acetate, scales.

AlF$_3$, triclinic; TlCl, cubic; Tl$_2$Cl$_3$, yellow, hexagonal; TlCNS, tetragonal needles; Tl$_3$PO$_4$, needles; Tl$_4$V$_2$O$_7$; Tl$_4$Fe(CN)$_6$ · 2 H$_2$O, yellow, triclinic.

GeO$_2$, orthorhombic; GeS, orthorhombic, monoclinic; GeS$_2$, white; PbCl$_2$, orthorhombic needles; PbBr$_2$, orthorhombic.

Sb$_2$O$_5$, yellow.

Literature

General Reference Books

(1) American Chemical Society: Critical Solution Temperatures, Advances in Chemistry Series No. 31. American Chemical Society Special Issue Sales, 1155 Sixteenth Str., N. W., Washington 6, D. C.
(2) *Beilsteins Handbuch der organischen Chemie*. Berlin-Göttingen-Heidelberg: Springer. 1957—1961.
(3) FURMAN, N. H., editor: Scott's Standard Methods of Chemical Analysis, 6th ed. New York: Van Nostrand. 1962.
(4) *Gmelin-Institut der Max-Planck-Gesellschaft: Gmelins Handbuch der anorganischen Chemie*, 8th ed. Weinheim: Verlag Chemie. 1930.
(5) *Ibid.*, Magnetic Materials, 2nd supplement to Part D of System No. 59, Iron. Weinheim: Verlag Chemie. 1959.
(6) HAUSNER, H. H.: Modern Materials, 2 Vols. New York: Academic Press. 1959—1960.
(7) HILLEBRAND, W. F., G. E. F. LUNDELL, H. A. BRIGHT, and J. I. HOFFMAN: Applied Inorganic Analysis, 2nd ed. New York: John Wiley. 1953.
(8) HODGMAN, C. D., editor: Handbook of Chemistry and Physics. Cleveland, Ohio: The Chemical Rubber Company.
(9) HORWITZ, W., editor: Official Methods of the Association of Official Agricultural Chemists, 9th ed. Washington: Assoc. of Official Agricultural Chemists. 1960.
(10) LANGE, N. A., editor: Handbook of Chemistry. Sandusky, Ohio: Handbook Publishers.
(11) MELLOR, J. W.: Comprehensive Treatise on Inorganic and Theoretical Chemistry. New York: Longmans, Green and Co. 1922—1961.
(12) RYSHKEWITCH, E.: Oxide Ceramics. New York: Academic Press. 1960.

Theory of Chemical Analysis

(13) BENEDETTI-PICHLER, A. A.: Essentials of Quantitative Analysis. New York: Ronald Press. 1956.
(14) BENEDETTI-PICHLER, A. A.: Theory and Practice of Sampling for Chemical Analysis, in W. G. BERL: Physical Methods of Chemical Analysis. New York: Academic Press. 1956.
(15) CHARLOT, G.: *Théorie et méthode nouvelles d'analyse qualitative*, 3rd ed. Paris: Masson et Cie. 1949.
(16) EMELÉUS, H. J., and J. S. ANDERSON: Modern Aspects of Inorganic Chemistry. New York: Van Nostrand. 1938.
(17) HÄGG, G.: *Theoretische Grundlagen der analytischen Chemie*. Basel: Birkhäuser. 1950.
(18) HAMMETT, L. P.: Solutions of Electrolytes, 2nd ed. New York: McGraw-Hill. 1936.
(19) HILDEBRAND, J. H., and R. L. SCOTT: Regular Solutions. New York: Prentice-Hall. 1962.

(20) HOGNESS, T. R., and W. C. JOHNSON: Qualitative Analysis and Chemical Equilibrium, 4th ed. New York: Holt. 1954.
(21) KOLTHOFF, I. M., and E. B. SANDELL: Textbook of Quantitative Inorganic Analysis, 3rd ed. New York: Macmillan. 1952.
(22) SMITH, T. B.: Analytical Processes. A Physico-Chemical Interpretation, 2nd ed. London: Edw. Arnold. 1940.
(23) YOE, J. H.: Chemical Principles. New York: John Wiley. 1937.

Reagents

(24) American Chemical Society: Reagent Chemicals, ACS Specifications, Washington, D. C.: American Chemical Society. 1950.
(25) British Drug Houses: The Reagents for Delicate Analysis Including Spot Tests, 7th ed. London: British Drug Houses. 1939.
(26) Hopkin & Williams, Ltd.: Organic Reagents for Metals and for Certain Acid Radicals, 4th ed. London: Hopkin & Williams. Ltd. 1943.
(27) MERCK, E.: *Prüfung der chemischen Reagenzien auf Reinheit.* Darmstadt: Merck. 1931.
(28) MERCK, E.: Merck Index, 7th ed., Rahway, N. J.: E. Merck. 1960.
(29) NIEUWENBURG, C. J. VAN, W. BÖTTGER, F. FEIGL, A. S. KOMAROVSKY, and N. STRAFFORD: Tables of Reagents for Inorganic Analysis. London: H. K. Lewis. *(Premier rapport de la commission internationale de réactions et réactifs analytiques nouveau de l'union internationale de chimie.)* Leipzig: Akademische Verlagsgesellschaft. 1938.
(30) ROSIN, J.: Reagent Chemicals and Standards, 2nd ed. New York: Van Nostrand. 1946.
(31) U. S. Pharmacopoeial Convention: Pharmacopoeia of the United States, Washington, D. C.
(32) WELCHER, F. J.: Organic Analytical Reagents, 4 Vols. New York: Van Nostrand. 1947.

Standard Tests and Procedures of Qualitative Analysis

(33) ALLEN, P. W., editor: Techniques of Polymer Characterization. New York: Academic Press. 1959.
(34) BARBER, H. H., and T. I. TAYLOR: Semimicro Qualitative Analysis. New York: Harper. 1953.
(35) CAMPBELL, N.: Qualitative Organic Chemistry. New York: Van Nostrand. 1939.
(36) CHARLOT, G., DENISE BÉZIER, and R. GAUGUIN: *Analyse qualitative rapide des cations.* Paris: Dunod. 1950; translated by R. E. OESPER: Rapid Detection of Cations. New York: Chemical Publishing Co. 1954.
(37) FRESENIUS, W., and G. JANDER, editors: *Handbuch der analytischen Chemie,* 2nd part, Vols. I to IX. Berlin-Göttingen-Heidelberg: Springer. 1944—1956.
(38) KAMM, O.: Qualitative Organic Analysis, 2nd ed. New York: John Wiley. 1932.
(39) LINSTEAD, R. P., and B. C. L. WEEDON: A Guide to Qualitative Organic Chemical Analysis. New York: Academic Press. 1956.
(40) McALPINE, R. K., and B. A. SOULE: Qualitative Chemical Analysis. New York: Van Nostrand. 1933.
(41) McELVAIN, S. M.: Characterization of Organic Compounds. New York: Macmillan. 1953.

(42) McGookin, A.: Qualitative Organic Analysis and the Scientific Method. New York: Reinhold. 1955.
(43) Meldrum, W. B., and E. W. Flosdorf: Qualitative Analysis of Inorganic Materials. New York: American Book Co. 1938.
(44) Middleton, A. R., and J. W. Willard: Semimicro Qualitative Analysis. New York: Prentice-Hall. 1939.
(45) Morrison, G. H., and H. Freiser: Solvent Extraction in Analytical Chemistry. New York: John Wiley. 1957.
(46) Mulliken, S. F.: The Identification of Pure Organic Compounds, 4 Vols. New York: John Wiley. 1904—1942.
(47) Nieuwenburg, C. J. van, and J. W. Ligten, Qualitative chemische Analyse. Wien: Springer. 1959.
(48) Noyes, A. A.: Qualitative Chemical Analysis of Inorganic Substances, 9th ed. New York: Macmillan. 1922.
(49) Noyes, A. A., and W. C. Bray: A System of Qualitative Analysis for the Rare Elements. New York: Macmillan. 1927.
(50) Radley, J. A., and J. Grant: Fluorescence Analysis in Ultraviolet Light, 2nd ed. New York: Van Nostrand. 1935.
(51) Shriner, R. L., R. C. Fuson, and D. V. Curtin: The Systematic Identification of Organic Compounds, 4th ed. New York: John Wiley. 1956.
(52) Siggia, S.: Quantitative Organic Analysis Via Functional Groups. New York: John Wiley. 1949.
(53) Swift, E. H.: A System of Chemical Analysis (Qualitative and Semi-Quantitative) for the Common Elements. New York: Prentice-Hall. 1939.
(54) Tables for Identification of Organic Compounds. Cleveland, Ohio: Chemical Rubber Co.
(55) Treadwell, F. P.: Kurzes Lehrbuch der analytischen Chemie, I. Qualitative Analyse, 21st ed. Wien: Deuticke. 1948—1949.
(56) Treadwell, F. P., and W. T. Hall: Analytical Chemistry, I. Qualitative Analysis, 9th ed. New York: John Wiley. 1937.
(57) Treadwell, F. P., and V. Meyer: Tabellen zur qualitativen Analyse, 7th ed. Wien: Deuticke. 1912.
(58) Treybal, R. E.: Liquid Extraction. New York: McGraw-Hill. 1951.
(59) Veibel, S.: The Identification of Organic Compounds, 2nd English ed. Copenhagen: G. E. C. Gad. 1960.
(60) Vogel, A. I.: Text Book of Macro and Semi Micro Qualitative Inorganic Analysis, 4th ed. London: Longmans, Green and Co. 1954.
(61) Vortmann, G., and R. Lieber: Qualitative chemische Analyse nach dem Schwefelnatriumgang. Wien: Haim. 1933.
(62) West, P. W., M. M. Vick, and A. L. Le Rosen: Qualitative Analysis and Analytical Separations (without H_2S). New York: Macmillan. 1953.

Chromatography and Ion Exchange

(63) Block, R. J., E. L. Durrum, and G. Zweig: A Manual of Paper Chromatography and Paper Electrophoresis, 2nd ed. New York: Academic Press. 1958.
(64) Brimley, R. C., and F. C. Barrett: Practical Chromatography. New York: Reinhold. 1953.
(65) Cassidy, H. G.: Adsorption and Chromatography. New York: Interscience. 1951.
(66) Cramer, F.: Papierchromatographie. Weinheim: Verlag Chemie. 1952.

(67) HESSE, G.: *Adsorptionsmethoden im chemischen Laboratorium.* Berlin: de Gruyter. 1943.
(68) LEDERER, E., and M. LEDERER: Chromatography, 2nd ed. New York: Van Nostrand. 1957.
(69) LINSKENS, H. F., editor: *Papierchromatographie in der Botanik.* Berlin-Göttingen-Heidelberg: Springer. 1955.
(70) NACHOD, F. C., editor: Ion Exchange, Theory and Application. New York: Academic Press. 1949.
(71) POLLARD, F. H., and J. F. W. MCOMIE: Chromatographic Methods in Inorganic Analysis. New York: Academic Press. 1953.
(72) SAMUELSON, O.: Ion Exchangers in Analytical Chemistry. New York: John Wiley. 1952.
(73) SMITH, O. C.: Inorganic Chromatography. New York: Van Nostrand. 1953.
(73a) STAHL, E., editor: *Dünnschicht-Chromatographie* (a laboratory manual). Berlin-Göttingen-Heidelberg: Springer. 1962.
(74) STRAIN, H. H.: Chromatographic Adsorption Analysis. New York: Interscience. 1942.
(75) WILLIAMS, R. T., and R. L. M. SYNGE: Partition Chromatography. Cambridge: University Press. 1951.
(76) WILLIAMS, T. I.: An Introduction to Chromatography. New York: Chemical Publishing Co. 1947.

Instrumental Methods

(77) BERL, W. G., editor: Physical Methods in Chemical Analysis, Vol. 3. New York: Academic Press. 1956.
(78) BRODE, W. R.: Chemical Spectroscopy, 2nd ed. New York: John Wiley. 1946.
(79) HÁMOS, L. v.: X-Ray Microanalyzer Camera. Göteborg: Elanders Boktryckeri Actiebolag. 1953.
(80) KLUG, H. P., and L. E. ALEXANDER: X-Ray Diffraction Procedures. New York: John Wiley. 1954.
(81) LÁNG, L., editor: Absorption Spectra in the Ultraviolet and Visible Region, 3 Vols. New York: Academic Press. 1961.
(81a) NAKANISHI, K.: Infrared Absorption Spectroscopy. San Francisco: Holden-Day. 1962.
(82) NYBURG, S. C.: X-Ray Analyses of Organic Structures. New York: Interscience. 1961.

Chemical Microscopy

(83) BARKER, T. V.: Systematic Crystallography. London: Murby and Co. 1930.
(84) BLASS, F. D.: An Introduction to the Methods of Optical Crystallography. New York: Holt, Rinehart and Winston. 1961.
(85) CHAMBERS, R., in C. F. MCCLUNG: Handbook of Microscopical Technique, 1st ed. New York: Hoeber. 1929.
(86) CHAMBERS, R., and M. J. KOPAC: Handbook of Microscopical Technique, 3rd ed. New York: Hoeber. 1950.
(87) CHAMOT, É. M.: The Microscopy of Small Arms Primers. Ithaca, N. Y. 1922.
(88) CHAMOT, É. M., and C. W. MASON: Handbook of Chemical Microscopy, Vol. I, 3rd ed. New York: John Wiley. 1958.
(89) CONN, G. K. T., and F. J. BRADSHAW, editors: Polarized Light in Metallography. New York: Academic Press. 1952.

(90) EL-BADRY, H. M.: Micromanipulators and Micromanipulation. *(Monographien aus dem Gebiete der qualitativen Mikroanalyse*, Vol. III.*)* Wien: Springer. 1963.
(91) DONNAY, J. D. H., and W. NOVACKI: Crystal Data. Geological Society of America. 1954.
(92) FONBRUNE, P. DE: *Technique de micromanipulation.* Paris: Masson et Cie. 1949.
(93) FRY, W. H.: Petrographic Methods for Soil Laboratories. Tech. Bull. No. 344. Washington, D. C.: U. S. Department of Agriculture. 1933.
(94) Groth's Encyclopedia of Chemical and Physical Crystallography, revised ed., University Park, Pa., Groth Institute, Pennsylvania State University (in preparation).
(95) HAITINGER, M.: *Die Fluoreszenzanalyse in der Mikrochemie.* Wien: Haim. 1937.
(96) HARTSHORNE, N. H., and A. STUART: Crystals and the Polarizing Microscope, 2nd ed. New York: Longmans, Green and Co. 1950.
(97) KERR, P. F.: Optical Mineralogy, 3rd ed. New York: McGraw-Hill. 1959.
(98) KOFLER, L., ADELHEID KOFLER, and A. MAYRHOFER: *Mikroskopische Methoden in der Mikrochemie.* Wien: Haim. 1936.
(99) LARSEN, E. S., and H. BERMAN: Microscopic Determination of the Non-Opaque Minerals. U. S. Geol. Survey Bull. 848 (1934).
(100) MARTIN, L. C., and B. K. JOHNSON: Practical Microscopy. New York: Chemical Publishing Co. 1951.
(101) MURDOCH, J.: Microscopical Determination of Opaque Minerals. New York: John Wiley. 1916.
(102) OLLIVER, C. W.: The Intelligent Use of the Microscope. New York: Chemical Publishing Co. 1953.
(103) OTTO, L.: *Der Mikromanipulator und seine Hilfsgeräte.* Berlin: Verlag Technik. 1954. See also (1133).
(104) PORTER, MARY W., and R. C. SPILLER: The Barker Index of Crystals, 3rd ed. Cambridge: Heffner and Sons. 1951—1959.
(105) SCHAEFFER, H. F.: Microscopy for Chemists. New York: Van Nostrand. 1953.
(106) SHILLABER, C. P.: Photomicrography in Theory and Practice. New York: John Wiley. 1944.
(107) SMITH, H. G.: Minerals and the Microscope, 4th ed. New York: Macmillan. 1956. (An introduction for beginners.)
(108) STOVES, J. L.: Fibre Microscopy. New York: Van Nostrand. 1958.
(109) UYTENBOGAARDT, W.: Tables for Microscopic Identification of Ore Minerals. Princeton: University Press. 1951.
(110) WAHLSTROM, E. E.: Optical Crystallography, 2nd ed. New York: John Wiley. 1943.
(111) WALLIS, T. E.: Analytical Microscopy, 2nd ed. Boston: Little, Brown and Co. 1957.
(112) WINCHELL, A. N.: The Optical Properties of Organic Compounds, 2nd ed. New York: Academic Press. 1954.
(113) WINCHELL, A. N., and H. WINCHELL: Microscopic Characters of Artificial Inorganic Solid Substances or Artificial Minerals, 3rd ed. New York: Academic Press. 1961.
(114) WINCHELL, A. N., and H. WINCHELL: Elements of Optical Mineralogy; I. Principles and Methods, 4th ed.; II. Descriptions of Minerals; III. Determinative Tables, 2nd ed. New York: John Wiley. 1951.

Slide Tests and Spot Tests

(115) BEHRENS, H.: A Manual of Microchemical Analysis. London: Macmillan. 1894.
(116) BEHRENS, H., and P. D. C. KLEY: *Mikrochemische Analyse*, 4th ed. Leipzig und Hamburg: Voss. 1921.
(117) BOŘICKÝ, E.: *Elemente einer neuen chemisch-mikroskopischen Mineral- und Gesteinsanalyse*, reprint from *Archiv der naturwissenschaftlichen Landesdurchforschung von Böhmen*, Prague, 1877. Translated into English by N. H. WINCHELL, 19th Annual Report of the Geological and Natural History Survey of Minnesota, Minneapolis. 1892.
(118) CHAMOT, É. M., and C. W. MASON: Handbook of Chemical Microscopy, Vol. II, 2nd ed. New York: John Wiley. 1940.
(119) DUVAL, C.: *Traité de micro-analyse minérale*, 4 Vols. Paris: Presses Scientifiques Internationales. 1951.
(120) FEIGL, F.: *Qualitative Analyse mit Hilfe von Tüpfelreaktionen*. Leipzig: Akademische Verlagsgesellschaft. 1931.
(121) FEIGL, F., and R. E. OESPER: Qualitative Analysis by Spot Tests, Inorganic and Organic Applications, 3rd ed. New York: Elsevier. 1948.
(122) FEIGL, F., and R. E. OESPER: Spot Tests, Vol. 2, Organic Applications. Houston: Elsevier. 1954.
(123) GEILMANN, W.: *Bilder zur qualitativen Mikroanalyse anorganischer Stoffe*, 3rd ed. Weinheim: Verlag Chemie. 1960.
(124) HARTING, P. (translated from the Dutch original by F. W. THEILE): *Theorie und allgemeine Beschreibung des Mikroskopes*. Braunschweig: Vieweg. 1866.
(125) HAUSHOFER, K.: *Anleitung zur Erkennung verschiedener Elemente mittels mikroskopischer Reaktionen*. Braunschweig: Vieweg. 1885.
(126) HELWIG, A.: *Das Mikroskop in der Toxikologie*. Mainz. 1865.
(127) HINRICHS, C. G.: First Course in Microchemical Analysis. St. Louis, Mo.: Carl Gustav Hinrichs; New York and Leipzig: Lemske and Buechner; London: H. Grevel & Co.; Paris: H. le Soudier. 1904.
(128) HUYSSE, A. C.: *Atlas zum Gebrauch bei der mikrochemischen Analyse*. Leiden: Brill. 1932.
(129) KLEMENT, C., and A. RENARD: *Réactions microchimiques à cristaux et leur application en analyse qualitative*. Brussels: Manceaux. 1886.
(130) KRAMER, G.: *Mikroanalytische Nachweise anorganischer Ionen, Ausführung und Reaktionsbilder*. Leipzig: Akademische Verlagsgesellschaft. 1937.
(131) LONGO, R. E.: *Mikroanalisis inorganica*, 2nd ed. Buenos Aires: Ciordia y Rodriguez. 1951.
(132) MALJAROFF, K. L.: *Qualitative anorganische Mikroanalyse*. Berlin: Verlag Technik. 1953.
(133) MARTINI, A., and S. SCHAMIS: *Nuevo método para el reconocimiento microquímico de los cationes mas comunes en las mezclas complejas*. Buenos Aires: Trab. pres. al Segundo Congresa de Química. 1924.
(134) POZZI-ESCOT, M. E.: *Analyse microchimique et spectroscopic*. Paris: Gauthier-Villars. 1899.
(135) STEPHENSON, C. H., and E. C. PARKER: Some Microchemical Tests for Alkaloids. Philadelphia: Lippincott. 1921.
(136) WENGER, P. E., and R. DUCKERT: *Réactifs pour l'analyse qualitative minérale, 2ème rapport*. Basel: Wepf. 1945.

(137) WENGER, P. E., and YVONNE RUSCONI: *Réactifs pour l'analyse qualitative minérale, 4ème rapport*. Paris: Sedes. 1950.
(138) WORMLEY, TH. G.: The Micro-Chemistry of Poisons, 2nd ed. Philadelphia: Lippincott. 1885.

Micro Analysis and Microtechnique

(139) ALIMARIN, I. P., and M. N. PETRIKOVA: Inorganic Ultramicro Analysis. Moscow: Academy of Sciences. 1960.
(140) BELCHER, R., and C. L. WILSON: Inorganic Microanalysis, Qualitative and Quantitative. A short Elementary Course, 2nd ed. London: Longmans, Green and Co. 1957.
(141) BENEDETTI-PICHLER, A. A.: Introduction to the Microtechnique of Inorganic Analysis. New York: John Wiley. 1942.
(142) BENEDETTI-PICHLER, A. A.: *Waagen und Wägen*. (*Handbuch der mikrochemischen Methoden*, Vol. 1, edited by F. HECHT and M. K. ZACHERL, Part 2.) Wien: Springer. 1959.
(143) BENEDETTI-PICHLER, A. A., and W. F. SPIKES: Introduction to the Microtechnique of Inorganic Qualitative Analysis. Douglaston, N. Y.: Microchemical Service. 1935.
(144) BRISCOE, H. V. A., and P. F. HOLT: Inorganic Microanalysis. London: Longmans, Green and Co. 1950.
(145) CHERONIS, N. D., editor: Submicrogram Experimentation. New York: Interscience. 1961.
(146) CHERONIS, N. D., and J. B. ENTRIKIN: Semimicro Qualitative Organic Analysis. The Systematic Identification of Organic Compounds, 2nd ed. New York: Interscience. 1957.
(147) CHERONIS, N. D., with A. R. RONZIO and T. S. MA: Micro and Semimicro Methods. (Technique of Organic Chemistry, edited by A. WEISSBERGER, Vol. VI.) New York: Interscience. 1954.
(148) DONAU, J.: *Die Arbeitsmethoden der Mikrochemie*. Stuttgart: Franckh'sche Verlagshandlung. 1913.
(149) EMICH, F.: *Lehrbuch der Mikrochemie*. Wiesbaden: Bergmann. 1911.
(150) EMICH, F.: *Lehrbuch der Mikrochemie*, 2nd ed. München: Bergmann. 1926.
(151) EMICH, F.: *Methoden der Mikrochemie*, in ABDERHALDEN's *Handbuch der biologischen Arbeitsmethoden*, Abt. 1, Teil 3. Wien und Berlin: Urban und Schwarzenberg. 1921.
(152) EMICH, F.: *Mikrochemisches Praktikum*, 1st ed. München: Bergmann. 1924.
(153) EMICH, F.: *Umsetzungen sehr kleiner Stoffmengen*, in STÄHLER-TIEDE-RICHTER's *Handbuch der Arbeitsmethoden in der anorganischen Chemie*, Vol. 2, Part 2. Berlin: de Gruyter. 1925.
(154) EMICH, F., and F. SCHNEIDER: Microchemical Laboratory Manual. New York: John Wiley. 1932.
(155) GORBACH, G.: *Mikrochemisches Praktikum (Anleitungen für die chemische Laboratoriumspraxis*, Band 7). Berlin-Göttingen-Heidelberg: Springer. 1956.
(156) HECHT, F., and M. K. ZACHERL, editors: *Handbuch der mikrochemischen Methoden*. Wien: Springer. 1954.
(157) KIRK, P. L.: Quantitative Ultramicroanalysis. New York: John Wiley. 1950.
(158) KLEIN, G., and R. STREBINGER: *Fortschritte der Mikrochemie (1915—1926)*. Wien: Deuticke. 1928.

(159) KOFLER, L., and ADELHEID KOFLER: *Mikromethoden zur Kennzeichnung organischer Stoffe und Stoffgemische.* Innsbruck: Wagner. 1948.
(160) KOFLER, L., ADELHEID KOFLER, and MARIA BRANDSTÄTTER: *Thermo-Mikro-Methoden,* 3rd ed. Weinheim: Verlag Chemie. 1954.
(161) LÉVY, R.: *Microanalyse organique élémentaire qualitative et quantitative.* Paris: Masson et Cie. 1961.
(162) MALISSA, H., and A. A. BENEDETTI-PICHLER: *Anorganische qualitative Mikroanalyse (Monographien aus dem Gebiete der qualitativen Mikroanalyse,* Vol. I). Wien: Springer. 1958.
(163) MCCRONE, W. C.: Fusion Methods in Chemical Microscopy. New York: Interscience. 1956.
(164) NIEDERL, J. B., and J. A. SOZZI: *Microanálisis elemental orgánico.* Buenos Aires: (Arcos 2073, Buenos Aires 28) 1958.
(165) REID, D. B.: Rudiments of Chemistry, 3rd ed. Edinburgh. 1848.
(166) SCHNEIDER, F.: Organic Qualitative Analysis. New York: John Wiley. 1946.
(167) SCHNEIDER, F. L.: Practical Organic Qualitative Micro Analysis. Cognition and Recognition. *(Monographien aus dem Gebiete der qualitativen Mikroanalyse,* Vol. II*).* Wien: Springer. 1964.
(168) SCHOORL, N.: *Beiträge zur mikrochemischen Analyse,* a collection of papers published in Z. analyt. Chem., Vols. 46, 47, 48 (1907—1909). Wiesbaden. 1909.
(169) SHORT, M. N.: Microscopic Determination of the Ore Minerals, Geological Survey Bulletin 825, Washington, D. C.: Superintendent of Documents. 1931.
(170) STRONG, J., H. V. NEHER, A. H. WHITFORD, C. H. CARTWRIGHT, and R. HAYWARD: Procedures in Experimental Physics. New York: Prentice-Hall. 1938.
(171) TOEPLER, A.: Vol. 157—158 of OSTWALD's *Klassiker der exakten Wissenschaften.* Leipzig: Engelmann. 1906.
(172) WEISZ, H.: Microanalysis by the Ring Oven Technique. New York: Pergamon Press. 1961.
(173) WILSON, C. L.: An Introduction to Microchemical Methods for Senior Students of Chemistry. New York: Chemical Publishing Co. 1938.

Miscellaneous Applications of Microtechnique

(174) AMELINK, F.: *Schema zur mikrochemischen Identifikation von Alkaloiden.* Amsterdam: Centen's. 1934.
(175) CORRINGTON, J. D.: Exploring with Your Microscope. New York: McGraw-Hill. 1957.
(176) DE WILD, A. M.: The Scientific Examination of Pictures. London: Bell and Sons. 1929.
(177) ESAU, KATHERINE: Plant Anatomy. New York: John Wiley. 1953.
(178) GARNER, W.: Industrial Microscopy. London: Pitman. 1932.
(179) GRAFF, J. H.: Pulp and Paper Microscopy. Appleton, Wisc.: Inst. Paper Chemistry. 1942.
(180) HANAUSEK, T. F., A. L. WINTON, and KATEE BARBER WINTON: The Microscopy of Technical Products. New York: John Wiley. 1907.
(181) HARRIS, K. L., editor: Microscopic-Analytical Methods in Food and Drug Control. Washington, D. C.: Food and Drug Technical Bulletin No. 1, Superintendent of Documents. 1960.

(182) HENRICI, A. T., C. W. EMMONS, C. E. SKINNER, and H. M. TSUCHIYA: Molds, Yeasts, and Actinomycetes, 2nd ed. New York: John Wiley. 1947.
(183) HERZOG, A.: *Mikrophotographischer Atlas der technisch wichtigsten Faserstoffe*. München: Obernetter. 1908.
(184) HEYN, A. N. J.: Fiber Microscopy. New York: Interscience. 1954.
(185) JOHANSEN, D. A.: Plant Microtechnique. New York: McGraw-Hill. 1940.
(186) KIRK, P. L.: Crime Investigation. New York: Interscience. 1953.
(187) KLINGER, P., and W. KOCH: *Beiträge zur metallkundlichen Analyse*. Düsseldorf: Verlag Stahleisen. 1949.
(188) KOCH, P. A., in H. SOMMER and F. WINKLER: *Die Prüfung der Textilien*. Berlin-Göttingen-Heidelberg: Springer. 1960.
(189) LEIFSON, E.: Atlas of Bacterial Flagellation. New York: John Wiley. 1959.
(190) LINDSLEY, L. C.: Industrial Microscopy. Richmond, Va.: Byrd. 1929.
(191) MAUERSBERGER, H. R., editor: Matthews' Textile Fibers, 6th ed. New York: John Wiley. 1954.
(192) MAYRHOFER, A.: *Mikrochemie der Arzneimittel und Gifte*, 2 Parts. Berlin: Urban und Schwarzenberg. 1928.
(193) MAYRHOFER, A.: *Qualitative mikrochemische Methoden zur Untersuchung der Heilmittel*, in ABDERHALDEN's *Handbuch der biologischen Arbeitsmethoden*, Abt. IV, Teil 7 C. Berlin: Urban und Schwarzenberg. 1929.
(194) MOLISCH, H.: *Mikrochemie der Pflanze*, 3rd ed. Jena: Fischer. 1923.
(195) PARRY, J. W.: Spices, Their Morphology, Histology, and Chemistry. New York: Chemical Publishing Co. 1962.
(196) PAULSEN: *Botanisk Mikrokemi*. Copenhagen: 1918.
(197) PÖSCHL, V.: *Technische Mikroskopie*. Stuttgart: Union Deutsche Verlagsgesellschaft. 1927. (Excellent Illustrations.)
(198) ROSENTHALER, L.: *Toxikologische Mikroanalyse*. Berlin: Borntraeger. 1935.
(199) SANDELL, E. B.: Colorimetric Determination of Traces of Metals. New York: Interscience. 1950.
(200) SCHNEIDER, A.: Microanalysis of Powdered Vegetable Drugs. Philadelphia: Blakiston's Son. 1920.
(201) SCHNEIDER, A.: Microbiology and Microanalysis of Foods. Philadelphia: Blakiston's Son. 1920.
(202) SCHNEIDER, H., and A. ZIMMERMANN: *Botanische Mikrotechnik*. Jena: Fischer. 1922.
(203) STERN, A. C., editor: Air Pollution, 2 Vols. New York: Academic Press. 1962.
(204) STOVES, J. L.: Fibre Microscopy. London: Natl. Trade Press. 1957; New York: Van Nostrand. 1958.
(205) TUNMANN, O., and L. ROSENTHALER: *Pflanzenmikrochemie*. Berlin: Borntraeger. 1931.
(206) VESCE, V. C.: Classification and Microscopic Identification of Organic Pigments, in MATIELLO: Protective and Decorative Coatings, Vol. 2. New York: John Wiley. 1942.
(207) WHIPPLE, G. C., revised by G. M. FAIR and M. C. WHIPPLE: The Microscopy of Drinking Water, 4th ed. New York: John Wiley. 1927.

Mineralogy

See also Chemical Microscopy.

(208) BRUSH, G. J., and S. L. PENFIELD: Manual of Determinative Mineralogy, 16th ed. New York: John Wiley. 1926.
(209) CAMERON, E. N.: Ore Microscopy. New York: John Wiley. 1961.

(210) DANA, J. D., C. PALACHE, H. BERMAN, and C. FRONDEL: The System of Mineralogy, 7th ed. 2 Vols. New York: John Wiley. 1951.
(211) DANA, E. S., and C. HURLBUT, Jr.: Minerals and How to Study Them, 3rd ed. New York: John Wiley. 1949.
(212) LEWIS, J. V., and A. C. HAWKINS: A Manual of Determinative Mineralogy, 4th ed. New York: John Wiley. 1931.

Journals

Acta Chemica Scandinavica
(300) 3, 630 (1949): J. N. OSPENSON.

American Journal of Botany
(310) 30, 477 (1943): B. ESTHER STRUCKMEYER.

American Mineralogist
(320) 43, 606 (1958): W. W. VIRGIN, Jr., and C. J. MASSONI.

Analitikeskoi Kimii
(330) 10, 251 (1955): I. P. ALIMARIN and M. N. PETRIKOWA.

Analytica Chimica Acta
(340) 3, 15 (1949): F. FEIGL and L. BAUMFELD.
(341) 3, 629 (1949): H. FLASCHKA.
(342) 19, 437 (1958): D. GOLDSTEIN and C. STARK-MEYER.

Analytical Chemistry
(Analytical Edition of Industrial and Engineering Chemistry, 1929—1946)
(400) 2, 177 (1930): E. R. CALEY.
(401) 2, 309 (1930): A. A. BENEDETTI-PICHLER.
(402) 3, 266 (1931): F. E. BLACET and P. A. LEIGHTON.
(403) 4, 336 (1932): A. A. BENEDETTI-PICHLER.
(404) 5, 272 (1933): F. E. BLACET, G. D. MAC DONALD, and P. A. LEIGHTON.
(405) 6, 334 (1934): F. E. BLACET and G. D. MAC DONALD.
(406) 7, 25 (1935): H. SCHAPIRO.
(407) 7, 218 (1935): B. L. CLARKE and H. W. HERMANCE.
(408) 9, 44 (1937): F. E. BLACET and D. H. VOLMAN.
(409) 9, 149 (1937): A. A. BENEDETTI-PICHLER.
(410) 9, 292 (1937): B. L. CLARKE and H. W. HERMANCE.
(411) 9, 483 (1937): A. A. BENEDETTI-PICHLER.
(412) 9, 496 (1937): A. C. SHEAD.
(413) 9, 589 (1937): A. A. BENEDETTI-PICHLER and J. R. RACHELE.
(414) 10, 47 (1938): H. K. ALBER and C. J. RODDEN.
(415) 10, 107 (1938): A. A. BENEDETTI-PICHLER and J. T. BRYANT.
(416) 10, 224 (1938): C. VAN BRUNT.
(417) 10, 348 (1938): H. K. ALBER.
(418) 10, 591 (1938): B. L. CLARKE and H. W. HERMANCE.
(419) 10, 662 (1938): A. C. SHEAD.

(420) **11**, 117 (1939): A. A. BENEDETTI-PICHLER, W. R. CROWELL, and C. DONAHUE.
(421) **11**, 294 (1939): B. S. ALSTODT and A. A. BENEDETTI-PICHLER.
(422) **11**, 403 (1939): P. L. KIRK and C. S. GIBSON.
(423) **11**, 409 (1939): J. R. BOWMAN.
(424) **12**, 233 (1940): A. A. BENEDETTI-PICHLER and J. R. RACHELE.
(425) **12**, 305 (1940): H. K. ALBER and J. T. BRYANT.
(426) **12**, 764 (1940): H. K. ALBER.
(427) **12**, 777 (1940): A. MARION.
(428) **13**, 127 (1941): W. G. BATT and H. K. ALBER.
(429) **13**, 494 (1941): A. A. MORTON and J. F. MAHONEY.
(430) **13**, 498 (1941): A. A. MORTON and J. F. MAHONEY.
(431) **13**, 656 (1941): H. K. ALBER.
(432) **14**, 278 (1942): W. W. RAZIM.
(433) **14**, 813 (1942): A. A. BENEDETTI-PICHLER and M. CEFOLA.
(434) **15**, 135 (1943): H. YAGODA.
(435) **15**, 227 (1943): A. A. BENEDETTI-PICHLER and M. CEFOLA.
(436) **15**, 648 (1943): C. R. GARCÍA.
(437) **17**, 187 (1945): ANNE G. LOSCALZO and A. A. BENEDETTI-PICHLER.
(438) **17**, 593 (1945): J. D. H. DONNAY and W. A. O'BRIEN.
(439) **18**, 81 (1946): D. SMITH and SHIRLEY A. EHRHARDT.
(440) **19**, 77 (1947): G. SHEPHERD.
(441) **19**, 355 (1947): P. L. KIRK, R. S. ROSENFELD, and J. D. HANAHAN.
(442) **20**, 976 (1948): G. C. CROSSMON.
(443) **20**, 1122 (1948): P. L. KIRK and M. DANIELSON.
(444) **20**, 1241 (1948): H. H. ANDERSON.
(445) **21**, 632 (1949): M. J. BABCOCK.
(446) **21**, 700 (1949): F. W. CHAPMAN, Jr., G. G. MARVIN, and S. YOUNG TYREE, Jr.
(447) **22**, 600 (1950): A. O. GETTLER, C. J. UMBERGER, and L. GOLDBAUM.
(448) **22**, 628 (1950): R. S. TIPSON.
(449) **22**, 892 (1950): E. H. GILMORE, MARIE MENAUL, and V. SCHNEIDER.
(450) **23**, 196 (1951): R. D. CADLE.
(451) **23**, 545 (1951): A. I. MEDALIA and R. W. STOENNER.
(452) **24**, 576 (1952): B. K. SEELY.
(453) **24**, 870 (1952): P. W. WEST and L. GRANATELLI.
(454) **26**, 1515 (1954): W. PRIMAK and P. DAY.
(455) **26**, 1829 (1954): J. P. LODGE.
(456) **27**, 93 (1955): B. K. SEELY.
(457) **27**, 704 (1955): J. L. MONKMAN.
(458) **27**, 865 (1955): A. J. FRANKLIN and S. E. VOLTZ.
(459) **28**, 1586 (1956): D. G. GRABAR and RITA HAESSLY.
(460) **29**, 167 (1957): S. T. ZENCHELSKY and J. S. SHOWELL.
(461) **29**, 167 (1957): R. C. BACKUS.
(462) **29**, 169 (1957): E. D. BLACK, J. D. MARGERUM, and G. M. WYMAN.
(463) **29**, 45 A (June, 1957): L. T. HALLETT.
(464) **29**, 860 (1957): BETTY J. STEINBACH and T. R. P. GIBB, Jr.
(465) **29**, 861 (1957): E. C. FIEBIG, E. L. SPENCER, and R. N. MCCOY.
(466) **29**, 1239 (1957): H. B. BRADLEY.
(467) **30**, 593 (1958): I. FANKUCHEN.
(468) **31**, 84 A (January, 1959): Norton Company.

(469) **31**, 148 (1959): D. E. Laskowski and O. W. Adams.
(470) **31**, 238 (1959): Barbara J. Tufts.
(471) **31**, 242 (1959): Barbara J. Tufts.
(472) **31**, 456 (1959): D. F. H. Wallach, D. M. Surgenor, J. Soderberg, and E. Delano.
(473) **31**, 947 (1959): P. W. West and A. K. Mukherji.
(474) **31**, 1124 (1959): J. S. Matthews and N. D. Coggeshall.
(475) **31**, 1287 (1959): J. E. Stewart.
(476) **32**, 25 A (February, 1960): staff report.
(477) **32**, 19 R (1960): S. Dal Nogare.
(478) **32**, 87 R (1960): G. Coven and R. Cox.
(479) **32**, 225 R (1960): R. C. Hirt.
(480) **32**, 229 R (1960): B. F. Scribner.
(481) **32**, 238 R (1960): R. C. Gore.
(482) **32**, 240 R (1960): H. A. Liebhafsky, E. H. Winslow, and H. Pfeiffer.
(483) **33**, 1559 (1961): L. D. Metcalfe.
(484) **34**, 143 A (February, 1962): R. H. Müller.
(485) **34**, 175 (1962): F. Davis, C. A. Dubbs, and W. S. Adams.
(486) **34**, 242 (1962): M. Sparagana and W. B. Mason.
(487) **34**, 448 (1962): C. Szonyi and J. D. Graske.
(488) **34**, 880 (1962): C. D. Felton.
(489) **34**, 966 (1962): E. M. Dodson.
(490) **34**, 1077 (1962): J. P. Faris and J. W. Warton.
(491) **34**, 1183 (1962): L. Spialter and M. Ballester.
(492) **34**, 1319 (1962): M. H. Swann and M. L. Adams.
(493) **34**, 1346 (1962): R. Wasicky.
(494) **34**, 1370 (1962): A. M. Liebman.
(495) **35**, 621 (1963): T. O. Ziebold and R. E. Ogilvie.

Annalen der Chemie

(550) **351**, 426 (1907): F. Emich.

Annales de chimie analytique et de chimie appliquée

(560) **8**, 130 (1926): R. Meurice.

Archiv für das Eisenhüttenwesen

(570) **28**, 785 (1957): W. Koch, H. Malissa, and D. Ditges.

Archiv für experimentelle Pathologie und Pharmakologie

(580) **135**, 188 (1928): H. Eitel.

Australian Chemical Institute, Journal & Proceedings

(590) **4**, 26 (1937): F. P. Dwyer.
(591) **5**, 32 (1938): F. P. Dwyer.

Berichte der deutschen chemischen Gesellschaft

(600) **9**, 217 (1876): V. Wartha.
(601) **16**, 2234 (1883): E. Fischer.

(602) 42, 1126 (1909): H. v. WARTENBERG.
(603) 46, 255 (1913): O. BINDER and R. F. WEINLAND.
(604) 51, 1739 (1918): F. PANETH.
(605) 56, 1338 (1923): J. VON BRAUN, W. GMELIN, and A. SCHULTHEISS.
(606) 61, 1654 (1928): G. JAEGER.
(607) 73, 1388 (1940): L. KOFLER and R. WANNENMACHER.

Chemist-Analyst

(620) 32, 4 (1943): F. FEIGL and H. A. SUTER.
(621) 36, 38 (1947): A. F. GILMAN, Jr.
(622) 49, 4 (1960): M. FUJIMOTO.
(623) 49, 20 (1960): R. DELHEZ.
(624) 49, 113 (1960): F. L. HAHN.
(625) 49, 114 (1960): J. T. STOCK and M. A. FILL.
(626) 50, 80 (1961): P. J. HOWE, J. S. HILL, and J. D. SLATER.

Comptes rendus de l'academie des sciences, Paris

(640) 100, 605 (1885): LECOQ DE BOISBAUDRAN.
(641) 176, 1012, 1187 (1923): A. POLICARD.
(642) 180, 538 (1925): L. HERRERA.

Die Chemie — Zeitschrift für angewandte Chemie

(650) 31, 50 (1918): C. G. SCHWALBE.
(651) 39, 461 (1926): A. STOCK.
(652) 42, 954 (1929): A. A. BENEDETTI-PICHLER.
(653) 55, 244 (1942): R. FISCHER.

Economic Geology and the Bulletin of the Society of Economic Geologists

(660) 26, 415 (1933): M. H. HAYCOK.

Endeavour

(665) 10, 188 (1951): MARY W. PORTER.

Farben-Zeitung

(670) 31, 1456 (1926): C. P. VAN HOEK-HILVERSUM.

Gerberei Collegium

(672) 1916, 16: W. MOELLER.

Halle aux cuirs

(675) 1925, 23, 39, 73: C. GENOT.

Industrial and Engineering Chemistry

(680) 9, 969 (1917): E. M. CHAMOT and H. I. COLE.
(681) 10, 48 (1918): E. M. CHAMOT and H. I. COLE.
(682) 15, 725 (1923): E. B. SPEAR and H. A. ENDRES.

Jahrbuch der Mineralogie

(690) **1876**, 1215: W. N. HARTLEY.
(691) **1877**, 1251: W. N. HARTLEY.

Journal of the American Chemical Society

(695) **21**, 417 (1899): V. LENHER and J. S. C. WELLS.
(696) **27**, 104 (1905): T. W. RICHARDS.
(697) **33**, 718 (1911): L. J. CURTMANN and P. ROTHBERG.
(698) **39**, 1148 (1917): N. F. HALL.
(699) **39**, 2186 (1917): O. R. SWEENEY.
(700) **47**, 2625 (1925): H. S. BOOTH and N. E. SCHREIBER.
(701) **62**, 3165 (1940): C. A. HUTCHISON and H. L. JOHNSTON.

Journal of the Biological Photographers Association

(708) **8**, 115 (1940): G. L. ROYER and MARIE E. WISSEMANN.

Journal of Chemical Education

(710) **11**, 624 (1934): J. DUNNING and P. PRATT with O. E. LOWMAN.
(711) **34**, 381 (1957): A. A. BENEDETTI-PICHLER, F. SCHNEIDER, and O. F. STEINBACH.
(712) **34**, 383 (1957): R. E. FRANK.
(713) **35**, 453 (1958): R. T. CONLEY.

Journal of the Chemical Society, London

(720) **93**, 1442 (1908): O. BRILL and C. DE B. EVANS.
(721) **125**, 1946 (1924): L. REEVE.
(722) **138**, 362 (1935): A. LAWSON and E. W. BALSON.
(723) **1940**, 1258: CHRISTINA C. MILLER and A. J. LOWE.
(724) **1941**, 786: CHRISTINA C. MILLER.

Journal of the Franklin Institute

(730) **182**, 19 (1916): G. K. BURGESS.

Journal of the Institution of Petroleum Technologists

(735) **34**, 331 (1948): E. GLYNN and L. GRUNBERG.

Journal of the Oil and Colour Chemists' Association

(740) **9**, 255 (1926): J. PARRISH.

Journal of the Optical Society of America

(745) **23**, 299 (1933): S. B. HENDRICKS and M. E. JEFFERSON.
(746) **45**, 740 (1955): W. THORNBURG.

Journal für praktische Chemie

(750) (1) **74**, 341 (1858): J. LÖWE.

Journal of Research of the National Bureau of Standards

(755) **57**, 137 (1956): A. R. GLASCOW, Jr., and G. ROSS.
(760) **10**, 378 (1933): H. G. TAYLOR and J. M. WALDRAM.
(761) **34**, 207 (1957): P. R. ROWLAND and G. W. WHITING.

Laboratory Methods

(765) **55**, 151 (1957): G. L. KELLY, H. STEINMETZ, and W. G. MCGONNAGLE.

Klinische Wochenschrift

(766) **32**, 988 (1954): G. NÖLLER.

Kolloidchemische Beihefte

(767) **23**, 309 (1927): J. MIKA.

Metall und Erz

(768) **12**, 189 (1929): G. GRANIGG.

Microchemical Journal

(770) **2**, 3 (1958): A. A. BENEDETTI-PICHLER.
(771) **2**, 43 (1958): N. D. CHERONIS.
(772) **2**, 205 (1958): M. CEFOLA.
(773) **3**, 285 (1959): J. KRC, Jr.
(774) **3**, 323 (1959): A. A. BENEDETTI-PICHLER.
(775) **3**, 433 (1959): N. D. CHERONIS.
(776) **3**, 515 (1959): L. FINE and E. A. WYNNE.
(777) **4**, 423 (1960): N. D. CHERONIS.
(778) **4**, 459 (1960): G. T. CHANG and A. A. BENEDETTI-PICHLER.
(779) **5**, 331 (1961): A. A. BENEDETTI-PICHLER.
(780) **5**, 525 (1961): N. D. CHERONIS.

Mikrochemie, **1—24** (1923—1938)

Mikrochemie vereinigt mit Mikrochimica Acta, **25—40** (1938—1953)

(850) **1**, 4 (1923): F. FEIGL.
(851) **2**, 138 (1924): F. STEIDLER.
(852) **2**, 197 (1924): A. POLICARD.
(853) Emich-Festschrift 152 (1930): H. HETTERICH.
(854) Emich-Festschrift 243 (1930): P. RÂY and P. B. SARKAR.
(855) Emich-Festschrift 275 (1930): A. SOLTYS.
(856) **8**, 77 (1930): F. L. HAHN.
(857) **9**, 31 (1931): F. L. HAHN.
(858) **9**, 385 (1931): H. J. BRENNEIS.
(859) **10**, 313 (1931): F. L. HAHN.
(860) **10**, 380 (1932): H. HETTERICH.
(861) **11**, 167 (1932): A. O. GETTLER, J. B. NIEDERL, and A. A. BENEDETTI-PICHLER.
(862) **14**, 219 (1933—1934): H. K. ALBER.
(863) **15**, 271 (1934): A. A. BENEDETTI-PICHLER and W. F. SPIKES.
(864) **17**, 165 (1935): F. FEIGL, J. V. SANCHEZ, and R. ZAPPERT.

(865) **17**, 279 (1935): W. F. WHITMORE and H. SCHNEIDER.
(866) Molisch-Festschrift 3, 36 (1936): A. A. BENEDETTI-PICHLER and W. F. SPIKES.
(867) **18**, 272 (1935): F. FEIGL and O. FREHDEN.
(868) **18**, 289 (1935): B. L, CLARKE and H. W. HERMANCE.
(869) **19**, 1 (1935): A. A. BENEDETTI-PICHLER and J. R. RACHELE.
(870) **19**, 239 (1935): A. A. BENEDETTI-PICHLER and J. R. RACHELE.
(871) **21**, 98 (1937): A. K. RUSSANOW.
(872) **21**, 133 (1936): J. W. YOUNG.
(873) **21**, 268 (1937): A. A. BENEDETTI-PICHLER and W. F. SPIKES.
(874) **21**, 300 (1937): A. A. BENEDETTI-PICHLER.
(875) **26**, 29 (1939): J. I. ADAMS, A. A. BENEDETTI-PICHLER, and J. T. BRYANT.
(876) **26**, 143 (1939): D. L. JOHNSON and C. L. SHREWSBURY.
(877) **26**, 182 (1939): E. BONTINCK.
(878) **27**, 47 (1939): G. BECK.
(879) **27**, 249 (1939): W. F. WHITMORE and C. A. WOOD.
(880) **28**, 1 (1939): W. F. WHITMORE and C. A. WOOD.
(881) **31**, 263 (1943): G. SKALOS.
(882) **31**, 309 (1943): G. GORBACH.
(883) **33**, 281 (1948): O. KÖNIG, W. R. CROWELL, and A. A. BENEDETTI-PICHLER.
(884) **33**, 300 (1948): O. KÖNIG and W. R. CROWELL.
(885) **33**, 303 (1948): O. KÖNIG and W. R. CROWELL.
(886) **33**, 316 (1948): R. FISCHER and G. KARASEK.
(887) **33**, 385 (1948): J. ERDÖS.
(888) **34**, 39 (1948): A. A. BENEDETTI-PICHLER.
(889) **34**, 319 (1949): R. FISCHER and E. NEUPAUER. Tables of data.
(890) **34**, 382 (1949): J. LINDNER.
(891) **34**, 395 (1949): H. MALISSA.
(892) **35**, 34 (1950): H. MALISSA.
(893) **35**, 135 (1950): S. S. BURKE.
(894) **35**, 213 (1951): H. MALISSA.
(895) **35**, 266 (1951): H. MALISSA.
(896) **36/37**, 224 (1950): M. C. ALVAREZ QUEROL and C. L. WILSON.
(897) **36/37**, 296 (1950): R. FISCHER.
(898) **38**, 33 (1951): H. MALISSA.
(899) **38**, 50 (1951): J. GILLIS.
(900) **38**, 100 (1951): P. W. WEST and W. C. HAMILTON.
(901) **38**, 342 (1951): R. FISCHER.
(902) **38**, 471 (1951): J. S. WIBERLEY, R. K. SIEGFRIEDT, and A. A. BENEDETTI-PICHLER.
(903) **40**, 141 (1952): H. M. EL-BADRY and C. L. WILSON.
(904) **40**, 245 (1952): T. S. MA and R. TEN EYCK SCHENK.

Mikrochimica Acta, **1—3,** (1937—1938), **1953—1962**

Mikrochimica et Ichnoanalytica Acta, **1963**

(910) **1**, 266 (1937): H. V. A. BRISCOE and JANET W. MATTHEWS.
(911) **2**, 9, 287 (1937): G. BECK.
(912) **3**, 30 (1938): C. DUVAL and P. FAUCONNIER.
(913) **3**, 243 (1938): S. AUGUSTI.
(914) **1953**, 305: ALICE LACOURT, G. SOMMEREYNS, C. FRANCOTTE, and N. DELANDE.

(915) **1953**, 332: G. SOMMEREYNS.
(916) **1954**, 140: H. WEISZ.
(917) **1954**, 376: H. WEISZ.
(918) **1954**, 460, 785: H. WEISZ.
(919) **1955**, 134: H. JURÁNY.
(920) **1955**, 821: R. M. RUSH and L. B. ROGERS.
(921) **1956**, 422: J. MITCHELL, Jr., and ADA L. RYLAND.
(922) **1956**, 667: H. WEISZ.
(923) **1956**, 1225: H. WEISZ.
(924) **1956**, 1565: B. FLASCHENTRÄGER and M. S. ZEIN.
(925) **1956**, 1705: J. KÖRBL.
(926) **1956**, 1729: CHARLOTTE L. BROWN and P. L. KIRK.
(927) **1956**, 1783: J. KOLŠEK.
(928) **1956**, 1856: H. WEISZ and F. SCOTT.
(929) **1957**, 341: F. FEIGL, V. GENTIL, and C. STARK-MAYER.
(930) **1957**, 390: B. FLASCHENTRÄGER, SAMIHA M. ABDEL-WAHAB, and G. HABIB LABIB.
(931) **1957**, 417: W. KNÖDEL and H. WEISZ.
(932) **1957**, 427: MARIA BRANDSTÄTTER-KUHNERT and H. GRIMM.
(933) **1957**, 501: J. W. SHELL, C. F. POE, and N. F. WITT.
(934) **1957**, 527: E. WIESENBERGER.
(935) **1957**, 567: A. A. BENEDETTI-PICHLER and H. E. SCHNEIDER.
(936) **1957**, 640: P. SENISE. The tolidine is dissolved in 10% acetic acid; private communication.
(937) **1957**, 714: P. L. KIRK and CHARLOTTE L. BROWN.
(938) **1957**, 720: CHARLOTTE L. BROWN and P. L. KIRK.
(939) **1957**, 726: F. FEIGL, J. R. AMARAL, and V. GENTIL.
(940) **1957**, 736: P. LUIS and A. CORAZZA.
(941) **1957**, 751: H. BALLCZO and H. WEISZ.
(942) **1958**, 201: R. BELCHER, R. HARRISON, and W. I. STEPHEN.
(943) **1958**, 248, 253: G. SANDRI.
(944) **1958**, 305: B. M. TURNER.
(945) **1958**, 337: F. FEIGL and J. R. AMARAL.
(946) **1958**, 353: L. W. BRADFORD and J. W. BRACKETT.
(947) **1958**, 411: CH. R. WILLMS and W. M. HARDING.
(948) **1958**, 577: V. M. BHUCHAR.
(949) **1958**, 630: O. MANNS and S. PFEIFER.
(950) **1959**, 32: H. WEISZ.
(951) **1959**, 36: H. WEISZ, M. B. ČELOP, and V. V. ALMAŽAN.
(952) **1959**, 87: W. COLLOM and P. L. KIRK.
(953) **1959**, 314: H. BALLCZO.
(954) **1959**, 357: G. ACKERMANN.
(955) **1959**, 406: G. SCHMIDT.
(956) **1959**, 432: A. C. SHEAD.
(957) **1959**, 541: P. LUIS.
(958) **1959**, 657: A. C. SHEAD.
(959) **1959**, 801: J. A. JAECKER and F. SCHNEIDER.
(960) **1960**, 31: J. W. SHELL, N. F. WITT, and C. F. POE.
(961) **1960**, 592: V. ANGER and G. FISCHER.
(962) **1960**, 650: F. L. HAHN.
(963) **1960**, 830: J. MAHON and A. A. BENEDETTI-PICHLER.

(964) **1960**, 946: E. WIESENBERGER.
(965) **1960**, 967: ARLEEN PIERCE, R. LOESCH, and F. SCHNEIDER.
(966) **1961**, 11: R. WEISS.
(967) **1961**, 65: J. CHURÁČEK.
(968) **1961**, 140, 145, 149: H. REIMERS.
(969) **1961**, 899: G. PRAZAK.
(970) **1962**, 421: A. C. SHEAD.
(971) **1962**, 490, 498: A. C. SHEAD.
(972) **1962**, 529: J. J. PEIFER.
(973) **1962**, 830: H. WEISZ.
(974) **1962**, 922: H. WEISZ.
(975) **1962**, 1165: P. W. WEST, A. J. LLACER, and C. CIMERMAN.
(976) **1963**, 104: J. P. CRISLER, N. F. WITT, and MARJORIE H. CRISLER.
(977) **1963**, 316: J. P. CRISLER, N. F. WITT, and MARJORIE H. CRISLER.
(978) **1963**, 416: E. R. DU FRESNE.

Monatshefte für Chemie

(1000) **22**, 670 (1901): F. EMICH.
(1001) **23**, 76 (1902): F. EMICH.
(1002) **25**, 545 (1904): J. DONAU.
(1003) **25**, 915 (1904): J. DONAU.
(1004) **28**, 825 (1907): F. EMICH and J. DONAU.
(1005) **29**, 959 (1908): J. DONAU.
(1006) **34**, 949 (1913): J. DONAU.
(1007) **38**, 219 (1917): F. EMICH.
(1008) **39**, 775 (1918): F. EMICH.
(1009) **42**, 411 (1921): H. SCHEUCHER.
(1010) **43**, 129 (1922): A. FUCHS.
(1011) **43**, 405 (1922): F. LANYAR and L. ZECHNER.
(1012) **46**, 261 (1925): F. EMICH.
(1013) **46**, 265 (1925): J. VOGEL.
(1014) **50**, 745 (1928): F. EMICH.
(1015) **53/54**, 312 (1929): F. EMICH *et al.*
(1016) **53/54**, 335 (1929): F. EMICH and H. HÄUSLER.
(1017) **58**, 399 (1931): E. SCHALLY.
(1018) **64**, 385 (1934): E. SCHALLY and F. NAGL.
(1019) **65**, 153 (1935): J. HARAND.

Nature

(1050) **134**, 809 (1934): Y. D. BERNAL and D. CROWFOOT.
(1051) **135**, 305 (1935): Y. D. BERNAL and D. CROWFOOT.
(1052) **172**, 809 (1953): J. E. EDSTRÖM.
(1053) **174**, 128 (1954): J. E. EDSTRÖM and H. HYDÉN.
(1054) **179**, 628 (1957): S. H. U. BOWIE and K. TAYLOR.
(1055) **180**, 50 (1957): H. H. ALLEN.
(1056) **183**, 1423 (1959): V. E. COSLETT.

Die Naturwissenschaften

(1060) **38**, 287 (1951): P. DECKER.

Neues Jahrbuch für Mineralogie, Geologie und Paläontologie, Abhandlungen
(1065) 89, 149 (1956): W. LINDENBERG.

Proceedings of the Physical Society, London
(1068) 44, 511 (1932): A. FERGUSON and J. S. KENNEDY.

Recueil des travaux des Pays-Bas et de la Belgique
(1070) II, 16, 369 (1898): M. H. HEMMES.

The Laboratory (Fisher Scientific Co.)
(1072) 31, 48 (1963).

The Review of Scientific Instruments
(1075) 28, 256 (1957): J. S. COURTNEY-PRATT and C. M. HUGGINS.

Science
(1080) 73, 344 (1931): D. DU BOIS.

Sitzungsberichte der Akademie der Wissenschaften in Wien. Abteilung I
(1085) 111, 171 (1902): O. RICHTER.
(1086) 130, 383 (1922): H. BRUNSWIK.
(1087) 142, 339 (1933): M. HAITINGER.

Skandinavisches Archiv für Physiologie
(1090) 20, 279 (1908): A. KROGH.

Spectrochimica Acta
(1092) 12, 276 (1958): D. A. CLARK and A. P. BOER.

Technical Studies in the Field of Fine Arts
(1092) 1, 1 (1932): R. G. GETTENS.
(1093) 2, 185 (1934): R. G. GETTENS.
(1094) 7, 200 (1938): R. G. GETTENS; data on painting materials.
(1095) 8, 12 (1939): R. G. GETTENS; data on painting materials.

The Analyst
(1100) 57, 2, 107 (1932): J. C. MABY.
(1101) 63, 467 (1938): JANET W. MATTHEWS.

The New Scientist, London
(1105) October 31, 1957: L. W. CODD and W. T. MOORE.

Transactions of the Institution of Mining Engineers, London
(1108) June 1937: H. V. A. BRISCOE, JANET W. MATTHEWS, P. F. HOLT, and PHYLLIS M. SANDERSON.

U. S. National Bureau of Standards Publications

(1110) Bulletin 3, 345 (1905): G. K. BURGESS.
(1111) Bulletin 11, 591 (1915): G. K. BURGESS and WILTTENBERG.
(1112) Research Paper, RS 1809 (1947): B. W. SCRIBNER and W. K. WILSON.

U. S. Department of Agriculture Publications

(1120) Bur. Chem. Bull. **122**, 97 (1909); **137**, 189 (1911): B. J. HOWARD and C. H. STEPHENSON.

U. S. Geological Survey Publications

(1125) Bulletin 679 (1921): E. S. LARSEN, The Microscopic Determination of the Nonopaque Minerals.

Vierteljahresschrift für gerichtliche Medizin

(1130) 3. Folge, **59**, 233 (1920): G. STRASSMANN.

Zeitschrift des Vereines deutscher Ingenieure

(1133) **94**, 754 (1952): L. OTTO.

Zeitschrift für analytische Chemie

(1135) **28**, 374 (1889): E. LÉGER.
(1136) **36**, 195 (1897): L. STAHRE.
(1137) **46**, 658 (1907): N. SCHOORL.
(1138) **47**, 18 (1908): O. LUTZ.
(1139) **54**, 493 (1915): F. EMICH.
(1140) **54**, 500 (1915): F. EMICH.
(1141) **56**, 1 (1917): F. EMICH.
(1142) **62**, 284 (1922): R. KEMPF.
(1143) **70**, 257 (1927): A. A. BENEDETTI-PICHLER.
(1144) **73**, 54 (1928): H. FISCHER.
(1145) **74**, 191 (1928): A. A. BENEDETTI-PICHLER, review.
(1146) **75**, 395 (1928): A. A. BENEDETTI-PICHLER, review.
(1947) **76**, 216 (1929): A. A. BENEDETTI-PICHLER, review.
(1148) **76**, 443 (1929): A. A. BENEDETTI-PICHLER, review.
(1149) **77**, 130 (1929): A. A. BENEDETTI-PICHLER. review.
(1150) **79**, 94 (1929): P. RĂY.
(1151) **80**, 247, (1930): H. MEISSNER.
(1152) **82**, 113 (1930): F. L. HAHN.
(1153) **82**, 241 (1930): H. K. ALBER, review.
(1154) **86**, 69 (1931): A. A. BENEDETTI-PICHLER and F. SCHNEIDER.
(1155) **86**, 114 (1931): H. ALBER and MARIA v. RENZENBERG.
(1156) **89**, 121 (1932): E. EEGRIWE.
(1157) **90**, 87 (1932): H. ALBER.
(1158) **99**, 402 (1934): I. M. KORENMAN.
(1159) **102**, 102 (1935): R. W. FELDMANN.

Zeitschrift für anorganische und allgemeine Chemie

(1200) **147**, 156 (1925): F. HABER and J. JAENICKE.
(1201) **175**, 383 (1928): F. PANETH and K. PETERS.

(1202) 199, 77 (1931): W. GEILMANN and K. BRÜNGER.
(1203) 201, 347, 353 (1931): L. WOLF and W. JUNG.

Zeitschrift für Elektrochemie

(1210) 28, 89 (1922): A. GÜNTHER-SCHULZE.

Zeitschrift für physikalische Chemie

(1215) 76, 491 (1911): J. L. ANDRAE.

Zeitschrift für den physikalischen und chemischen Unterricht

(1220) 21, 17 (1908): V. DVOŘÁK.
(1221) 21, 281 (1908): A. WEINHOLD.

Zeitschrift für wissenschaftliche Mikroskopie

(1225) 38, 1 (1921): K. SPANGENBERG.
(1226) 39, 316 (1923): H. BRUNSWIK.

Zentralblatt für Mineralogie, Geologie und Paläontologie

(1230) A, 1929, 251: H. MORITZ.

Reports

(1240) Calco Technical Bulletin No. 770: Microscopical Techniques for the Study of Dying: G. L. ROYER, C. MARESH, and ANNA M. HARDING.
(1241) Committee for the Study of New Analytical Reagents, International Union of Pure and Applied Chemistry, Paris, May 1937.

Theses

(1250) E. FORCHE, Leipzig, 1938.
(1251) R. N. BOOS, Master's Thesis, New York University, 1940.
(1252) L. BRANCONI, Master's Thesis, New York University, 1940.
(1253) K. D. FLEISCHER, Master's Thesis, New York University, 1940.
(1254) G. C. T. CHANG, Master's Thesis, Brooklyn College, 1961.

Unpublished Experiments

(1260) Dr. KUNZ ALFONS, Budapest, during a visit to EMICH's Institute in 1926.
(1261) ANNE G. LOSCALZO at the Washington Square College of N. Y. U., 1940.
(1262) Dr. O. F. STEINBACH at Queens College, 1955.
(1263) SHAO-HSUN LY, unpublished experiments.

Private Communications

(1270) Suggestion of Dr. GULBRAND LUNDE during a visit to EMICH's Institute in summer 1926.
(1271) C. VAN BRUNT, General Electric Company, Schenectady, October 1943.
(1272) Dr. R. TEN EYCK SCHENK, Washington Square College of N. Y. U., 1950.
(1273) THOMAS P. SCHREIBER, Senior Research Physicist, Chemistry Department, General Motors Technical Center, Warren, Michigan.

Demonstrations of Microgram Technique by Dr. M. Cefola

(1280) Meeting of the Metropolitan Microchemical Society of New York. Washington Square College, January 11, 1940.
(1281) Meeting of the New York Section of the American Chemical Society. Hotel Pennsylvania, April 5, 1940.
(1282) In Service Training Lecture, Division of Laboratories, Department of Hospitals. Washington Square College, December 20, 1940.

Meetings

(1283) MARY L. WILLARD, Eastern Analytical Symposium, New York, Nov. 16, 1962.

Addresses

(1290) Alfred Fritsch, Laborgerätebau, Hauptstraße 542, Idar-Oberstein, West Germany.
(1291) Canal Industrial Corporation, Bethesda, Md.
(1292) Intercontinental Electronics Corporation, Minneola, N. Y.

Subject Index

Absorption spectrophotometry 296
Acetate 357
Air, supply of 97
Akro technique 130
Alloys, analysis of 403
—, dissolution of 321, 403
Alkali metals, tests for 325–328
Alkaline earths, tests for 329–331
Aluminum, tests for 352
Ammonium, tests for 365
Analysis *see* Identification
Analyzer 49, 50
Angle, interfacial 247
—, profile 248
—, —, list of 250
Anions, classification with barium nitrate 424
—, — with silver nitrate 425
—, — with sulfuric acid 301, 309
—, sensitivity of classification tests 307
—, systematic testing for 422
Anisotropic matter 47, 52
Antimony, tests for 374
Apparatus and scale of work 11
— — surface forces 15
— — volatility 14
—, cleaning of, centigram scale 99
—, — —, general 68
—, — —, microgram scale 209
—, — —, milligram scale 124
—, general, on centigram scale 82
—, —, — gram scale 69
—, —, — microgram scale 194–202
—, —, — milligram scale 124
—, selection of 228
Aperture 24
— diaphragm 25
Arsenic, tests for 373
Artifacts, identification of 242
Ashing of tissue 242
Autoradiography 254
Axial figures 61
Azide, tests for 367

Barium, tests for 331
Batch identification by *schlieren* 45
Bead, dimensions and weight of 205
— tests 315
— —, list of 316
— —, performance of 70, 191
Becke line 42, 43
Beilstein test 81
Bertrand lens 49, 62
Beryllium, test for 329
Biological matter 240
Bismuth, tests for 375
Blanks 324
Body color 244
Boiling Point, determination on microgram scale 215
— —, — — milligram scale 102
— Range, determination on microgram scale 215
Books, classified list 463
Boron and borate, tests for 351
Bromate, tests for 392
Bromide, tests for 391
Bromine, tests for 391
Bunsen flame, zones of 81

Cadmium, tests for 351
Calcium, tests for 330
Capillaries, examination of contents of 162, 164
—, preparation of 99
—, sealing of 161
—, working in 159
Capillary cone 198
— pipets, calibration of 125
— —, preparation of 100
— siphon 114
Carbides 356
Carbon monoxide 356
—, tests for 299, 355
Carbonate, tests for 356
Carius oxidation 166, 200
Centrifugal pipet 126

Centrifuge 84
—, use of 86
— tube, angle of taper 13
— —, dimensions and scale of work 13
Cerium, tests for 332
Cesium, tests for 328
Chlorate, tests for 390
Chloride, tests for 388
Chlorine, tests for 300, 387
Chromium, from cleaning solution 229
—, tests for 336
Citrate, tests for 359
Classification tests and general technique 268
Cleaning, general 68, 229
— on centigram scale 99
— — microgram scale 209, 219
— — milligram scale 124
Cobalt, tests for 341
Color 54, 244
— observation, general 16
— — under microscope 17, 175
— of inorganic substances, table 432
—, orders of 54
Coloriscopic capillary 175
Columbium *see* Niobium
Compensators 52, 53, 57
Condenser 33
—, use of 37
— rod, microgram scale 200
Cone *see* Centrifuge tube
Confirmatory tests 6, 324–395
— — by luminescence 193
— — in beads 191
— — on centigram scale 98
— — — exchange resins 182
— — — gram scale 75
— — — microgram scale 217, 218, 221
— — — milligram scale 123
— — — paper 128–136, 177ff.
— — — slide 136–149
— — — textile fibers 183, 188, 191
— — — wires 187
Conflagrating inorganic substances, list of 458
Congo fiber 185
Contraction pipet 166
Controls 324
Copper, tests for 347
Critical solution temperature 239

Critical temperature 236, 238
Cross-hair eyepiece 49
Cross hairs 32
— —, checking of 52
Crystal classes, criteria of 251
Crystallographical optical analysis 292, 296
Crystals, identification 243
—, list of cubic substances 433
—, — — hexagonal substances 435
Cyanate, test for 360
Cyanide, tests for 359

Decantation on the centigram scale 86, 87
— — — gram scale 72
— — — microgram scale 211
— — — milligram scale 113
— — — submilligram scale 150, 151, 152, 160, 165
Density of liquids, determination of 262
— — solids, determination of 264
Directions *see* Working directions
Distillation in capillaries 163, 170
— on centigram scale 90, 101
— — gram scale 74
— — microgram scale 213
— — milligram scale 118
Double refraction 47
— —, sign of 62
Drills, drilling (micro) 232, 233
Drying in capillary 168, 171
— on slide 141, 152, 156–157
Dust particles, sampling 230

Electrodes 168
Electrolysis 187
— in the centifuge cone 115
— on milligram scale 115
— — microgram scale 186
— — slide 186
Electrolytic slide 186
Electron diffraction 292
— probe 293
Emulsions, breaking of 73
Europium, tests for 333
Evaporation on centigram scale 90
— — fibers 189
— — gram scale 70
— — microgram scale 212
— — milligram scale 117

Evaporation on paper 178, 179, 184
— — slide 140, 154, 155
—, prevention of 14, 15, 194
Experimentation *see* Technique
Exploding substances, list of inorganic 459
— —, — — organic 460
— —, test for 269
Explosives 272
Extinction angle 56
— position 53
—, types of 56
Extraction of solids on centigram scale 88
— — — — gram scale 73
— — — — milligram scale 115
— — — — paper 178, 179, 181
— — — — submilligram scale 155, 157
— — liquids on centigram scale 88
— — — — gram scale 73
— — — — milligram scale 116
Eyepiece 31
— micrometer, calibration of 37

Fat, test for 242
Ferricyanide, tests for 361
Ferrocyanide, tests for 361
Ferromagnetism, test for 254
Fibers 184
—, identification of 240, 242
—, spot tests on 218
—, working on 178, 183, 188–191
Field diaphragm 25
— of vision 36, 38
Filter paper, impregnation with reagents 130
— —, working on 128
Filtration in capillaries 169, 173, 174
— — paper 178
— of hot solutions in capillaries 169
— on centigram scale 86, 87
— — gram scale 71
— — milligram scale 113
— — the slide 150, 151, 152
Final review of findings 428
Flame colorations, list of 314
— test 313
— —, performance of 79
Flash figures 62
Fluorescence 246

Fluoride, tests for 385
Fluorine 385
Forceps 202, 232
Formate, tests for 357
Fractional distillation on centigram scale 91, 101
— — — microgram scale 213
— — — milligram scale 120
Fractionation by melting in capillary 171
— — — on slide 151
Front lenses 24
Fusions on the centigram scale 83
— — — gram scale 70
— — — microgram scale 204
— — — milligram scale 111
— — — submilligram scale 152, 179 (footnote)
— with potassium pyrosulfate 112, 323
— — sodium carbonate 111, 323
— — — peroxide 111, 324
— — sodium-potassium carbonate 323

Gallium, tests for 353
Gas, expelling of, on the microgram scale 213
—, — —, — — milligram scale 108, 280
—, liberation and testing for, on centigram scale 96, 303
—, — — — — —, — gram scale 303
—, — — — — —, — milligram scale 123, 280, 308
—, — — — — —, — submilligram scale 158, 308
— reaction cell 96
—, sensitivity of tests for 307
—, test for evolution of 309
Gaseous reagent, treating with, on centigram scale 83
— —, — —, — gram scale 69
— —, — —, — microgram scale 210
— —, — —, — milligram scale 106
Germanium, tests for 363
Gold, tests for 348
Goniometer 248
Grinding 231
Gypsum plate 53, 57

Halogens, tests for 300, 385–395
Hard materials, list of 260

Hardness tests 258
Heating *see also* Ignition
— block 109
— element (μg scale) 201
— in capillaries 162, 165, 166, 167
— — cones under pressure 110
— on centigram scale 83
— — gram scale 70
— — microgram scale 211, 213
— — milligram scale 108
— up to 350° C 269
Hexagonal crystals, criteria 252
— —, list of substances 435
History of sample 234
Hook of glass 304
— — platinum 128
Hydrazine, tests for 366
Hydrazoic acid, tests for 367
Hydrofluoric acid, treatment with 322, 323
— —, — —, on milligram scale 112
Hydrogen, test for 299, 304
— peroxide, tests for 376
Hydroxylamine, tests for 366
Hypochlorite, tests for 389
Hypophosphate, tests for 370
Hypophosphite, tests for 369

Identification, exercises 225–227
— of artifacts 242
— — crystals 243
— — inorganic substances 272, 291, 301
— — liquids 236
— — organic substances 273, 294, 299
— — organized (biological) matter 240
— — solids 244
—, record of 235
—, systematic procedure of 228, 234, 268, 317, 428
Ignition above 300° C 274
— below 350° C 269
— in air 77, 274, 283
— — chlorine 285
— — closed tube 75, 274
— — gas stream 276
— — hydrogen 284, 324
— — — sulfide 285
— — inert gas 75, 282
— — open tube 274
— — oxygen 283

Ignition on charcoal block 78, 275
Illumination 34
—, critical 27
— in microscopy 35
—, oblique 42, 43
Immersion method 39, 43
Inclusions, sampling of 232, 233
Indium, tests for 354
Interference color 54
— —, determination of order 58
Iodate, tests for 394
Iodide, tests for 393
Iodine, tests for 300, 393
Iridium, tests for 346
Iron, as contamination on platinum 229
—, tests for 340
Isotropic substances 52

Journal articles 472

Laboratory work *see* Technique
Lanthanides 332
L. C. *see* Limiting concentration
Lead, tests for 365
L. I. *see* Limit of identification
Lignin, test for 242
Limit of identification 4, 5, 324
— — — of color tests 17
Limiting concentration 4, 5, 324
— proportions 4, 5, 324
Lithium, tests for 325
Litmus fiber 184
Loop of glass 305
— — platinum 127
Loss by evaporation 14
— — spreading of drops 16
— with residual liquid left in apparatus 15
L. P. *see* Limiting proportions
Luminescence 193, 246

Magnesium, tests for 329
Magnetism, test for 254
Magnification 23
Majors 2
Manganese, tests for 338
Manipulators 195, 196, 231
Measuring capillary (μg scale) 200
— (volumes) on centigram scale 82
— — — gram scale 68
— — — microgram scale 206
— — — milligram scale 105, 125, 127

Subject Index

Melting point 295
— —, determination in capillary 170, 173
— — of eutectic mixture 295
— — — inorganic substances 438
— — on slide 269
— — under microscope 269
Mercury, tests for 351
Metals, analysis of 403
—, dissolution of 321, 403
Metaphosphate 371
Mica plate 59
Micro- *see* item following the prefix micro
Microburner, preparation of 100
Microcone 105, 106
—, observation in 124
Micrometer eyepiece 31
— —, calibration of 37
— scale 39
Micromethods, definition 3
Micropipet 198
—, inserting into capillary 205
—, — — chamber 203
—, mounting 202
Microprojection 26
Microscope 20
—, binocular 29
—, cleaning of 28
—, condenser 33, 37
—, field of vision of 36, 38
—, focusing 30, 32
— for microgram work 194
—, illumination 29, 30, 34, 35, 43
—, inspection of 28, 33, 50
— lamps 28
—, monocular 29
—, petrographic 49
—, polarizing 49
—, selection of 26
— slide *see* slide
—, stage 33
—, — rotating 49
—, — —, centering of 51
—, testing of 28, 33, 50
—, working distance of 25, 38
Microscopical preparations, preserving of 65
— tests, preserving of 65
Microsonde 293
Minors 3

Mixing in capillaries 161, 165
— on centigram scale 82
— — gram scale 69
— — microgram scale 210
— — milligram scale 106
Moderately soluble inorganic substances, list of 461
Moist chamber 198
— —, assembling of 202
Molybdenum, tests for 337

New substances 1
Nickel, tests for 342
Nicol prism 48
Nicols 48
—, crossed 50
Niobium, tests for 335
Nitrate, tests for 368
Nitrite, tests for 368
Nitrogen, tests for 300, 365
Nonmetallic substances, dissolution of 321
— —, test for anions 427

Objectives 32
Ocular *see* Eyepiece
Odor, testing for 256
Optical emission spectrochemical analysis 294
Organic elemental analysis 299
Organized matter, identification 240
Osmium, tests for 345
Oxalate, tests for 358
Oxidants, tests for 426
Oxygen, tests for 375
Ozone, tests for 376

Palladium, tests for 344
Paper, impregnation with reagents 130
—, work in 128, 177
Perborate, tests for 302, 377, 382
Percarbonate, tests for 302, 377, 382
Perchlorate, tests for 390
Periodate, tests for 395
Permanent mounts 65
Permanganate, test for 339
Peroxide, tests for 302, 376, 382
Persulfate, tests for 302, 377, 381
Petrographic microscope 49, 50
pH, adjustment of, on centigram scale 82

pH, adjustment of, on gram scale 69
—, — —, — smaller scales 19, 20
Phosphate from cleaning solutions 229
—, tests for 371
Phosphide, tests for 369
Phosphite, tests for 370
Phosphorus 369
Photomicrography 26
Pipet holder (μg scale) 197
Plastic ware 229
Platinum, tests for 346
— metals, glow test 343
pL. C. 5
Pleochroism 61
pL. I. 5
Plunger control 197
Polarized light 47
Polarizer 48, 50
Polarizing microscope 49
— —, testing and adjusting 50
— —, data easily determined with 251
Polaroid 48
Polars 48
—, crossed 51
Position of extinction 53
Potassium, tests for 326
Preliminary inspection 235
— treatment 6
Preserving microscopic tests 65
Procedure of analysis, selection of 18
Profile angles, list of 250
— —, measuring 56
Projection 26
Properties, effect of size upon 9
—, observation of 16
Proteins, tests for 242
Pseudomorphs 143, 144, 148
Purification see Distillation, Fractional distillation, Fractional Melting, Recrystallization, Sublimation
Pyrophosphate 371

Qualitative analysis, definition 2
— —, limitations 1
— —, method of 6
— —, principal steps 6
— —, purpose and task 1
Quantitative estimations on microgram scale 212, 216
— — — milligram scale 149
Quartz wedge 53, 57

Radiation, test for 253
Rare earths, tests for 332, 333, 334, 335, 336
Reagent container (μg scale) 199
Reagents, adding, on centigram scale 82
—, —, — gram scale 69
—, —, — microgram scale 209
—, —, — milligram scale 105
—, —, — submilligram scale 152
—, care and use 68, 228
—, — — — on milligram scale 104
—, listing of 432
Recrystallization, in capillary 161, 168, 171
— on slide 139, 151
Reductants, tests for 426
Refraction see Refractive index
Refractive index, Becke test 42
— —, determination of 39
— —, — —, in anisotropic matter 57, 60
— —, double variation method 41
— —, oblique illumination 42
— —, standard liquids 40
— —, — solids 39
— —, use in identification 260
— — via schlieren observation 45
Report on analysis 428
Retardation plates 49, 52, 53, 57
Rhenium, tests for 340
Rhodium, tests for 344
Rubidium, tests for 327
Ruthenium, tests for 343

Sample, acid insoluble 322
—, definition 3
—, dissolution 320
—, minimum size of 7
Sampling 230
—, dust 230
—, ferromagnetic particles 233
—, inclusions 232, 233
Saturation chamber 108
Scale of work, indication of 3
— — —, theoretical lower limit 8
Scandium, test for 331
Scheme of separation, classical or H_2S 396
— — — of Noyes and Bray 404
Schlieren and observation 44, 45
Selenate, tests for 383

Selenide, tests for 302, 306, 382
Selenite plate 53, 57
—, tests for 383
Selenium 382
Sensitivity 4, 19
Separation by mechanical sorting 219
—, chemical 395
—, —, of anions 424
—, —, — arsenic and copper groups 149
—, —, — arsenic from antimony 222
—, —, — cations 396
—, —, on fibers 178, 183
—, —, — microgram scale 222
—, —, — milligram scale 149
—, —, — paper 128, 177
—, —, upon slide 153, 156
Silhouette angle 56
Silicate and silicon, tests 362
Silver, tests for 347
Slide, chemical work upon 150–161
— tests, examples 139–148
— —, general 136–138
— —, preserving of 65
Sodium, tests for 326
Solubility 285
—, definition, inorganic 287
—, —, organic 290
—, determination of, inorganic 288
—, — —, organic 290
— of small particles 10, 287
—, rules for inorganic substances 288
Specific gravity see Density
— surface area 9
Specificity 4
Spindrying in capillary 171
Spot tests on centigram scale 98
— — — microgram scale 218
— — — (sub) milligram scale 128 to 136
— —, sensitivity of 129
Stability on heating, test 269
Stage micrometer 37
— of microscope 33
— rotating 49
— —, centering of 51
Standards of refractive index 39, 40
Starch, test for 242
Steam bath 109
Storing of work on centigram scale 99
— — — — gram scale 75

Storing of work on microgram scale 219
— — — — milligram scale 124
Strontium, tests for 330
Study of chemical behavior 224
Sublimation from slide to slide 157
— in capillary 170
— on centigram scale 92
— — gram scale 74
— — milligram scale 122
Subliming inorganic substances, list of 457
Substances not described 1
Sulfate, tests for 381
Sulfide fiber 185
—, tests for 302, 378
Sulfite, tests for 303, 379
Sulfur, tests for 300, 377
Surface color 245
— force, effects of 15

Tantalum, tests for 336
— group, separation 410
Tartrate, tests for 358
Technique and scale of work 11
— — surface forces 15
—, basic rules 67
— of centigram scale 81
— — gram scale 68
— — microgram scale 193
— — milligram scale 104
— — picogram scale 9
— — submilligram scale 128–193
— — working with microscopic drops 9
Tellurium, tests for 302, 306, 383
Test solutions 430
Testing chemical behavior 224
—, non destructive 244–268
Thallium, tests for 355
Thiocyanate, tests for 361
Thionate, tests for 381
Thiosulfate, tests for 380
Thorium, tests for 335
Tin, tests for 364
Titanium, tests for 333
Traces, definition 3
Transfer of liquids and slurries, centigram scale 86, 87
— — — — —, gram scale 70, 71
— — — — —, microgram scale 208

Transfer of liquids and slurries, milligram scale 113, 114
— — solids, centigram scale 82, 88
— — —, gram scale 69, 72
— — —, microgram scale 203, 209, 219
— — —, milligram scale 115
— — —, submilligram scale 160
Transition phenomena in polarized light 63
— points above 300° C 274ff., 438
— — below 350° C 269, 438
Tungsten *see* Wolfram 337
Turmeric linen 188

Ultrasonic jack hammer 233
Unknowns, preparation of 431
Uranium, tests for 338

Vanadium, tests for 335
Vibration directions, determination 55
— — of polars 50
— — — specimen 53, 55
Vibrator 73
Volatile substances and size of apparatus 14

Washing in capillaries 166, 167
— on centigram scale 87
— — gram scale 71
— — microgram scale 211
— — milligram scale 115
— — paper 179
— — submilligram scale 153
Water, tests for 376
— bath (micro scale) 109
Weighing on centigram scale 82
— — gram scale 68
— — microgram scale 203
— — milligram scale 104
Wolfram, tests for 337
Working directions, transposition from scale to scale 8, 9
— distance (microscopical) 25
— technique *see* Technique

X-ray diffraction 292
— emission spectrography 292, 296

Ytterbium, tests for 333
Yttrium, tests for 332

Zinc, tests for 349
Zirconium, tests for 334